Singularly Perturbed Problems: Asymptotic Analysis and Approximate Solution

Singularly Perturbed Problems: Asymptotic Analysis and Approximate Solution

Editor

Valery Y. Glizer

MDPI • Basel • Beijing • Wuhan • Barcelona • Belgrade • Manchester • Tokyo • Cluj • Tianjin

Editor
Valery Y. Glizer
ORT Braude College of
Engineering
Israel

Editorial Office
MDPI
St. Alban-Anlage 66
4052 Basel, Switzerland

This is a reprint of articles from the Special Issue published online in the open access journal *Axioms* (ISSN 2075-1680) (available at: https://www.mdpi.com/journal/axioms/special_issues/singularly_perturbed_problems).

For citation purposes, cite each article independently as indicated on the article page online and as indicated below:

LastName, A.A.; LastName, B.B.; LastName, C.C. Article Title. *Journal Name* **Year**, *Volume Number*, Page Range.

ISBN 978-3-0365-6029-8 (Hbk)
ISBN 978-3-0365-6030-4 (PDF)

Contents

About the Editor

Valery Y. Glizer

Valery Y. Glizer received his Ph.D. degree in mathematics and physics in 1980. From 2007–2020, he served as an associate/full professor at the Department of Mathematics in the ORT Braude College of Engineering, Karmiel, Israel. Currently, Prof. Glizer serves as a researcher at The Galilee Research Center for Applied Mathematics, ORT Braude College of Engineering, Karmiel, Israel. Prof. Glizer is an author/coauthor of more than 240 scientific publications, including the monographs "Singular Linear–Quadratic Zero-Sum Differential Games and H_∞ Control Problems: Regularization Approach", Birkhauser (2022); "Controllability of Singularly Perturbed Linear Time Delay Systems", Birkhauser (2021) and "Robust Controllability of Linear Systems", Nova Science (2012). He is an associate editor of a number of international journals. He has served as the chairman of IPCs for more than 10 international conferences and has been a member of IPCs of more than 80 international conferences. Prof. Glizer was awarded for the best paper at the 2011 International Conference on Applied and Engineering Mathematics of the World Congress of Engineering (WCE 2011), London, UK. His main areas of research interest are: asymptotic methods, dynamic games, hybrid systems, optimal control, robust control, singular perturbations, systems' theory and time delay systems.

Preface to "Singularly Perturbed Problems: Asymptotic Analysis and Approximate Solution"

This book collects papers which were published in the Special Issue "Singularly Perturbed Problems: Asymptotic Analysis and Approximate Solution" of the *Axioms* journal. These papers represent various aspects of singular perturbation theory and its applications.

In their contribution, Margarita Besova and Vasiliy Kachalov develop the axiomatic approach in the analytic theory of singular perturbations in the frame of topological algebras. This allows the authors to present the main concepts of the singular perturbation analytical theory with maximal generality.

Dana Bibulova, Burkhan Kalimbetov and Valeriy Safonov consider a singularly perturbed integral–differential equation with a rapidly oscillating inhomogeneity and with a rapidly decreasing kernel of the integral operator of Fredholm type. For this equation, the authors construct and justify the regularized (in the sense of S.A. Lomov) asymptotic solution.

The contribution by Abduhafiz Bobodzhanov, Valeriy Safonov and Vasiliy Kachalov is devoted to the analysis of one singularly perturbed integral equation with weakly and rapidly varying kernels. The authors study the influence of the weakly varying integral kernel on the asymptotic solution of this equation. Additionally, using the method of holomorphic regularization, the authors consider the problem of constructing pseudo-analytic solutions of singularly perturbed problems.

The paper by Yuli D. Chashechkin and Artem A. Ochirov applies the theory of singular perturbation to study the propagation of two-dimensional periodic perturbations, including capillary and gravitational surface waves, in a viscous continuously stratified fluid.

Vasile Drăgan studies a stochastic linear–quadratic optimal control problem, the dynamics of which are described by a system of singularly perturbed Itô differential equations with two fast time scales. The author derives the proper stabilizing asymptotic solution from the algebraic Riccati equation associated with this problem. Using this asymptotic solution, the author designs the suboptimal control with the gain matrices not depending upon the small parameters.

In their contribution, Alexander Eliseev and Tatjana Ratnikova consider a singularly perturbed Cauchy problem for a two-dimensional differential equation with a "simple" turning point. Using the regularization method of S.A. Lomov, the authors construct and justify the asymptotic solution to the problem under consideration. Additionally, under proper additional conditions on the considered problem, the authors show that the series, representing the asymptotic solution, converges and that its sum represents the exact solution to the singularly perturbed Cauchy problem.

The contribution by Valery Y. Glizer considers a singularly perturbed linear time-dependent controlled system with multiple point-wise delays and distributed delays in the state and control variables. It is assumed that the delays are small, in the order of a small positive multiplier for a part of the derivatives in the system. Both types of the considered system, standard and nonstandard, are analyzed. For each of these types, two much simpler, parameter-free subsystems (the slow and fast ones) are associated with the original system. The author establishes that the proper kinds of controllability of the slow and fast subsystems yield the complete Euclidean space controllability of the original system for all sufficiently small values of the singular perturbation parameter.

Burkhan Kalimbetov and Valeriy Safonov consider a system with rapidly oscillating coefficients. This system includes an integral operator with an exponentially varying kernel. The authors develop an algorithm for the regularization method (in the sense of S.A. Lomov) for this system. They also analyze the influence of the integral term on the asymptotic behavior of the solution to the original system.

The paper by Galina Kurina considers an initial value problem for a class of singularly perturbed systems in the case where the matrix of the coefficients for the state variable is singular (the critical case). The author applies the orthogonal projector method for the construction and justification the asymptotic solution to the considered problem.

In their paper, Galina Kurina and Margarita Kalashnikova apply the direct scheme method for the asymptotic solution of a weakly nonlinearly perturbed linear–quadratic optimal control problem with three-tempo state variables. Asymptotic expansions for the optimal control, optimal trajectory and optimal value of the minimized functional of the considered problem are derived and justified. Monotonic (non-increasing) behavior is established for the asymptotic expansion of the optimal value of the functional with respect to the order of this expansion.

Tatiana Ratnikova studies the singularly perturbed Cauchy problem for a parabolic equation in the case of violation of stability conditions of the limit–operator spectrum. This case is due to the presence of a "simple" turning point in the equation. Using the Lomov's regularization method, the author constructs a uniform asymptotic solution to the considered problem and proves the asymptotic convergence of the regularized series.

The contribution by Olga Tsekhan is devoted to undertaking an analysis of complete controllability of a linear time-invariant singularly perturbed system, with multiple commensurate non-small delays in the slow state variables. An extension of the Chang-type time-scale separation of a singularly perturbed system to the considered time delay system is carried out. Based on this time-scale separation of the original system, sufficient conditions are obtained for its complete controllability. These conditions are independent of the parameter of singular perturbation, while they provide the complete controllability of the original system for all sufficiently small values of this parameter.

Vladimir Turetsky and Valery Y. Glizer consider a finite-horizon zero-sum linear–quadratic differential game, modeling a pursuit–evasion problem. In the game's cost functional, the cost of the control of the minimizing player is much smaller than the cost of the control of the maximizing player and the cost of the state variable. This smallness is due to a positive small multiplier (a small parameter) for the quadratic form of the minimizing player's control in the cost functional. Parameter-free sufficient conditions for the existence of the game's solution, valid for all sufficiently small values of the parameter, are presented. The boundedness with respect to the small parameter of the time realizations of the players' optimal state feedback controls, along the corresponding game's trajectory, is established. The best achievable game value from the minimizing player's viewpoint is derived. A relation is established between solutions of the original game and the game that is obtained from the original one by replacing the small parameter with zero.

The paper by Robert Vrabel considers the problem of asymptotic behavior of the solutions for one class of non-resonant, singularly perturbed linear Neumann boundary value problems. The approach, proposed by the author for analysis of asymptotic behavior of the solution to such problems, is based on the study of an integral equation associated with this problem.

I express my sincere gratitude to all of the authors for their contributions to this book.

Valery Y. Glizer
Editor

axioms MDPI

Article

Axiomatic Approach in the Analytic Theory of Singular Perturbations

Margarita Besova † **and Vasiliy Kachalov** *,†

Department of Mathematical Sciences, National Research University "MPEI", ul. Krasnokazarmennaya 14, 111250 Moscow, Russia; besova.margarita@yandex.ru

* Correspondence: vikachalov@rambler.ru; Tel.: +7-495-362-78-74

† These authors contributed equally to this work.

Received: 18 November 2019; Accepted: 25 December 2019; Published: 16 January 2020

Abstract: Introduced by S.A. Lomov, the concept of a pseudoanalytic (pseudoholomorphic) solution laid the foundation for the development of the singular perturbation analytical theory. In order for this concept to work in case of linear problems, an apparatus for the theory of exponential type vector spaces was developed. When considering nonlinear singularly perturbed problems, an algebraic approach is currently used. This approval is based on the properties of algebra homomorphisms for holomorphic functions with various numbers of variables, as a result of which it is possible to obtain pseudoholomorphic solutions. In this paper, formally singularly perturbed equations are considered in topological algebras, which allows the authors to formulate the main concepts of the singular perturbation analytical theory from the standpoint of maximal generality.

Keywords: ε-regular function; invariants of equations and systems; ε-pseudoregular solution; essentially singular manifold

MSC: 34E15

1. Introduction

The basic concept of the singular perturbation analytic theory is the concept pseudoholomorphic solution, i.e., such a solution, which can be presented as a series in powers of a small parameter that converges in the usual sense (and not asymptotically). The nature of this convergence is determined by the topology of the spaces in which the investigated problems are considered. As a rule, spaces of holomorphic functions (of one or several variables) are used. In this regard, it was possible to formulate the main principles for the theory of singularly perturbed differential equations and systems—under fairly general assumptions that they possess holomorphics in small parameter first integrals [1,2]. Moreover, a connection between the first integrals and homomorphisms of algebras of holomorphic functions with various numbers of variables was established. The pseudoholomorphic solutions themselves are obtained as a result of applying the implicit function theorem. In the presented paper, all of these constructions will be carried out in topological algebras for formally singularly perturbed equations.

2. Algebraic and Analytic Aspects of the Theory of Singular Perturbations

Let $\mathcal{J}_a$ be a complete topological commutative algebra with unit e and let $X, Y_1, \ldots, Y_k, \ldots$ be a sequence of open sets $\mathcal{J}_a$. Let us denote by $\mathcal{A}_0, \mathcal{A}_1, \ldots, \mathcal{A}_k, \ldots$ the spaces of functions continuous on the sets $X, X \times Y_1, \ldots, X \times Y_1 \times \ldots \times Y_k, \ldots$ respectively with their values in $\mathcal{J}_a$. Let us formulate the block I of necessary conditions:

(1°) If the sequence $\{x_i\}_{i=0}^{\infty}$ is a bounded set [3] in $\mathcal{J}_a$, then the series $x_0 + \varepsilon x_1 + \ldots + \varepsilon^i x_i + \ldots$ converges at $|\varepsilon| < 1$.

(2°) If the sequence $\{h_{i,k}\}_{i=1}^{\infty} \subset \mathcal{A}_k$ is such that the series

$$\sum_{i=0}^{\infty} \varepsilon^i h_{i,k}(x, y_1, \ldots, y_k) \tag{1}$$

converges on each set $T \times T_1 \times \ldots \times T_k$, where T is an arbitrary compact from X; T_1 is an arbitrary compact from $Y_1, \cdots$; T_k is an arbitrary compact set from Y_k in some neighborhood of the value $\varepsilon = 0$, the function $\Phi \in \mathcal{A}_0$ and it can be extended to all $\mathcal{J}_a$, then we have

$$\Phi\left(\sum_{i=0}^{\infty} \varepsilon^i h_{i,k}\right) = \Phi(h_{0,k}) + \sum_{i=1}^{\infty} \varepsilon^i g_{i,k}$$

and the last row with coefficients from $\mathcal{A}_k$ is convergent.

Definition 1. *The function $f(x, y_1, \ldots, y_k, \varepsilon) \in \mathcal{A}_k$ represented by (1), is called ε-regular.*

(3°) If the system

$$\begin{cases} F_1(x, y_1, \ldots, y_k, \varepsilon) = q_1, \\ \ldots\ldots\ldots\ldots\ldots\ldots\ldots \\ F_k(x, y_1, \ldots, y_k, \varepsilon) = q_k \end{cases}$$

with ε-regular left-hand sides is uniquely solvable with respect to $\{y_1, \ldots, y_k\}$ for $\varepsilon = 0$ in some neighborhood of the point $x_0 \in X$, then it is also uniquely solvable in some neighborhood of the same point and thus functions $y_m(x, \varepsilon) \in \mathcal{A}_0$ $(m = \overline{1,k})$ are ε-regular.

Remark 1. *The conditions of the block I are satisfied if $\mathcal{J}_a = \mathbb{C}$, $X, Y_1, Y_2, \ldots$ are simply connected regions, $\mathcal{A}_0, \mathcal{A}_1, \ldots$ are spaces of holomorphic functions on $X, X \times Y_1, X \times Y_1 \times Y_2, \ldots$ respectively.*

In order to formulate the conditions of block II, we give some definitions.

Definition 2. *s-product of tuples $\boldsymbol{\varphi} = \{\varphi_1, \ldots, \varphi_k\}$ and $\boldsymbol{\psi} = \{\psi_1, \ldots, \psi_k\}$ is a function $\boldsymbol{\varphi} \circledS \boldsymbol{\psi} = \varphi_1\psi_1 + \ldots + \varphi_k\psi_k$.*

Definition 3. *Let $f \in \mathcal{A}_k$, $\varphi_i(x) \in \mathcal{A}_0$, $i = \overline{1,k}$. The composition f and $\boldsymbol{\varphi} = \{\varphi_1, \ldots, \varphi_k\}$ is determined by the formula $f \circ \boldsymbol{\varphi} = f(x, \varphi_1(x), \ldots, \varphi_k(x))$ as usual.*

Block of conditions II:

(1°) All algebras $\mathcal{A}_0, \mathcal{A}_1, \ldots, \mathcal{A}_k, \ldots$ contain constant functions and linear functions. We consider the embeddings $\mathcal{A}_0 \subset \mathcal{A}_1 \subset \ldots \subset \mathcal{A}_k \subset \ldots$ together with topologies to be obvious.

(2°) On all spaces $\mathcal{A}_0, \mathcal{A}_1, \ldots, \mathcal{A}_k, \ldots$, a linear operation ∂_0 is defined such that $\partial_0 p = 0$, where p is a constant function, $\partial_0 x = e$ and $\partial_0 f = 0$ if $f \in \mathcal{A}_k$ and does not depend on x. On each space $\mathcal{A}_k$ $(k = 1, 2, \ldots)$, linear operations $\{\partial_i\}_{i=1}^{k}$ are defined and they comply with the following laws:

(a) $\partial_i p = 0$, $i = \overline{1,k}$, where $p \in \mathcal{A}_k$ is a constant function;
(b) $\partial_i y_i = e$, $i = \overline{1,k}$;
(c) if the function $f \in \mathcal{A}_k$ does not depend on y_m, then $\partial_i f = 0$ for $i \neq m$.

(3°) The operations $\{\partial_i\}_{i=0}^{\infty}$ form a commutative ring.

(4°) An operation d is introduced, and it satisfies the following rules:

(a) $d \equiv \partial_0$ on $\mathcal{A}_0$;
(b) $d(f \circ g) = \partial_0 f \cdot \partial_0 g$ $\forall f, g \in \mathcal{A}_0$;
(c) if $f \in \mathcal{A}_k$, $\varphi_i(x) \in \mathcal{A}_0$ $(i = \overline{1,k})$, $\boldsymbol{\varphi} = \{\varphi_1, \ldots, \varphi_k\}$, then $d(f \circ \boldsymbol{\varphi}) = \boldsymbol{\partial} f \circledS \boldsymbol{\partial\varphi}$, where $\boldsymbol{\partial} f \equiv \{\partial_0 f, \partial_1 f, \ldots, \partial_k f\}$, $\partial_0 \boldsymbol{\varphi} = \{e, \partial_0\varphi_1, \ldots, \partial_0\varphi_k\}$ are tuples of length $(k+1)$.

(5°) For every natural number k in the algebra $\mathcal{A}_k$, there exist a lot of tuples $\mathbf{f} = \{f_1(x, y_1, \ldots, y_k), \ldots, f_k(x, y_1, \ldots, y_k)\}$ such that the operator $D_k^{\mathbf{f}} = \mathbf{f} \circledS \boldsymbol{\partial}$, where $\boldsymbol{\partial} = \{\partial_1, \ldots, \partial_k\}$, with a specially defined domain $\mathcal{D}(D_k^{\mathbf{f}})$ is surjective and has the inverse $J_k^{\mathbf{f}}$, which has the following property: for arbitrary compact sets $T \subset X, T_1 \subset Y_1, \ldots, T_k \subset Y_k$ there is a number $C > 0$ such that, for an arbitrary function $\varphi(x) \in \mathcal{A}_0$, the set $\Gamma_k^{\varphi} = \{C^{-n}(J_k^{\mathbf{f}}\partial_0)^n \varphi(x, y_1, \ldots, y_k) \in T \times T_1 \times \ldots \times T_k\}_{n=1}^{\infty}$ is bounded in $\mathcal{J}_a$.

Let us consider the case $k = 1$. We investigate the following equation:

$$\varepsilon dy_1 = F(x, y_1), \tag{2}$$

in which $F \in \mathcal{A}_1$ and ε is a small complex parameter. The function $y_1(x) \in \mathcal{A}_0$ satisfying the initial condition

$$y_1(x^0, \varepsilon) = y_1^0, \tag{3}$$

where $x^0 \in X, y_1^0 \in Y_1$, is required to be found.

Definition 4. *The invariant of Equation (2) is the function $U(x, y_1, \varepsilon) \in \mathcal{A}_1$, which turns into a constant on the solution $y_1(x, \varepsilon)$ of this equation.*

Theorem 1. *When the blocks of conditions I and II are satisfied, then Equation (2) has ε-regular invariants.*

Proof of Theorem 1. If $U(x, y, \varepsilon)$ is an invariant of the Equation (2), then, as it follows from Definition 4, we have

$$\varepsilon \partial_0 U + D_1^F U = 0, \tag{4}$$

where $D_1^F = F\partial_1$.

We seek a solution of Equation (4) in the form of a series in powers of ε:

$$U(x, y_1, \varepsilon) = U_0(x, y_1) + \varepsilon U_1(x, y_1) + \ldots + \varepsilon^n U_n(x, y_1) + \ldots \tag{5}$$

for the coefficients of the equation above the following series of equations holds:

$$\begin{aligned} &D_1^F U_0 = 0, \\ &D_1^F U_1 = -\partial_0 U_0, \\ &\ldots\ldots\ldots\ldots\ldots \\ &D_1^F U_n = -\partial_0 U_{n-1}. \\ &\ldots\ldots\ldots\ldots\ldots \end{aligned} \tag{6}$$

As a solution to the first equation of this series, we take an arbitrary function $\varphi(x) \in \mathcal{A}_0$. To satisfy the condition (5°) of block II, we assume that the domain of the surjective operator D_1^F consists of functions from $\mathcal{A}_1$ that vanish when $y_1 = y_1^0 \ \forall x \in X$, and the inverse operator J_1^F is such that, for any compact sets $T \subset X$, $T_1 \subset Y_1$, there exists a number $C > 0$ such that, for an arbitrary function $\varphi(x) \in \mathcal{A}_0$ set $\Gamma_1^{\varphi} = \{C^{-n}(J_1^F\partial_0)^n\varphi,\ (x, y_1) \in T \times T_1\}_{n=1}^{\infty}$ is limited in $\mathcal{J}_a$.

As a result, all equations of the series (6), starting from the second, are uniquely solvable:

$$U(x, y_1, \varepsilon) = \varphi - \varepsilon(J_1^F\partial_0)\varphi + \ldots + (-1)^n\varepsilon^n(J_1^F\partial_0)^n\varphi + \ldots \tag{7}$$

and this series converges in some neighborhood of the value $\varepsilon = 0$ on the set $T \times T_1$. Theorem 1 is proved. □

Remark 2. *As it comes out from the form of series (7), we can consider $U(x, y_1, \varepsilon)$ for each fixed ε as the image of the linear operator $H_\varepsilon : \mathcal{A}_0 \to \mathcal{A}_1$ given by the formula*

$$H_\varepsilon = I - \varepsilon(J_1^F \partial_0) + \ldots + (-1)^n \varepsilon^n (J_1^F \partial_0)^n + \ldots,$$

where I is the identity operator. Thus, $U = H_\varepsilon[\varphi]$.

Theorem 2. *$\{H_\varepsilon\}$ forms a ε-regular family for homomorphisms of the algebra $\mathcal{A}_0$ into the algebra $\mathcal{A}_1$.*

Proof of Theorem 2. Let U and V be invariants of the Equation (2). Obviously, then there exists a function Φ such that $V = \Phi(U)$, and therefore $H_\varepsilon[\varphi(x)] = \Phi(H_\varepsilon[x])$. If in this equality we put $y_1 = y_1^0$, then $\varphi(x) = \Phi(x)\ \forall x \in X$, therefore

$$H_\varepsilon[\varphi(x)] = \varphi(H_\varepsilon[x]). \tag{8}$$

The equality (8) is called the commutation relation.

Now, let $\varphi_1(x), \varphi_2(x) \in \mathcal{A}_0$; then,

$$H_\varepsilon[\varphi_1\varphi_2] = (\varphi_1\varphi_2)(H_\varepsilon[x]) = \varphi_1(H_\varepsilon[x])\varphi_2(H_\varepsilon[x]) = H_\varepsilon[\varphi_1]H_\varepsilon[\varphi_2],$$

where $H_\varepsilon : \mathcal{A}_0 \to \mathcal{A}_1$ is a homomorphism. Theorem 2 is proved. □

For the concepts given below, we need a definition introduced by S.A. Lomov for the notion of the essentially singular manifold [4].

Definition 5. *Let $\varphi(x) \in \mathcal{A}_0$, $\varphi(x_0) = 0$, $\Phi \in \mathcal{A}_0$, let it allow continuation to all $\mathcal{J}_a$, and let T_0 be some compact from X containing the point x_0. The set $Q^+(\varphi, \Phi, T_0) = \{q : \Phi(\varphi(x)/\varepsilon),\ x \in T_0,\ \varepsilon > 0\}$ is called an essentially singular variety generated by the point $\varepsilon = 0$. Moreover, we say that it has the correct structure if*

$$Q^+ = \bigcup_{m=1}^{\infty} \Pi_m,$$

where $\Pi_1 \subset \Pi_2 \subset \ldots$ is an increasing compact system.

We introduce the concept of ε-pseudoregularity necessary for studying the analytic properties of a solution of $y(x, \varepsilon)$.

Definition 6. *The solution to the problems (2), (3) is called ε-pseudoregular if $y_1(x, \varepsilon) = \widetilde{Y}(x, \varphi(x)/\varepsilon, \varepsilon)$, in which $\varphi(x) \in \mathcal{A}_0$; the function $\widetilde{Y}(x, \eta, \varepsilon)$ is ε-regular for all $(x, \eta) \in T_0 \times G$ where T_0 is some compact set containing the point x_0, G is some unlimited set from $\mathcal{J}_a$.*

Theorem 3. *If the essentially singular manifold $Q^+(\varphi, \Phi, T_0)$ is a bounded set in $\mathcal{J}_a$ and the equation*

$$(J_1^F \partial_0)\varphi = \varphi(x)/\varepsilon \tag{9}$$

has a unique solution of the form $y_1 = Y_{1,0}(x, q)\big|_{q=\Phi(\varphi(x)/\varepsilon)}$ such that the function $Y_{1,0}(x, q)$ coincides with the contraction to the set $T_0 \times Q^+$ of some function from $\mathcal{A}_1$, then problems (2), (3) have a ε-pseudoregular solution.

Proof of Theorem 3. For the invariant represented by the Formula (7), we compose the equality

$$(J_1^F \partial_0)\varphi - \varepsilon(J_1^F \partial_0)^2\varphi + \ldots + (-1)^{n-1}\varepsilon^{n-1}(J_1^F \partial_0)^n\varphi + \ldots = \varphi(x)/\varepsilon,$$

which defines the solution to the problems (2), (3). We apply the function Φ to its left-hand and right-hand sides and, using the condition (2°) of block I, we obtain the following equality:

$$\Phi((J_1^F \partial_0)\varphi) + \varepsilon\Psi(x, y_1, \varepsilon) = q, \tag{10}$$

where $\Psi(x, y_1, \varepsilon)$ is some ε-regular function.

Let the small parameter $\varepsilon > 0$ in Equation (2) be such that the following expression holds:

$$\{q : q = \Phi(\varphi(x)/\varepsilon),\ x \in T_0\} = \Pi_m$$

for some natural number m (depending on ε). In accordance with the theorem conditions and the condition (3°) of block I, Equation (10) is solvable in some neighborhood σ_{xq} of each point $(x, q) \in T_0 \times \Pi_m$ and has a solution $y_1 = Y_1(x, q, \varepsilon)$ that is ε-regular in a neighborhood of $|\varepsilon| < \varepsilon_{xq}$, where $\varepsilon_{xq} > 0$ and is determined by this neighborhood. From the cover $\{\sigma_{xq}\}$ of the compact set $T_0 \times \Pi_m$, we choose the finite subcover $\{\sigma_{xq}\}_{i=1}^N$. Then, $y_1 = Y_1(x, q, \varepsilon)$ will be a ε-regular function in the smallest neighborhood of the point $\varepsilon = 0$ defined by a finite subcover; the function $y_1 = Y_1(x, \Phi(\varphi(x)/\varepsilon), \varepsilon)$ will give a ε-pseudoregular solution to the problem (2), (3) on the part $\widetilde{T}_0 \subset T_0$ such that the set $\{(x, q) : x \in \widetilde{T}_0,\ q = \Phi(\varphi(x)/\varepsilon)\} \subset T_0 \times \Pi_m$. The theorem is proved. □

3. Invariants and ε-Pseudoregular Solutions of Systems of Equations

We take into the consideration the system of equations

$$\begin{cases} \varepsilon dy_1 = F_1(x, y_1, \ldots, y_k), \\ \cdots\cdots\cdots\cdots\cdots\cdots\cdots \\ \varepsilon dy_k = F_k(x, y_1, \ldots, y_k), \end{cases} \tag{11}$$

the right-hand sides of which belong to the algebra $\mathcal{A}_k$. It is required to find its solution $\mathbf{y}(x, \varepsilon) = \{y_1(x, \varepsilon), \ldots, y_k(x, \varepsilon)\}$ satisfying the initial conditions

$$y_1(x^0, \varepsilon) = y_1^0,\ \ldots,\ y_k(x^0, \varepsilon) = y_k^0. \tag{12}$$

We rewrite system (11) by introducing the following denotation:

$$\begin{aligned} &\mathbb{F}(x, \mathbf{y}) = \{F_1(x, \mathbf{y}), \ldots, F_k(x, \mathbf{y})\}, \\ &\mathbf{y}^0 = \{y_1^0, \ldots, y_k^0\}. \end{aligned}$$

Thus, we have

$$\begin{aligned} &\varepsilon d\mathbf{y} = \mathbb{F}(x, \mathbf{y}), \\ &\mathbf{y}(x^0, \varepsilon) = \mathbf{y}^0 \end{aligned} \tag{13}$$

to be the initial investigated problem.

Definition 7. *The function $U(x, \mathbf{y}, \varepsilon) \in \mathcal{A}_k$ is called the invariant of the system (11) if it turns into a constant on the solution $\mathbf{y}(x, \varepsilon)$.*

We formulate a theorem similar to Theorem 1.

Theorem 4. *The system (11) has ε-regular invariants.*

Proof of Theorem 4. The proof is carried out according to the same scheme as in the case of a single equation. □

Definition 8. *The solution of the* $\mathbf{y}(x,\varepsilon)$ *problem* (13) *is called* ε*-pseudoregular if* $\mathbf{y}(x,\varepsilon) = \widetilde{\mathbb{Y}}(x,\boldsymbol{\varphi}(x)/\varepsilon,\varepsilon)$*, in which* $\boldsymbol{\varphi}(x) = \{\varphi_1(x),\ldots,\varphi_k(x)\}$*,* $\varphi_i(x) \in \mathcal{A}_0$ $(i=\overline{1,k})$ *and the function* $\widetilde{\mathbb{Y}}(x,\boldsymbol{\eta},\varepsilon)$ *in which* $\boldsymbol{\eta} = (\eta_1,\ldots,\eta_k)$*, is* ε*-regular for all* $(x,\eta_1,\ldots,\eta_k) \in T_0 \times G_1 \times \ldots \times G_k$ *where* T_0 *is some compact set from* X *containing* x_0 *and* G_i $(i=\overline{1,k})$ *are unbounded sets from* $\mathcal{J}_a$*.*

Theorem 5. *Let the following conditions be fulfilled:*

(1°) *The functions* Φ_i *and* φ_i *are such that the essentially singular manifolds* $Q_i^+(\varphi_i,\Phi_i,T_0)$ $(i=\overline{1,k})$ *are bounded sets in* $\mathcal{J}_a$*.*

(2°) *The equation* $D_k^{\mathbb{F}} V = e$ *has the solutions* $\{V_1(x,\mathbf{y}),\ldots,V_k(x,\mathbf{y})\}$ *such that system*

$$\begin{cases} \partial_0\varphi_1 \cdot V_1(x,\mathbf{y})) = \varphi_1(x)/\varepsilon, \\ \cdots\cdots\cdots\cdots\cdots\cdots \\ \partial_0\varphi_k \cdot V_k(x,\mathbf{y})) = \varphi_k(x)/\varepsilon \end{cases}$$

has the only solution

$$\mathbf{y} = \mathbb{Y}_0(x,\mathbf{q})\Big|_{\substack{\mathbf{q}=(q_1,\ldots,q_k) \\ q_i=\Phi_i(\varphi_i(x)/\varepsilon)}},$$

and each component $Y_{0,i}(x,\mathbf{q})$ $(i=\overline{1,k})$ *of it coincides with the restriction to the set* $T_0 \times Q_1^+ \times \ldots \times Q_k^+$ *of some function from* $\mathcal{A}_k$*.*

Then, the solution $\mathbf{y}(x,\varepsilon)$ *of the problem* (13) *is* ε*-pseudoregular.*

Proof of Theorem 5. We write the equalities for the invariants of the system (11) in the following form:

$$\begin{cases} \partial_0\varphi_1 \cdot V_1(x,\mathbf{y}) - \varepsilon(J_k^{\mathbb{F}}\partial_0)^2\varphi_1 + \varepsilon^2(J_k^{\mathbb{F}}\partial_0)^3\varphi_1 - \ldots = \varphi_1(x)/\varepsilon, \\ \cdots\cdots\cdots\cdots\cdots\cdots\cdots\cdots\cdots\cdots \\ \partial_0\varphi_k \cdot V_k(x,\mathbf{y}) - \varepsilon(J_k^{\mathbb{F}}\partial_0)^2\varphi_k + \varepsilon^2(J_k^{\mathbb{F}}\partial_0)^3\varphi_k - \ldots = \varphi_k(x)/\varepsilon. \end{cases} \tag{14}$$

In order for this system to determine the solution of the problem (11), (12) (or (13)), we assume (see condition (5°) of block II) that $\mathcal{D}(D_k^{\mathbb{F}})$ consists of functions that vanish when $\mathbf{y} = \mathbf{y}^0$ for any $x \in X$ and $V_i(x,\mathbf{y}^0) = 0$, $i=\overline{1,k}$.

We apply the functions $\Phi_1,\ldots,\Phi_k$ to the equations of the system (14), respectively. Then, in accordance with the condition (2°) of block I, we obtain the system

$$\begin{cases} \Phi_1(\partial_0\varphi_1 \cdot V_1(x,\mathbf{y})) + \varepsilon\Psi_1(x,\mathbf{y},\varepsilon) = q_1, \\ \cdots\cdots\cdots\cdots\cdots\cdots\cdots\cdots \\ \Phi_k(\partial_0\varphi_k \cdot V_k(x,\mathbf{y})) + \varepsilon\Psi_k(x,\mathbf{y},\varepsilon) = q_k. \end{cases} \tag{15}$$

Let the small parameter $\varepsilon > 0$ in the system (13) be such that

$$\{q_i : q_i = \Phi_i(\varphi(x)/\varepsilon),\ x \in T_0\} = \Pi_{m_i},\ \ i=\overline{1,k}$$

for natural $m_1,\ldots,m_k$. By the condition of the (2°) Theorem 5, the system (15) for $\varepsilon = 0$ has a unique solution $\mathbf{y} = \mathbb{Y}_0(x,\mathbf{q})$ and, therefore, in accordance with the condition (3°) of block I, this system is solvable in some neighborhood $\sigma_{x\mathbf{q}}$ of each point $(x,\mathbf{q}) \in T_0 \times \Pi_{m_1} \times \ldots \times \Pi_{m_k}$, and its solution $\mathbb{Y}(x,\mathbf{q},\varepsilon)$ is ε-regular there for $|\varepsilon| < \varepsilon_{x\mathbf{q}}$. After that, from the cover $\{\sigma_{x\mathbf{q}}\}$ of the compact set $T_0 \times \Pi_{m_1} \times \ldots \times \Pi_{m_k}$, we choose a finite subcover and $\mathbb{Y}(x,\mathbf{q},\varepsilon)$ will be ε-regular in the minimal neighborhood from the neighborhood of $\varepsilon = 0$ corresponding to a finite subcover. As in the proof of Theorem 3, we choose $\widetilde{T}_0 \subset T_0$, a compact set on which there exists a ε-regular solution

$$\mathbf{y}(x,\varepsilon) = \mathbb{Y}(x,\mathbf{q},\varepsilon)\Big|_{\substack{q_i=\Phi_i(\varphi_i(x)/\varepsilon)\\ (i=\overline{1,k})}}.$$

The theorem is proved. □

4. Concrete Implementations of the Theory

In this section of the article, we assume that $\mathcal{J}_a = \mathbb{C}$, $X = P_0 \equiv \{z \in \mathbb{C} : |z - z^0| < r_0\}$, $Y_i = P_i \equiv \{w_i \in \mathbb{C} : |w_i - w_i^0| < r_i\}, i = \overline{1,k}$. We shall use the following denotations $\mathbf{w} = \{w_1, \ldots, w_k\}$, $\mathbf{w}^0 = \{w_1^0, \ldots, w_k^0\}$, $\mathbb{P}^k = P_1 \times \ldots \times P_k$ a polycircle of $\mathbb{C}^k$.

Let $\mathcal{A}_0$ be the algebra of holomorphic functions in the P_0 circle of the variable z; let $\mathcal{A}_1$ be the algebra of holomorphic functions in the $P_0 \times P_1$ bicircle of the variables (z, w_1), ...; let $\mathcal{A}_k$ be the algebra of holomorphic functions of the variables $(z, w_1, \ldots, w_k)$ in the polycircle $P_0 \times \mathbb{P}^k$. It is clear that, if $\partial_0 = \partial_z, \partial_1 = \partial_{w_1}, \ldots, \partial_k = \partial_{w_k}$, then all the conditions of block I and the conditions (1°)—(4°) of block II are satisfied. In the concepts given below, we show that the condition (5°) also holds under fairly general assumptions.

Thus, we investigate the Cauchy problem for $\varepsilon > 0$:

$$\begin{aligned} &\varepsilon\frac{d\mathbf{w}}{dz} = \mathbb{F}(z,\mathbf{w}), \;\; z \in \widetilde{P}_0 = \{z \in \mathbb{C} : |z - z^0| < \widetilde{r}_0, \; 0 < \widetilde{r}_0 < r_0\}, \\ &\mathbf{w}(z^0,\varepsilon) = \mathbf{w}^0, \end{aligned} \tag{16}$$

where $\mathbb{F}(z,\mathbf{w}) = \{F_1(z,\mathbf{w}), \ldots, F_k(z,\mathbf{w})\}$, $F_i(z,\mathbf{w}) \in \mathcal{A}_k$ for $i = \overline{1,k}$.

From the nonlinear system (16), we come to the linear equation of its integrals (invariants):

$$\varepsilon\partial_z\mathbb{U} + D_k^{\mathbb{F}}\mathbb{U} = 0. \tag{17}$$

Here, $D_k^{\mathbb{F}} = F_1\partial_{w_1} + \ldots + F_k\partial_{w_k}$ is the linear partial differential operator of the first order in partial derivatives: $\mathbb{U} = \{U^{[1]}, \ldots, U^{[k]}\}$, where $\{U^{[i]}\}_{i=1}^k$ is the system of independent integrals.

First of all, we present an integral method for solving inhomogeneous linear differential equations of the first order with partial derivatives [5].

Let Λ be a holomorphically smooth surface in $\mathbb{C}^k$ and we need to solve the initial problem

$$\begin{aligned} &D_k^{\mathbb{F}}V = f, \;\; f \in \mathcal{A}_k, \\ &V\big|_{\mathbf{w}\in\Lambda} = 0. \end{aligned} \tag{18}$$

Let us suppose that the surface Λ is given by the coordinates $\widetilde{\mathbf{w}} = \{\widetilde{w}_1, \ldots, \widetilde{w}_{k-1}\}$ and, namely, $\Lambda = \{\mathbf{w} \in \mathbb{C}^k : w_i = \lambda(\widetilde{\mathbf{w}}), i = \overline{1,k}\}$, where $\lambda_i(\widetilde{\mathbf{w}})$ are functions holomorphic in some region $\mathbb{C}^{k-1}$. Next, we compose the equation system for the characteristic equation

$$\frac{d\mathbf{w}}{ds} = \mathbb{F}(z,\mathbf{w}), \tag{19}$$

in which $s \in \mathbb{C}$ is an independent variable, and z acts as a parameter. Let $\mathbf{w} = \mathbf{g}(z, \widetilde{\mathbf{w}}, s)$ be a solution to the system (19) with the initial condition

$$\mathbf{w}\big|_{s=0} = \boldsymbol{\lambda}(\widetilde{\mathbf{w}}),$$

where $\boldsymbol{\lambda} = \{\lambda_1, \ldots, \lambda_k\}$.

The existence and uniqueness theorem guarantees the unique solvability of the system $\mathbf{g}(z, \widetilde{\mathbf{w}}, s) = \mathbf{w}$ relative to $\widetilde{\mathbf{w}}$ and s: $s = S(z, \mathbf{w})$, $\widetilde{\mathbf{w}} = \widetilde{\mathbb{W}}(z, \mathbf{w})$. We denote the operator of replacing variables $(s, \widetilde{\mathbf{w}})$ by the variable $\mathbf{w}$ by $R(z)$ and the backward replacement operator is denoted by $R^{-1}(z, s)$:

$$R(z)[\chi(z,s,\widetilde{\mathbf{w}})] = \chi(z,S(z,\mathbf{w}),\widetilde{\mathbb{W}}(z,\mathbf{w})),$$
$$R^{-1}(z,s)[\theta(z,\mathbf{w})] = \theta(z,\mathbf{g}(z,\widetilde{\mathbf{w}},s)).$$

Then, as you know, if the phase trajectories in the system of characteristics are transversal (not tangent) to the surface Λ, then a solution to the Cauchy problem (17) exists, is unique, and is expressed by the following formula:

$$V(z,\mathbf{w}) = \int_0^s f(z,\mathbf{g}(z,\widetilde{\mathbf{w}},s_1))ds_1 \Big|_{\substack{s=S(z,\mathbf{w}), \\ \widetilde{\mathbf{w}}=\widetilde{\mathbb{W}}(z,\mathbf{w})}}. \tag{20}$$

We return to Equation (18). We have

$$\mathbb{U}(z,\mathbf{w},\varepsilon) = \mathbb{U}_0(z,\mathbf{w}) + \varepsilon\mathbb{U}_1(z,\mathbf{w}) + \ldots + \varepsilon^n\mathbb{U}_n(z,\mathbf{w}) + \ldots, \tag{21}$$

and, as this takes place

$$\begin{aligned} &D_k^{\mathbb{F}}\mathbb{U}_0(z,\mathbf{w}) = 0, \\ &D_k^{\mathbb{F}}\mathbb{U}_1(z,\mathbf{w}) = -\partial_z\mathbb{U}_0(z,\mathbf{w}), \\ &\ldots\ldots\ldots\ldots\ldots\ldots \\ &D_k^{\mathbb{F}}\mathbb{U}_n(z,\mathbf{w}) = -\partial_z\mathbb{U}_{n-1}(z,\mathbf{w}). \\ &\ldots\ldots\ldots\ldots\ldots\ldots \end{aligned} \tag{22}$$

As a solution to the first equation of this series, we take the vector function $\mathbb{U}_0 = \{\varphi_1(z),\ldots,\varphi_k(z)\}$, $\varphi_i(z) \in \mathcal{A}_0$ for $i = \overline{1,k}$. The solution to the second equation of the series (22) is the vector function $\mathbb{U}_1 = \{-\partial_z\varphi_1 V^{[1]},\ldots,-\partial_z\varphi_k V^{[k]}\}$ where $\{V^{[1]},\ldots,V^{[k]}\}$ are functionally independent solutions of the equation $D_k^{\mathbb{F}}V = 1$ and such that $V^{[i]}(z,\mathbf{w}^0) = 0\ \forall z \in P_0$, $i = \overline{1,k}$. We find solutions to other equations using Formula (20), assuming that $\mathbf{w}^0 \in \Lambda$:

$$\begin{aligned} &\mathbb{U}_2(z,\mathbf{w}) = -R(z)\int_0^s R^{-1}(z,s_1)\partial_z\mathbb{U}_1(z,\mathbf{w})ds_1, \\ &\mathbb{U}_3(z,\mathbf{w}) = R(z)\int_0^s ds_1 R^{-1}(z,s_1)\partial_z R(z)\int_0^{s_1} R^{-1}(z,s_2)\partial_z\mathbb{U}_1(z,\mathbf{w})ds_2, \\ &\ldots\ldots\ldots\ldots\ldots\ldots \\ &\mathbb{U}_n(z,\mathbf{w}) = (-1)^{n-1}R(z)\int_0^s ds_1 R^{-1}(z,s_1)\partial_z R(z)\int_0^{s_1} ds_2 R^{-1}(z,s_2)\partial_z \ldots \\ &\qquad \ldots R(z)\int_0^{s_{n-2}} R^{-1}(z,s_{n-1})\partial_z\mathbb{U}_1(z,\mathbf{w})ds_{n-1}. \\ &\ldots\ldots\ldots\ldots\ldots\ldots \end{aligned} \tag{23}$$

Next, to each natural $n \geq 2$, we associate $(n-1)$ concentric circles $C_m = \{z : |z - z^0| = t_m\}$ where

$$t_m = \widetilde{r}_0 + \frac{r - \widetilde{r}_0}{n-1}m, \quad m = \overline{1,n-1}$$

and $\widetilde{r}_0 < r < r_0$.

These circles are situated at the same distance from each other:

$$\rho = t_{n-1} - t_{n-2} = \ldots = t_2 - \widetilde{r}_0 = \frac{r - \widetilde{r}_0}{n-1}.$$

We use the equalities (23) with the Cauchy integral formula:

$$\begin{aligned}
\mathbb{U}_2(z,\mathbf{w}) &= -\tfrac{1}{2\pi i}R(z)\int_0^s R^{-1}(z,s_1)\oint_{C_1}\tfrac{dz_1}{(z_1-z)^2}\mathbb{U}_1(z_1,\mathbf{w})ds_1,\\
\mathbb{U}_3(z,\mathbf{w}) &= \tfrac{1}{(2\pi i)^2}R(z)\int_0^s ds_1R^{-1}(z,s_1)\oint_{C_1}\tfrac{dz_1}{(z_1-z)^2}R(z_1)\int_0^{s_1}R^{-1}(z_1,s_2)\oint_{C_2}\tfrac{dz_2}{(z_2-z_1)^2}\mathbb{U}_1(z_2,\mathbf{w})ds_2,\\
&\cdots\cdots\cdots\cdots\cdots\cdots\cdots\cdots\cdots\cdots\\
\mathbb{U}_n(z,\mathbf{w}) &= \tfrac{(-1)^{n-1}}{(2\pi i)^{n-1}}R(z)\int_0^s ds_1R^{-1}(z,s_1)\oint_{C_1}\tfrac{dz_1}{(z_1-z)^2}R(z_1)\int_0^{s_1}ds_2R^{-1}(z_1,s_2)\oint_{C_2}\tfrac{dz_2}{(z_2-z_1)^2}\cdots\\
&\quad \ldots R(z_{n-2})\int_0^{s_{n-2}}R^{-1}(z_{n-2},s_{n-1})\oint_{C_{n-1}}\tfrac{dz_{n-1}}{(z_{n-1}-z_{n-2})^2}\mathbb{U}_1(z_{n-1},\mathbf{w})ds_{n-1}.\\
&\cdots\cdots\cdots\cdots\cdots\cdots\cdots\cdots\cdots\cdots
\end{aligned} \tag{24}$$

We represent $\mathbb{U}_n(z,\mathbf{w})$ in the following form:

$$\mathbb{U}_n(z,\mathbf{w}) = \frac{(-1)^{n-1}}{(2\pi i)^{n-1}}\int_0^s ds_1\int_0^{s_1}ds_2\ldots\int_0^{s_{n-2}}ds_{n-1}\oint_{C_1}\frac{dz_1}{(z-z_1)^2}\cdots\oint_{C_{n-2}}\frac{dz_{n-2}}{(z_{n-3}-z_{n-2})^2}\cdot$$
$$\cdot\oint_{C_{n-1}}\frac{R(z)R^{-1}(z,s_1)R(z_1)R^{-1}(z_1,s_2)\ldots R(z_{n-2})R^{-1}(z_{n-2},s_{n-1})\mathbb{U}_1(z_{n-1},\mathbf{w})dz_{n-1}}{(z_{n-2}-z_{n-1})^2}.$$

Let $\|\cdot\|_k$ be the norm in $\mathbb{C}^k$; then, for all $z\in\widehat{P}_0=\{z\in\mathbb{C}:|z-z_0|<r_0\}$ and all $\mathbf{w}$ from some subregion $\widehat{\mathbb{P}}_k$ of the polycircle $\mathbb{P}_k$, the following inequality takes place:

$$\|\mathbb{U}_n(z,\mathbf{w})\|_k \le \left|\int_0^s ds_1\int_0^{s_1}ds_2\ldots\int_0^{s_{n-2}}ds_{n-1}\right|\frac{1}{(2\pi)^{n-1}}H_{n-1}\|\mathbb{U}_1(z,\mathbf{w})\|_k,$$

where

$$\begin{aligned}
H_{n-1} &= \oint_{C_1}\tfrac{|dz_1|}{|z-z_1|^2}\cdots\oint_{C_{n-1}}\tfrac{|dz_{n-1}|}{|z_{n-2}-z_{n-1}|^2} =\\
&= \int_0^{2\pi}\tfrac{t_1d\alpha}{t_1^2+|z|^2-2t_1|z|\cos\alpha}\cdots\int_0^{2\pi}\tfrac{t_{n-1}d\alpha}{t_{n-1}^2+t_{n-2}^2-2t_{n-1}t_{n-2}\cos\alpha} =\\
&= \tfrac{(2\pi)^{n-1}t_1t_2\ldots t_{n-1}}{(t_1-|z|^2)(t_2^2-t_1^2)\ldots(t_{n-1}^2-t_{n-2}^2)} \le \tfrac{(2\pi)^{n-1}r_0^{n-1}(n-1)^{n-1}}{2^{n-2}\widetilde{r}_0^{\,n-1}(r-\widetilde{r}_0)^{n-1}}.
\end{aligned}$$

As we have

$$\left|\int_0^s ds_1\int_0^{s_1}ds_2\ldots\int_0^{s_{n-2}}ds_{n-1}\right| = \frac{|s|^{n-1}}{(n-1)!},$$

then

$$\|\mathbb{U}_n(z,\mathbf{w})\|_k \le \frac{r_0^{n-1}(n-1)^{n-1}}{2^{n-2}\widetilde{r}_0^{\,n-1}(r-\widetilde{r}_0)^{n-1}(n-1)!}\|\mathbb{U}_1(z,\mathbf{w})\|_k,$$

and from that the convergence of the series (21) on any compact set from the set $\widehat{P}_0\times\widehat{\mathbb{P}}_k$ follows.

Thus, it is proved that the components of the vector $\mathbb{U}(z,\mathbf{w},\varepsilon)$ form an independent system of integrals (invariants) and are holomorphic (ε-regular) at the point $\varepsilon=0$. It is also clear that there is a statement similar to Theorem 5 on the existence of a pseudoholomorphic (ε-pseudoregular) solution of the Cauchy problem (16). Without loss of generality, we assume that $z_0=0$.

Theorem 6. *Let the entire functions* $\{\Phi_1, \ldots, \Phi_k\}$ *and the functions* $\{\varphi_1(z), \ldots, \varphi_k(z)\}$ *which are holomorphic in the circle* P_0 *be such that the essentially singular manifolds* $\{Q_1^+(\varphi_1, \Phi_1, T_0), \ldots, Q_k^+(\varphi_k, \Phi_k, T_0)\}$ *created by the functions described above where* T_0 *is some segment of the real axis, the left end of which coincides with the origin and belongs to the circle* $\widehat{P}_0$*, are sets bounded in* $\mathbb{C}$*; and the system of equations*

$$\begin{cases} \varphi_1'(z)V_1(z, \mathbf{w}) = \varphi_1(z)/\varepsilon, \\ \ldots\ldots\ldots\ldots\ldots\ldots\ldots \\ \varphi_k'(z)V_k(z, \mathbf{w}) = \varphi_k(z)/\varepsilon, \end{cases}$$

in which $\{V_1(z, \mathbf{w}), \ldots, V_k(z, \mathbf{w})\}$ *are independent solutions of the equation* $D_k^{\mathbb{F}} V = 1$*, has a solution of the form*

$$\mathbf{w} = \mathbb{W}_0(z, \mathbf{q})\Bigg|_{\mathbf{q}=\{\Phi_1(\varphi_1(z)/\varepsilon), \ldots, \Phi_k(\varphi_k(z)/\varepsilon)\}},$$

each component $W_{0,i}(z, \mathbf{q})$ $(i = \overline{1,k})$ *of it is holomorphic on the set* $T_0 \times Q_1^+ \times \ldots \times Q_k^+$*. Then, the solution* $\mathbf{w}(z, \varepsilon)$ *of the initial problem* (16) *is pseudoholomorphic at the point* $\varepsilon = 0$ *(ε-pseudoregular).*

It should be noted [6] that this solution can be continued in a pseudoholomorphic way for a fixed $\varepsilon > 0$ from some segment $[0, \widetilde{T}_0]$ (see the end of the proof of Theorem 5) by segment $[0, T_0]$.

5. Conclusions

Further development of the axiomatic approach in the analytical singular perturbation theory will allow us to consider a more general class of equations with a small parameter, in particular, an analogue of nonlinear differential equations in partial derivatives (for example, equations of the Navier–Stokes type, etc.). This is very urgent since the range of problems leading to singularly perturbed problems is constantly expanding. In this sense, the "Dyson argument" that appeared in theoretical physics is quite indicative—the solutions of the equations arising in astrophysics can depend holomorphically on the gravitational constant only after isolating a revealing the essentially singular manifold [7].

As for the current state of the singular perturbation of theory, the asymptotic approach prevails there. In our opinion, when solving most singularly perturbed equations, the following methods are used: the Vasilieva–Butuzov–Nefedov boundary function method [8,9], the Maslov method [10], the Lomov regularization method [4,11], and the Bogolyubov–Krylov–Mitropolsky average method [12,13]. In the case of more specific situations, these methods are combined and new approaches to the asymptotic integration are proposed [14].

Author Contributions: Conceptualization and writing—review and editing by M.B. and V.K. All authors have read and agreed to the published version of the manuscript.

Funding: This research received no external funding.

Conflicts of Interest: The authors declare no conflict of interest.

References

1. Kachalov, V.I. On a method of holomorphic regularization of singularly perturbed problems. *Russ. Math. (Izv. VUZov)* **2017**, *61*, 44–50. [CrossRef]
2. Kachalov, V.I. On holomorphic regularization of strongly nonlinear singularly perturbed problems. *Ufa Math. J.* **2018**, *10*, 35–42. [CrossRef]
3. Rudin, W. *Functional Analysis*; McGraw-Hill Book Company: New York, NY, USA, 1973.
4. Lomov, S.A.; Lomov, I.S. *Fundamentals of the Mathematical Theory of Boundary Layer*; Moscow State University: Moscow, Russia, 2011.
5. Samoilenko, A.M.; Krivosheya, S.A.; Perestyuk, N.A. *Differential Equations*; Vysshaya Shkola: Moscow, Russia, 1989.

6. Bobodzhanov, A.A.; Safonov, V.F.; Kachalov, V.I. Asymptotic and Pseudoholomorphic Solutions of Singularly Perturbed Differential and Integral Equations in the Lomov's Regularization Method. *Axioms* **2019**, *8*, 27. [CrossRef]
7. Krivoruchenko, M.I.; Nadyozhin, D.K.; Yudin, A.V. Hydrostatic Equilibrium of Stars without Electroneutrality constraint. *Phys. Rev. D* **2018**, *97*, 083016. [CrossRef]
8. Vasilieva, A.B.; Butuzov, V.F. *Asymptotic Expansions of Solutions of Singularly Perturbed Equations*; Nauka: Moscow, Russia, 1989.
9. Vasilieva, A.B.; Butuzov, V.F.; Nefedov, N.N. Contrast structures in singularly perturbed problems. *Fundam. Appl. Math.* **1998**, *4*, 799–851.
10. Maslov, V.P. *Asymptotic Methods and Perturbation Theory*; Nauka: Moscow, Russia, 1988.
11. Lomov, S.A. *Introduction to the General Theory of Singular Perturbations*; Nauka: Moscow, Russia, 1989.
12. Bogolyubov, N.N; Mitropolsky, Yu.A. *Asymptotic Methods in the Theory of Nonlinear Oscillations*; Publ. H. of the Academy of Sciences of the USSR: Moscow, Russia, 1963.
13. Mitropolsky, Y.A. *Averaging Method in Nonlinear Mechanics*; Naukova Dumka: Kiev, Ukraine, 1971.
14. Glizer, V.Y. Conditions of Functional Null Controllability for Some Types of Singularly Perturbed Nonlineal Systems with Delays. *Axioms* **2019**, *8*, 80. [CrossRef]

Article

Regularized Asymptotic Solutions of a Singularly Perturbed Fredholm Equation with a Rapidly Varying Kernel and a Rapidly Oscillating Inhomogeneity

Dana Bibulova [1], Burkhan Kalimbetov [2,*] and Valeriy Safonov [3]

[1] Department of Higher Mathematics, South Kazakhstan University Named after M. Auezov, Tauke-Khan Ave., 5, Shymkent 160000, Kazakhstan; danass86@mail.ru
[2] Department of Mathematics, Akhmed Yassawi University, B. Sattarkhanov 29, Turkestan 161200, Kazakhstan
[3] Department of Higher Mathematics, National Research University «MPEI», Krasnokazarmennaya 14, 111250 Moscow, Russia; singsaf@yandex.ru
* Correspondence: burkhan.kalimbetov@ayu.edu.kz

Citation: Bibulova, D.; Kalimbetov, B.; Safonov, V. Regularized Asymptotic Solutions of a Singularly Perturbed Fredholm Equation with a Rapidly Varying Kernel and a Rapidly Oscillating Inhomogeneity. *Axioms* **2022**, *11*, 141. https://doi.org/10.3390/axioms11030141

Academic Editors: Hans J. Haubold and Valery Y. Glizer

Received: 15 December 2021
Accepted: 10 March 2022
Published: 18 March 2022

Publisher's Note: MDPI stays neutral with regard to jurisdictional claims in published maps and institutional affiliations.

Copyright: © 2022 by the authors. Licensee MDPI, Basel, Switzerland. This article is an open access article distributed under the terms and conditions of the Creative Commons Attribution (CC BY) license (https://creativecommons.org/licenses/by/4.0/).

Abstract: This article investigates an equation with a rapidly oscillating inhomogeneity and with a rapidly decreasing kernel of an integral operator of Fredholm type. Earlier, differential problems of this type were studied in which the integral term was either absent or had the form of a Volterra-type integral. The presence of an integral operator and its type significantly affect the development of an algorithm for asymptotic solutions, in the implementation of which it is necessary to take into account essential singularities generated by the rapidly decreasing kernel of the integral operator. It is shown in tise work that when passing the structure of essentially singular singularities changes from an integral operator of Volterra type to an operator of Fredholm type. If in the case of the Volterra operator they change with a change in the independent variable, then the singularities generated by the kernel of the integral Fredholm-type operators are constant and depend only on a small parameter. All these effects, as well as the effects introduced by the rapidly oscillating inhomogeneity, are necessary to take into account when developing an algorithm for constructing asymptotic solutions to the original problem, which is implemented in this work.

Keywords: singular perturbation; integro-differential equation; rapidly oscillating inhomogeneity; regularization; asymptotic convergence

MSC: 34K26; 45J05

1. Introduction

Integro-differential equations

$$L_\varepsilon y(t,\varepsilon) \equiv \varepsilon \frac{dy}{dt} - a(t)y - \int_0^\alpha e^{\frac{1}{\varepsilon}\int_s^\alpha \mu(\theta)d\theta} K(t,s)y(s,\varepsilon)ds =$$

$$= h_1(t) + h_2(t)e^{\frac{i\beta(t)}{\varepsilon}}, \ y(0,\varepsilon) = y^0, \ t \in [0,T]$$

(1)

with rapidly changing Volterra-type kernels ($\alpha = t$) have been studied from various positions in a number of works (see, for example, [1] and its bibliography). The problems were considered on the construction of a regularized asymptotics for the solution of problem type (1) in the case of stability of the operator $a(t)$ and the spectral value $\mu(t)$ of the kernel of the integral operator [2–7]. As for the integro-differential equations (1) with rapidly changing kernels of the Fredholm type ($\alpha = T$), it was assumed that the results obtained for the Volterra equations are automatically extended to equations of the Fredholm type. However, when considering the simplest case of scalar Equation (1) for (see, for example, [8–14]) it turned out that the spectral value does not participate in the regularization of problem (1) (in contrast to the case $\alpha = t$), and the coefficients of the elements of the space of resonance-free solutions (in the terminology of S.A. Lomov [15–17] depend on the exponentials

$$\sigma_1 = \exp\{\tfrac{1}{\varepsilon}\int_0^1 a(\theta)d\theta\}, \sigma_2 = \exp\{\tfrac{i}{\varepsilon}\int_0^1 \beta'(\theta)d\theta\}, \sigma_3 = \exp\{\tfrac{1}{\varepsilon}\int_0^1 \mu(\theta)d\theta\},$$

$$\sigma_4 = \exp\{\tfrac{i}{\varepsilon}\beta(0)\}, (\beta'(t) > 0 \ \forall t \in [0,1])$$

bounded for $\varepsilon \to +0$, if $a(t) < 0$ and $\mu(t) < 0, \beta'(t) > 0$ $(t \in [0,1]), \beta(t)$ is a real function). Therefore, the regularization of problem (1) and the theory of normal and unique solvability of the corresponding iterative problems do not fit into the previously developed scheme for equations of the Volterra type and should be revised, taking into account those changes which are introduced by the Fredholm operator. Fredholm-type integro-differential equations with slow and rapidly changing kernels have been studied in [18]. In recent years, the main attention of researchers has been focused on the development of asymptotic solutions for integro-differential equations with rapidly oscillating coefficients in the presence of rapidly oscillating inhomogeneities [19]. Therefore, in this paper, an attempt is made to create an algorithm for constructing an asymptotic solution of problem (1) at $\varepsilon \to +0$. Without a loss of generality, we can assume that $\alpha = T = 1$, and we will proceed to the study of the problem (1).

Thus, in this paper we consider the following Cauchy problem:

$$L_\varepsilon y(t,\varepsilon) \equiv \varepsilon\frac{dy}{dt} - a(t)y - \int_0^1 e^{\frac{1}{\varepsilon}\int_s^1 \mu(\theta)d\theta}K(t,s)y(s,\varepsilon)ds = \\ = h_1(t) + h_2(t)e^{\frac{i\beta(t)}{\varepsilon}}, y(0,\varepsilon) = y^0, t \in [0,1] \quad (2)$$

with the Fredholm type of integral operator.

2. Regularization of the Problem (1)

The problem (1) will be considered under the following conditions:

(1) $a(t), \mu(t), \beta(t) \in C^\infty([0,1], \mathbb{R}), h_1(t), h_2(t) \in ([0,1], \mathbb{C}), K(t,s) \in C^\infty(\{0 \le s \le t \le 1\}, \mathbb{C})$;

(2) $a(t) \ne \mu(t), \mu(t) < 0, a(t) < 0 \forall t \in [0,1]$.

Let us denote $\lambda_1(t) \equiv a(t), \lambda_2(t) \equiv i\beta'(t), \lambda_3(t) \equiv \mu(t)$ and call the set $\{\lambda_j(t)\}$ the spectrum of problem (1). We introduce the regularizing variables

$$\tau_j = \frac{1}{\varepsilon}\int_0^t \lambda_j(\theta)d\theta = \frac{\psi_j(t)}{\varepsilon}, j = 1,2$$

along the points of the spectrum $\lambda_1(t)$ and $\lambda_2(t)$ of the problem (1) (in this case, as will be shown below, the variable $\tau_3 = \varepsilon^{-1}\int_0^t \lambda_3(\theta)d\theta$ does not participate in the regularization). For the "extension" $\tilde{y}(t,\tau,\varepsilon)$ we obtain the following problem:

$$L_\varepsilon \tilde{y}(t,\tau,\varepsilon) \equiv \varepsilon\frac{\partial \tilde{y}}{\partial t} + \sum_{j=1}^{2} \lambda_j(t)\frac{\partial \tilde{y}}{\partial \tau_j} - \lambda_1(t)\tilde{y} - \int_0^1 e^{\frac{1}{\varepsilon}\int_s^1 \lambda_3(\theta)d\theta}K(t,s)\tilde{y}(s,\tfrac{\psi(s)}{\varepsilon},\varepsilon)ds = \\ = h_1(t) + h_2(t)e^{\tau_2}\sigma_4, \ \tilde{y}(0,0,\varepsilon) = y^0, \ t \in [0,1] \quad (3)$$

where $\tau = (\tau_1, \tau_2), \psi = (\psi_1, \psi_2)$. The function $\tilde{y}(t,\tau,\varepsilon)$ satisfies the necessary regularization condition: $\tilde{y}(t,\frac{\psi(t)}{\varepsilon},\varepsilon) \equiv y(t,\varepsilon)(y(t,\varepsilon)$ is the exact solution to problem (1)). However, problem (3) cannot be considered completely regularized, since the integral term

$$J\tilde{y} \equiv J\left(\tilde{y}(t,\tau,\varepsilon)|_{t=s,\tau=\psi(s)/\varepsilon}\right) = \int_0^1 e^{\frac{1}{\varepsilon}\int_s^1 \lambda_3(\theta)d\theta}K(t,s)\tilde{y}(s,\frac{\psi(s)}{\varepsilon},\varepsilon)ds. \quad (4)$$

has not been regularized in it. For its regularization, as is known, it is necessary to introduce a class M_ε, that is asymptotically invariant with respect to the operator J (see [15], pp. 62–64).

Definition 1. *We say that a vector function $y(t,\tau,\sigma)$ belongs to the space U, if it is represented by a sum of the form*

$$y(t,\tau) \equiv y(t,\tau,\sigma) = y_0(t,\sigma) + \sum_{j=1}^{2} y_j(t,\sigma)e^{\tau_j} \tag{5}$$

where the functions $y_j(t,\sigma)$ are polynomials in $\sigma = (\sigma_1,\ldots,\sigma_4)$ with coefficients from the class $C^\infty([0,1],\mathbb{C})$, i.e.,

$$y_j(t,\sigma) = \Sigma_{|m|=0}^{N_j} y_j^{(m_1,\ldots,m_4)}(t)\sigma_1^{m_1}\cdots\sigma_4^{m_4},$$

$$y_j^{(m_1,\ldots,m_4)}(t) \in C^\infty([0,1],\mathbb{C}),\quad 0 \le |m| \equiv m_1+\cdots+m_4,\ N_i < \infty,\quad j=0,1,2.$$

We take as M_ε the class $U|_{\tau=\frac{\psi(t)}{\varepsilon}}$. It should be shown that the image $Jy(t,\tau)$ on functions (4) can be represented in the form of a series

$$\sum_{k=0}^{\infty} \varepsilon^k \left(\sum_{j=1}^{2} z_j^{(k)}(t,\sigma)e^{\tau_j} + z_0^{(k)}(t,\sigma) \right)|_{\tau=\frac{\psi(t)}{\varepsilon}},$$

converging asymptotically to Jy at $\varepsilon \to +0$ (uniformly with respect to $t \in [0,1]$). Substituting (5) in, we will have

$$Jy(t,\tau,\sigma) = \int_0^1 K(t,s)y_0(s,\sigma)e^{\frac{1}{\varepsilon}\int_s^1 \lambda_3(\theta)d\theta}ds + \sum_{j=1}^{2}\int_0^1 K(t,s)y_j(s,\sigma)e^{\frac{1}{\varepsilon}\int_s^1 \lambda_3(\theta)d\theta + \frac{1}{\varepsilon}\int_0^s \lambda_j(\theta)d\theta}ds.$$

We take the integrals here by parts:

$$J_0(t,\varepsilon) = \int_0^1 e^{\frac{1}{\varepsilon}\int_s^1 \lambda_3(\theta)d\theta}K(t,s)y_0(s,\sigma)ds = \varepsilon\int_0^1 \frac{K(t,s)y_0(s,\sigma)}{-\lambda_3(s)}d(\exp\left(\frac{1}{\varepsilon}\int_s^1 \lambda_3(\theta)d\theta\right)) =$$

$$= \varepsilon[\frac{K(t,1)y_0(1,\sigma)}{-\lambda_3(1)} - \frac{K(t,0)y_0(0,\sigma)}{-\lambda_3(0)}\sigma_3] - \varepsilon\int_0^1 \exp\left(\frac{1}{\varepsilon}\int_s^1 \lambda_3(\theta)d\theta\right)\frac{\partial}{\partial s}(\frac{K(t,s)y_0(s,\sigma)}{-\lambda_3(s)})ds =$$

$$= \sum_{\nu=0}^{\infty}(-1)^\nu\varepsilon^{\nu+1}[(I_0^\nu(K(t,s)y_0(s,\sigma)))_{s=1} - (I_0^\nu(K(t,s)y_0(s,\sigma)))_{s=0}\sigma_3], \tag{5a}$$

$$J_j(t,\varepsilon) = \int_0^1 \exp\left(\frac{1}{\varepsilon}\int_s^1 \lambda_3(\theta)d\theta + \frac{1}{\varepsilon}\int_0^s \lambda_j(\theta)d\theta\right)K(t,s)y_j(s,\sigma)ds \equiv$$

$$\equiv \sigma_3\int_0^1 \exp\left(\frac{1}{\varepsilon}\int_0^s(\lambda_j(\theta)-\lambda_3(\theta))d\theta\right)K(t,s)y_j(s,\sigma)ds \equiv$$

$$\equiv \sigma_3\varepsilon\int_0^1 \frac{K(t,s)y_j(s,\sigma)}{\lambda_j(s)-\lambda_3(s)}d(\exp\left(\frac{1}{\varepsilon}\int_0^s(\lambda_j(\theta)-\lambda_3(\theta))d\theta\right)) =$$

$$= \varepsilon\sigma_3[\frac{K(t,1)y_j(1,\sigma)}{\lambda_j(1)-\lambda_3(1)}\frac{\sigma_j}{\sigma_3} - \frac{K(t,0)y_j(0,\sigma)}{(\lambda_j(0)-\lambda_3(0))}]-$$

$$-\varepsilon\sigma_3\int_0^1 \exp\left(\frac{1}{\varepsilon}\int_0^s(\lambda_j(\theta)-\lambda_3(\theta))d\theta\right)\frac{\partial}{\partial s}\left(\frac{K(t,s)y_j(s,\sigma)}{\lambda_j(s)-\lambda_3(s)}\right)ds =$$

$$= \sum_{\nu=0}^{\infty}(-1)^{\nu}\varepsilon^{\nu+1}\left[(I_j^{\nu}(K(t,s)y_j(s,\sigma)))_{s=1}\sigma_j - (I_j^{\nu}(K(t,s)y_j(s,\sigma)))_{s=0}\sigma_3\right] \tag{5b}$$

where $j = 1,2$ and the operators are introduced:

$$\begin{aligned} I_0^0 &= \frac{1}{-\lambda_3(s)}, I_0^{\nu} = \frac{1}{-\lambda_3(s)}\frac{\partial}{\partial s}I_0^{\nu-1}, (\nu \geq 1), \\ I_j^0 &= \frac{1}{\lambda_j(s)-\lambda_3(s)}, I_j^{\nu} = \frac{1}{\lambda_j(s)-\lambda_3(s)}\frac{\partial}{\partial s}I_j^{\nu-1}, (\nu \geq 1, j = 1,2). \end{aligned} \tag{5c}$$

It is easy to show (see, for example, [20], pp. 291–294) that the series (5a, b) converge asymptotically (at $\varepsilon \to +0$) to the corresponding integrals $J_j(t,\varepsilon)$ (uniformly with respect to $t \in [0,1]$), and hence the image $Jy(t,\tau)$ is represented as series $\sum_{k=0}^{\infty}\varepsilon^k(\sum_{j=1}^{2} z_j^{(k)}(t,\sigma)e^{\tau_j} + z_0^{(k)}(t,\sigma)g)|_{\tau=\frac{\psi(t)}{\varepsilon}}$, also converging asymptotically to $Jy(t,\tau)$ uniformly with respect to $t \in [0,1]$). Thus, it is shown that the class $M_\varepsilon = U|_{\tau=\frac{\psi(t)}{\varepsilon}}$ is asymptotically invariant with respect to the operator J.

Now let $\tilde{y}(t,\tau,\varepsilon)$ be an arbitrary continuous in $(t,\tau) \in [0,1] \times \Pi(\Pi = \{\tau : \mathrm{Re}\tau_j \leq 0, j = 1,2\}$ function represented by the series

$$\tilde{y}(t,\tau,\varepsilon) = \sum_{k=0}^{\infty}\varepsilon^k y_k(t,\tau,\sigma), y_k(t,\tau,\sigma) \in U \tag{6}$$

converging asymptotically at $\varepsilon \to +0$ (uniformly with respect to $t \in [0,1]$). Substituting (6) into (4) and collecting the coefficients at the same degrees of ε, we obtain the series

$$J\tilde{y}(t,\tau,\varepsilon) = \sum_{k=0}^{\infty}\varepsilon^k Jy_k(t,\tau,\sigma) = \sum_{r=0}^{\infty}\varepsilon^r\sum_{s=0}^{r}R_{r-s}y_s(t,\tau,\sigma)$$

converging asymptotically to $J\tilde{y}$ for $\varepsilon \to +0$ (uniformly in $t \in [0,1]$). Here $R_\nu : U \to U$ (the operators of order in ε) are of the following form:

$$\begin{aligned} &R_0 y(t,\tau,\sigma) \equiv 0, \\ &R_1(t,\tau,\sigma) = \left[\frac{K(t,1)y_0(1,\sigma)}{-\lambda_3(1)} - \frac{K(t,0)y_0(0,\sigma)}{-\lambda_3(0)}\sigma_3\right] + \\ &+\sum_{j=1}^{2}\left[\frac{K(t,1)y_j(1,\sigma)}{\lambda_j(1)-\lambda_3(1)}\sigma_j - \frac{K(t,0)y_j(0,\sigma)}{\lambda_j(0)-\lambda_3(0)}\sigma_3\right], \end{aligned} \tag{6a}$$

$$\begin{aligned} R_{\nu+1}y(t,\tau,\sigma) &= (-1)^{\nu}[(I_0^{\nu}(K(t,s)y_0(s,\sigma)))_{s=1} - (I_0^{\nu}(K(t,s)y_0(s,\sigma)))_{s=0}\sigma_3] + \\ &+\sum_{j=1}^{2}(-1)^{\nu}\left[(I_j^{\nu}(K(t,s)y_j(s,\sigma)))_{s=1} - (I_j^{\nu}(K(t,s)y_j(s,\sigma)))_{s=0}\sigma_3\right] \end{aligned} \tag{6b}$$

where I_j^{ν} are the operators (5c), $j = \overline{0,2}, \nu \geq 0$, introduced above, and $y(t,\tau,\sigma)$ is the function (5).

Definition 2. *By a formal extension of an operator J, we mean an operator $\tilde{J}$, acting on any continuous in $(t,\tau) \in [0,1] \times \Pi$ function $\tilde{y}(t,\tau,\varepsilon)$ of the form (6) according to the law*

$$\tilde{J}\tilde{y} \equiv \tilde{J}\left(\sum_{k=0}^{\infty}\varepsilon^k y_k(t,\tau,\sigma)\right) = \sum_{r=0}^{\infty}\varepsilon^r\sum_{s=0}^{r}R_{r-s}y_s(t,\tau,\sigma). \tag{7}$$

Equality (7) is the basis for the definition of the operator $\tilde{J}$, extended with respect to the integral operator J. Despite the fact that the extension $\tilde{J}$ of the operator J is defined formally, it is quite *possible to use it (see Theorem 3 below) when constructing an asymptotic solution of finite order in ε. Now we can write the problem completely regularized with respect to the original (1):*

$$L_\varepsilon \tilde{y}(t,\tau,\sigma,\varepsilon) \equiv \varepsilon \tfrac{\partial \tilde{y}}{\partial t} + \textstyle\sum_{j=1}^{2} \lambda_j(t) \tfrac{\partial \tilde{y}}{\partial \tau_j} - \lambda_1(t)\tilde{y} - \tilde{J}\tilde{y} =$$
$$= h_1(t) + h_2(t)e^{\tau_2}\sigma_4, \tilde{y}(t,\tau,\varepsilon)|_{t=0,\tau=0} = y^0 \qquad (8)$$

where the operator $\tilde{J}$ has the form (7).

3. Iterative Problems and Their Solvability in the Space U

Substituting (6) into (8) and equating the coefficients at the same degrees of ε, we obtain the following iterative problems:

$$Ly_0(t,\tau,\sigma) \equiv \textstyle\sum_{j=1}^{2} \lambda_j(t) \tfrac{\partial y_0}{\partial \tau_j} - \lambda_1(t)y_0 =$$
$$= h_1(t) + h_2(t)e^{\tau_2}\sigma_4, y_0(0,0) = y^0; \qquad (9_0)$$

$$Ly_1(t,\tau,\sigma) = -\frac{\partial y_0}{\partial t} + R_1 y_0, y_1(0,0) = 0; \qquad (9_1)$$

$$Ly_2(t,\tau,\sigma) = -\frac{\partial y_1}{\partial t} + R_1 y_1 + R_2 y_0, y_2(0,0) = 0; \qquad (9_2)$$

$$\ldots$$

$$Ly_k(t,\tau,\sigma) = -\frac{\partial y_{k-1}}{\partial t} + R_k y_0 + \ldots + R_1 y_{k-1},\ y_k(0,0) = 0, k \geq 1. \qquad (9_k)$$

Each of the iterative problems (9_k) has the form

$$Ly(t,\tau,\sigma) \equiv \sum_{j=1}^{2} \lambda_j(t) \frac{\partial y}{\partial \tau_j} - \lambda_1(t)y = H(t,\tau,\sigma), y(0,0,\sigma) = y_* \qquad (10)$$

where $H(t,\tau,\sigma) = H_0(t,\sigma) + \sum_{j=1}^{2} H_j(t,\sigma)e^{\tau_j}$. We introduce in the space U a scalar product (for each $t \in [0,1]$ and σ) :

$$< y(t,\tau,\sigma), z(t,\tau,\sigma) > \equiv < \textstyle\sum_{j=1}^{2} y_j(t,\sigma)e^{\tau_j} + y_0(t,\sigma), \sum_{j=1}^{2} z_j(t,\sigma)e^{\tau_j} +$$

$$+ z_0(t,\sigma) > \stackrel{def}{=} \textstyle\sum_{j=0}^{2} (y_j(t,\sigma), z_j(t,\sigma))$$

where $(*,*)$ is the usual scalar product in $\mathbb{C}$. Let us prove the following statement.

Theorem 1. *Let $H(t,\tau) \in U$, and conditions (1) and (2) be satisfied. Then, for the solvability of Equation (10) in the space U , it is necessary and sufficient that*

$$< H_1(t,\tau,\sigma), e^{\tau_1} > \equiv 0 \Leftrightarrow H_1(t,\sigma) \equiv 0, \forall t \in [0,1]. \qquad (11)$$

Proof. Defining the solution of the Equation (10) in the form of function (5), we obtain the identity

$$\sum_{j=1}^{2} [\lambda_j(t) - \lambda_1(t)] y_j(t,\sigma)e^{\tau_j} - \lambda_1(t)y_0(t,\sigma) = H_0(t,\sigma) + \sum_{j=1}^{2} H_j(t,\sigma)e^{\tau_j}.$$

Equating here separately the free terms and the coefficients at the exponentials e^{τ_j}, we will have

$$-\lambda_1(t)y_0(t,\sigma) = H_0(t,\sigma), \tag{12_0}$$

$$[\lambda_j(t) - \lambda_1(t)]y_j(t,\sigma) = H_j(t,\sigma), j = 1,2. \tag{12_j}$$

Since $\lambda_1(t) \neq 0 \forall t \in [0,1]$, that the Equation (12_0) has a unique solution

$$y_0(t,\sigma) = -\frac{H_0(t,\sigma)}{\lambda_1(t,\sigma)}. \tag{13}$$

Since $\lambda_1(t)$ is a real function and $\lambda_2(t) = i\beta'(t)$ is purely imaginary, the Equation (12_2) has a unique solution in the space $C^\infty([0,1],\mathbb{C})$. For the solvability of the Equation (12_1) in the space $C^\infty([0,1],\mathbb{C})$ it is necessary and sufficient for identity (11) to hold. Thus, Theorem 1 is proved. □

Remark 1. *It follows from the Equalities* (12_0)*–(13) that under conditions (1) and (2) and condition (11), Equation (10) has the following solution in the space* U*:*

$$y(t,\tau,\sigma) = y_0(t,\sigma) + \alpha_1(t,\sigma)e^{\tau_1} + y_2(t,\sigma)e^{\tau_2} \tag{14}$$

where $\alpha_1(t,\sigma) \in C^\infty([0,1],\mathbb{C})$ *is an arbitrary function,*

$$y_0(t,\sigma) = -\lambda_1^{-1}(t)H_0(t,\sigma), \quad y_2(t,\sigma) = (\lambda_2(t) - \lambda_1(t))^{-1}H_2(t,\sigma).$$

Thus, the solution (14) of the Equation (10) is determined ambiguously in the space U*. Let now* $y_* \in \mathbb{C}$ *be a fixed constant vector. Consider the following problem:*

$$\begin{aligned} &y(0,0,\sigma) = y_*, \\ &< -\tfrac{\partial y}{\partial t} + R_1 y + Q(t,\tau,\sigma), e^{\tau_1} >\equiv 0, \forall t \in [0,1] \end{aligned} \tag{15}$$

where $Q(t,\tau,\sigma) = Q_0(t,\sigma) + \sum_{j=1}^{2} Q_i(t,\sigma)e^{\tau_j}$ *is the well-known vector function of the space* U*, and* R_1 *is the order operator described above (see (6a)). Let us prove the following statement.*

Theorem 2. *Let conditions (1) and (2) be satisfied and the vector function* $H(t,\tau,\sigma) \in U$ *satisfies the orthogonality conditions (11). Then the problem (10) under additional conditions (15) has a unique solution in the space* U*.*

Proof. Since the condition (11) is satisfied, Equation (10) has a solution in the space U in the form of the function (14), where $\alpha_1(t,\sigma) \in C^\infty([0,1],\mathbb{C})$ is an arbitrary function. Submitting (14) to the condition $y(0,0) = y_*$, we have

$$y_* = y_0(0,\sigma) + \alpha_1(0,\sigma) + \frac{H_2(0,\sigma)}{\lambda_2(0) - \lambda_1(0)} \Leftrightarrow$$

$$\Leftrightarrow \alpha_1(0,\sigma) = y_* + \frac{H_0(0,\sigma)}{\lambda_1(0)} - \frac{H_2(0,\sigma)}{\lambda_2(0) - \lambda_1(0)}.$$

Let us now subordinate (14) to the second condition (15):

$$-\frac{\partial y_0}{\partial t} + R_1 y_0 + Q(t,\tau) = -\dot{y}_0(t,\sigma) - \dot{\alpha}_1(t,\sigma)e^{\tau_1} - \dot{y}_2(t,\sigma)e^{\tau_2} +$$

$$+\left[\frac{K(t,1)y_0(1,\sigma)}{-\lambda_3(1)}\sigma_3 - \frac{K(t,0)y_0(0,\sigma)}{-\lambda_3(0)}\right] + \left[\frac{K(t,1)\alpha_1(1,\sigma)}{\lambda_1(1) - \lambda_3(1)}\sigma_1 - \frac{K(t,0)\alpha_1(0,\sigma)}{\lambda_1(0) - \lambda_3(0)}\sigma_3\right] +$$

$$+\left[\frac{K(t,1)y_2(1,\sigma)}{\lambda_2(1)-\lambda_3(1)}\sigma_2-\frac{K(t,0)y_2(0,\sigma)}{\lambda_2(0)-\lambda_3(0)}\sigma_3\right]+Q(t,\tau).$$

Considering that here the expressions in square brackets do not contain an exponent e^{τ_1}, we perform scalar multiplication in the second equality (15). This gives

$$-\dot{\alpha}_1(t,\sigma)+Q_1(t)=0\Leftrightarrow\alpha_1(t,\sigma)=\int_0^t Q_1(\theta)d\theta+y_*+\frac{H_0(0,\sigma}{\lambda_1(0)}-\frac{H_2(0,\sigma)}{\lambda_2(0)-\lambda_1(0)}$$

and hence, we construct the solution (14) of the problem (10) in the space U in a unique way. Theorem 2 is proved. □

4. Construction of the Solution to the First Iterative Problem

Let us apply Theorem 1 to iterative problems (9_k). Since the right-hand side $h_1(t)+h_2(t)e^{\tau_2}\sigma_4$ of the Equation (9_0) satisfies condition (11), the solution $y_0(t,\tau)\in U$ of the first iterative problem (9_0) has the form

$$y_0(t,\tau,\sigma)=\frac{h_1(t)}{-\lambda_1(t)}+\alpha_1^{(0)}(t,\sigma)e^{\tau_1}+\frac{h_2(t)}{\lambda_2(t)-\lambda_1(t)}e^{\tau_2}\sigma_4 \quad (16)$$

where $\alpha_1^{(0)}(t,\sigma)\in C^\infty[0,1]$ is an arbitrary function. Submitting this solution to the initial condition $y_0(0,0,\sigma)=y^0$, we find

$$\begin{gathered}\frac{h_1(0)}{-\lambda_1(0)}+\alpha_1^{(0)}(0,\sigma)+\frac{h_2(0)}{\lambda_2(0)-\lambda_1(0)}\sigma_4=y^0\Leftrightarrow\\ \Leftrightarrow\alpha_1^{(0)}(0,\sigma)=y^0+\frac{h_1(0)}{\lambda_1(0)}-\frac{h_2(0)}{\lambda_2(0)-\lambda_1(0)}\sigma_4.\end{gathered} \quad (17)$$

For the final calculation of the function $\alpha_1^{(0)}(t,\sigma)$, it is necessary to write down conditions (11) for the next iterative problem (9_1). Since $R_1y_0(t,\tau)$ does not contain an exponent, then, under the orthogonality conditions (11), it can be omitted and an equality can be obtained $\dot{\alpha}_1^{(0)}(t,\sigma)=0$, which, taking into account the initial condition (17), leads to an unambiguous calculation of the function

$$\alpha_1^{(0)}(t,\sigma)=y^0+\frac{h_1(0)}{\lambda_1(0)}-\frac{h_2(0)}{\lambda_2(0)-\lambda_1(0)}\sigma_4=\text{const}$$

and hence to an unambiguous calculation of the solution (16) of the first iterative problem (9_0) in the space U.

Remark 2. *The solution of the following problem (9_1) is determined from the system*

$$\begin{gathered}Ly_1(t,\tau,\sigma)=-\frac{\partial y_0}{\partial t}+R_1y_0,y_1(0,0)=0,\\ <-\frac{\partial y_1}{\partial t}+R_1y_1+R_2y_0,e^{\tau_1}>\equiv 0\forall t\in[0,1].\end{gathered} \quad (18)$$

As in the previous case, the expression R_1y_1 and R_2y_0 does not contain an exponent e^{τ_1}, therefore, under orthogonality conditions (18), they can be omitted, and then the solution $y_1(t,\tau)\in U$ of the iterative problem (9_1) will be determined from the system

$$Ly_1(t,\tau,\sigma)=-\frac{\partial y_0}{\partial t}+R_1y_0,y_1(0,0)=0,$$

$$<-\frac{\partial y_1}{\partial t},e^{\tau_1}>\equiv 0\forall t\in[0,1].$$

The same situation takes place for all subsequent iterative problems $(9_k)(k \geq 2)$. Thus, the influence of the Fredholm-type integral operator in (1) affects only the formation of particular solutions of equations for functions $\alpha_1^{(k)}(t,\sigma)$, while in Volterra systems the kernel $K(t,s)$ of the integral operator participates in the formation of common solutions for these functions.

5. Justification of the Asymptotic Convergence of Formal Solutions to the Exact Solutions

Applying Theorems 1 and 2 to iterative problems (9_k), we can uniquely calculate their solutions $y_k(t,\tau,\sigma)$ in the space U. Denote the N-th partial sum of series (6) by $S_N(t,\tau,\sigma)$, and through $y_{\varepsilon N}(t) = S_N(t, \frac{\psi(t)}{\varepsilon}, \varepsilon)$ is the restriction of this sum at $\tau = \frac{\psi(t)}{\varepsilon}$. It is easy to prove the following assertion (see, for example, [15], pp. 37–40).

Lemma 1. *Let conditions (1) and (2) be satisfied. Then the function $y_{\varepsilon N}(t)$ is a formal asymptotic solution of the problem (1) of order N, that is, it satisfies the problem*

$$\varepsilon\frac{dy_{\varepsilon N}}{dt} - a(t)y_{\varepsilon N} - \int_0^1 \exp\left(\frac{1}{\varepsilon}\int_s^1 \mu(\theta)d\theta\right)K(t,s)y_{\varepsilon N}(s)ds = \\ = h_1(t) + h_2(t)e^{\frac{i\beta(t)}{\varepsilon}} + \varepsilon^{N+1}F_N(t,\varepsilon), y_{\varepsilon N}(0) = y^0 \tag{19}$$

where $||F_N(t,\varepsilon)||_{C[0,1]} \leq \bar{F}(\bar{F} > 0$ is a constant independent of ε at $\varepsilon \in (0,\varepsilon_0], \varepsilon_0$ is small enough).

To prove the Theorem on the estimate of the remainder term, we first consider the integro-differential equation

$$\varepsilon\frac{dz}{dt} = a(t)z + \int_0^1 \exp\left(\frac{1}{\varepsilon}\int_s^1 \mu(\theta)d\theta\right)K(t,s)z(s,\varepsilon)ds + H(t,\varepsilon), z(0,\varepsilon) = 0 \tag{20_0}$$

and try to estimate the norm of its solution $z(t,\varepsilon)$ in terms of the norm of the right-hand side $H(t,\varepsilon)$. The function $Y(t,s,\varepsilon) = e^{\frac{1}{\varepsilon}\int_s^t a(\theta)d\theta}$ is the fundamental Cauchy solution for a homogeneous equation $\varepsilon\dot{z} = a(t)z$. Under conditions (1) and (2) it is uniformly bounded, i.e., $||Y(t,s,\varepsilon)|| \leq c_0 = \text{const}$ for all $(t,s,\varepsilon) \in \{0 \leq s \leq t \leq 1, \varepsilon > 0\}$. Let us convert the Equation (20_0), using $Y(t,s,\varepsilon)$; we obtain the equivalent integral equation

$$z(t,\varepsilon) = \frac{1}{\varepsilon}\int_0^t e^{\frac{1}{\varepsilon}\int_x^t a(\theta)d\theta}\left(\int_0^1 \left(e^{\frac{1}{\varepsilon}\int_s^1 \mu(\theta)d\theta}\right)K(x,s)z(s,\varepsilon)ds\right)dx + \frac{1}{\varepsilon}\int_0^t e^{\frac{1}{\varepsilon}\int_x^t a(\theta)d\theta}H(x,\varepsilon)dx.$$

Denoting $H_1(t,\varepsilon) \equiv \int_0^t e^{\frac{1}{\varepsilon}\int_x^t a(\theta)d\theta}H(x,\varepsilon)dx$ and changing the order of integration in the iterated integral, we obtain the following integral equation of the Fredholm type:

$$z(t,\varepsilon) = \int_0^1 \exp\left(\frac{1}{\varepsilon}\int_s^1 \mu(\theta)d\theta\right)G(t,s,\varepsilon)z(s,\varepsilon)ds + \frac{H_1(t,\varepsilon)}{\varepsilon} \tag{20}$$

where $G(t,s,\varepsilon) = \frac{1}{\varepsilon}\int_0^t e^{\frac{1}{\varepsilon}\int_x^t a(\theta)d\theta}K(x,s)dx$. Let us show that the kernel $G(t,s,\varepsilon)$ of this equation is uniformly bounded for $0 \leq s,t \leq 1$, i.e., which the following statement holds.

Lemma 2. *Let conditions (1) and (2) be satisfied. Then the kernel $G(t,s,\varepsilon)$ is uniformly bounded, i.e., $|G(t,s,\varepsilon)| \leq M$ for all $(s,t,\varepsilon) \in [0,1] \times [0,1] \times (0,+\infty)$.*

Proof. Using the operation of integration by parts, we have

$$G(t,s,\varepsilon) = \frac{1}{\varepsilon}\int_0^t e^{\frac{1}{\varepsilon}\int_x^t a(\theta)d\theta}K(x,s)dx = \int_0^t \frac{K(x,s)}{-a(x)}d_x e^{\frac{1}{\varepsilon}\int_x^t a(\theta)d\theta} =$$

$$= \frac{K(x,s)}{-a(x)}e^{\frac{1}{\varepsilon}\int_x^t a(\theta)d\theta}\Big|_{x=0}^{x=t} - \int_0^t e^{\frac{1}{\varepsilon}\int_x^t a(\theta)d\theta}\frac{\partial}{\partial x}\left(\frac{K(x,s)}{-a(x)}\right)dx =$$

$$= \left[\frac{K(t,s)}{-a(t)} - \frac{K(0,s)}{-a(0)} e^{\frac{1}{\varepsilon}\int_0^t a(\theta)d\theta} \right] - \int_0^t e^{\frac{1}{\varepsilon}\int_x^t a(\theta)d\theta} \frac{\partial}{\partial x}\left(\frac{K(x,s)}{-a(x)} \right) dx.$$

Hence, it is clear that under conditions (1) and (2) the kernel $G(t,s,\varepsilon)$ is uniformly bounded, i.e., $|G(t,s,\varepsilon)| \le M$ for all $0 \le s,t \le 1, \varepsilon > 0$. The Lemma 2 is proved.

We now turn to the proof of the correct solvability of Equation (20). To do this, we will try to estimate the norm of the resolvent $R(t,s,\varepsilon)$ of the kernel $\tilde{K}(t,s,\varepsilon) = \exp\left(\frac{1}{\varepsilon}\int_s^1 \mu(\theta)d\theta\right)$ $G(t,s,\varepsilon)$ of integral Equation (20). Let us denote $\chi = \min\limits_{t\in[0,1]} \operatorname{Re}(-\mu(t))$ and estimate the iterated kernels of the integral operator of this system. By Lemma 2, for all $0 \le s,t \le 1$ and $\varepsilon > 0$ we have

$$|\tilde{K}_1(t,s,\varepsilon)| \equiv |\tilde{K}(t,s,\varepsilon)| \le M;$$

$$|\tilde{K}_2(t,s,\varepsilon)| \equiv |\int_0^1 \tilde{K}(t,x,\varepsilon)\tilde{K}_1(x,s,\varepsilon)dx| \equiv$$

$$\equiv |\int_0^1 \exp\left(\frac{1}{\varepsilon}\int_x^1 \mu(\theta)d\theta\right) G(t,x,\varepsilon) \exp\left(\frac{1}{\varepsilon}\int_s^1 \mu(\theta)d\theta\right) G(x,s,\varepsilon)dx| \le$$

$$\le M^2 \int_0^1 \exp\left(\frac{1}{\varepsilon}\int_x^1 \operatorname{Re}\mu(\theta)d\theta\right)dx \le M^2 \int_0^1 \exp\left(-\frac{\chi(1-x)}{\varepsilon}\right)dx =$$

$$\le M^2\varepsilon \frac{\exp\left(-\frac{\chi(1-x)}{\varepsilon}\right)}{x}\Bigg|_{x=0}^{x=1} = \frac{M^2\varepsilon}{\chi}(1 - e^{\frac{\chi}{\varepsilon}}) \le \frac{M^2}{\varepsilon},$$

$$|\tilde{K}_3(t,s,\varepsilon)| \equiv |\int_0^1 \tilde{K}(t,x,\varepsilon)\tilde{K}_2(x,s,\varepsilon)dx| \le \int_0^1 |\tilde{K}(t,s,\varepsilon)| \cdot |\tilde{K}_2(x,s,\varepsilon)|dx \le$$

$$\le \frac{M^2}{\chi}\varepsilon \int_0^1 \left(\frac{1}{\varepsilon}\int_x^1 \operatorname{Re}\mu(\theta)d\theta\right)|G(t,x,\varepsilon)|dx \le \frac{M^3}{\chi}\varepsilon \int_0^1 \exp\left(-\frac{\chi(1-x)}{\varepsilon}\right)dx \le \frac{M^3\varepsilon^2}{\chi^2}.$$

Suppose now that, for $n = r \ge 1$, the estimate

$$|\tilde{K}_r(t,s,\varepsilon)| \le \frac{M^r\varepsilon^{r-1}}{\chi^{r-1}}, 0 \le s,t \le 1, \varepsilon > 0$$

holds. Let us show that this estimate is also true for $n = r+1$. Indeed,

$$|\tilde{K}_{r+1}(t,s,\varepsilon)| \equiv \int_0^1 |\tilde{K}(t,x,\varepsilon)\tilde{K}_r(x,s,\varepsilon)dx| \le \int_0^1 |\tilde{K}(t,x,\varepsilon)| \cdot |\tilde{K}_r(x,s,\varepsilon)|dx \le$$

$$\le \tfrac{M^r\varepsilon^{r-1}}{\chi^{r-1}} \textstyle\int_0^1 |\tilde{K}(t,x,\varepsilon)|dx = \tfrac{M^{r+1}\varepsilon^{r-1}}{\chi^{r-1}} \tfrac{\varepsilon}{\chi} e^{-\frac{\chi(1-x)}{\varepsilon}} |_{x=0}^{x=1} =$$

$$= \tfrac{M^{r+1}\varepsilon^r}{\chi^r}\left(1 - e^{-\frac{\chi}{\varepsilon}}\right) \le \tfrac{M^{r+1}\varepsilon^r}{\chi^r}\ (0 \le s,t \le 1, \varepsilon > 0).$$

So, for all $0 \le s,t \le 1, \varepsilon > 0$ we have proved the estimate

$$|\tilde{K}_n(t,s,\varepsilon)| \le \frac{M^n\varepsilon^{n-1}}{\chi^{n-1}} (n = 1,2,3,\ldots).$$

But then the resolvent

$$R(t,s,\varepsilon) \equiv \tilde{K}_1(t,s,\varepsilon) + \tilde{K}_2(t,s,\varepsilon) + \cdots + \tilde{K}_n(t,s,\varepsilon) + \cdots \equiv \sum_{n=1}^{\infty} \tilde{K}_n(t,s,\varepsilon)$$

majorized by a number series

$$\sum_{n=1}^{\infty} \frac{M^n \varepsilon^{n-1}}{\chi^{n-1}} \equiv M \sum_{n=1}^{\infty} \left(\frac{M\varepsilon}{\chi}\right)^{n-1} = \frac{M}{1 - \frac{M\varepsilon}{\chi}}$$

converging absolutely for $0 < \varepsilon < \frac{\chi}{M}$. This means that the series for the resolvent converges absolutely and uniformly in $(s,t) : 0 \le s,t \le 1$ for all $\varepsilon \in (0, \frac{\chi}{2M}]$. In this case, we have the estimate

$$|R(t,s,\varepsilon)| \le \frac{M}{1 - \frac{M\varepsilon}{\chi}} \le 2M,$$

at $(s,t,\varepsilon) : 0 \le s,t \le 1, 0 < \varepsilon \le \varepsilon_0$ (where $\varepsilon_0 > 0$ is small enough). Consequently, for $\varepsilon \in (0, \varepsilon_0]$ Equation (20) (and hence the equivalent Equation (20_0)) is uniquely solvable in the class $C^1([0,1], \mathbb{C})$ and its solution is represented in the form

$$z(t,\varepsilon) = \frac{1}{\varepsilon} H_1(t,\varepsilon) + \frac{1}{\varepsilon} \int_0^1 R(t,s,\varepsilon) H_1(t,s,\varepsilon) ds$$

for any right-hand side $H_1(t,\varepsilon) \equiv \int_0^t Y(t,x,\varepsilon) H(x,\varepsilon) dx$. From this we derive the estimate

$$||z(t,\varepsilon)||_{C[0,1]} \le \frac{1}{\varepsilon} ||H_1(t,\varepsilon)||_{C[0,1]} + \frac{1}{\varepsilon} 2M ||H_1(t,\varepsilon)|| \le$$

$$\le \frac{1}{\varepsilon} \left(||H(t,\varepsilon)||_{C[0,1]} + 2Mc_0 ||H(t,\varepsilon)||_{C[0,1]} \right) \le \bar{c}_0 \frac{||H(t,\varepsilon)||_{C[0,1]}}{\varepsilon}$$

where $\bar{c}_0 = c_0(1 + 2M) > 0$ is a constant independent of $\varepsilon \in (0, \varepsilon_0]$. The following statement is proved. □

Lemma 3. *Let conditions (1) and (2) be satisfied. Then, for sufficiently small $\varepsilon (0 < \varepsilon \le \varepsilon_0)$, the Equation (20_0) is uniquely solvable in the class $C^1([0,1], \mathbb{C})$ and its solution satisfies the estimate*

$$||z(t,\varepsilon)||_{C[0,1]} \le \frac{\bar{c}_0}{\varepsilon} ||H(t,\varepsilon)||_{C[0,1]}$$

where the constant $\bar{c}_0 > 0$ does not depend on $\varepsilon (0 < \varepsilon \le \varepsilon_0]$.

Remark 3. *Correct solvability of the integral system (20) means that the integral operator $\int_0^1 \exp\left(\frac{1}{\varepsilon} \int_s^1 \mu(\theta) d\theta\right) G(t,s,\varepsilon) z(s,\varepsilon) ds$ has no eigenvalues in the space $C([0,1], \mathbb{C})$ (for sufficiently small $\varepsilon > 0$).*

We apply Lemma 3 to prove the following statement.

Theorem 3. *Let conditions (1) and (2) be satisfied. Then the problem (1) is uniquely solvable in the class $C^1([0,1], \mathbb{C})$ and its solution $y(t,\varepsilon)$ satisfies the estimate*

$$||y(t,\varepsilon) - y_{\varepsilon N}(t)||_{C[0,1]} \le c_N \varepsilon^{N+1}, N = 0,1,2,\ldots$$

where $y_{\varepsilon N}(t)$ is the narrowing (for $\tau = \frac{\psi(t)}{\varepsilon}$), N-th partial sum of the series (6) (with coefficients $y_k(t,\tau) \in U$ satisfying the iterative problems (9_k)), and the constant $c_N > 0$ does not depend on ε at $\varepsilon \in (0, \varepsilon_0] (\varepsilon_0 > 0$ is small enough).

Proof. The problem (1) is uniquely solvable, since it is reduced to the problem (20_0) by a change $y - y^0 = z$. By Lemma 1, for the difference $\Delta_N(t,\varepsilon) = y(t,\varepsilon) - y_{\varepsilon N}(t)$, we obtain the equation

$$\varepsilon\frac{\Delta_N}{dt} = a(t)\Delta_N(t,\varepsilon) + \int_0^1 \exp\left(\frac{1}{\varepsilon}\int_s^1 \mu(\theta)d\theta\right)K(t,s)\Delta_N(s,\varepsilon)ds - \varepsilon^{N+1}F_N(t,\varepsilon), \Delta_N(t,\varepsilon) = 0.$$

It has the form of the problem (20) with inhomogeneity $H(t,\varepsilon) \equiv -\varepsilon^{N+1}F_N(t,\varepsilon)$. By Lemma 3, we have the estimate

$$||\Delta_N(t,\varepsilon)||_{C[0,1]} \equiv ||y(t,\varepsilon) - y_{\varepsilon N}(t)||_{C[0,1]} \le \frac{\bar{c}_0}{\varepsilon}\varepsilon^{N+1}||F_N(t,\varepsilon)||_{C[0,1]} \le \bar{c}_0\bar{F}_N\varepsilon^N \equiv \bar{c}_{N-1}\varepsilon^N$$

and, therefore, for $\Delta_{N+1}(t,\varepsilon) = y(t,\varepsilon) - y_{\varepsilon,N+1}(t)$ will have the estimate

$$||\Delta_{N+1}(t,\varepsilon)||_{C[0,1]} \equiv ||(y(t,\varepsilon) - y_{\varepsilon N}(t)) - \varepsilon^{N+1}y_{N+1}(t,\tfrac{\psi(t)}{\varepsilon})||_{C[0,1]} \le \bar{c}_N\varepsilon^{N+1}.$$

Hence, we obtain that

$$\bar{c}_N\varepsilon^{N+1} \ge ||y(t,\varepsilon) - y_{\varepsilon N}(t)||_{C[0,1]} - \varepsilon^{N+1}||y_{N+1}(t,\frac{\psi(t)}{\varepsilon})||_{C[0,1]}$$

or $||y(t,\varepsilon) - y_{\varepsilon N}(t)||_{C[0,1]} \le c_N\varepsilon^{N+1}$, where $c_N = \bar{c}_N + \bar{y}_N > 0$, $||y_{N+1}(t,\frac{\psi(t)}{\varepsilon})||_{C[0,1]} \le \bar{y}_N$, and the constant c_N does not depend on $\varepsilon \in (0,\varepsilon_0]$, where $\varepsilon_0 > 0$ is small enough. The Theorem 3 is proved. □

According to this Theorem 3, the leading term of the asymptotics of the solution the problem (1) has the form (see Formula (16))

$$\begin{aligned} y_{\varepsilon 0}(t,\sigma) &= \tfrac{h_1(t)}{-\lambda_1(t)} + \alpha_1^{(0)}(t,\sigma)e^{\frac{1}{\varepsilon}\int_0^t a(\theta)d\theta} + \tfrac{h_2(t)}{\lambda_2(t)-\lambda_1(t)}e^{\frac{i}{\varepsilon}\int_0^t \beta'(\theta)d\theta}\sigma_4 = \\ &= \tfrac{h_1(t)}{-\lambda_1(t)} + \left[y^0 + \tfrac{h_1(0)}{\lambda_1(0)} - \tfrac{h_2(0)}{\lambda_2(0)-\lambda_1(0)}e^{\frac{i}{\varepsilon}\beta(0)}\right]e^{\frac{1}{\varepsilon}\int_0^t a(\theta)d\theta} + \tfrac{h_2(t)}{\lambda_2(t)-\lambda_1(t)}e^{\frac{i}{\varepsilon}\beta(t)}. \end{aligned} \quad (21)$$

It is clearly seen here how the rapidly oscillating inhomogeneity affects the asymptotic behavior of the solution to Equation (1), but the contribution of the integral operator $\int_0^1 e^{\frac{1}{\varepsilon}\int_s^T \mu(\theta)d\theta}K(t,s)y(s,\varepsilon)ds$ to it is not found; therefore, we calculate the next term of the asymptotics.

Substituting the solution to the problem in the right-hand side, we obtain the following equation:

$$\begin{aligned} Ly_1(t,\tau,\sigma) &= -\frac{\partial y_0}{\partial t} + R_1y_0 = \\ &= -\frac{\partial}{\partial t}\left(-\frac{h_1(t)}{\lambda_1(t)} + \alpha_1^{(0)}(t,\sigma)e^{\tau_1} + \frac{h_2(t)}{\lambda_2(t)-\lambda_1(t)}e^{\tau_2}\sigma_4\right) + R_1y_0 = \\ &= \left(\frac{h_1(t)}{\lambda_1(t)}\right)^{\bullet} - \dot{\alpha}_1^{(0)}(t,\sigma)e^{\tau_1} - \left(\frac{h_2(t)}{\lambda_2(t)-\lambda_1(t)}\right)^{\bullet}e^{\tau_2}\sigma_4 - \\ &-\left[\frac{K(t,1)h_1(1,\sigma)}{\lambda_3(1)\lambda_1(1)}\sigma_3 - \frac{K(t,0)h_1(0,\sigma)}{\lambda_3(0)\lambda_1(0)}\right] + \left[\frac{K(t,1)\alpha_1^{(0)}(1,\sigma)}{\lambda_1(1)-\lambda_3(1)}\sigma_1^2 - \frac{K(t,0)\alpha_1^{(0)}(0,\sigma)}{\lambda_1(0)-\lambda_3(0)}\sigma_3\right] + \\ &+\left[\frac{K(t,1)h_2(1,\sigma)}{[\lambda_2(1)-\lambda_3(1)]^2}\sigma_2^2\sigma_4 - \frac{K(t,0)h_2(0,\sigma)}{[\lambda_2(0)-\lambda_3(0)]^2}\sigma_3\sigma_4\right] = \left(\frac{h_1(t)}{\lambda_1(t)}\right)^{\bullet} - \dot{\alpha}_1^{(0)}(t,\sigma)e^{\tau_1} - \\ &-\left(\frac{h_2(t)}{\lambda_2(t)-\lambda_1(t)}\right)^{\bullet}e^{\tau_2}\sigma_4 + \frac{K(t,0)h_1(0,\sigma)}{\lambda_3(0)\lambda_1(0)} + \frac{K(t,1)\alpha_1^{(0)}(1,\sigma)}{\lambda_1(1)-\lambda_3(1)}\sigma_1^2 + \frac{K(t,1)h_2(1,\sigma)}{[\lambda_2(1)-\lambda_3(1)]^2}\sigma_2^2\sigma_4 - \end{aligned}$$

$$-\left[\frac{K(t,1)h_1(1,\sigma)}{\lambda_3(1)\lambda_1(1)}+\frac{K(t,0)\alpha_1^{(0)}(0,\sigma)}{\lambda_1(0)-\lambda_3(0)}+\frac{K(t,0)h_2(0,\sigma)}{[\lambda_2(0)-\lambda_3(0)]^2}\sigma_4\right]\sigma_3.$$

Defining the solution of this equation as an element

$$y_1(t,\tau)=y_0^{(1)}(t,\sigma)+\sum_{j=1}^{2}y_j^{(1)}(t,\sigma)e^{\tau_j}$$

of the space U, we arrive at the following equality:

$$\Sigma_{j=1}^2[\lambda_j(t)-\lambda_1(t)]y_j^{(1)}(t,\sigma)e^{\tau_j}-\lambda_1(t)y_0^{(1)}(t,\sigma)=$$

$$=\left(\frac{h_1(t)}{\lambda_1(t)}\right)^{\bullet}-\dot{\alpha}_1^{(0)}(t,\sigma)e^{\tau_1}-\left(\frac{h_2(t)}{\lambda_2(t)-\lambda_1(t)}\right)^{\bullet}e^{\tau_2}\sigma_4+$$

$$+\frac{K(t,0)h_1(0,\sigma)}{\lambda_3(0)\lambda_1(0)}+\frac{K(t,1)\alpha_1^{(0)}(1,\sigma)}{\lambda_1(1)-\lambda_3(1)}\sigma_1^2+\frac{K(t,1)h_2(1,\sigma)}{[\lambda_2(1)-\lambda_3(1)]^2}\sigma_2^2\sigma_4-$$

$$-\left[\frac{K(t,1)h_1(1,\sigma)}{\lambda_3(1)\lambda_1(1)}+\frac{K(t,0)\alpha_1^{(0)}(0,\sigma)}{\lambda_1(0)-\lambda_3(0)}+\frac{K(t,0)h_2(0,\sigma)}{[\lambda_2(0)-\lambda_3(0)]^2}\sigma_4\right]\sigma_3.$$

Equating here separately the free terms and the coefficients at the exponentials e^{τ_j}, we will have

$$-\lambda_1(t)y_0^{(1)}(t,\sigma)=\left(\frac{h_1(t)}{\lambda_1(t)}\right)^{\bullet}+\frac{K(t,0)h_1(0,\sigma)}{\lambda_3(0)\lambda_1(0)}+\frac{K(t,1)\alpha_1^{(0)}(1,\sigma)}{\lambda_1(1)-\lambda_3(1)}\sigma_1^2+$$

$$+\frac{K(t,1)h_2(1,\sigma)}{[\lambda_2(1)-\lambda_3(1)]^2}\sigma_2^2\sigma_4-\left[\frac{K(t,1)h_1(1,\sigma)}{\lambda_3(1)\lambda_1(1)}+\frac{K(t,0)\alpha_1^{(0)}(0,\sigma)}{\lambda_1(0)-\lambda_3(0)}+\frac{K(t,0)h_2(0,\sigma)}{[\lambda_2(0)-\lambda_3(0)]^2}\sigma_4\right]\sigma_3.$$

$$0\cdot y_1^{(1)}(t,\sigma)=-\dot{\alpha}_1^{(0)}(t,\sigma),$$

$$[\lambda_2(t)-\lambda_1(t)]y_2^{(1)}(t,\sigma)=-\left(\frac{h_2(t)}{\lambda_2(t)-\lambda_1(t)}\right)^{\bullet}\sigma_4.$$

Since the orthogonality condition $\dot{\alpha}_1^{(0)}(t,\sigma)\equiv 0$ is satisfied, these equations have solutions in the form of functions:

$$y_0^{(1)}(t,\sigma)=-\frac{1}{\lambda_1(t)}\left\{\left(\frac{h_1(t)}{\lambda_1(t)}\right)^{\bullet}+\frac{K(t,0)h_1(0,\sigma)}{\lambda_3(0)\lambda_1(0)}+\frac{K(t,1)\alpha_1^{(0)}(1,\sigma)}{\lambda_1(1)-\lambda_3(1)}\sigma_1^2+\right.$$

$$\left.+\frac{K(t,1)h_2(1,\sigma)}{[\lambda_2(1)-\lambda_3(1)]^2}\sigma_2^2\sigma_4-\left[\frac{K(t,1)h_1(1,\sigma)}{\lambda_3(1)\lambda_1(1)}+\frac{K(t,0)\alpha_1^{(0)}(0,\sigma)}{\lambda_1(0)-\lambda_3(0)}+\frac{K(t,0)h_2(0,\sigma)}{[\lambda_2(0)-\lambda_3(0)]^2}\sigma_4\right]\sigma_3\right\},$$

$$y_2^{(1)}(t,\sigma)=-\frac{\left(\frac{h_2(t)}{\lambda_2(t)-\lambda_1(t)}\right)^{\bullet}}{\lambda_2(t)-\lambda_1(t)}\sigma_4$$

and $y_1^{(1)}(t,\sigma) = \alpha_1^{(1)}(t,\sigma) \in C^\infty[0,1]$ is an arbitrary function. Thus, the solution to the problem (9_1) will be as follows:

$$y_1(t,\tau,\sigma) = -\frac{1}{\lambda_1(t)}\left\{\left(\frac{h_1(t)}{\lambda_1(t)}\right)^{\bullet} + \frac{K(t,0)h_1(0,\sigma)}{\lambda_3(0)\lambda_1(0)} + \frac{K(t,1)\alpha_1^{(0)}(1,\sigma)}{\lambda_1(1)-\lambda_3(1)}\sigma_1^2 + \frac{K(t,1)h_2(1,\sigma)}{[\lambda_2(1)-\lambda_3(1)]^2}\sigma_2^2\sigma_4 -\right.$$

$$\left. -\left[\frac{K(t,1)h_1(1,\sigma)}{\lambda_3(1)\lambda_1(1)} + \frac{K(t,0)\alpha_1^{(0)}(0,\sigma)}{\lambda_1(0)-\lambda_3(0)} + \frac{K(t,0)h_2(0,\sigma)}{[\lambda_2(0)-\lambda_3(0)]^2}\sigma_4\right]\sigma_3\right\} + \alpha_1^{(1)}(t,\sigma)e^{\tau_1} - \frac{\left(\frac{h_2(t)}{\lambda_2(t)-\lambda_1(t)}\right)^{\bullet}}{\lambda_2(t)-\lambda_1(t)}e^{\tau_2}\sigma_4$$

where $\alpha_1^{(1)}(t,\sigma) \in C^\infty[0,1]$ is an arbitrary function that is calculated in the process of solving the next iterative problem (9_2). As a result, we obtain an asymptotic solution of the first order:

$$y_{\varepsilon 1}(t) = \frac{h_1(t)}{-\lambda_1(t)} + \left[y^0 + \frac{h_1(0)}{\lambda_1(0)} - \frac{h_2(0)}{\lambda_2(0)-\lambda_1(0)}\sigma_4\right]e^{\frac{1}{\varepsilon}\int_0^t a(\theta)d\theta} + \frac{h_2(t)}{\lambda_2(t)-\lambda_1(t)}e^{\frac{i}{\varepsilon}\int_0^t \beta'(\theta)d\theta}\sigma_4 -$$

$$-\frac{\varepsilon}{\lambda_1(t)}\left\{\left(\frac{h_1(t)}{\lambda_1(t)}\right)^{\bullet} + \frac{K(t,0)h_1(0,\sigma)}{\lambda_3(0)\lambda_1(0)} + \frac{K(t,1)\alpha_1^{(0)}(1,\sigma)}{\lambda_1(1)-\lambda_3(1)}\sigma_1^2 + \frac{K(t,1)h_2(1,\sigma)}{[\lambda_2(1)-\lambda_3(1)]^2}\sigma_2^2\sigma_4 -\right.$$

$$\left. -\left[\frac{K(t,1)h_1(1,\sigma)}{\lambda_3(1)\lambda_1(1)} + \frac{K(t,0)\alpha_1^{(0)}(0,\sigma)}{\lambda_1(0)-\lambda_3(0)} + \frac{K(t,0)h_2(0,\sigma)}{[\lambda_2(0)-\lambda_3(0)]^2}\sigma_4\right]\sigma_3\right\} + \varepsilon\alpha_1^{(1)}(t,\sigma)e^{\frac{1}{\varepsilon}\int_0^t a(\theta)d\theta} -$$

$$-\varepsilon\frac{\left(\frac{h_2(t)}{\lambda_2(t)-\lambda_1(t)}\right)^{\bullet}}{\lambda_2(t)-\lambda_1(t)}\sigma_4 e^{\frac{i}{\varepsilon}\int_0^t \beta'(\theta)d\theta}$$

from which it is seen that the kernel of the integral operator affects only the formation of particular solutions of iterative problems (9_k) and particular solutions of equations for the functions $\alpha_1^{(k)}(t,\sigma)$.

In conditions of solvability of the type (11), as already mentioned above, the integral operator does not participate. This is the main difference between integro-differential equations of Fredholm type from equations of Volterra type, where the kernel of the integral operator significantly affects the construction of the general solution of the equations for functions $\alpha_1^{(k)}(t,\sigma)$ (see, for example, [20]).

6. Conclusions

Since the terms of order ε in $y_{\varepsilon 1}(t)$ uniformly tend to zero, when $\varepsilon \to +0$, then the behavior of the exact solution of the problem (1) as the small parameter tends to zero completely is determined by its main term of asymptotics (21): after leaving the point $y = y^0$ at $t = 0$, the exact solution $y(t,\varepsilon)$ of the problem (1) (for $t > 0$ and $\varepsilon \to +0$) will perform fast oscillations around the "degenerate solution" $\bar{\bar{y}}(t) = \frac{h_1(t)}{-\lambda_1(t)}$, not tending for any limit.

Author Contributions: All authors contributed evenlly. All authors have read and agreed to the published version of the manuscript.

Funding: This work was supported by grant No. AP05133858 of the Ministry of Education and Science of the Republic of Kazakhstan.

Institutional Review Board Statement: Not applicable.

Informed Consent Statement: Not applicable.

Data Availability Statement: Not applicable.

Conflicts of Interest: The funders had no role in the design of the study; in the collection, analyses, or interpretation of data; in the writing of the manuscript, or in the decision to publish the results.

References

1. Bobodzhanov, A.A.; Safonov, V.F. *Singularly Perturbed Integral, Integro-Differential Equations with Rapidly Varying Kernels and Equations with Diagonal Degeneration of the Kernel;* Publishing House "Sputnik +": Moscow, Russia, 2017.
2. Yeliseev, A. On the Regularized Asymptotics of a Solution to the Cauchy Problem in the Presence of a Weak Turning Point of the Limit Operator. *Axioms* **2020**, *9*, 86. [CrossRef]
3. Eliseev, A.G.; Kirichenko P.V. A solution of the singularly perturbed Cauchy problem in the presence of a «weak» turning point at the limit operator. *SEMR* **2020**, *17*, 51–60. [CrossRef]
4. Eliseev, A.G. The regularized asymptotics of a solution of the Cauchy problem in the presence of a weak turning point of the limit operator. *Sb. Math.* **2021**, *212*, 1415–1435. [CrossRef]
5. Lomov, S.A.; Eliseev, A.G. Asymptotic integration of singularly perturbed problems. *Russ. Math. Surv.* **1988**, *43*, 1–63. [CrossRef]
6. Besova, M.I.; Kachalov, V.I. On holomorphic regularization of nonlinear singularly perturbed boundary value problems. *Vestn. MEI/Bull. MPEI* **2018**, *4*, 152–156. [CrossRef]
7. Besova, M.I.; Kachalov, V.I. On a nonlinear differential equation in a Banach space. *SEMR* **2021**, *18*, 332–337. [CrossRef]
8. Bobodzhanova, M.A. Substantiation of the regularization method for nonlinear integro-differential equations with a zero operator of the differential part. *Vestn. MEI/Bull. MPEI* **2011**, *6*, 85–95.
9. Bobodzhanova, M.A. Singularly perturbed integro-differential systems with a zero operator of the differential part. *Vestn. MEI/Bull. MPEI* **2010**, *6*, 63–72.
10. Besova, M.I.; Kachalov, V.I. Analytical aspects of the theory of Tikhonov systems. *Mathematics* **2022**, *10*, 72. [CrossRef]
11. Nefedov, N.N.; Nikitin, A.G. The Cauchy problem for a singularly perturbed integro-differential Fredholm equation. *Comput. Math. Math. Phys.* **2007**, *47*, 629–637. [CrossRef]
12. Eliseev, A.G.; Ratnikova, T.A. Shaposhnikova, D.A. On an initialization problem. *Math. Notes* **2020**, *108*, 286–291. [CrossRef]
13. Vasil'eva, A.B.; Butuzov, V.F.; Nefedov, N.N. Singularly perturbed problems with boundary and internal layers. *Proc. Steklov Inst. Math.* **2010**, *268*, 258–273. [CrossRef]
14. Abramov, V.S.; Bobodzhanov, A.A.; Bobodzhanova, M.A. A method of normal forms for nonlinear singularly perturbed systems in case of intersection of eigenvalues of limit operator. *Russian Math. (Iz. VUZ)* **2019**, *63*, 1–9. [CrossRef]
15. Lomov, S.A. *Introduction to the General Theory of Singular Perturbations;* Science: Moscow, Russia, 1981.
16. Lomov, S.A.; Lomov, I.S. *Foundations of Mathematical Theory of Boundary Layer;* Izdatelstvo MSU: Moscow, Russia, 2011.
17. Ryzhikh, A.D. Asymptotic solution of a linear differential equation with a rapidly oscillating coefficient. *Tr. Mosk. Energeticheskogo Instituta* **1978**, *357*, 92–94.
18. Safonov, V.F.; Bobodzhanov, A.A. *Singularly Perturbed Integro-Differential Equations of Fredholm Type and Systems with Inner Transition Layers;* Publishing House "Sputnik +": Moscow, Russia, 2018.
19. Bobodzhanov, A.A.; Kalimbetov, B.T.; Safonov, V.F. Generalization of the regularization method to singularly perturbed integro-differential systems of equations with rapidly oscillating inhomogeneity. *Axioms* **2021**, *10*, 40. [CrossRef]
20. Safonov, V.F.; Bobodzhanov, A.A. *Course of Higher Mathematics. Singularly Perturbed Equations and the Regularization Method: Textbook;* Publishing House of MPEI: Moscow, Russia, 2012.

axioms

MDPI

Article

Asymptotic and Pseudoholomorphic Solutions of Singularly Perturbed Differential and Integral Equations in the Lomov's Regularization Method

Abduhafiz Bobodzhanov †, Valeriy Safonov and Vasiliy Kachalov *,†,‡

National Research University "MPEI", 111250 Moscow, Russia; bobojanova@yandex.ru (A.B.); SafonovVF@yandex.ru (V.S.)

* Correspondence: vikachalov@rambler.ru; Tel.: +8-495-362-71-31

† Sections devoted to singularly perturbed integral equations were written by Bobodzhanov A.A. and Safonov V.F., the sections connected with holomorphic regularization were written by Kachalov V.I.

‡ Current address: National Research University "MPEI", Ul. Krasnokazarmennaya 14, 111250 Moscow, Russia; Tel.: +8-495-362-71-31.

Received: 9 December 2018; Accepted: 21 February 2019; Published: 1 March 2019

Abstract: We consider a singularly perturbed integral equation with weakly and rapidly varying kernels. The work is a continuation of the studies carried out previously, but these were focused solely on rapidly changing kernels. A generalization for the case of two kernels, one of which is weakly, and the other rapidly varying, has not previously been carried out. The aim of this study is to investigate the effects introduced into the asymptotics of the solution of the problem by a weakly varying integral kernel. In the second part of the work, the problem of constructing exact (more precise, pseudo-analytic) solutions of singularly perturbed problems is considered on the basis of the method of holomorphic regularization developed by one of the authors of this paper. The power series obtained with the help of this method for the solutions of singularly perturbed problems (in contrast to the asymptotic series constructed in the first part of this paper) converge in the usual sense.

Keywords: singularly perturbed; integral equations; regularization of the integral; weakly and rapidly changing kernel; holomorphic integrals; family of homomorphisms; asymptotic and pseudoholomorphic solutions

1. Introduction

In the first part of this work, we consider a singularly perturbed equation in which integral operators contain both weakly and rapidly changing kernels. The problem of constructing a regularized asymptotic solution for this problem, uniformly applicable over the entire time interval under consideration, was previously solved but only for rapidly varying kernels (see, for example References [1–4]). A generalization for the case of two kernels, one of which is weakly, and the other rapidly varying, has not previously been carried out. The aim of the present study is to investigate the effects introduced into the asymptotics of the solution by a weakly varying kernel. Notice that this problem was not considered from the point of view of other methods of asymptotic integration (for example, using the methods of References [5–7]).

The second part of our paper is devoted to the construction of approximate solutions of singularly perturbed problems using the method of holomorphic regularization [8,9]. The analysis of asymptotic methods for solving singularly perturbed problems shows that the solutions of such problems depend in two ways on a small parameter: regularly and singularly. This dependence is especially vividly demonstrated by the method of regularization of Lomov. Moreover, regularized series representing solutions of singularly perturbed problems can converge in the usual sense. In this connection, it became necessary to study a special class of functions—pseudoholomorphic functions. This very important part of the complex analysis is designed to substantiate the main provisions of the so-called analytic theory of singular perturbations. On the other hand, the relevance of the theory is also supported by the fact that pseudoholomorphic functions, in contrast to holomorphic functions, are determined when the conditions of the implicit function theorem are violated.

The concept of a pseudoanalytic (pseudoholomorphic) function and the associated concept of an essentially singular manifold are of a general mathematical nature, although they arose in the framework of the regularization method for singular perturbations. First of all, they reflect the new concept of a pseudoholomorphic solution of singularly perturbed problems, i.e., such a solution, which is representable in the form of a series converging in the usual (but not asymptotic) sense in powers of a small parameter. We must also take into account the fact that the modern mathematical theory of the boundary layer [1], along with the Vasilyeva–Butuzov–Nefedov boundary-function method [5] and the method of barrier functions [10], widely uses the notion of a pseudoholomorphic solution. The importance of considering singularly perturbed problems from the standpoint of the method of pseudoholomorphic solutions is illustrated by applications (see, for example, References [11,12]).

2. An Equivalent Integro-Differential System and Its Regularization

We consider the singularly perturbed equation

$$\varepsilon y(t,\varepsilon)=\int_0^t e^{\frac{1}{\varepsilon}\int_s^t \mu(\theta)d\theta}K_2(t,s)y(s,\varepsilon)ds+\int_0^t K_1(t,s)y(s,\varepsilon)ds+h(t), t\in[0,T]. \tag{1}$$

Differentiating Equation (1) with respect to t, will have

$$\varepsilon^2\left(\frac{dy(t,\varepsilon)}{dt}\right)=\int_0^t\left(\mu(t)e^{\frac{\int_s^t\mu(\theta)\,d\theta}{\varepsilon}}K_2(t,s)\,y(s,\varepsilon)+\varepsilon\cdot e^{\frac{\int_s^t\mu(\theta)\,d\theta}{\varepsilon}}\left(\frac{\partial}{\partial t}K_2(t,s)\right)y(s,\varepsilon)\right)ds+$$
$$+\varepsilon\cdot K_2(t,t)\,y(t,\varepsilon)+\varepsilon\cdot\int_0^t\left(\frac{\partial}{\partial t}K_1(t,s)\right)y(s,\varepsilon)\;ds+\varepsilon\cdot K_1(t,t)\,y(t,\varepsilon)+\varepsilon\cdot\frac{d}{dt}h(t),$$

or

$$\begin{aligned}&\varepsilon^2\frac{dy}{dt}=(K_1(t,t)+K_2(t,t))\,\varepsilon y+\mu(t)\,z+\\&+\int_0^t e^{\frac{\int_s^t\mu(\theta)\,d\theta}{\varepsilon}}\frac{\partial}{\partial t}K_2(t,s)\,\varepsilon y(s,\varepsilon)\;ds+\int_0^t\frac{\partial}{\partial t}K_1(t,s)\,\varepsilon y(s,\varepsilon)\;ds+\varepsilon\cdot\frac{d}{dt}h(t),\end{aligned}\tag{2}$$

where $z(t,\varepsilon)=\int_0^t e^{\frac{1}{\varepsilon}\int_s^t\mu(\theta)d\theta}K_2(t,s)\,y(s,\varepsilon)\,ds$. By differentiating this function with respect to t, we also obtain

$$\varepsilon\frac{dz}{dt}=\mu(t)\cdot z+\int_0^t\left(\varepsilon\cdot e^{\frac{\int_s^t\mu(\theta)\,d\theta}{\varepsilon}}\left(\frac{\partial}{\partial t}K_2(t,s)\right)y(s,\varepsilon)\right)ds+\varepsilon\cdot K_2(t,t)\,y. \tag{3}$$

Finally, denoting by $\varepsilon y=v$, rewriting Equations (2) and (3) in the form

$$\begin{aligned}&\varepsilon\frac{dv}{dt}=(K_1(t,t)+K_2(t,t))\,v+\mu(t)\,z+\\&+\int_0^t e^{\frac{\int_s^t\mu(\theta)\,d\theta}{\varepsilon}}\frac{\partial}{\partial t}K_2(t,s)\,v(s,\varepsilon)\;ds+\int_0^t\frac{\partial}{\partial t}K_1(t,s)\,v(s,\varepsilon)\;ds+\varepsilon\cdot\dot h(t),\\&\varepsilon\frac{dz}{dt}=\mu(t)\cdot z+\int_0^t\left(e^{\frac{\int_s^t\mu(\theta)\,d\theta}{\varepsilon}}\left(\frac{\partial}{\partial t}K_2(t,s)\right)v(s,\varepsilon)\right)ds+K_2(t,t)\,v.\end{aligned}$$

We have obtained an integro-differential system of equations

$$\varepsilon\begin{pmatrix}\frac{dv}{dt}\\ \frac{dz}{dt}\end{pmatrix}=\begin{pmatrix}K_1(t,t)+K_2(t,t) & \mu(t)\\ K_2(t,t) & \mu(t)\end{pmatrix}\begin{pmatrix}v\\ z\end{pmatrix}+$$
$$+\int_0^t e^{\frac{\int_s^t \mu(\theta)\,d\theta}{\varepsilon}}\begin{pmatrix}\frac{\partial}{\partial t}K_2(t,s) & 0\\ \frac{\partial}{\partial t}K_2(t,s) & 0\end{pmatrix}\begin{pmatrix}v(s,\varepsilon)\\ z(s,\varepsilon)\end{pmatrix}ds+\int_0^t\begin{pmatrix}\frac{\partial}{\partial t}K_1(t,s) & 0\\ 0 & 0\end{pmatrix}\begin{pmatrix}v(s,\varepsilon)\\ z(s,\varepsilon)\end{pmatrix}ds+$$
$$+\varepsilon\begin{pmatrix}\dot h(t)\\ 0\end{pmatrix},\ \begin{pmatrix}v(0,\varepsilon)\\ z(0,\varepsilon)\end{pmatrix}=\begin{pmatrix}h(0)\\ 0\end{pmatrix},$$

or

$$\begin{gathered}\varepsilon\frac{dw}{dt}=A(t)w+\int_0^t B(t,s)\,w(s,\varepsilon)ds+\\ +\int_0^t e^{\frac{1}{\varepsilon}\int_s^t\mu(\theta)d\theta}G(t,s)\,w(s,\varepsilon)ds+\varepsilon H(t),w(0,\varepsilon)=w^0\equiv\begin{pmatrix}h(0)\\ 0\end{pmatrix},\end{gathered}\tag{4}$$

where $w=\{v,z\}$, matrixes $A(t)$, $A_1(t)$, $B(t,s)$, $G(t,s)$, and the vector function $H(t)$ have the form

$$A(t)=\begin{pmatrix}K_1(t,t)+K_2(t,t) & \mu(t)\\ K_2(t,t) & \mu(t)\end{pmatrix},\ B(t,s)=\begin{pmatrix}\frac{\partial K_1(t,s)}{\partial t} & 0\\ 0 & 0\end{pmatrix},$$
$$G(t,s)=\begin{pmatrix}\frac{\partial K_2(t,s)}{\partial t} & 0\\ \frac{\partial K_2(t,s)}{\partial t} & 0\end{pmatrix},\ H(t)=\begin{pmatrix}\dot h(t)\\ 0\end{pmatrix},\ w^0\equiv\begin{pmatrix}h(0)\\ 0\end{pmatrix}.$$

The roots of the characteristic equation of matrix $A(t)$:

$$\lambda^2-(\mu(t)+K_1(t,t)+K_2(t,t))\,\lambda+\mu(t)K_1(t,t)=0$$

form the spectrum $\sigma(A(t))=\{\lambda_1(t),\lambda_2(t)\}$ of the matrix $A(t)$. We assume that the following conditions hold:

1) $h(t),\mu(t)\in C^\infty([0,T],\mathbb{C}),\ K_j(t,s)\in C^\infty(0\le s\le t\le T,\mathbb{C}),j=1,2;$
2) $\mu(t)\neq 0,\operatorname{Re}\mu(t)\le 0,\ \lambda_j(t)\neq 0,\ \operatorname{Re}\lambda_j(t)\ \forall t\in[0,T],\ j=1,2.$

We denote by $\lambda_3(t)\equiv\mu(t)$ and (according to the method [13] of Lomov) we introduce regularizing variables

$$\tau_j=\frac{1}{\varepsilon}\int_0^t\lambda_j(\theta)d\theta\equiv\frac{\psi_j(t)}{\varepsilon},\ j=1,2,3.\tag{5}$$

For the extension $\tilde w=\{v(t,\tau,\varepsilon),z(t,\tau,\varepsilon)\}$, we get the following system:

$$\begin{gathered}\frac{\partial\tilde w}{\partial t}+\sum_{j=1}^3\lambda_j(t)\frac{\partial\tilde w}{\partial\tau_j}-A(t)\tilde w-\int_0^t B(t,s)\,\tilde w(s,\tfrac{\psi(s)}{\varepsilon},\varepsilon)ds-\\ -\int_0^t e^{\frac{1}{\varepsilon}\int_s^t\lambda_3(\theta)d\theta}G(t,s)\,\tilde w(s,\tfrac{\psi(s)}{\varepsilon},\varepsilon)ds=\varepsilon H(t),\tilde w(t,\tau,\varepsilon)|_{t=0,\tau=0}=w^0,\end{gathered}\tag{6}$$

where $\tau=(\tau_1,\tau_2,\tau_3)$, $\psi=(\psi_1,\psi_2,\psi_3)$. However, Equation (6) cannot be considered completely regularized, since the integral operator

$$J\tilde w=\int_0^t B(t,s)\,\tilde w(s,\frac{\psi(s)}{\varepsilon},\varepsilon)ds+\int_0^t e^{\frac{1}{\varepsilon}\int_s^t\lambda_3(\theta)d\theta}G(t,s)\,\tilde w(s,\frac{\psi(s)}{\varepsilon},\varepsilon)ds$$

has not been regularized. To regularize the operator $J\tilde w$, we introduce a class $M_\varepsilon=U|_{\tau=\frac{\psi(t)}{\varepsilon}}$, asymptotically invariant with respect to the operator J (see Reference [13], p. 62). In this case, we take as the space U the vector-valued functions representable by the sums of the form

$$w(t,\tau)=\sum_{j=1}^3 w_j(t)e^{\tau_j}+w_0(t),w_j(t)\in C([0,T],\mathbb{C}^2),j=\overline{0,3}.\tag{7}$$

We must show that the image $Jw(t,\tau)$ of the functions of the form of Equation (7) can be represented in the form of a series

$$Jw(t,\tau) = \sum_{k=0}^{\infty} \varepsilon^k (\sum_{j=1}^{3} w_1^{(k)}(t)e^{\tau_j} + w_0^{(0)}(t))|_{\tau=\frac{\psi(t)}{\varepsilon}},$$

converging asymptotically to Jw (as $\varepsilon \to +0$) and that this convergence is uniform with respect to $t \in [0,T]$. Substituting Equation (7) into $Jw(t,\tau)$, we obtain

$$\begin{array}{l} Jw(t,\tau) = \int_0^t B(t,s) \left(\sum_{j=1}^3 w_j(s)e^{\frac{1}{\varepsilon}\int_0^s \lambda_j(\theta)d\theta} + w_0(s)\right) ds+ \\ + \int_0^t e^{\frac{1}{\varepsilon}\int_s^t \lambda_3(\theta)d\theta} G(t,s) \left(\sum_{j=1}^3 w_j(s)e^{\frac{1}{\varepsilon}\int_0^s \lambda_j(\theta)d\theta} + w_0(s)\right) ds \equiv \\ \equiv \int_0^t B(t,s)\, w_0(s)ds + \int_0^t e^{\frac{1}{\varepsilon}\int_s^t \lambda_3(\theta)d\theta} G(t,s)\, w_0(s)ds+ \\ + \sum_{j=1}^3 \int_0^t B(t,s)\, w_j(s)e^{\frac{1}{\varepsilon}\int_0^s \lambda_j(\theta)d\theta} ds+ \\ + \sum_{j=1}^3 \int_0^t G(t,s)\, w_j(s)e^{\frac{1}{\varepsilon}\int_0^s \lambda_j(\theta)d\theta + \frac{1}{\varepsilon}\int_s^t \lambda_3(\theta)d\theta} ds \equiv \\ \equiv \int_0^t B(t,s)\, w_0(s)ds + e^{\frac{1}{\varepsilon}\int_0^t \lambda_3(\theta)d\theta} \int_0^t G(t,s)\, w_3(s)ds+ \\ + \int_0^t e^{\frac{1}{\varepsilon}\int_s^t \lambda_3(\theta)d\theta} G(t,s)\, w_0(s)ds + \sum_{j=1}^3 \int_0^t B(t,s)\, w_j(s)e^{\frac{1}{\varepsilon}\int_0^s \lambda_j(\theta)d\theta} ds+ \\ + \sum_{k=1}^2 \int_0^t G(t,s)\, w_k(s)e^{\frac{1}{\varepsilon}\int_0^s \lambda_k(\theta)d\theta + \frac{1}{\varepsilon}\int_s^t \lambda_3(\theta)d\theta} ds. \end{array} \tag{7a}$$

Applying the operation of integration by parts, we find that

$$\begin{array}{l} \int_0^t e^{\frac{1}{\varepsilon}\int_s^t \lambda_3(\theta)d\theta} G(t,s)\, w_0(s)ds = -\varepsilon \int_0^t \frac{G(t,s)w_0(s)}{\lambda_3(s)} de^{\frac{1}{\varepsilon}\int_s^t \lambda_3(\theta)d\theta} = \\ = \varepsilon \left[\frac{G(t,0)w_0(0)}{\lambda_3(0)} e^{\frac{1}{\varepsilon}\int_0^t \lambda_3(\theta)d\theta} - \frac{G(t,t)w_0(t)}{\lambda_3(t)}\right] + \\ + \varepsilon \int_0^t e^{\frac{1}{\varepsilon}\int_s^t \lambda_3(\theta)d\theta} \frac{\partial}{\partial s}\left(\frac{G(t,s)w_0(s)}{\lambda_3(s)}\right) ds = \\ = \sum_{m=0}^{\infty} \varepsilon^{m+1} [(I_3^m (G(t,s)\, w_0(s)))_{s=0} e^{\frac{1}{\varepsilon}\int_0^t \lambda_3(\theta)d\theta} - (I_3^m (G(t,s)\, w_0(s)))_{s=t}]; \end{array}$$

$$\begin{array}{l} \int_0^t B(t,s)\, w_j(s)e^{\frac{1}{\varepsilon}\int_0^s \lambda_j(\theta)d\theta} ds = \varepsilon \int_0^t \frac{B(t,s)w_j(s)}{\lambda_j(s)} de^{\frac{1}{\varepsilon}\int_0^s \lambda_j(\theta)d\theta} = \\ = \varepsilon \left[\frac{B(t,t)w_j(t)}{\lambda_j(t)} e^{\frac{1}{\varepsilon}\int_0^t \lambda_j(\theta)d\theta} - \frac{B(t,0)w_j(0)}{\lambda_j(0)}\right] - \\ - \int_0^t \frac{\partial}{\partial s}\left(\frac{B(t,s)w_j(s)}{\lambda_j(s)}\right) e^{\frac{1}{\varepsilon}\int_0^s \lambda_j(\theta)d\theta} ds = \\ = \sum_{m=0}^{\infty} (-1)^m \varepsilon^{m+1} [\left(I_j^m (B(t,s)\, w_j(s))\right)_{s=t} e^{\frac{1}{\varepsilon}\int_0^t \lambda_j(\theta)d\theta} - \left(I_j^m (B(t,s)\, w_j(s))\right)_{s=0}]; \end{array}$$

$$\begin{array}{l} \int_0^t G(t,s)\, w_k(s)e^{\frac{1}{\varepsilon}\int_0^s \lambda_k(\theta)d\theta + \frac{1}{\varepsilon}\int_s^t \lambda_3(\theta)d\theta} ds = e^{\frac{1}{\varepsilon}\int_0^t \lambda_3(\theta)d\theta} \int_0^t e^{\frac{1}{\varepsilon}\int_0^s [\lambda_k(\theta)-\lambda_3(\theta)]d\theta} G(t,s)\, w_k(s)ds = \\ = \varepsilon e^{\frac{1}{\varepsilon}\int_0^t \lambda_3(\theta)d\theta} \int_0^t \frac{G(t,s)w_k(s)}{\lambda_k(s)-\lambda_3(s)} de^{\frac{1}{\varepsilon}\int_0^s [\lambda_k(\theta)-\lambda_3(\theta)]d\theta} = \\ = \varepsilon e^{\frac{1}{\varepsilon}\int_0^t \lambda_3(\theta)d\theta} \{[\int_0^t \frac{G(t,t)w_k(t)}{\lambda_k(t)-\lambda_3(t)} e^{\frac{1}{\varepsilon}\int_0^t [\lambda_k(\theta)-\lambda_3(\theta)]d\theta} - \frac{G(t,0)w_k(0)}{\lambda_k(0)-\lambda_3(0)}] - \\ - \int_0^t e^{\frac{1}{\varepsilon}\int_0^s [\lambda_k(\theta)-\lambda_3(\theta)]d\theta} \frac{\partial}{\partial s}\left(\frac{G(t,s)w_k(s)}{\lambda_k(s)-\lambda_3(s)}\right) ds\} = \\ = \varepsilon [\int_0^t \frac{G(t,t)w_k(t)}{\lambda_k(t)-\lambda_3(t)} e^{\frac{1}{\varepsilon}\int_0^t \lambda_k(\theta)d\theta} - \frac{G(t,0)w_k(0)}{\lambda_k(0)-\lambda_3(0)} e^{\frac{1}{\varepsilon}\int_0^t \lambda_3(\theta)d\theta}] - \\ - \varepsilon e^{\frac{1}{\varepsilon}\int_0^t \lambda_3(\theta)d\theta} \int_0^t e^{\frac{1}{\varepsilon}\int_0^s (\lambda_k(\theta)-\lambda_3(\theta))d\theta} \frac{\partial}{\partial s}\left(\frac{G(t,s)w_k(s)}{\lambda_k(s)-\lambda_3(s)}\right) ds = \\ = \sum_{m=0}^{\infty} (-1)^m \varepsilon^{m+1} [(I_{k3}^m (G(t,s)\, w_k(s)))_{s=t} e^{\frac{1}{\varepsilon}\int_0^t \lambda_k(\theta)d\theta} - \\ - (I_{k3}^m (G(t,s)\, w_k(s)))_{s=0} e^{\frac{1}{\varepsilon}\int_0^t \lambda_3(\theta)d\theta}], \end{array}$$

where operators are introduced:

$$\begin{array}{l} I_j^0 = \frac{1}{\lambda_j(s)}, I_j^m = \frac{1}{\lambda_j(s)} \frac{\partial}{\partial s} I_j^{m-1}, m \geq 1, j = 1,2,3; \\ I_{k3}^0 = \frac{1}{\lambda_k(s)-\lambda_3(s)}, I_{k3}^m = \frac{1}{\lambda_k(s)-\lambda_3(s)} \frac{\partial}{\partial s} I_{k3}^{m-1}, m \geq 1, k = 1,2. \end{array} \tag{8}$$

Consequently, for the operator $Jw(t,\tau)$ there is a decomposition

$$\begin{array}{l} Jw(t,\tau) \equiv \int_0^t B(t,s)\,w_0(s)ds + e^{\frac{1}{\varepsilon}\int_0^s \lambda_3(\theta)d\theta}\int_0^t G(t,s)\,w_3(s)ds+ \\ +\sum_{m=0}^{\infty}\varepsilon^{m+1}[(I_3^m(G(t,s)\,w_0(s)))_{s=0}e^{\frac{1}{\varepsilon}\int_0^t \lambda_3(\theta)d\theta} - (I_3^m(G(t,s)\,w_0(s)))_{s=t}] \\ +\sum_{m=0}^{\infty}(-1)^m\varepsilon^{m+1}\sum_{j=1}^{3}[\left(I_j^m(B(t,s)\,w_j(s))\right)_{s=t}e^{\frac{1}{\varepsilon}\int_0^t \lambda_j(\theta)d\theta} - \\ -\left(I_j^m(B(t,s)\,w_j(s))\right)_{s=0}]+ \\ +\sum_{m=0}^{\infty}(-1)^m\varepsilon^{m+1}\sum_{k=1}^{2}[(I_{k3}^m(G(t,s)\,w_k(s)))_{s=t}e^{\frac{1}{\varepsilon}\int_0^t \lambda_k(\theta)d\theta} - \\ -\left(I_{k3}^m(G(t,s)\,w_k(s))\,e^{\frac{1}{\varepsilon}\int_0^t \lambda_3(\theta)d\theta}\right)_{s=0}]. \end{array} \tag{9}$$

It is not hard to show (see Reference [14]) that the series on the right-hand side of Equation (9) converges to $Jw(t,\varepsilon)$ (as $\varepsilon \to +0$) uniformly with respect to $t \in [0,T]$. We introduce operators of order (on ε) $R_\nu : U \to U$:

$$\begin{array}{l} R_0 w(t,\tau) \equiv \int_0^t B(t,s)\,w_0(s)ds + e^{\tau_3}\int_0^t G(t,s)\,w_3(s)ds, \\ R_1 w(t,\tau) = \frac{G(t,0)w_0(0)}{\lambda_3(0)}e^{\tau_3} - \frac{G(t,t)w_0(t)}{\lambda_3(t)}+ \\ +\sum\limits_{j=1}^{3}\left[\frac{B(t,t)w_j(t)}{\lambda_j(t)}e^{\tau_j} - \frac{B(t,0)w_j(0)}{\lambda_j(0)}\right]+ \\ +\sum\limits_{k=1}^{2}\left[\frac{(G(t,t)w_k(t))}{\lambda_k(t)-\lambda_3(t)}e^{\tau_k} - \frac{(G(t,0)w_k(0))}{\lambda_k(0)-\lambda_3(0)}e^{\tau_3}\right], \end{array} \tag{10}$$

$$\begin{array}{l} R_{m+1}w(t,\tau) = [(I_3^m(G(t,s)\,w_0(s)))_{s=0}e^{\tau_3} - (I_3^m(G(t,s)\,w_0(s)))_{s=t}]+ \\ +(-1)^m\sum\limits_{j=1}^{3}[\left(I_j^m(B(t,s)\,w_j(s))\right)_{s=t}e^{\tau_j} - \left(I_j^m(B(t,s)\,w_j(s))\right)_{s=0}]+ \\ +(-1)^m\sum\limits_{k=1}^{2}[(I_{k3}^m(G(t,s)\,w_k(s)))_{s=t}e^{\tau_k} - (I_{k3}^m(G(t,s)\,w_k(s))\,e^{\tau_3})_{s=0}], \\ m \geq 1, \tau = \frac{\psi(t)}{\varepsilon}. \end{array}$$

Then, the image $Jw(t,\tau)$ can be written in the form

$$Jw(t,\tau) = R_0 w(t,\tau) + \sum_{m=0}^{\infty}\varepsilon^{m+1}R_{m+1}w(t,\tau), \tag{11}$$

where $\tau = \frac{\psi(t)}{\varepsilon}$. We now extend the operator J on the series of the form

$$\tilde{w}(t,\tau,\varepsilon) = \sum_{k=0}^{\infty}\varepsilon^k w_k(t,\tau) \tag{12}$$

with coefficients $w_k(t,\tau) \in U, k \geq 0$. The formal extension $\tilde{J}$ of the operator J on the series of the form of Equation (12) is called the operator

$$\tilde{J}\tilde{w}(t,\tau,\varepsilon) \overset{def}{=} \sum_{\nu=0}^{\infty}\varepsilon^\nu\sum_{s=0}^{\nu}R_{\nu-s}w_s(t,\tau). \tag{13}$$

In spite of the fact that the extension in Equation (13) of the operator J is defined formally, it is quite possible to use it (see Theorem 3 below) in constructing an asymptotic solution of a finite order in ε. Now, it is easy to write out the regularized (with respect to Equation (1)) problem:

$$\frac{\partial\tilde{w}}{\partial t} + \sum_{j=1}^{3}\lambda_j(t)\frac{\partial\tilde{w}}{\partial\tau_j} - A(t)\tilde{w} - \tilde{J}\tilde{w} = \varepsilon H(t), \tilde{w}(t,\tau,\varepsilon)|_{t=0,\tau=0} = w^0. \tag{14}$$

3. The Solvability of Iterative Problems and the Asymptotic Convergence of Formal Solutions to the Exact Ones

Substituting the series of Equation (12) into Equation (14) and equating the coefficients for the same powers of ε, we obtain the following iteration problems:

$$L_0 w_0(t,\tau) \equiv \sum_{j=1}^{3} \lambda_j(t) \frac{\partial w_0}{\partial \tau_j} - A(t) w_0 - R_0 w_0 = 0, w_0(0,0) = w^0; \tag{15a}$$

$$L_0 w_1(t,\tau) = -\frac{\partial w_0}{\partial t} + R_1 w_0 + H(t), w_1(0,0) = 0; \tag{15b}$$

$$L_0 w_2(t,\tau) = -\frac{\partial w_1}{\partial t} + R_1 w_1 + R_2 w_0;\ w_2(0,0) = 0; \tag{15c}$$

$$\ldots$$

$$\begin{array}{l} L_0 w_k(t,\tau) = -\frac{\partial w_{k-1}}{\partial t} + R_1(t) w_{k-1} + R_2 w_{k-2} + \\ + \ldots + R_k w_0, w_k(0,0) = 0, k \geq 1, \end{array} \tag{15d}$$

where $R_0 w(t,\tau) \equiv R_0\left(\sum_{j=1}^{3} w_1(t) e^{\tau_j} + w_0(t)\right) = \int_0^t B(t,s) w_0(s) ds + e^{\tau_3} \int_0^t G(t,s) w_3(s) ds.$

Turning to the formulation of theorems on the normal and unique solvability of the iterative problems of Equations (15a)–(15d), we denote by

$$\varphi_j(t) \equiv \begin{pmatrix} \varphi_j^1(t) \\ \varphi_j^2(t) \end{pmatrix} = \begin{pmatrix} \lambda_j(t) - \mu(t) \\ K_2(t,t) \end{pmatrix}, j = 1,2,$$

the eigenvectors of the matrix $A(t)$. As the eigenvectors $\chi_j(t)$ of the matrix $A^*(t)$ we take the columns of the matrix $\left(\Phi^{-1}(t)\right)^* \equiv (\chi_1(t), \chi_2(t))$, where $\Phi(t) = (\varphi_1(t), \varphi_2(t))$ is the matrix whose columns are the eigenvectors of the matrix $A(t)$. Therefore, if $\varphi_j(t)$ is $\lambda_j(t)$-eigenvector of the matrix $A(t)$, then $\chi_j(t)$ is an $\bar{\lambda}_j(t)$-eigenvector of the matrix $A^*(t)$, and the systems $\{\varphi_j(t)\}$ and $\{\chi_k(t)\}$ are biorthonormal (see Reference [14], pp. 81–83), that is,

$$(\varphi_j(t), \chi_k(t)) \equiv \delta_{jk} = \begin{cases} 1, & j = k, \\ 0, & j \neq k \end{cases} \quad (j,k = 1,2).$$

Each of the iterative systems of Equation (15d) has the form

$$L_0 w(t,\tau) \equiv \sum_{j=1}^{3} \lambda_j(t) \frac{\partial w}{\partial \tau_j} - A(t) w - R_0 w = P(t,\tau), \tag{16}$$

where $P(t,\tau) = \sum_{j=1}^{3} P_j(t) e^{\tau_j} + P_0(t) \in U$. We prove the following assertion.

Theorem 1. *Suppose that the conditions (1)–(2) are satisfied and $P(t,\tau) \in U$. Then, the system of Equation (16) is solvable in the space U if and only if*

$$(P_j(t), \chi_j(t)) \equiv 0\ \forall t \in [0,T], j = 1,2. \tag{17}$$

Proof. We will determine the solution of the system of Equation (16) as the sum of Equation (7). Substituting Equation (7) into Equation (16) and equating separately the coefficients of e^{τ_j} and the free terms, we have

$$(\lambda_k(t) I - A(t)) w_k(t) = P_k(t),\ k = 1,2, \tag{18a}$$

$$(\lambda_3(t) I - A(t)) w_3(t) - \int_0^t G(t,s) w_3(s) ds = P_3(t), \tag{18b}$$

$$-A(t)\,w_0(t) - \int_0^t B(t,s)\,w_0(s)\,ds = P_0(t). \tag{18c}$$

For the systems of Equation (18a) to be solvable in space $C^\infty([0,T],\mathbb{C}^2)$, it is necessary and sufficient that the identities of Equation (17) hold (see, for example, Reference [14], p. 84). Moreover, these systems have a solution in the form of vector functions

$$w_k(t) = \alpha_k(t)\,\varphi_k(t) + \sum_{s=1,s\neq k}^{2} \frac{(P_k(t),\chi_s(t))}{\lambda_k(t)-\lambda_s(t)}\varphi_s(t),\ k=1,2,$$

where $\alpha_k(t) \in C^\infty([0,T],\mathbb{C}^1)$ are arbitrary functions. Since $\lambda_3(t) \notin \sigma(A(t))$ and $0 \notin \sigma(A(t))$, the systems of Equations (18b) and (18c) can also be rewritten in the form

$$\begin{array}{l} w_3(t) - \int_0^t (\lambda_3(t)I - A(t))^{-1}G(t,s)\,w_3(s)\,ds = (\lambda_3(t)I - A(t))^{-1}P_3(t), \\ w_0(t) + \int_0^t A^{-1}(t)\,B(t,s)\,w_0(s)\,ds = -A^{-1}(t)\,P_0(t). \end{array} \tag{19}$$

These Volterra integral systems have kernels belonging to the class $C^\infty([0,T],\mathbb{C}^{2\times 2})$, so they have unique solutions in the space $C^\infty([0,T],\mathbb{C}^2)$. The theorem is proved. □

Remark 1. *It follows from the proof of Theorem 1 that if the conditions of Equation (17) are satisfied, then the system of Equation (17) has the following solution in the space* U *:*

$$\begin{array}{l} w(t,\tau) = \sum_{k=1}^{2}\left[\alpha_k(t)\,\varphi_k(t) + \sum_{s=1,s\neq k}^{2} p_{ks}(t)\,\varphi_s(t)\right]e^{\tau_k} + w_3(t)\,e^{\tau_3} + w_0(t), \\ \left(p_{ks}(t) \equiv \frac{(P_k(t),\chi_s(t))}{\lambda_k(t)-\lambda_s(t)},\ k,s=1,2\right), \end{array} \tag{20}$$

where $\alpha_k(t) \in C^\infty([0,T],\mathbb{C}^1)$ *are arbitrary functions, and vector-valued functions* $w_3(t), w_0(t)$ *are solutions of the integral systems of Equation (19).*

We now consider the system of Equation (16) under additional conditions

$$\begin{array}{l} w(0,0) = w^*, \\ < -\frac{\partial w}{\partial t} + R_1 w + Q(t,\tau), \chi_j(t)e^{\tau_j} >\equiv 0,\ j=1,2, \end{array} \tag{21}$$

where $Q(t,\tau) = \sum_{j=1}^{3} Q_j(t)e^{\tau_j} + Q_0(t)$ are known functions of class $U, w^* \in \mathbb{C}^2$ is a known constant vector, the operator R_1 is defined by the equality of Equation (10), and by the $<,>$ we denote the inner product (for each $t \in [0,T]$) in space U :

$$< p(t,u), q(t,u) >\equiv< \sum_{j=1}^{3} p_j(t)e^{\tau_j} + p_0(t), \sum_{j=1}^{3} q_j(t)e^{\tau_j} + q_0(t) > \stackrel{def}{=} \sum_{k=0}^{3}(p_k(t),q_k(t)),$$

where $(,)$ is an ordinary inner product in $\mathbb{C}^2$. The following assertion holds true.

Theorem 2. *Suppose that the conditions (1)–(2) hold and the vector function* $P(t,\tau) \in U$ *satisfies the conditions of Equation (17). Then, the system of Equation (16) under additional conditions of Equation (21) is uniquely solvable in* U.

Proof. Since the conditions of Equation (17) are satisfied, the system of Equation (16) has a solution for Equation (20) in the space U, where $\alpha_j(t)$ are arbitrary functions for now. Subordinating Equation (18) to the initial condition $w(0,0) = w^*$, we obtain the equality

$$\alpha_1(0)\,\varphi_1(0) + p_{12}(0)\,\varphi_2(0) + \alpha_2(0)\,\varphi_2(0) + p_{21}(0)\,\varphi_1(0) = w^*,$$

where $w^* = w_* - w_3(0) - w_0(0)$. Multiplying both sides of this equation scalarly in turn by $\chi_1(0)$ and $\chi_2(0)$, taking into account the biorthonormality of the eigenvector systems $\{\varphi_j(t)\}, \{\chi_k(t)\}$, we have

$$\alpha_1(0) = (w^*, \chi_1(0)) - p_{21}(0), \alpha_2(0) = (w^*, \chi_2(0)) - p_{12}(0). \tag{22}$$

We now calculate the expression $-\frac{\partial w}{\partial t} + R_1 w + Q(t,\tau)$. Taking into account Equation (21) and the form of the operator $R_1 w(t,\tau)$, we have (here and everywhere below, a fatty dot denotes differentiation with respect to t.)

$$\begin{array}{l} -\frac{\partial w}{\partial t} + R_1 w + Q(t,\tau) = -\sum_{k=1}^{2} (\alpha_k(t)\varphi_k(t) + p_{ks}(t)\varphi_s(t))^{\bullet} e^{\tau_k} - \\ -\dot{w}_3(t) e^{\tau_3} - \dot{w}_0(t) + \frac{G(t,0)w_0(0)}{\lambda_3(0)} e^{\tau_3} - \frac{G(t,t)w_0(t)}{\lambda_3(t)} + \\ +\sum_{j=1}^{3}\left[\frac{B(t,t)w_j(t)}{\lambda_j(t)} e^{\tau_j} - \frac{B(t,0)w_j(0)}{\lambda_j(0)}\right] + \\ +\sum_{k=1}^{2}\left[\frac{G(t,t)w_k(t)}{\lambda_k(t)-\lambda_3(t)} e^{\tau_k} - \frac{G(t,0)w_k(0)}{\lambda_k(0)-\lambda_3(0)} e^{\tau_3}\right] + \sum_{j=1}^{3} Q_j(t) e^{\tau_j} + Q_0(t). \end{array}$$

When writing the conditions of Equation (21) in this expression, it is necessary to preserve only terms containing exponentials e^{τ_1} and e^{τ_2}, that is, Equation (21) is equivalent to the conditions

$$\begin{array}{l} < -\sum\limits_{k=1}^{2}\left(\alpha_k(t)\varphi_k(t) + \sum\limits_{s=1,s\neq k}^{2} p_{ks}(t)\varphi_s(t)\right)^{\bullet} e^{\tau_k} + \\ +\sum\limits_{k=1}^{2}\left(\frac{B(t,t)}{\lambda_k(t)} + \frac{G(t,t)}{\lambda_k(t)-\lambda_3(t)}\right)\left(\alpha_k(t)\varphi_k(t) + \sum\limits_{s=1,s\neq k}^{2} p_{ks}(t)\varphi_s(t)\right) e^{\tau_k} + \\ +\sum\limits_{k=1}^{2} Q_j(t) e^{\tau_k}, \chi_j(t) e^{\tau_j} > \equiv 0, j = 1,2, \end{array}$$

or

$$\begin{array}{l} (-(\alpha_1(t)\varphi_1(t) + p_{12}(t)\varphi_2(t))^{\bullet} + \left(\frac{B(t,t)}{\lambda_1(t)} + \frac{G(t,t)}{\lambda_1(t)-\lambda_3(t)}\right)(\alpha_1(t)\varphi_1(t) + p_{12}(t)\varphi_2(t)) + \\ +Q_1(t), \chi_1(t)) \equiv 0, \\ (-(\alpha_2(t)\varphi_2(t) + p_{21}(t)\varphi_1(t))^{\bullet} + \left(\frac{B(t,t)}{\lambda_2(t)} + \frac{G(t,t)}{\lambda_2(t)-\lambda_3(t)}\right)(\alpha_2(t)\varphi_2(t) + p_{21}(t)\varphi_1(t)) + \\ +Q_2(t), \chi_2(t)) \equiv 0. \end{array}$$

Performing inner multiplication here, we obtain differential equations

$$\dot{\alpha}_1(t) + \left(\dot{\varphi}_1(t) - \left(\frac{B(t,t)}{\lambda_1(t)} + \frac{G(t,t)}{\lambda_1(t)-\lambda_3(t)}\right)\varphi_1(t), \chi_1(t)\right)\alpha_1(t) = g_1(t),$$

$$\dot{\alpha}_2(t) + \left(\dot{\varphi}_2(t) - \left(\frac{B(t,t)}{\lambda_2(t)} + \frac{G(t,t)}{\lambda_2(t)-\lambda_3(t)}\right)\varphi_2(t), \chi_2(t)\right)\alpha_2(t) = g_2(t),$$

where $g_j(t)$ are known scalar functions, $j = 1,2$. Adding the initial conditions of Equation (22) to these equations, we find uniquely the functions $\alpha_j(t)$ in the solution of Equation (20) of the system of Equation (16), and therefore, we construct a solution of this system in the space U in a unique way. The theorem is proved. □

Applying Theorems 1 and 2 to iterative problems, we uniquely determine their solutions in space U and construct the series of Equation (12). As in Reference [2], we prove the following assertion.

Theorem 3. *Assume that the conditions (1)–(2) are satisfied for the system of Equation (2). Then, for* $\varepsilon \in (0,\varepsilon_0]$ *($\varepsilon_0 > 0$ is sufficiently small) the system of Equation (2) has a unique solution* $w(t,\varepsilon) \in C^1([0,T],\mathbb{C}^2)$*; and here we have the estimate*

$$||w(t,\varepsilon) - w_{\varepsilon N}(t)||_{C[0,T]} \leq c_N \varepsilon^{N+1}, N = 0,1,2,...,$$

where $w_{\varepsilon N}(t)$ *is the restriction (for* $\tau = \frac{\psi(t)}{\varepsilon}$*) N-partial sum of the series of Equation (12) (with coefficients* $w_k(t,\tau) \in U$*, satisfying the iterative problems of Equation (15d)), the constant* $c_N > 0$ *does not depend on* ε *at* $\varepsilon \in (0, \varepsilon_0]$.

Since $y(t,\varepsilon) = \frac{1}{\varepsilon} v(t,\varepsilon)$, the series

$$\frac{1}{\varepsilon}\sum_{k=0}^{\infty} v_k\left(t, \frac{\psi(t)}{\varepsilon}\right) \equiv \frac{1}{\varepsilon} v_0\left(t, \frac{\psi(t)}{\varepsilon}\right) + v_1\left(t, \frac{\psi(t)}{\varepsilon}\right) + \varepsilon v_2\left(t, \frac{\psi(t)}{\varepsilon}\right) + \ldots$$

is an asymptotic solution (for $\varepsilon \to +0$) of the original problem of Equation (1), that is, the estimate

$$||y(t,\varepsilon) - \sum_{k=-1}^{N} \varepsilon^k v_{k+1}\left(t, \frac{\psi(t)}{\varepsilon}\right)||_{C[0,T]} \le C_N \varepsilon^{N+1},\ N = -1, 0, 1, \ldots, \quad (23)$$

is correct, where the constant $C_N > 0$ does not depend on $\varepsilon \in (0, \varepsilon_0]$.

Conclusion 1. The influence of the weakly varying integral kernel $K_0(t,s)$ on the asymptotic of the solution of the problem of Equation (1) consists of two factors: Firstly, the kernel $K_0(t,s)$ participates in the formation of the matrix $A(t)$ and its eigenvectors and eigenvalues, secondly, it participates in the construction of the limit operator L_0, which leads to an additional integral system $w_0(t) + \int_0^t A^{-1}(t) B(t,s) w_0(s) ds = -A^{-1}(t) P_0(t)$ in the solvability of conditions Equation (17) of iterative problems.

4. The Limit Transition in the Problem of Equation (1). Solving the Initialization Problem

It follows from Equation (23) that the exact solution of the problem of Equation (1) is represented in the form

$$\begin{aligned} &y(t,\varepsilon) = \tfrac{1}{\varepsilon} v_0\left(t, \tfrac{\psi(t)}{\varepsilon}\right) + v_1\left(t, \tfrac{\psi(t)}{\varepsilon}\right) + \varepsilon F(t,\varepsilon), \\ &||F(t,\varepsilon)||_{C^n} \le \bar{F} = \text{const}\ (\forall (t,\varepsilon) \in [0,T] \times (0,\varepsilon_0]), \end{aligned} \quad (24)$$

therefore, in order to study the passage to the limit (for $\varepsilon \to +0$) in the solution of the problem of Equation (1), it is necessary to find the solutions of the two iteration problems of Equation (15d) ($k = 0, 1$) under the conditions of Equation (18) for the solvability of the third problems of Equation (15c). We start with the problem of Equation (15a):

$$\begin{aligned} &L_0 w_0(t,\tau) \equiv \sum_{j=1}^{3} \lambda_j(t) \tfrac{\partial w_0}{\partial \tau_j} - A(t) w_0 - R_0 w_0 = 0, w_0(0,0) = w^0 \\ &\left(R_0 w(t,\tau) = \int_0^t B(t,s) w_0(s) ds + e^{\tau_3} \int_0^t G(t,s) w_3(s) ds\right). \end{aligned} \quad (15a)$$

Since the right-hand side of the system of Equation (15a) $P^{(0)}(t,\tau) = \sum_{j=1}^{3} P_j^{(0)}(t) e^{\tau_j} + P_0^{(0)}(t)$ is identically zero, it has (according to Theorem 1) a solution

$$w_0(t,\tau) = \sum_{k=1}^{2} \alpha_k^{(0)}(t) \varphi_k(t) e^{\tau_k} + w_3^{(0)}(t) e^{\tau_3} + w_0^{(0)}(t),$$

where the vector functions $w_3^{(0)}(t), w_0^{(0)}(t)$ satisfy the equations

$$\begin{aligned} &w_3^{(0)}(t) - \int_0^t (\lambda_3(t) I - A(t))^{-1} G(t,s) w_3^{(0)}(s) ds = 0, \\ &w_0^{(0)}(t) + \int_0^t A^{-1}(t) B(t,s) w_0(s) ds = 0. \end{aligned}$$

These equations are homogeneous, and therefore, they have the unique solutions $w_3^{(0)}(t) = w_0^{(0)}(t) \equiv 0$, and the solution of the system of Equation (15a) is written in the form

$$w_0(t,\tau) = \sum_{k=1}^{2} \alpha_k^{(0)}(t)\,\varphi_k(t)\,e^{\tau_k}. \tag{25}$$

Let $\chi_k(t) = \{\chi_k^1(t), \chi_k^2(t)\}$, $k = 1,2$. Subordinating Equation (24) to the initial condition $w_0(0,0) = w^0$, we find the values

$$\alpha_k^{(0)}(0) = \left(w^0, \chi_k(0)\right) = h(0)\,\bar{\chi}_k^1(0),\ k = 1,2. \tag{26}$$

For the final computation of the functions $\alpha_k^{(0)}(t)$, we pass to the next iteration problem

$$L_0 w_1(t,\tau) = -\sum_{k=1}^{2} \left(\alpha_k^{(0)}(t)\,\varphi_k(t)\right)^{\bullet} e^{\tau_k} + R_1 w_0 + H(t), w_1(0,0) = 0, \tag{15b}$$

where

$$\begin{aligned} R_1 w_0 &= R_1\left(\textstyle\sum_{k=1}^{2} \alpha_k^{(0)}(t)\,\varphi_k(t)\,e^{\tau_k}\right) = \\ &= \textstyle\sum_{j=1}^{3}\left[\frac{B(t,t)\alpha_j^{(0)}(t)\varphi_k(t)}{\lambda_j(t)} e^{\tau_j} - \frac{B(t,0)\alpha_j^{(0)}(0)\varphi_j(0)}{\lambda_j(0)}\right] + \\ &+ \textstyle\sum_{k=1}^{2}\left[\frac{G(t,t)\alpha_k^{(0)}(t)\varphi_k(t)}{\lambda_k(t)-\lambda_3(t)} e^{\tau_k} - \frac{G(t,0)\alpha_k^{(0)}(0)\varphi_k(0)}{\lambda_k(0)-\lambda_3(0)} e^{\tau_3}\right]. \end{aligned}$$

Keeping, as in Theorem 2, only the terms containing exponentials e^{τ_1} and e^{τ_2}, we write down conditions of Equation (17) in the form (see Equation (26)):

$$\begin{aligned} &\dot{\alpha}_k^{(0)}(t) = \left(\frac{B(t,t)\,\varphi_k(t)}{\lambda_k(t)} + \frac{G(t,t)\,\varphi_k(t)}{\lambda_k(t) - \lambda_3(t)} - \dot{\varphi}_k(t), \chi_k(t)\right)\alpha_k^{(0)}(t), \\ &\alpha_k^{(0)}(0) = h(0)\,\bar{\chi}_k^1(0), k = 1,2, \end{aligned}$$

from which we find that

$$\alpha_k^{(0)}(t) = h(0)\,\bar{\chi}_k^1(0)\,e^{\int_0^t q_k(\theta)d\theta},\ k = 1,2, \tag{27}$$

where it is denoted: $q_k(t) \equiv \left(\frac{B(t,t)\varphi_k(t)}{\lambda_k(t)} + \frac{G(t,t)\varphi_k(t)}{\lambda_k(t)-\lambda_3(t)} - \dot{\varphi}_k(t), \chi_k(t)\right), k = 1,2$. Thus, the solution of the problem of Equation (15a) is found in the form of Equation (25), where the functions $\alpha_k^{(0)}(t)$ are Equation (27). Similarly, we can find the solution of the problem of Equation (15b). However, having in mind to solve the initialization problem in the future, we must put $v_0\left(t, \frac{\psi(t)}{\varepsilon}\right) \equiv 0$ in Equation (24). This identity holds if and only if $\alpha_k^{(0)}(t) \equiv 0\ (k = 1,2) \Leftrightarrow h(0) = 0$ (remember that $v_0\left(t, \frac{\psi(t)}{\varepsilon}\right) = \sum_{k=1}^{2} \alpha_k^{(0)}(t)\varphi_k^1(t)\,e^{\frac{\psi_k(t)}{\varepsilon}}$, $\varphi_j^1(t) = \lambda_j(t) - \mu(t), j = 1,2$ and see Equation (27)), we will therefore carry out further calculations for $h(0) = 0$. In this case, $w_0(t,\tau) \equiv 0, R_1 w_0 \equiv 0$, and the problem of Equation (15b) takes the form

$$L_0 w_1(t,\tau) \equiv \sum_{j=1}^{3} \lambda_j(t)\,\frac{\partial w_1}{\partial \tau_j} - A(t) w_1 - R_0 w_1 = H(t), w_1(0,0) = 0.$$

Since here $P^{(1)}(t,\tau) = H(t)\ \left(P_j^{(1)}(t) \equiv 0, j = 1,2,3, P_0^{(1)}(t) = H(t)\right)$, in formula

$$w_1(t,\tau) = \sum_{k=1}^{2}\left[\alpha_k^{(1)}(t)\,\varphi_k(t) + \sum_{s=1, s\neq k}^{2} p_{ks}^{(1)}(t)\,\varphi_s(t)\right] e^{\tau_k} + w_3^{(1)}(t)\,e^{\tau_3} + w_0^{(1)}(t)$$

for the solution of the problem of Equation (15b) functions $p_{ks}^{(1)}(t) \equiv 0\ (k,s=1,2)$, functions $w_3^{(1)}(t)$ and $w_0^{(1)}(t)$ are solutions of the integral equations

$$\begin{array}{l} w_3^{(1)}(t) - \int_0^t (\lambda_3(t) I - A(t))^{-1} G(t,s)\, w_3^{(1)}(s)\, ds = 0 \Leftrightarrow w_3^{(1)}(t) \equiv 0, \\ w_0^{(1)}(t) + \int_0^t A^{-1}(t) B(t,s)\, w_0^{(1)}(s)\, ds = -A^{-1}(t) H(t), \end{array} \tag{28}$$

therefore, the solution of the problem will be as follows:

$$w_1(t,\tau) = \sum_{k=1}^{2} \alpha_k^{(1)}(t)\, \varphi_k(t)\, e^{\tau_k} + w_0^{(1)}(t), \tag{29}$$

where $\alpha_k^{(1)}(t)$, for the time being, are arbitrary functions, $k=1,2$, and the vector-valued function $w_0^{(1)}(t)$ is a solution of the system of Equation (28). Subordinating Equation (29) to the initial condition $w_1(0,0)=0$, we obtain

$$\begin{array}{l} \sum_{k=1}^2 \alpha_k^{(1)}(0)\, \varphi_k(0) = -w_0^{(1)}(0) \equiv A^{-1}(0) H(0) \Rightarrow \\ \alpha_k^{(1)}(0) = \left(A^{-1}(0) H(0), \chi_k(0)\right) = \left(H(0), A^{-1}(0)\chi_k(0)\right) = \\ = \left(H(0), \bar{\lambda}_k(0)\chi_k(0)\right) = \lambda_k(0)\left(H(0), \chi_k(0)\right), \end{array}$$

i.e.,

$$\begin{cases} \alpha_1^{(1)}(0) = \lambda_1(0)\, \dot{h}(0)\, \bar{\chi}_1^1(0), \\ \alpha_2^{(1)}(0) = \lambda_2(0)\, \dot{h}(0)\, \bar{\chi}_2^1(0). \end{cases} \tag{30}$$

For the final calculation of the solution of Equation (29) of the problem of Equation (15b), let us pass to the following problem (note that $w_0 \equiv 0$):

$$L_0 w_2(t,\tau) = -\frac{\partial w_1}{\partial t} + R_1 w_1, \quad w_2(0,0) = 0. \tag{15c}$$

Substituting here the function of Equation (29), we obtain the system

$$\begin{array}{l} L_0 w_2(t,\tau) = -\sum_{k=1}^2 \left(\alpha_k^{(1)}(t)\, \varphi_k(t)\right)^{\bullet} e^{\tau_k} + \\ + \frac{G(t,0) w_0^{(1)}(0)}{\lambda_3(0)} e^{\tau_3} - \frac{G(t,t) w_0^{(1)}(t)}{\lambda_3(t)} + \\ + \sum_{j=1}^3 \left[\frac{B(t,t)\alpha_j^{(1)}(t)\varphi_j(t)}{\lambda_j(t)} e^{\tau_j} - \frac{B(t,0)\alpha_j^{(1)}(0)\varphi_j(0)}{\lambda_j(0)} \right] + \\ + \sum_{k=1}^2 \left[\frac{G(t,t)\alpha_k^{(1)}(t)\varphi_k(t)}{\lambda_k(t)-\lambda_3(t)} e^{\tau_k} - \frac{G(t,0)\alpha_k^{(1)}(0)\varphi_k(0)}{\lambda_k(0)-\lambda_3(0)} e^{\tau_3} \right] = 0. \end{array}$$

Keeping here, as in Theorem 2, only terms containing exponentials e^{τ_1} and e^{τ_2}, we write the conditions of Equation (17) for the solvability of this system in the form

$$\begin{array}{l} \dot{\alpha}_k^{(1)}(t) = \left(\frac{B(t,t)\varphi_k(t)}{\lambda_k(t)} + \frac{G(t,t)\varphi_k(t)}{\lambda_k(t)-\lambda_3(t)} - \dot{\varphi}_k(t), \chi_k(t) \right) \alpha_k^{(1)}(t), \\ \alpha_1^{(1)}(0) = \lambda_1(0)\, \dot{h}(0)\, \bar{\chi}_1^1(0), \alpha_2^{(1)}(0) = \lambda_2(0)\, \dot{h}(0)\, \bar{\chi}_2^1(0), \end{array}$$

from which we uniquely find the functions $\alpha_k^{(1)}(t)$:

$$\alpha_k^{(1)}(t) = \lambda_k(0)\, \dot{h}(0)\, \bar{\chi}_k^1(0)\, e^{\int_0^t q_k(\theta) d\theta}, k = 1,2,$$

and therefore, we uniquely construct the solution of Equation (29) of the problem of Equation (15b). In this case, the equality holds (remember that $w_0(t,\tau) \equiv 0$)

$$w(t,\varepsilon) = \varepsilon \left(\sum_{k=1}^{2} \lambda_k(0)\,\dot{h}(0)\,\bar{\chi}_k^1(0)\, e^{\int_0^t q_k(\theta)d\theta} \varphi_k(t)\, e^{\frac{\psi_k(t)}{\varepsilon}} + w_0^{(1)}(t) \right) + \varepsilon^2 F_1(t,\varepsilon) \Rightarrow$$

$$\Rightarrow y(t,\varepsilon) = \sum_{k=1}^{2} \lambda_k(0)\,\dot{h}(0)\,\bar{\chi}_k^1(0)\, e^{\int_0^t q_k(\theta)d\theta} \varphi_k^1(t)\, e^{\frac{\psi_k(t)}{\varepsilon}} + v_0^{(1)}(t) + \varepsilon f_1(t,\varepsilon), \tag{31}$$

where $w_0^{(1)}(t) = \left\{ v_0^{(1)}(t), z_0^{(1)}(t) \right\}$ is the solution of the integral system

$$\begin{cases} v_0^{(1)}(t) + \int\limits_0^t \frac{\left(\frac{\partial}{\partial t} K_1(t,s)\right) v_0^{(1)}(s)}{K_1(t,t)} ds = -\frac{\frac{d}{dt} h(t)}{K_1(t,t)}, \\ z_0^{(1)}(t) - \left(\int\limits_0^t \frac{K_2(t,t)\left(\frac{\partial}{\partial t} K_1(t,s)\right) v_0^{(1)}(s)}{\mu(t) K_1(t,t)} ds \right) = \frac{K_2(t,t)\left(\frac{d}{dt} h(t)\right)}{\mu(t) K_1(t,t)}. \end{cases} \tag{32a}$$

It follows from Equation (31) that when $\operatorname{Re} \lambda_k(t) < 0$ $(\forall t \in [0,T], k = 1,2)$ there is a passage to the limit

$$\left\| y(t,\varepsilon) - v_0^{(1)}(t) \right\|_{C[\delta,T]} \to 0 \ (\varepsilon \to +0),$$

where $\delta \in (0,T)$ is an arbitrary fixed constant, and $w_0^{(1)}(t) = \left\{ v_0^{(1)}(t), z_0^{(1)}(t) \right\}$. However, in our case, there can be purely imaginary eigenvalues $(\operatorname{Re} \lambda_k(t) \equiv 0)$, so the indicated limit transition does not hold. The following problem is posed: to find a class $\Sigma = \{h(t), K_1(t,s), K_2(t,s)\}$ of initial data of Equation (1) for which the passage to the limit

$$\left\| y(t,\varepsilon) - v_0^{(1)}(t) \right\|_{C[0,T]} \to 0 \ (\varepsilon \to +0), \tag{$*$}$$

takes place on the whole segment $[0,T]$, including the boundary layer zone. This task is called *the initialization problem.* . It is clear from Equation (31) that the limit transition $(*)$ occurs if and only if $\dot{h}(0) = 0$, therefore, the following result follows from Equation (32a).

Theorem 4. *Suppose that the conditions (1)–(2) are satisfied. Then, the passage to the limit $(*)$ holds if and only if $h(0) = \dot{h}(0) = 0$ (here, $v_0^{(1)}(t)$ is the solution of the first equation of the system of Equation (32a)).*

Conclusion 2. Thus, the initialization class Σ has the form $\Sigma = \left\{ h(t) : h(0) = \dot{h}(0) = 0 \right\}$. Here, the kernels $K_j(t,s)$ can be arbitrary, provided that conditions (1)–(2) are satisfied.

Example 1. *Consider the equation*

$$\varepsilon y(t,\varepsilon) = \int\limits_0^t e^{-\frac{1}{\varepsilon}(t-s)} \left(e^{-s} - 1\right) y(s,\varepsilon)\, ds + \int\limits_0^t \left(-2e^{-s} y(s,\varepsilon)\right) ds + t^2. \tag{32b}$$

Here, $h(t) = t^2$, $\mu(t) = -1$, $K_1(t,s) = -2e^{-s}$, $K_2(t,s) = e^{-s} - 1$. *The characteristic equation of the matrix* $A(t) = \begin{bmatrix} -e^{-t} - 1 & -1 \\ e^{-t} - 1 & -1 \end{bmatrix}$ *has two roots* $\lambda_1(t) = -2$, $\lambda_2(t) = -e^{-t}$. *Using the algorithm developed above, we find that*

$$v_0^{(1)}(t) + \int\limits_0^t \frac{\left(\frac{\partial}{\partial t} K_1(t,s)\right) v_0^{(1)}(s)}{K_1(t,t)} ds = -\frac{\frac{d}{dt} h(t)}{K_1(t,t)} \Leftrightarrow v_0^{(1)}(t) = te^t.$$

Since $\dot{h}(0) = 0$, the main term of the asymptotic of the solution of our Equation (32b) coincides with $v_0^{(1)}(t)$ (see Equation (31)). By Theorem 4, there is a passage to the limit:

$$\left\|y(t,\varepsilon) - te^t\right\|_{C[0,T]} \to 0\,(\varepsilon \to +0)\,.$$

We note that the function $v_0^{(1)}(t) = te^t$ is a solution of the integral equation $\int\limits_0^t (-2e^{-s}y(s))\,ds + t^2 = 0$, which is degenerative with respect to Equation (1). If only $h(0) = 0$, but $\dot{h}(0) \neq 0$, then from Equation (31), we would have obtained that

$$y(t,\varepsilon) = v_1\left(t, \frac{\psi(t)}{\varepsilon}\right) + \varepsilon F(t,\varepsilon)\,,$$

and the function $v_1\left(t, \frac{\psi(t)}{\varepsilon}\right)$ contains exponents $e^{\frac{-2t}{\varepsilon}}$ and $e^{\frac{1}{\varepsilon}(e^{-t}-1)}$, which prevent uniform convergence of the solution $y(t,\varepsilon)$ on the whole interval $[0,T]$ to the limit function. In this case, uniform convergence will occur only outside the boundary layer $[\delta, T]$ $(\delta \in (0,T))$.

The analysis of asymptotic methods for solving singularly perturbed problems shows that the solutions of such problems depend in two ways on a small parameter: regularly and singularly. This dependence is especially vividly demonstrated by the method of regularization of Lomov. Moreover, regularized series representing solutions of singularly perturbed problems can converge in the usual sense. In this connection, it became necessary to study a special class of functions—pseudoholomorphic functions. This very important part of the complex analysis is designed to substantiate the main provisions of the so-called analytic theory of singular perturbations. On the other hand, the relevance of the theory is also dictated by the fact that pseudoholomorphic functions, in contrast to holomorphic functions, are determined when the conditions of the implicit function theorem are violated. The concept of a pseudoanalytic (pseudoholomorphic) function and the associated concept of an essentially singular manifold are of a general mathematical nature, although they arose in the framework of the regularization method for singular perturbations. First of all, they reflect the new concept of a pseudoholomorphic solution of singularly perturbed problems, i.e., such a solution, which is representable in the form of a series converging in the usual (but not asymptotic) sense in powers of a small parameter. We must also take into account the fact that the modern mathematical theory of the boundary layer [13], along with the Vasilyeva–Butuzov–Nefedov's boundary-function method [5,6], widely uses the concept of a pseudoholomorphic solution. The following sections of our work are devoted to the construction of exactly such solutions [15].

5. Pseudoholomorphic Functions in the Theory of Singular Perturbations. Basic Concepts and Statements

We consider the set of functions $F(z,w,\varepsilon)$, where $w = (w_1,\ldots,w_k)$, $F = (F_1,\ldots,F_k)$, holomorphic in a polydisc $D = D_{z_0} \times D_{w_0} \times D_0$, in which

$$D_{z_0} = \{z:\ |z - z_0|\ < R_0\}, D_{w_0} = \{w:\ |w_j - w_{0,j}|\ < R_j,\ j = \overline{1,k}\}, D_0 = \{\varepsilon:\ |\varepsilon|\ < \varepsilon_0\}.$$

Definition 1. *A function $w(z,\varepsilon)$, defined implicitly by the equation*

$$F(z,w,\varepsilon) = 0, \tag{33}$$

is said to be pseudoholomorphic at a point of $\varepsilon = 0$ of rank r, if the following conditions are satisfied:

1^0. *$F(z_0, w_0, 0) = 0$;*

2^0. *$\partial_{w_j} F_i\big|_{\varepsilon=0} = 0\ \forall (z,w) \in D_{z_0} \times D_{w_0}$, $i = \overline{1,r}$, $j = \overline{1,k}$;*

3^0. *$\det \|f_{ij}\| \neq 0\ \forall (z,w) \in D_{z_0} \times D_{w_0}$, where $f_{ij} = \partial^2_{\varepsilon w_j} F_i\big|_{\varepsilon=0}$, $i = \overline{1,r}$, $j = \overline{1,k}$; $f_{ij} = \partial_{w_j} F_i\big|_{\varepsilon=0}$, $i = \overline{r+1,k}$, $j = \overline{1,k}$.*

4^0. *$w(z,\varepsilon)$ is unbounded in any sufficiently small neighborhood of a point $\varepsilon = 0$ and there exists a set $E_0 \subset D_0$, for which the point $\varepsilon = 0$ is a limit point and such that it is bounded on a set $T_{z_0} \times E_0$, where T_{z_0} is a compact that belongs D_{z_0} and contains a point z_0.*

From definition , it follows that

$$\begin{aligned} F_i(z,w,\varepsilon) &= \varphi_i(z) - \varepsilon U_{i,1}(z,w) - \ldots - \varepsilon^n U_{i,n}(z,w) - \ldots, \quad i = \overline{1,r}; \\ F_i(z,w,\varepsilon) &= U_{i,0}(z,w) + \varepsilon U_{i,1}(z,w) + \ldots + \varepsilon^n U_{i,n}(z,w) + \ldots, \quad i = \overline{r+1,k}, \end{aligned} \tag{34}$$

and these series converge uniformly on any compact set $D_{z_0} \times D_{w_0}$ in some neighborhood of the point $\varepsilon = 0$ (depending on the compact).

We compose the following system of equations:

$$\begin{cases} U_{1,1}(z,w) = \varphi_1(z)/\varepsilon, \\ \quad\ldots \\ U_{r,1}(z,w) = \varphi_r(z)/\varepsilon, \\ U_{r+1,0}(z,w) = 0, \\ \quad\ldots \\ U_{k,0}(z,w) = 0, \end{cases} \tag{35}$$

which will be used in the future. We shall call Equation (35) *the main system*.

Suppose that the entire functions $\Psi_1,\ldots,\Psi_r$ of one variable with the asymptotic values $a_1,\ldots,a_r$ are such that the sets $\omega_i = \{q_i : \ q_i = \Psi_i(\varphi_i(z)/\varepsilon)\} \subset \mathbb{C}_{q_i}$ are bounded if $z \in T_{z_0}$ and $\varepsilon \in E_0$, where T_{z_0} and E_{z_0} are sets satisfying the condition 4^0 of the Definitions 1. We also assume that the points a_i close these sets: $\bar{\omega}_i = \omega_i \cup \{a_i\}$, $i = \overline{1,r}$. We introduce the notations: $\Psi = (\Psi_1,\ldots,\Psi_r)$, $\varphi = (\varphi_1,\ldots,\varphi_r)$, $a = (a_1,\ldots,a_r)$.

Definition 2. *The set $\Omega(\Psi,\varphi,T_{z_0},E_0) = \omega_1 \times \ldots \times \omega_r \subset \mathbb{C}_{q_1} \times \ldots \times \mathbb{C}_{q_2}$ is called an essentially singular manifold, generated by the functions Ψ and φ on the set $T_{z_0} \times E_0$; we call the set $\bar{\Omega}(\Psi,\varphi,T_{z_0},E_0) = \bar{\omega}_1 \times \ \ldots \times \bar{\omega}_r$ an extended essentially singular manifold.*

Let us formulate sufficient conditions for the existence of a pseudoholomorphic function. For this, along with the system of Equation (35), we consider the system

$$\begin{cases} U_{1,1}(z,w) = q_1, \\ \quad\ldots \\ U_{r,1}(z,w) = q_r, \\ U_{r+1,0}(z,w) = 0, \\ \quad\ldots \\ U_{k,0}(z,w) = 0. \end{cases} \tag{36}$$

Theorem 5. *If a functionw $= W_0(z,q)$ that is a solution of the system of Equation (36) is holomorphic on a compact $\bar{Q} = T_{z_0} \times \bar{\Omega}(\Psi,\varphi,T_{z_0},E_0)$ and maps it to a polydisk D_{w_0}, then the function $w(z,\varepsilon)$, implicitly defined by Equation (33), is pseudoholomorphic at the point $\varepsilon = 0$.*

Proof. We represent the vector of Equation (33) in the form of a system as follows:

$$\begin{cases} U_{1,1}(z,w) + \varepsilon U_{1,2}(z,w) + \ldots + \varepsilon^n U_{1,n+1}(z,w) + \ldots = \varphi_1(z)/\varepsilon, \\ \ldots \\ U_{r,1}(z,w) + \varepsilon U_{r,2}(z,w) + \ldots + \varepsilon^n U_{r,n+1}(z,w) + \ldots = \varphi_r(z)/\varepsilon, \\ U_{r+1,0}(z,w) + \varepsilon U_{r+1,1}(z,w) + \ldots + \varepsilon^n U_{r+1,n}(z,w) + \ldots = 0, \\ \ldots \\ U_{k,0}(z,w) + \varepsilon U_{k,1}(z,w) + \ldots + \varepsilon^n U_{k,n}(z,w) + \ldots = 0, \end{cases} \tag{37}$$

and calculate the values of the functions $\Psi_1, \ldots, \Psi_r$ from the left and right parts of the first r equations:

$$\Psi_1(U_{1,1}(z,w) + \varepsilon U_{1,2}(z,w) + \ldots + \varepsilon^n U_{1,n+1}(z,w) + \ldots) = \Psi_1(\varphi_1(z)/\varepsilon),$$
$$\ldots$$
$$\Psi_r(U_{r,1}(z,w) + \varepsilon U_{r,2}(z,w) + \ldots + \varepsilon^n U_{r,n+1}(z,w) + \ldots) = \Psi_r(\varphi_r(z)/\varepsilon),$$

and then in the left-hand sides of these equations we distinguish the main terms:

$$\begin{aligned} \Psi_1(U_{1,1}(z,w)) + \varepsilon V_1(z,w,\varepsilon) &= \Psi_1(\varphi_1(z)/\varepsilon), \\ &\ldots \\ \Psi_r(U_{r,1}(z,w)) + \varepsilon V_r(z,w,\varepsilon) &= \Psi_r(\varphi_r(z)/\varepsilon). \end{aligned} \tag{38}$$

Using the notations introduced earlier, we rewrite the system of Equation (36):

$$\begin{cases} \Psi_1(U_{1,1}(z,w)) + \varepsilon V_1(z,w,\varepsilon) = q_1, \\ \ldots \\ \Psi_r(U_{r,1}(z,w)) + \varepsilon V_r(z,w,\varepsilon) = q_r, \\ U_{r+1,0}(z,w) = 0, \\ \ldots \\ U_{k,0}(z,w) = 0. \end{cases} \tag{39}$$

When $\varepsilon = 0$, the system of Equation (39) has a solution $w = W_0(z,q)$, holomorphic on a set $\bar{Q}$, that which maps to a compact, belonging to D_{w_0}, and therefore, in accordance with the implicit function theorem, in some neighborhood σ_{zq} of each point $(z,q) \in \bar{Q}$ this system has a solution w that is holomorphic at the point $\varepsilon = 0 : W(z,q,\varepsilon) = \sum_{n=0}^{\infty} \varepsilon^n W_n(z,q)$. From the covering $\{\sigma_{zq}\}$ of a compact set $\bar{Q}$, we choose a finite subcover, then the function $W(z,q,\varepsilon)$ will be holomorphic uniformly on $\bar{Q}$ in a neighborhood $|\varepsilon| < \varepsilon_1$, where ε_1 is the smallest number of the corresponding finite subcoverings. The boundedness of the function $w(z,\varepsilon) = W(z, \Psi_1(\varphi_1(z)/\varepsilon), \ldots, \Psi_r(\varphi_r(z)/\varepsilon), \varepsilon)$ for $\varepsilon \to 0$ ($\varepsilon \in E_0$) follows from the fact that the point (z,ε) belongs to an extended essentially singular manifold $\hat{\Omega}(\Psi, \varphi, T_{z_0}, E_0)$. The theorem is proved. □

Remark 2. *It follows from Theorem 5 that a pseudoholomorphic function decomposes into a power series with coefficients that depend in a singular way on* ε*:*

$$W(z,\varepsilon) = \sum_{n=0}^{\infty} \varepsilon^n W_n(z, \Psi_1(\varphi_1(z)/\varepsilon), \ldots, \Psi_r(\varphi_r(z)/\varepsilon)) \tag{40}$$

and this series converges for $|\varepsilon| < \varepsilon_1$ *(*$\varepsilon \in E_0$*) uniformly on* T_{z_0}*.*

6. *-Pseudoholomorphic Functions

In applications, for example, in the mathematical theory of the boundary layer [3], we have to impose less restrictive conditions on pseudomorphic functions.

Definition 3. *A *-transformation of a function* $F(z,w,\varepsilon) = (F_1,...,F_k)$*, defined by the equalities of Equation (34), is a vector-valued function of* $(k+3)$ *variables:*

$$F_*(z,w,\varepsilon,\varepsilon_*) = (F_{*1}(z,w,\varepsilon,\varepsilon_*),\ldots,F_{*k}(z,w,\varepsilon,\varepsilon_*)),$$

where the components with numbers $i=\overline{1,r}$ *have the form*

$$F_{*i}(z,w,\varepsilon,\varepsilon_*) = \varphi_i(z) - \varepsilon_* U_{i,1}(z,w) - \ldots - \varepsilon_*\varepsilon^{n-1}U_{1,n}(z,w) - \ldots,$$

(that is, they are obtained from $F_i(z,w,\varepsilon)$ *by replacing* ε^n *by* $\varepsilon_*\varepsilon^{n-1}, n=1,2,\ldots$*), and when* $i=\overline{r+1,k}$ *they remain unchanged:* $F_{*i}(z,w,\varepsilon,\varepsilon_*) \equiv F_i(z,w,\varepsilon)$*.*

Obviously, the function $F_*(z,w,\varepsilon,\varepsilon_*)$ is holomorphic in a polydisc $D \times D_{0*}$, where $D_{0*} = \{\varepsilon_* : |\varepsilon_*| < \varepsilon_0\}$, and the equation $F_*(z,w,\varepsilon,\varepsilon_*) = 0$ implicitly defines a function$w = w_*(z,\varepsilon,\varepsilon_*)$ for which the equality $w(z,\varepsilon) = w_*(z,\varepsilon,\varepsilon_*)$ holds true.

Definition 4. *A function* $w(z,\varepsilon)$ *is said to be *-pseudoholomorphic, if the function* $w_*(z,\varepsilon,\varepsilon_*)$ *is holomorphic with respect to the second variable at the point* $\varepsilon=0$ *uniformly with respect to*$z \in T_{z_0}$ *for each fixed* $\varepsilon_* \in E_0$*.*

Theorem 6. *If a function* $W_0(z,q)$ *is holomorphic on a set* $Q = T_{z_0} \times \Omega(\Psi,\varphi,T_{z_0},E_0)$ *and maps it to a polydisk* D_{w_0}*, then the function* $w(z,\varepsilon)$ *is * -pseudoholomorphic at a point* $\varepsilon = 0$*.*

Proof. We fix $\varepsilon_* \in E_0$, then choose arbitrarily $z \in T_{z_0}$, and let $q_* = \Psi(\varphi(z)/\varepsilon_*)$. It is clear that for the system

$$\begin{cases} \Psi_1(U_{1,1}(z,w)) + \varepsilon V_1(z,w,\varepsilon) = q_{1*}, \\ \ldots \\ \Psi_r(U_{r,1}(z,w)) + \varepsilon V_r(z,w,\varepsilon) = q_{r*}, \\ F_{r+1}(z,w,\varepsilon) = 0, \\ \ldots \\ F_k(z,w,\varepsilon) = 0 \end{cases} \tag{41}$$

the conditions of the implicit function theorem are satisfied, and since the set of all such q_* compacts (for a fixed $\varepsilon_*$and $z \in T_{z_0}$), the proof is completed in the same way as in the previous theorem. □

Corollary 1. *Thus, the solution of the system of Equation (41) can be represented in the form of a series in powers of* ε *:*

$$w(z,\varepsilon,\varepsilon_*) = \sum_{n=0}^{\infty} \varepsilon^n W_n(z,\Psi_1(\varphi_1(z)/\varepsilon_*),\ldots,\Psi_r(\varphi_r(z)/\varepsilon_*)) \tag{42}$$

which converges uniformly on T_{z_0} *at* $|\varepsilon| < \varepsilon_1$*, where* $\varepsilon_1 > 0$ *and depends on* ε_**. In addition, from the proof of Theorem 6, it follows that if* $\varepsilon_* = \varepsilon$ *(*ε *is fixed and belongs to the circle of convergence of this series), then uniform convergence will be observed even on a narrower set* $T_{z0*} \subset T_{z_0}$ *(*$z_0 \in T_{z_{0*}}$*).*

The main question that arises in connection with the notion of *-pseudoholomorphy is the following: when can a *-pseudoholomorphic function be extended to the whole compact T_{z_0}? The answer to this question will be given in the scalar case, i.e., when $n = r = 1$. Note that in this case

$$F(z,w,\varepsilon) = \varphi(z) - \varepsilon U_1(z,w) - \ldots - \varepsilon^n U_n(z,w) - \ldots \tag{43}$$

and $\partial_w U_1(z,w) \neq 0$ in the in bidisk $D_{z_0} \times D_{w_0}$.

Furthermore, we assume that the condition (R) is fulfilled: all the functions participating in the analysis take real values, when their arguments are real.

Let $\mathcal{A}(D_{z_0})$ and $\mathcal{A}(D_{z_0} \times D_{w_0})$, where $D_{w_0} = \{w : \ |w - w_0| \ < R\}$ be the algebras of holomorphic functions, respectively, in the domains D_{z_0} and $D_{z_0} \times D_{w_0}$. In connection with the condition (R), we will assume that z_0 and w_0 are real.

Theorem 7. *If $\{H_\varepsilon\}$ is a holomorphic at the point $\varepsilon = 0$ family of homomorphisms of an algebra $\mathcal{A}(D_{z_0})$ into an algebra $\mathcal{A}(D_{z_0} \times D_{w_0})$ such that $H_0 = I$ and the functions $\varphi(z)$, $\mathcal{F}(z, w, \varepsilon) \equiv H_\varepsilon[\varphi(z)]$ satisfy the condition (R), and the conditions of Theorem 6 on the compact set $T_{z_0} = [z_0, \ z_0 + \Delta] \subset D_{z_0}$ hold true, then the function $w(z, \varepsilon)$, implicitly defined by the equation $\mathcal{F}(z, w, \varepsilon) = 0$, admits a pseudoholomorphic extension to T_{z_0}.*

We preface the proof of Theorem 7 with the following lemma.

Lemma 1. *The mappings $H_\varepsilon : \mathcal{A}(D_{z_0}) \to \mathcal{A}(D_{z_0} \times D_{w_0})$ for each sufficiently small ε satisfy the commutation relation*

$$H_\varepsilon[\varphi(z)] = \varphi(H_\varepsilon[z]). \tag{44}$$

Proof. Indeed, since $\varphi(z) \in \mathcal{A}_{z_0}$, then $\varphi(z) = \sum_{k=0}^{\infty} c_k (z - z_0)^k$, and, therefore,

$$H_\varepsilon\left[\sum_{k=0}^{\infty} c_k (z - z_0)^k\right] = \sum_{k=0}^{\infty} c_k H_\varepsilon[(z - z_0)^k] = \sum_{k=0}^{\infty} c_k (H_\varepsilon[z - z_0])^k =$$

$$= \sum_{k=0}^{\infty} c_k (H_\varepsilon[z] - H_\varepsilon[z_0])^k = \sum_{k=0}^{\infty} c_k (H_\varepsilon[z] - z_0)^k = \varphi(H_\varepsilon[z]),$$

thus, Equation (44) is proved. □

Proof of Theorem 7. We differentiate Equation (12) with respect to z and w:

$$\partial_z H_\varepsilon[\varphi(z)] = \varphi'(H_\varepsilon[z])\partial_z H_\varepsilon[z],$$
$$\partial_w H_\varepsilon[\varphi(z)] = \varphi'(H_\varepsilon[z])\partial_w H_\varepsilon[z],$$

from which, it follows that

$$\varepsilon \mathcal{F}_z + f(z, w, \varepsilon)\mathcal{F}_w = 0, \tag{45}$$

where $f(z, w, \varepsilon) = -\varepsilon \partial_z H_\varepsilon[z]/\partial_w H_\varepsilon[z]$ is a holomorphic function at the point $\varepsilon = 0$, which differs from zero in the domain $D_{z_0} \times D_{w_0}$ for a sufficiently small ε. Equation (45) is the equation of integrals of the differential equation

$$\varepsilon \frac{dw}{dz} = f(z, w, \varepsilon), \tag{46}$$

and we seek its solution in the form of a series in powers of ε , assuming the operator ∂_z to be a subordinate operator $f\partial_w$. We have [8], for an arbitrary function $\varphi(z) \in \mathcal{A}_{z_0}$, that

$$\begin{aligned} \mathcal{F}(z, w, \varepsilon) \equiv H_\varepsilon[\varphi(z)] = \\ = \varphi(z) - \varepsilon \int_{w_0}^{w} \frac{\varphi'(z)dw_1}{f(z,w_1,\varepsilon)} + \varepsilon^2 \int_{w_0}^{w} \left(\frac{\partial}{\partial z} \int_{w_0}^{w_1} \frac{\varphi'(z)dw_2}{f(z,w_2,\varepsilon)}\right) \frac{dw_1}{f(z,w_1,\varepsilon)} - \dots . \end{aligned} \tag{47}$$

By uniqueness, the solution of the equation $\mathcal{F}(z, w, \varepsilon) = 0$ is the solution $\tilde{w}_1(z, \varepsilon)$ of the Cauchy problem for the differential Equation (46) with the initial condition $\tilde{w}_1(z_0, \varepsilon) = w_0$, which, in accordance with Theorem 7, is a *-pseudoholomorphic function in a neighborhood $|\varepsilon| \ < \varepsilon_1$ (see Corollary 1) and is defined on some interval $[z_0, \ z_0 + \Delta_1] \subset [z_0, \ z_0 + \Delta]$ (recall that Equation (46) is considered in the real domain). We will assume that the small parameter in Equation (46) satisfies the inequality $0 < \varepsilon < \varepsilon_1$, $\varphi'(z) < 0 \ \forall z \in [z_0, \ z_0 + \Delta]$. We show how in the real case we can find Δ_1. Thus, the series

$$\tilde{W}_1(z, \varepsilon, \varepsilon_*) = \sum_{n=0}^{\infty} \varepsilon^n W_n(z, \Psi(\varphi(z)/\varepsilon_*)), \tag{48}$$

where Ψ is an entire function, that satisfies Theorem 6, in the scalar case, converges uniformly on the interval $T_{z_0} = [z_0,\ z_0 + \Delta]$ (ε_* and ε are fixed!). Suppose also (without loss of generality) that an essentially singular manifold is a half-open interval $(p,\ \Psi(0))$, where p is the asymptotic value of the function Ψ, and hence the set $Q = T_{z_0} \times \Omega(\Psi, \varphi, T_{z_0}, E_0)$ is a rectangle. If $\varepsilon > \varepsilon_*$, then $w_1(z,\varepsilon) = \tilde{W}_1(z,\varepsilon,\varepsilon)$ it is defined on the entire segment T_{z_0} (ie $\Delta_1 = \Delta$), because the graph of the function $q = \varphi(z)$ completely belongs to Q. If $\varepsilon < \varepsilon_*$, then $w_1(z,\varepsilon) = \tilde{W}_1(z,\varepsilon,\varepsilon)$, where $z \in [z_0,\ z_0 + \Delta_1]$ and Δ_1 is found from the equation $\varphi(z_0 + \Delta_1)/\varepsilon = \varphi(z_0 + \Delta_1)/\varepsilon_*$. We now consider the Cauchy problem

$$\begin{array}{l} \varepsilon \frac{dw}{dz} = f(z,w,\varepsilon), \\ w(z_1,\varepsilon) = v_1, \end{array} \tag{49}$$

where $z_1 = z_0 + \Delta_1$, $v_1 = \tilde{w}_1(z_1,\varepsilon)$. The general integral of this equation can be represented in the form

$$\int_{v_1}^{w} \frac{\varphi'(z)dw_1}{f(z,w_1,\varepsilon)} - \varepsilon \int_{v_1}^{w} \left(\frac{\partial}{\partial z} \int_{v_1}^{w_1} \frac{\varphi'(z)dw_2}{f(z,w_2,\varepsilon)} \right) \frac{dw_1}{f(z,w_1,\varepsilon)} + \ldots = \frac{\varphi(z) - \varphi(z_1)}{\varepsilon}. \tag{50}$$

The solution $\tilde{w}_2(z,\varepsilon)$, obtained from it, is defined on the interval $[z_1,\ z_2]$, where $z_2 = z_1 + \Delta_2$ and Δ_2 is determined from the equation $\frac{\varphi(z_1+\Delta_2)-\varphi(z_1)}{\varepsilon} = \frac{\varphi(z_0+\Delta)}{\varepsilon_*}$. If $|\varphi'(z)| \le l\ \forall z \in T_{z_0}$, then in accordance with the Lagrange theorem we have

$$\Delta_2 \ge \frac{\varepsilon\varphi(z_0 + \Delta)}{\varepsilon_* l}. \tag{51}$$

Then, Equation (46) is considered with the initial condition $w(z_2,\varepsilon) = v_2$, when $v_2 = \tilde{w}_2(z_2,\varepsilon)$. A general integral analogous to Equation (50) is constructed, and so on. Since the estimate of Equation (51) is constant on an interval T_{z_0}, then in a finite number of steps the solution will be constructed on it. The Theorem is proved.

We give two examples of constructing pseudoholomorphic solutions in the real domain.

Example 2. *We consider the Cauchy problem for the scalar equation* ($n = r = 1$)

$$\begin{cases} \varepsilon y' = f(t,y), \ \ t \in [t_0, T], \\ y(t_0,\varepsilon) = y_0. \end{cases} \tag{52}$$

We assume that the function $f(t,y)$ admits a holomorphic extension to the bidisk $D_{t_0} \times D_{y_0}$, where $D_{t_0} = \{z : |z - t_0| < R_0,\ R_0 > T\}$, $D_{y_0} = \{w : |w - y_0| < R\}$, and is not equal to zero there. Then, the general integral has the form:

$$\begin{array}{c} \varphi(t) - \varepsilon\varphi'(t) \int_{y_0}^{y} \frac{dy_1}{f(t,y_1)} + \varepsilon \int_{y_0}^{y} \left(\frac{\partial}{\partial t} \int_{y_0}^{y_1} \frac{\varphi'(t)dy_2}{f(t,y_2)} \right) \frac{dy_1}{f(t,y_1)} - \\ -\varepsilon^2 \int_{y_0}^{y} \left(\frac{\partial}{\partial t} \int_{y_0}^{y_1} \left(\frac{\partial}{\partial t} \int_{y_0}^{y_2} \frac{\varphi'(t)dy_3}{f(t,y_3)} \right) \frac{dy_2}{f(t,y_2)} \right) \frac{dy_1}{f(t,y_1)} + \ldots = 0. \end{array}$$

*Hence, we obtain a *-pseudoholomorphic solution*

$$y(t,\varepsilon) = \sum_{n=0}^{\infty} \varepsilon^n Y_n \left(t, \frac{\varphi(z)}{\varepsilon} \right), \tag{53}$$

where $\varphi(t) \in \mathcal{A}_{t_0}$ and such that the conditions of Theorem 6 are satisfied.

We write out the formulas for the first terms of the series of Equation (53):

$$Y_1 = -\left.\frac{V_1}{V_2}\right|_{y=Y_0(t,\varphi(t)/\varepsilon)}, \quad Y_2 = -\left.\frac{V_{11}V_2^2 - 2V_{12}V_1V_2 + V_{22}V_1^2}{2V_2^2}\right|_{y=Y_0(t,\varphi(t)/\varepsilon)},$$

where

$$V_1 = -\int_{y_0}^{y}\left(\frac{\partial}{\partial t}\int_{y_0}^{y_1}\frac{\varphi'(t)dy_2}{f(t,y_2)}\right)\frac{dy_1}{f(t,y_1)};$$

$$V_2 = \frac{\varphi'(t)}{f(t,y)};$$

$$V_{11} = 2\int_{y_0}^{y}\left(\frac{\partial}{\partial t}\int_{y_0}^{y_1}\left(\frac{\partial}{\partial t}\int_{y_0}^{y_2}\frac{\varphi'(t)dy_3}{f(t,y_3)}\right)\frac{dy_2}{f(t,y_2)}\right)\frac{dy_1}{f(t,y_1)};$$

$$V_{12} = -\frac{1}{f(t,y)}\left(\frac{\partial}{\partial t}\int_{y_0}^{y}\frac{\varphi'(t)dy_1}{f(t,y_1)}\right);$$

$$V_{22} = -\frac{\varphi'(t)f_y'(t,y)}{f^2(t,y)}.$$

We recall that $y = Y_0(t,\varphi(t)/\varepsilon)$ is the bounded solution (for $\varepsilon \to +0$) of the equation

$$\varphi'(t)\int_{y_0}^{y}\frac{dy_1}{f(t,y_1)} = \frac{\varphi(t)}{\varepsilon}.$$

In particular, if $f(t,y) = y^2 - e^{2t}$, $t_0 = 0$, $y_0 = 0$, then

$$y(t,\varepsilon) = e^t th\frac{1-e^t}{\varepsilon} + \frac{\varepsilon}{2}th^2\frac{1-e^t}{\varepsilon} + \ldots .$$

Example 3. *Consider the Cauchy problem for Tikhonov's system [7] (here, $n = 2$, $r = 1$)*

$$\begin{cases} y' = g(t,y,v), \\ \varepsilon v' = f(t,y,v), \quad t \in [t_0,T], \end{cases} \qquad (54)$$
$$y(t_0,\varepsilon) = y_0, \quad v(t_0,\varepsilon) = v_0.$$

Denote by $\bar{y}(t)$ the solution of the limit problem $f(t,y,y') = 0$, $y(t_0) = 0$, and by $L = \partial_t + g\partial_y$ – the first-order linear partial differential operator. Then,

$$\begin{cases} \varphi(t) - \varepsilon\int_{v_0}^{v}\frac{\varphi'(t)dv_1}{f(t,y,v_1)} - \varepsilon^2\int_{v_0}^{v}\left(L\int_{v_0}^{v_1}\frac{\varphi'(t)dv_2}{f(t,y,v_2)}\right)\frac{dv_1}{f(t,y,v_1)} - \ldots = 0, \\ y - \bar{y}(t) - \varepsilon\int_{v_0}^{v}\frac{L(y-\bar{y}(t))dv_1}{f(t,y,v_1)} + \varepsilon^2\int_{v_0}^{v}\left(L\int_{v_0}^{v_1}\frac{L(y-\bar{y}(t))dv_2}{f(t,y,v_2)}\right)\frac{dv_1}{f(t,y,v_1)} - \ldots = 0 \end{cases}$$

*are independent first integrals of the system of Equation (54). Hence, we obtain a *-pseudoholomorphic solution of this system:*

$$y(t,\varepsilon) = \bar{y}(t) + \varepsilon\int_{v_0}^{v}\frac{L(y-\bar{y}(t))dv_1}{f(t,y,v_1)}\Bigg|_{\substack{y=\bar{y}(t)\\ v=V_0(t,\varphi(t)/\varepsilon)}} + \ldots;$$

$$v(t,\varepsilon) = V_0(t,\varphi(t)/\varepsilon) + \varepsilon\frac{f(t,y,v)}{\varphi'(t)}\left[\int_{v_0}^{v}\left(L\int_{v_0}^{v_1}\frac{\varphi'(t)dv_2}{f(t,y,v_2)}\right)\frac{dv_1}{f(t,y,v_1)} - \right.$$
$$\left. - \int_{v_0}^{v}\frac{L(y-\bar{y}(t))dv_1}{f(t,y,v_1)}\cdot\frac{\partial}{\partial y}\int_{v_0}^{v}\frac{\varphi'(t)dv_1}{f(t,y,v_1)}\right]\Bigg|_{\substack{y=\bar{y}(t)\\ v=V_0(t,\varphi(t)/\varepsilon)}} + \ldots .$$

Here, $v = V_0(t,\varphi(t)/\varepsilon)$ is the bounded solution (for $\varepsilon \to +0$) of the equation

$$\varphi'(t)\int_{v_0}^{v}\frac{dv_1}{f(t,\bar{y}(t),v_1)} = \frac{\varphi(t)}{\varepsilon}.$$

Conclusion 3. The algorithms developed in this paper allow one to theoretically substantiate two main approaches in the general theory of singular perturbations: an approach related to approximate (asymptotic) solutions, and an approach related to pseudoholomorphic (exact) solutions of such problems.

Author Contributions: Sections devoted to singularly perturbed integral equations were written by Bobodzhanov A.A. and Safonov V.F., the sections connected with holomorphic regularization were written by Kachalov V.I.

Funding: This research received no external funding.

Conflicts of Interest: The authors declare no conflict of interest.

References

1. Lomov, S.A.; Lomov, I.S. *Fundamentals of the Mathematical Theory of the Boundary Layer;* Moscow University Press: Moscow, Russia, 2011.
2. Bobodzhanov, A.A.; Safonov, V.F. Integral Volterra equations with rapidly varying kernels and their asymptotic integration. *Mat. Sb.* **2001**, *192*, 519–536. [CrossRef]
3. Bobodzhanov, A.A.; Safonov, V.F. *Singularly Perturbed Integral and Integrodifferential Equations with Rapidly Varying Kernels and Equations with Diagonal Degeneracy of the Kernals;* Sputnik: Moscow, Russia, 2017.
4. Bobodzhanov, A.A.; Safonov, V.F. Singularly perturbed non-linear integro-differential systems with rapidly changing kernels. *Mat. Sb.* **2002**, *72*, 654–664.
5. Vasilyeva, A.B.; Butuzov, V.F. *Asymptotic Expansions of Solutions Singularly Perturbed Equations;* Nauka: Moscow, Russia, 1973.
6. Vasilyeva, A.B.; Butuzov, V.F.; Nefedov N.N. Contrast structures in singularly perturbed problems. *J. Comput. Math. Math. Phys.* **2011**, *41*, 799–851.
7. Imanaliev, M. *Asymptotic Methods in the Theory of Singularly Perturbed Integro-Differential Equations;* Ilim: Frunze, Russia, 1972.
8. Kachalov, V.I. Holomorphic regularization of singularly perturbed problems. *Vestn. MEI* **2010**, *6*, 54–62.
9. Kachalov, V.I. On holomorphic regularization of singularly perturbed systems of differential equations. *J. Comput. Math. Math. Phy.* **57**, *4*, 64–71. [CrossRef]
10. Chang, K.; Howes, F. *Nonlinear Singularly Perturbed Boundary Value Problems. Theory and Application;* Mir: Moscow, Russia, 1988.
11. Rozov, N.K.; Kolesov, A.Y.; Glyzin, S.D. Theory of nonclassical relaxation oscillations in singularly perturbed systems with delay. *Mat. Sb.* **2002**, *72*, 654–664. (In Russian)
12. Davlatov, D.B. Singularly perturbed boundary value Dirichlet problem for a stationary system of linear elasticity theory. *Izv. Vuzov. Math.* **2008**, *12*, 7–16. [CrossRef]
13. Lomov, S.A. *Introduction to the General Theory of Singular Perturbations;* Nauka: Moscow, Russia, 1981.
14. Safonov, V.F.; Bobodzhanov, A.A. Course of higher mathematics. *Singularly Perturbed Equations and the Regularization Method: Textbook;* Izdatelstvo MEI: Moscow, Russia, 2012.
15. Kachalov, V.I.; Lomov, S.A. Pseudoanalytic solutions of singularly perturbed problems. *Rep. Rus. Acad. Sci.* **1994**, *334*, 694–695.

axioms

MDPI

Article

Periodic Waves and Ligaments on the Surface of a Viscous Exponentially Stratified Fluid in a Uniform Gravity Field

Yuli D. Chashechkin [1,*] and Artem A. Ochirov [2]

1 Laboratory of Fluid Mechanics, Ishlinsky Institute for Problems in Mechanics RAS, 119526 Moscow, Russia
2 Department of Intelligent Information Radiophysical Systems, Physics Faculty, P.G. Demidov Yaroslavl State University, 150003 Yaroslavl, Russia
* Correspondence: chakin@ipmnet.ru; Tel.: +7-(495)-434-01-92

Abstract: The theory of singular perturbations in a unified formulation is used, for the first time, to study the propagation of two-dimensional periodic perturbations, including capillary and gravitational surface waves and accompanying ligaments in the $10^{-4} < \omega < 10^{3}\ \mathrm{s}^{-1}$ frequency range, in a viscous continuously stratified fluid. Dispersion relations for flow constituents are given, as well as expressions for phase and group velocities for surface waves and ligaments in physically observable variables. When the wave-length reaches values of the order of the stratification scale, the liquid behaves as homogeneous. As the wave frequency approaches the buoyancy frequency, the energy transfer rate decreases: the group velocity of surface waves tends to zero, while the phase velocity tends to infinity. In limiting cases, the expressions obtained are transformed into known wave dispersion expressions for an ideal stratified or actually homogeneous fluid.

Keywords: waves; ligaments; gradient flow; singular solution; stratification; viscous fluid

Citation: Chashechkin, Y.D.; Ochirov, A.A. Periodic Waves and Ligaments on the Surface of a Viscous Exponentially Stratified Fluid in a Uniform Gravity Field. *Axioms* **2022**, *11*, 402. https://doi.org/10.3390/axioms11080402

Academic Editor: Valery Y. Glizer

Received: 11 July 2022
Accepted: 10 August 2022
Published: 15 August 2022

1. Introduction

Waves on the surface of rivers, seas and oceans became one of the main objects of theoretical research as the first fundamental equations of continuum mechanics and closed systems of equations were formulated [1–3]. Due to a large difference in the physical properties of the atmosphere and water [4], the theory of waves evolved with the approximation of the homogeneity and immutability of the density of the contacting media. B. Franklin in the 18th century observed the water–olive oil interface motion in a ship's lighting lamp and pointed out that the variability of density with depth should be taken into account when analyzing the wave phenomena in a liquid [5].

The results of the analysis of the first papers, which considered the variability of the density of the liquid in depth, were presented [6], but for some unknown reason, mathematicians and mechanics did not pay attention to these topics. For example, such a fundamental characteristic as the buoyancy frequency calculated in [7], which is the limiting frequency for propagating internal waves in a continuously stratified fluid, escaped from attention. Independently the buoyancy frequency was re-discovered as the natural oscillation frequency of probe balls drifting in a stratified atmosphere [8] and later it was associated with the local extreme in the spectrum of high-frequency pressure oscillations recorded by a microbarograph [9].

Due to the practical importance of the issue, for a long time, researchers have been concentrating their efforts on studying periodic phenomena in a liquid in the context of the force action of waves on obstacles and the wave resistance of bodies moving in a liquid. The scientists assumed the constancy of density and incompressibility of the liquid [10] to be the most important qualities. Nowadays the constant density approximation is still considered high priority in the description of wave processes [11].

In theoretical works much attention is paid to the study of nonlinear properties of waves. The first heuristic discovery—the theory of periodic nonlinear vortex waves [12]—is significant in wave theory development. The approach was extended in [13,14], then analyzed in [15] and it has recently been supplemented with new results [16,17]. Another large cycle of studies of nonlinear waves, initiated by observations of a solitary wave in a shipping channel [18], is successfully proceeding at the present time. The number of various model equations [19,20] for the description of nonlinear wave properties is continuously increasing [21].

Seminal works [22,23] occupied a special place among the first publications. In these studies, the parameters of infinitesimal waves were determined, the waves of finite amplitude were calculated and showed the existence of the wave transfer of matter (Stokes drift) using the methods of the regular perturbations theory. The mass transfer is caused by nonlinear effects, which are distinctly revealed in the deviation of the waveform from the ideal one, i.e., in a nonlinear wave, the crests become sharper, and the troughs become flatter. The limiting angle between the tangents to the right and left sides of the crest is calculated as $\alpha = 120°$ in [23], and as $\alpha = 90°$ in [13]. It depends on the nonlinearity parameter $\varepsilon = \omega^2\zeta_0/g$ [24] when describing the propagating potential waves of finite amplitude by means of Lambert's complex functions (ζ_0 and ω are the amplitude and frequency of the wave, g is the gravity acceleration).

Calculations of the viscous attenuation of waves in a deep liquid and a channel of finite depth were carried out by asymptotic methods in a linear and nonlinear formulation [6,25,26]. Such calculations continue to be conducted at the present time using various approximations [27–30].

The "boundary layer" ideas (i.e., a mathematical model of the flow adjacent to the surface of the liquid, in which the influence of viscosity is considerably revealed), which were formulated at the beginning of the 20th century, had a significant impact on the development of the theory of waves in a liquid. The approximation of the boundary layer, accompanied by the reformulation of the defining equations [31], stimulated the development of the search for new analytical methods of finding their solutions [32]. The analysis of the wave equations by the theory of regular perturbations methods showed that in a homogeneous liquid with kinematic viscosity ν, a surface gravitational wave with a frequency ω is accompanied by a boundary layer with a specific thickness $\delta_\omega = \sqrt{2\nu/\omega}$. Separate boundary layers are formed on the free surface and on the solid bottom of the channel through which the waves propagate [33].

The boundary layer greatly influences all the parameters of fluid flows, namely pressure distribution, velocity and substance transfer characteristics [34]. Corrections of the theory formulas [33] due to the viscous attenuation of waves were later calculated in the second order of the perturbation theory [35]. The analysis of the dispersion relation for surface gravitational waves and accompanying boundary layers in a viscous liquid, supplemented by calculations of the attenuation coefficient and the scale of spatial attenuation of the wave was carried out in [36].

A more complex model of a "double boundary layer", inside which there is a periodic boundary layer with a length scale $\delta_\omega = \sqrt{2\nu/\omega}$, was proposed to describe standing waves [37].

With the frequency increasing, the surface tension influences more the pattern of waves on the surface of the liquid. The surface tension coefficient σ is a parameter that determines the type of dispersion relation for short waves [38] and accompanying fine constituents [39]. The surface tension changes the dependences of the group and phase velocity, the attenuation coefficient and the velocity of matter transfer U_ρ on frequency ω, wave vector $\mathbf{k}$ or wavelength λ. The pattern of waves generated by a short-range localized source in a viscous liquid, taking into account the effects of surface tension, was calculated in [40].

The data of the first systematic experimental studies of surface waves, which were conducted in laboratory basins at the end of the 19th century [41], generally showed satisfactory agreement with the calculations of velocity in the liquid thickness [23]. The visualization of the velocity profile using a "marker" (i.e., a colored wake of a submerging particle) showed the existence of a drift in the near-surface and bottom layers in the direction of wave propagation and slow counterflow in the middle of the liquid layer [42]. Additional cleaning from dust and film of the target liquid surface and consideration of viscous attenuation in the calculations of the near-surface boundary layer significantly reduced the difference between the data of calculations and experiments in a laboratory channel [43]. The observations of rearrangement of the distribution pattern of initially homogeneous suspension in the field of standing gravity waves in a vertically oscillating basin are given in [44].

In the field of propagating gravitational-capillary waves ($5 < \omega < 50\,\text{Hz}$), the substance of the colored drop is distributed unevenly over the surface of the primary cavity [45], the same as in the case of a drop merging with a target liquid at rest [46]. In the evolutionary process a slowly drifting colored primary contact area, a near-surface jet with a vortex head and a sinking ring vortex remain in the distribution pattern of the substance. There is a pronounced fibrous structure in the distribution of the drop substance in the target fluid [45].

Widespread capillary waves, which are observed in various rivers, seas and oceans, are caused by weak wind gusts in calm [47] and are certainly present on the slopes of large gravitational waves [48] in the stormy open ocean [49] and coastal regions [50]. Capillary waves contribute to the formation of bubbles, drops and foam [51] and participate in the processes of generating sound packets when drops fall [52]. Drop impact sounds form rain noise [53], which is one of the main sources determining the acoustic background of the ocean [54].

The capillary ripples determine the roughness and the real area of the contact surface. The estimation of their influence on the transfer of momentum, energy and matter between the atmosphere and the hydrosphere is of scientific and practical interest. A continuously expanding list of scientific tasks supports the interest in studying the dynamics of capillary waves in a wide range of conditions.

The analysis of the additional dissipation of short gravitational and gravitational-capillary waves caused by "parasitic" capillary waves on the slopes of longer waves in a laboratory pool are given in [55]. The presence of a surfactant film dampens short capillary waves, which in turn amplifies decimeter waves [56]. The discussion of modern optical methods for studying capillary waves and the data of detailed experiments is carried out in [57]. A review of papers on the nonlinear interaction of coexisting waves of different frequencies within the framework of the theory of "wave turbulence" is presented in [58]. The basic mechanism of energy transfer in weak turbulence theory is validated experimentally in the gravity (four-wave interactions) and capillary (three-wave interactions) regimes. Advanced experiments enable the achievement of full spatiotemporal reconstruction of the wave field in a weakly or strong nonlinear regime, to infer wave statistics as well as waves' nonlinear dispersion relationship, to compare with theory [58]. The study in [59] discusses the influence of attenuation on nonlinear interactions of short waves and surface waves and the properties of "wave turbulence". In all cases, the density of the liquid is assumed to be homogeneous.

However, in real conditions, the influence of differences in atmospheric and hydrosphere temperatures, insolation, radiative heat transfer, variability in the concentration of dissolved and suspended particles and flows ensure the heterogeneity of the density profile throughout the depth of the liquid [60]. The same happens in a thin near-surface layer, where temperature changes rapidly with depth and forms a "cold" or "warm" film [61].

The given study of the properties of surface waves and accompanying fine components in a viscous stably stratified liquid is based on a complete system of fundamental equations, which was firstly collected in [62] and later was considered in [63,64]. In this paper, a reduced system of the fundamental equations in which the density variability is preserved without considering variations in temperature, salinity or pressure that cause it to change is used. This system of equations is analyzed by methods of the singular perturbations theory [65] taking into account the compatibility condition [66], enabling calculation of both the waves themselves and the family of accompanying fine constituents (i.e., ligaments).

Analysis of the solutions of the fundamental equations system in a continuously stratified fluid has shown that for periodic flows (with a frequency less than the buoyancy frequency), regular solutions describe waves. The wave frequency ω is related to the wave number $\mathbf{k}$ by an algebraic dispersion relation $\omega = \omega(\mathbf{k}, \mathbf{Ak})$, which may include the amplitude $\mathbf{A}$ [66,67].

Singular solutions describe ligaments perturbations—extended in some direction and fine in others—outlining wave beams. The transverse scale of the ligaments δ_ω^ν is determined by the kinematic viscosity of the liquid ν and the wave frequency ω: $\delta_\omega^\nu = \sqrt{\nu/\omega}$ (Stokes scale) or equivalent parameter with the buoyancy frequency N: $\delta_N^\nu = \sqrt{\nu/N}$. The number and positions of the ligaments in space are determined by the geometry of the source, the amplitude and frequency of its oscillations. Calculations of internal wave beams and accompanying ligaments in stratified liquids and gases are given in [68,69].

In this paper, a dispersion equation in a linear approximation for two-dimensional periodic perturbations on the surface of a viscous exponentially stratified liquid is constructed for the first time. The properties of its complete solutions are analyzed. Approximate solutions are obtained by perturbation theory. Regular solutions describe waves in the range from infra-low frequency to gravitational-capillary and capillary waves. The dispersion relations continuously transform into the well-known formulas of the theory of waves in a homogeneous viscous or ideal liquid. Singular solutions characterize ligaments—fine constituents that complement waves. Graphs of the dependence of the wavelengths and ligaments on the frequency, phase and group velocity of the constituents on the wavelength are given. The results can be used to solve problems of generation and propagation of surface waves with physically justified initial and boundary conditions and comparison with a high-resolution experiment.

2. Periodic Flows in a Viscous Exponentially Stratified Fluid

We consider the propagation of periodic perturbations over the surface of a viscous exponentially stratified fluid in a uniform gravity field. The undisturbed liquid filling the lower half-space is regarded to be incompressible; meanwhile the effects of thermal conductivity and diffusion are neglected.

We perform the study in a Cartesian coordinate system $Oxyz$ in which the plane Oxy coincides with the equilibrium position of a free surface of the liquid, and the axis Oz is directed vertically upwards against the direction of the gravity acceleration $\mathbf{g}$. The undisturbed density distribution over depth is exponential $\rho_0(z) = \rho_{00} \exp(-z/\Lambda)$. It is characterized by reference density $\rho_{00}(z_0)$ (i.e., the density value at the equilibrium level $z = z_0$), as well as the scale $\Lambda = |d \ln \rho/dz|^{-1}$, frequency $N = \sqrt{g/\Lambda}$ and buoyancy period $T_b = 2\pi/N$.

Mostly, changes of real fluids density are usually small and produce a small impact on the inertial properties of the flows. Nevertheless, the conservation of terms describing the stratification effects in the governing equations set is important since gravity acceleration is large. In this regard, it is useful to consider three types of medium: stratified fluids when buoyancy scale Λ, frequency N and period T_b which are included in the list of main parameters; then very weakly stratified fluids, when the scale of buoyancy substantially exceeds the values of other length scales of the problem (so called potentially homogeneous fluid) and actually homogeneous liquid whose density is assumed to be constant in the entire space [70]. Using a weak but variable density helps to save the rank of complete non-linear set and order of the linearized set of governing equations [66,71] and analyzes additional solutions which are lost in approximation of homogeneous fluid.

Depending on the magnitude of the density gradient created by the corresponding temperature or salinity distributions, it is customary to distinguish strongly and weakly stratified fluids, as well as potentially and actually homogeneous fluids. The values of the characteristic physical quantities are given in Table 1 [70].

Table 1. Values of characteristic physical quantities.

Parameter	Fluids			
	Stratified (SF)		Homogeneous (HF)	
	Strongly	Weakly	Potentially	Actually
Buoyancy frequency N, s^{-1}	1	0.01	0.00001	0.0
Period T_b	6.28 s	10.5 min	7.3 days	∞
Scale	9.8 m	100 km	10^8 km	∞
Viscous wave scale $L_N^{g\nu} = \sqrt[3]{g\nu}/N$, cm	2.14	200	$2 \cdot 10^5$	∞
Stokes microscale $\delta_N^{\nu} = \sqrt{\nu/N}$, cm	0.1	1.0	30	∞

Strong stratification ($T_b \sim 10\,\mathrm{s}$) is created in laboratory installations. Relatively weak stratification ($T_b \sim 10\,\mathrm{min}$) is observed in natural conditions. Figure 1 shows the density–depth dependencies for exponential stratification (solid line) and for linear stratification (dotted line) for a strongly stratified fluid $N = 1\,\mathrm{s}^{-1}$.

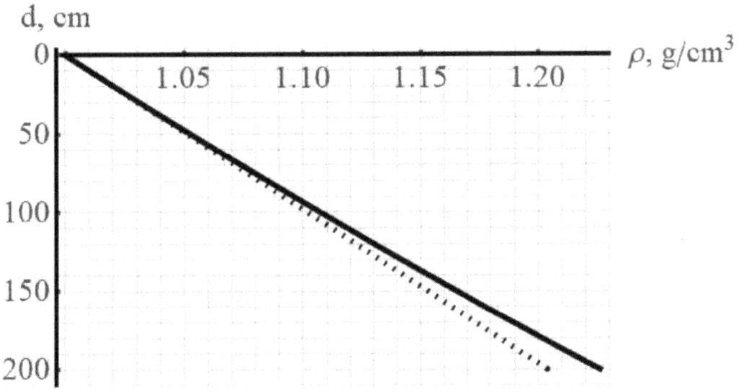

Figure 1. The density–depth dependencies for exponential stratification (solid line) and for linear stratification (dotted line) for strongly stratified fluid ($N = 1\,\mathrm{s}^{-1}$, $\nu = 0.01$ St, $\sigma = 72$ dyn/cm, $\rho_{00} = 1.0\,\mathrm{g/cm^3}$).

To compare the properties of exponentially and linearly stratified media, we analyze the changes in the density gradient with depth. For exponential stratification, the magnitude of the density gradient depends on the vertical coordinate z.

$$\frac{d\rho}{dz} = -\frac{\rho_{00}}{\Lambda} e^{-\frac{z}{\Lambda}} \quad (1)$$

For linear stratification, the density gradient does not depend on the depth:

$$\frac{d\rho}{dz} = -\frac{\rho_{00}}{\Lambda} \quad (2)$$

For small variations $z \ll \Lambda$ the dependencies of the density gradient are not distinguishable for exponential and linear stratification. The exponent can be represented as a Taylor series expansion over a small parameter z/Λ and expression (1) will be written as follows:

$$\frac{d\rho}{dz} = -\frac{\rho_{00}}{\Lambda} e^{-\frac{z}{\Lambda}} \simeq -\frac{\rho_{00}}{\Lambda}\left(1 - \frac{z}{\Lambda} + \frac{z^2}{2\Lambda^2}\right) \quad (3)$$

The numerical values for the stratification scale for a strongly stratified fluid are $\Lambda = 9.81 \cdot 10^2$ cm, and for a weakly stratified fluid $\Lambda = 9.81 \cdot 10^6$ cm. At depths of less than 10% of the stratification scale, it can be argued with a high degree of accuracy that the calculations in the model of a linearly stratified fluid and exponentially stratified fluid coincide.

We study below two-dimensional periodic flows of the form $A = A_0 \exp i(\mathbf{kx} - \omega t)$ with a positive definite frequency ω and a complex wave number $\mathbf{k}$, the imaginary part of which characterizes the spatial attenuation of the flow. The disturbances which are homogeneous in the direction of the transverse horizontal coordinate y are selected. They include the displacement of the free surface position $\zeta(x,t)$. The velocity $\mathbf{u} = u_x\mathbf{e}_x + u_z\mathbf{e}_z$ with horizontal and vertical velocity components u_x, u_z in an incompressible fluid (div$\mathbf{u} = 0$) is represented by derivatives of the stream function ψ:

$$u_x = \partial_z\psi, \qquad u_z = -\partial_x\psi \tag{4}$$

The expression for the density of the liquid $\rho = \rho_{00}(r_0(z) + s(x,z,t))$ replacing the equation of state in [66] includes the initial distribution $r_0(z)$ and the perturbation caused by the examined periodic flow $s(x,z,t)$.

Taking into account these assumptions, the system of continuity and Navier–Stokes equations is extremely simplified and takes the traditional form [62,63,69]:

$$z < \zeta : \begin{cases} \rho = \rho_{00}(r(z) + s(x,z,t)) \\ \rho\partial_t\mathbf{u} + \rho(\mathbf{u}\cdot\nabla)\mathbf{u} - \rho\nu\Delta\mathbf{u} = \rho\mathbf{g} - \nabla P \\ \partial_t\rho + \text{div}(\rho\mathbf{u}) = 0 \end{cases} \tag{5}$$

Kinematic and dynamic boundary conditions on a perturbed surface are traditional [62]:

$$z = \zeta : \begin{cases} \partial_t(z-\zeta) + \mathbf{u}\cdot\nabla(z-\zeta) = 0 \\ \boldsymbol{\tau}\cdot(\mathbf{n}\cdot\nabla\mathbf{u}) + \mathbf{n}(\boldsymbol{\tau}\cdot\nabla\mathbf{u}) = 0 \\ P - P_0 - \sigma\nabla\cdot\mathbf{n} - 2\rho\nu\mathbf{n}(\mathbf{n}\cdot\nabla\mathbf{u}) = 0 \end{cases} \tag{6}$$

$$\mathbf{n} = \frac{\nabla(z-\zeta)}{|\nabla(z-\zeta)|} = \frac{-\partial_x\zeta\mathbf{e}_x + \mathbf{e}_z}{\sqrt{1+(\partial_x\zeta)^2}}, \qquad \boldsymbol{\tau} = \frac{\mathbf{e}_x + \partial_x\zeta\mathbf{e}_z}{\sqrt{1+(\partial_x\zeta)^2}},$$

where P is the hydrodynamic pressure, $\sigma = \gamma uprho_{00}$ is the total coefficient of the surface tension of the liquid, $\mathbf{n}$ and $\boldsymbol{\tau}$ are the vectors of the external normal and tangent to the free surface of the liquid, respectively.

The studied problem includes the following dimensional parameters: density ρ and its gradient $\frac{d\rho}{dz}$, dynamic μ and density-normalized kinematic viscosity $\nu = \mu/\rho$, gravity acceleration g, surface tension coefficient σ and density-normalized surface tension coefficient $\gamma = \sigma/\rho$, amplitude A, frequency ω period $T_\omega = 2\pi/\omega$ and wavelength λ. The wave is also characterized by a wave vector $\mathbf{k}$, the modulus of which is related to the wavelength $\lambda = 2\pi/|\mathbf{k}|$.

In the transition to a mathematical description, physical quantities serve as the basis for the introduction of characteristic scales of length, time and speed. These scales are used in the physical interpretation of the results obtained, assessing the degree of influence of various physical factors.

The degree of relative influence of viscosity and gravity on fluid flows is characterized by the scale $\delta_g^\nu = \sqrt[3]{\nu^2/g}$ that appears in [71].

Traditionally, stratification is characterized by its own length scale which takes into account or neglect the effects of compressibility, the frequency $N = \sqrt{g/\Lambda}$, period $T_b = 2\pi/N$ and the inverse value of the buoyancy frequency $T_N = T_b/2\pi$. The stratification itself is characterized by a combinational viscous wave scale $L_N^{g\nu} = (g\nu)^{1/3}/N$ [72,73], as well as a length scale $\delta_N^\nu = \sqrt{\nu/N}$– a functional analogue of the dissipative Stokes microscale $\delta_\omega^\nu = \sqrt{\nu/\omega}$.

The dualism of the nature of surface waves, whose properties depend on the gravity acceleration g, normalized coefficients of surface tension γ and kinematic viscosity ν results in the introduction of several scales, including the capillary length $\delta_g^\gamma = \sqrt{\gamma/g}$. The inequality $\lambda < \delta_g^\gamma$ indicates the severity of surface force effects, and $\lambda >> \delta_g^\gamma$—indicates the severity of gravitational effects.

The structure of near-surface flows is also characterized by a known proper velocity scale $v_c = \gamma/\nu$ and an additional time scale of kinematic nature $T_{\nu g}^{\gamma} = v_c/g = \gamma/\nu g$.

Wave propagation is characterized by group $\mathbf{c}_g = \frac{\partial \omega}{\partial \mathbf{k}}$ and phase velocity $\mathbf{c}_{ph} = \frac{\omega \mathbf{k}}{\mathbf{k}^2}$.

Ratios of uniform scales form a set of dimensionless parameters, parts of which are used in further calculations.

2.1. Equations of Periodic Flows and Dispersion Relations for Plane Infinitesimal Waves

To simplify expressions, we accept the initial density distribution to be exponential, as it is customary in most models:

$$\rho_0(z) = \rho_{00} r(z) = \rho_{00} \exp(-z/\Lambda) \tag{7}$$

The change of variables proposed in [74] allows the transformation of the equations of motion for an arbitrary smooth density profile into equations with constant coefficients that determine the preferential choice of the exponential density profile.

Further calculations are carried out in the Boussinesq approximation, considering the smallness of absolute density variations in stratified media (in particular, a small value of the wavelength and buoyancy scale ratio $C = \lambda/\Lambda$). The density variations are neglected everywhere, except for the term with a large coefficient (i.e., the gravity acceleration $\mathbf{g}$). In this approximation, the system (5) takes the form:

$$\begin{cases} \partial_{tz}\psi + \partial_z\psi\partial_{xz}\psi - \partial_x\psi\partial_{zz}\psi - \nu\partial_z\Delta\psi + \partial_x P = 0 \\ g(r+s) - g - \partial_{tx}\psi + \partial_x\psi\partial_{xz}\psi - \partial_z\psi\partial_{xx}\psi - \nu\partial_x\Delta\psi + \partial_z P = 0 \\ \partial_t s + \partial_z\psi\partial_x s - \partial_x\psi\partial_z(r+s) = 0 \end{cases} \tag{8}$$

In a linear approximation the system (8) reduces to:

$$\begin{cases} \partial_{tz}\psi - \nu\partial_z\Delta\psi + \partial_x P = 0 \\ g(r+s) - g - \partial_{tx}\psi - \nu\partial_x\Delta\psi + \partial_z P = 0 \\ \partial_t s - \partial_x\psi\partial_z r = 0 \end{cases} \tag{9}$$

Cross-differentiation of spatial coordinates of the upper and middle equations of the system (9) allows getting rid of pressure:

$$\begin{cases} \partial_t\Delta\psi - g\partial_x s - \nu\Delta\Delta\psi = 0 \\ \partial_t s - \partial_x\psi\partial_z r = 0 \end{cases} \tag{10}$$

Subtraction of the second equation, differentiated by coordinate and multiplied by a coefficient g from the first equation of the system (10), differentiated by time, allows getting rid of the function $s(x,z,t)$:

$$\partial_{tt}\Delta\psi - \nu\partial_t\Delta\Delta\psi + \partial_{xx}\psi\partial_z r = 0 \tag{11}$$

Equation (11) takes the next form for an exponentially stratified fluid (7):

$$\partial_{tt}\Delta\psi - \nu\partial_t\Delta\Delta\psi + N^2\exp(-z/\Lambda)\partial_{xx}\psi = 0, \quad N = \sqrt{g/\Lambda} \tag{12}$$

The boundary conditions (6) for all variables of the infinitesimal waves $A = A_0\exp(i\mathbf{kx} - i\omega t)$ are traditionally carried away from the wave surface $z = \zeta$ to the unperturbed level $z = 0$ and take the form

$$z = 0 \begin{cases} \partial_t\zeta + \partial_x\psi = 0 \\ -\rho g\zeta + \rho_{00}P + 2\rho_{00}\nu\partial_{xz}\psi + \sigma\partial_{xx}\zeta = 0 \\ \partial_{zz}\psi - \partial_{xx}\psi = 0 \end{cases} \tag{13}$$

The small coefficient (kinematic viscosity ν), which ensures slow attenuation of the wave as it propagates in the liquid, allows the classification of the system (12) as a system of singularly perturbed equations and the application of the theory of singular perturbations [65] for its analysis. The compatibility condition, which requires the analysis of all the roots of the dispersion equation and the system (12) solutions, should be taken into account [66].

For surface waves with $\omega > N$, the solution of Equation (12) is sought in the form of plane waves:

$$\psi = \left(A_{+}e^{ik_x x - i\omega t} + A_{-}e^{-ik_x x - i\omega t}\right)\left(e^{k_z z} + \beta e^{k_l z}\right) \tag{14}$$

Here k_z corresponds to the regular solution of the dispersion equation, and k_l corresponds to the singular solution of the dispersion equation.

We obtain the shape of the deviation of the free surface of the liquid, the condition of the relationship between the amplitudes of the regular and singular component by substituting (14) into Equation (12) and boundary conditions (13):

$$\begin{aligned} \zeta &= \tfrac{k_x}{\omega}\left(A_{+}e^{ik_x x - i\omega t} - A_{-}e^{-ik_x x - i\omega t}\right)(1+\beta) \\ \beta &= -\tfrac{k_x^2 + k_z^2}{k_x^2 + k_l^2} \end{aligned} \tag{15}$$

as well as the dispersion relations:

$$\begin{cases} -N^2 k_x^2 + e^{z/\Lambda}\left(k_x^2 - k_{z,l}^2\right)\omega\left(ik_x^2\nu - ik_{z,l}^2\nu + \omega\right) = 0 \\ gk_x^2 - 3ik_x^2 k_{z,l}\nu\omega + k_{z,l}\left(ik_{z,l}^2\nu - \omega\right)\omega + k_x^4\gamma = 0, \qquad \gamma = \sigma/\rho_{00} \end{cases} \tag{16}$$

Calculated for the first time the expressions (16) are transformed into the relations for a viscous homogeneous fluid and for an ideal exponentially stratified fluid in limiting transitions $N \to 0$ and $\nu \to 0$ respectively. Similar dispersion relations for internal waves and ligaments in the thickness of a stratified fluid were presented in [75].

2.2. Solution of the Dispersion Equation

It is convenient to analyze the dispersion relations (13) in a dimensionless form. The time scale will be the parameter, which is inverse to the buoyancy frequency N, and the viscous wave scale $L_\nu = (\nu g)^{1/3}/N$ is chosen as the length scale. The ratios of the eigenscales of the problem $\delta = \left(\frac{\delta_g^\gamma}{\delta_N^\nu}\right)^2 = \frac{\gamma}{g} \cdot \frac{N}{v}$ and $\varepsilon = \frac{L_v}{\Lambda} = \frac{N\nu^{1/3}}{g^{2/3}}$ are used to construct new dimensionless parameters δ, ε involved in further calculations. Then, expressions (16) could be rewritten in a dimensionless form:

$$\begin{cases} -k_{*x}^2 + ie^{z/\Lambda}\left(k_{*x}^2 - k_{*z,l}^2\right)^2 \varepsilon\omega_* + e^{z/\Lambda}\left(k_{*x}^2 - k_{*z,l}^2\right)\omega_*^2 = 0 \\ k_{*x}^2 + k_{*x}^4\delta\varepsilon - 3ik_{*x}^2 k_{*z,l}\varepsilon^2\omega_* + ik_{*z,l}^3\varepsilon^2\omega_* - k_{*z,l}\varepsilon\omega_*^2 = 0 \\ \delta = N\gamma/\nu g, \qquad \varepsilon = N\nu^{1/3}/g^{2/3} \end{cases} \tag{17}$$

The upper dispersion relation in (17) has regular solutions, which we have denoted k_{*z}, and singular solutions, which are denoted k_{*l}. For a large number of real liquids, the parameter $\varepsilon \ll 1$ and regular solutions are found by direct decomposition over a small parameter ε:

$$k_{*z} = k_{*0z} + \varepsilon k_{*1z} + \varepsilon^2 k_{*2z} + \ldots \tag{18}$$

Substituting (18) into the upper expression in (17) we obtain up to the terms of the order $O(\varepsilon^2)$:

$$k_{*z} = \pm\frac{\sqrt{\omega_*^2 - e^{-z/\Lambda}}}{\omega_*}k_{*x} \pm \varepsilon\frac{ie^{-2z/\Lambda}k_{*x}^3}{2\omega_*^4\sqrt{\omega_*^2 - e^{-z/\Lambda}}} \tag{19}$$

Since the equations in the system (17) have the fourth degree, it is necessary to find two more solutions, which are singular perturbed ones in the form of decomposition [65]:

$$k_{*l} = \varepsilon^{-\eta}\left(k_{*0l} + \varepsilon k_{*1l} + \varepsilon^2 k_{*2l} + \ldots\right) \tag{20}$$

The parameter η in (20) is chosen in such a way that at the highest degree the main term of the decomposition persists. Substituting instead of k_{*l} in the upper expression in (17) $k_{*l} = \varepsilon^{-\eta} k_{*0l}$ and equating the exponents ε in different terms we get:

$$\begin{array}{l} 1 - 4\eta = -2\eta \\ 1 - 4\eta = 1 - 2\eta \\ 1 - 4\eta = 0 \\ 1 - 2\eta = -2\eta \\ 1 - 2\eta = 0 \\ -2\eta = 0 \end{array} \tag{21}$$

At the highest degree, the main term of the decomposition remains only at $\eta = 1/2$. Substituting the value $\eta = 1/2$ in (20) and then in (17) up to the terms of the order $O\left(\varepsilon^{3/2}\right)$ we get:

$$k_{*l} = \pm\frac{(1-i)\sqrt{\omega_*}}{\sqrt{2\varepsilon}} \pm \frac{(1+i)\left(e^{-z/\Lambda} + \omega_*^2\right)k_{*x}^2}{2\sqrt{2}\omega_*^{5/2}}\sqrt{\varepsilon} \tag{22}$$

Leaving only the main terms of ε in the lower dispersion relation (17), we obtain the dispersion equation for the regular part of the solution:

$$k_{*x} + \left(k_{*x}^3\delta - \omega_*\sqrt{\omega_*^2 - 1}\right)\varepsilon = 0 \tag{23}$$

For the singular part of the solution, leaving only the main terms of the expansion of ε, we obtain:

$$1 + k_{*x}^2\delta\varepsilon + \frac{(1+i)\left(1 - 2\omega_*^2\right)}{\sqrt{2\omega_*}}\varepsilon^{3/2} = 0 \tag{24}$$

The solution of the dispersion Equation (23) for the regular wave part is:

$$\begin{array}{c} k_{*x} = -\frac{2^{1/3}}{\alpha} + \frac{\alpha}{3\cdot 2^{1/3}\delta\varepsilon} \\ k_{*x} = \frac{1 \pm i\sqrt{3}}{2^{2/3}\alpha} - \frac{\left(1 \mp i\sqrt{3}\right)\alpha}{6\cdot 2^{1/3}\delta\varepsilon} \\ \alpha = \left(27\delta^2\varepsilon^3\omega_*\sqrt{\omega_*^2 - 1} + \sqrt{108\delta^3\varepsilon^3 + 729\delta^4\varepsilon^6\omega_*^2(\omega_*^2 - 1)}\right)^{1/3} \end{array} \tag{25}$$

From the condition of the physical realization of the solution (i.e., the damping of the flow with depth) it follows that only roots with $\mathrm{Re}(k_{*z}) > 0$ possess a physical meaning. Decomposing the solution (25) for the regular wave part into a Taylor series, we obtain:

$$\begin{array}{l} k_{*x} = \omega_*\varepsilon\sqrt{\omega_*^2 - 1} + O(\varepsilon^4) \\ k_{*x} = \pm\frac{i}{\sqrt{\delta\varepsilon}} - \omega_*\varepsilon\frac{\sqrt{\omega_*^2 - 1}}{2} + O\left(\varepsilon^{5/2}\right) \end{array} \tag{26}$$

With consideration of (19) and the condition $\mathrm{Re}(k_{*z}) > 0$, it can be seen that we implement only one root of (26) and finally for the regular wave solution we get:

$$\begin{array}{c} k_{*x} = -\frac{2^{1/3}}{\alpha} + \frac{\alpha}{3\cdot 2^{1/3}\delta\varepsilon} \\ k_{*z} = \frac{\sqrt{\omega_*^2 - e^{-z/\Lambda}}}{\omega_*}\left(-\frac{2^{1/3}}{\alpha} + \frac{\alpha}{3\cdot 2^{1/3}\delta\varepsilon}\right) + \varepsilon\frac{ie^{-2z/\Lambda}\left(-\frac{2^{1/3}}{\alpha} + \frac{\alpha}{3\cdot 2^{1/3}\delta\varepsilon}\right)^3}{2\omega_*^4\sqrt{\omega_*^2 - e^{-z/\Lambda}}} \end{array} \tag{27}$$

We perform similar transformations for singular ligament-solution. The solutions (24), which are found taking into account (22) and the conditions of the physical implementation of the roots $\mathrm{Re}(k_{*l}) > 0$ of the singular ligament, take the form:

$$\begin{aligned} k_{*x} &= \pm\sqrt{(\delta\varepsilon)^{-1}(((1+i)(2\omega_*^2-1)/\sqrt{2\omega_*})\varepsilon^{3/2}-1)} \\ k_{*l} &= \frac{(1-i)2\delta\omega_*^3+(1+i)\left(e^{-z/\Lambda}+\omega_*^2\right)\left(((1+i)\left(2\omega_*^2-1\right)/\sqrt{2\omega_*}\right)\varepsilon^{3/2}-1)}{2\sqrt{2}\omega_*^{5/2}\delta\sqrt{\varepsilon}} \end{aligned} \tag{28}$$

Expressions (27), (28) describe new dispersion relations for surface waves (27) and associated ligaments (28) in a viscous exponentially stratified fluid.

2.3. *Low Frequency Waves*

The solution for low frequency waves ($\omega < N$) is sought as the sum of the propagation of gravitational waves (waves with wave number components k_z) and fine perturbation (with wave number component k_s) as well:

$$\psi = \left(A_+e^{ik_xx-i\omega t} + A_-e^{-ik_xx-i\omega t}\right)\left(\alpha e^{ik_zz} + \beta e^{-ik_zz} + \chi e^{k_sz}\right) \tag{29}$$

Substituting (26) into (9) leads to dispersion relations:

$$\begin{cases} -N^2k_x^2 + e^{z/\Lambda}(k_x^2+k_z^2)\omega(ik_x^2\nu + ik_z^2\nu + \omega) = 0 \\ -N^2k_x^2 + ie^{z/\Lambda}(k_s^2-k_x^2)\omega(k_s^2\nu - k_x^2\nu + i\omega) = 0 \end{cases} \tag{30}$$

Substituting solution (29) into the boundary conditions (13), we obtain an expression for the shape of the free surface and additional relations. These additional relations determine the relationship between the amplitudes of the various components of periodic motion:

$$\zeta = \left(A_+e^{ik_xx-i\omega t} - A_-e^{-ik_xx-i\omega t}\right)(\alpha+\beta+\chi)k_x/\omega \tag{31}$$

$$(\alpha+\beta)\left(k_x^2-k_z^2\right) + \chi\left(k_x^2+k_s^2\right) = 0 \tag{32}$$

$$k_x^2\left(g+\gamma k_x^2\right)(\alpha+\beta+\chi) + \nu\omega\left(k_z(\alpha-\beta)\left(3k_x^2+k_z^2\right) + i\chi k_s\left(k_s^2-3k_x^2\right)\right) - \omega^2(ik_z(\alpha-\beta)-\chi k_s) = 0 \tag{33}$$

In a dimensionless form, the dispersion relations (30) take the form:

$$\begin{cases} -k_{*x}^2 + ie^{z/\Lambda}(k_{*x}^2-k_{*s}^2)^2\varepsilon\omega_* + e^{z/\Lambda}(k_{*x}^2-k_{*s}^2)\omega_*^2 = 0 \\ -k_{*x}^2 + ie^{z/\Lambda}\left(k_{*x}^2+k_{*z,l}^2\right)^2\varepsilon\omega_* + e^{z/\Lambda}\left(k_{*x}^2+k_{*z,l}^2\right)\omega_*^2 = 0 \end{cases} \tag{34}$$

The expressions for the surface component of periodic motion, up to the notation, coincide with the dispersion equations of surface waves discussed above. For low frequency waves, we will look for regular wave solutions in the form of expansion (18) and up to the terms of the order $O(\varepsilon^2)$:

$$k_{*z} = \pm\frac{\sqrt{e^{-z/\Lambda}-\omega_*^2}}{\omega_*}k_{*x} \mp \varepsilon\frac{ie^{-2z/\Lambda}k_{*x}^3}{2\omega_*^4\sqrt{\omega_*^2-e^{-z/\Lambda}}} \tag{35}$$

Similarly to surface waves case, we find singular solutions of the dispersion equation for low frequency waves using decomposition (20). Just as in the case of high frequency surface waves $\omega > N$, the exponent of the degree $\eta = 1/2$ is the only one that satisfies the condition of the main term of the decomposition at the highest degree k_l. Making a substitution up to the terms of the order $O\left(\varepsilon^{3/2}\right)$ we get:

$$k_{*l} = \pm\frac{(1+i)\sqrt{\omega_*}}{\sqrt{2\varepsilon}} \mp \frac{(1-i)\left(e^{-z/\Lambda}+\omega_*^2\right)k_{*x}^2}{2\sqrt{2}\omega_*^{5/2}}\sqrt{\varepsilon} \tag{36}$$

From the relation (32) we obtain that

$$\chi = (\alpha+\beta)\frac{k_z^2 - k_x^2}{k_x^2 + k_s^2} \tag{37}$$

$$\alpha+\beta+\chi = (\alpha+\beta)\frac{k_s^2 + k_z^2}{k_x^2 + k_s^2} \tag{38}$$

From the ratio (33) follows:

$$\alpha+\beta = -\frac{2i\left(k_s^2+k_x^2\right)kz\alpha\left(3k_x^2\nu+k_z^2\nu-i\omega\right)\omega}{ik_x^2(k_s^2+k_z^2)(g+k_x^2\gamma)-(k_s+ik_z)(-k_s^2k_x^2+3k_x^4+4ik_sk_x^2k_z+(k_s^2+k_x^2)k_z^2)\nu\omega-(k_s-ik_z)(ik_x^2+k_sk_z)\omega^2} \tag{39}$$

In the low viscosity approximation, we obtain that

$$\alpha+\beta \simeq 0, \quad \chi \simeq -(\alpha+\beta) = 0 \tag{40}$$

The relations (40) correspond to the situation when all the energy is concentrated in gravitational waves.

2.4. Periodic Flows on the Surface of a Viscous Exponentially Stratified Liquid

First of all, let us consider the dependence of the wavelength λ on the frequency of the wave motion ω. We define the wavelength as follows:

$$\lambda = 2\pi/\sqrt{\mathrm{Re}(k_x^2)+\mathrm{Im}(k_z^2)} \tag{41}$$

This method of determination of the wavelength is due to the fact that the imaginary part of the wave number k_x component and the real part of the wave number k_z component are responsible for the spatial attenuation of motion and are not impactful to wave propagation. Substituting the dispersion relations (27) in (32), we can construct the desired dependencies. Figure 2 shows the dependence of the wavelength on the frequency of wave motion $\lambda(\omega)$.

The velocity of movement of the phase front of structures (wave and ligament) and the rate of energy transfer are of particular interest. The phase front moves with the phase velocity of the wave (and its analogue for the singular solution):

$$\mathbf{c}_{ph} = \omega\mathbf{k}/|k|^2 \tag{42}$$

The energy is transferred with a group velocity c_{gr}, which is defined as follows:

$$\mathbf{c}_{gr} = \left\{\left(\frac{\partial k_x}{\partial\omega}\right)^{-1}, \left(\frac{\partial k_{z,l}}{\partial\omega}\right)^{-1}\right\} \tag{43}$$

Figure 3 shows the dependencies for the phase (dashed line) and group (solid line) velocities.

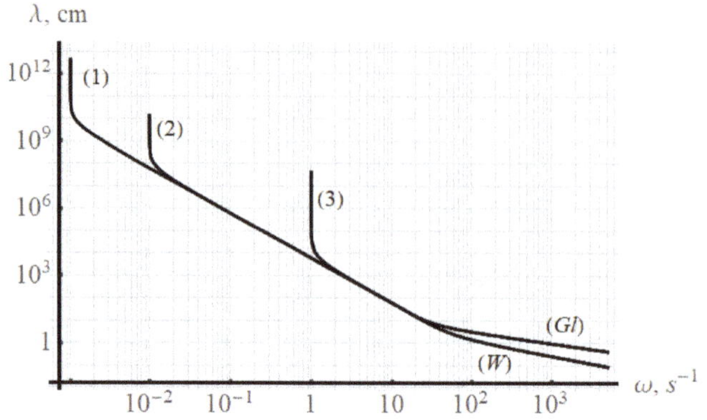

Figure 2. The dependence of the wavelength on the frequency of wave motion in a viscous exponentially stratified fluid for a periodic solution. The curves indicated by the letter (W) are constructed for a liquid with $\nu = 0.01$ St, $\sigma = 72$ dyn/cm, $\rho_{00} = 1\,\text{g/cm}^3$, and by the letter (Gl)—for a liquid with glycerin parameters ($\nu = 11.746$ St, $\rho_{00} = 1.26\,\text{g/cm}^3$, $\sigma = 64.7$ dyn/cm). The numbers indicate a different degree of stratification. Index (1) corresponds to a weak pycnocline $N = 0.001\ \text{s}^{-1}$, index (2) to a weakly stratified fluid $N = 0.01\ \text{s}^{-1}$, and index (3) to a strongly stratified fluid $N = 1\ \text{s}^{-1}$ for water and glycerin.

Figures 2 and 3 show that viscosity has a noticeable effect on capillary waves with a wavelength $\lambda < \delta_g^\gamma$, and the stratification influences the waves with frequencies close to the buoyancy frequency N. With the advent of stratification, a forbidden part of the spectrum appears for surface waves (with frequencies lower than the buoyancy frequency). With the tending to the stratification frequency, the group velocity goes to zero, and the phase velocity tends to infinity. A similar pattern is observed when electromagnetic waves propagate in waveguides. The size of the waveguide sets a certain critical size, beyond which the electromagnetic wave does not propagate in the waveguide channel. In stratified fluids, this mechanism is not due to external geometric constraints, but is determined by the characteristics of the medium (stratification).

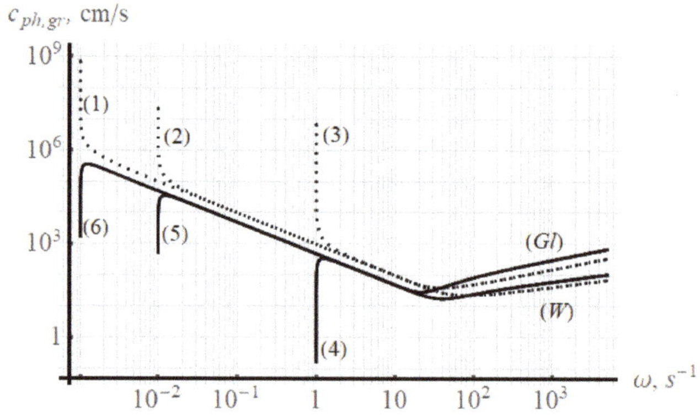

(a)

Figure 3. *Cont.*

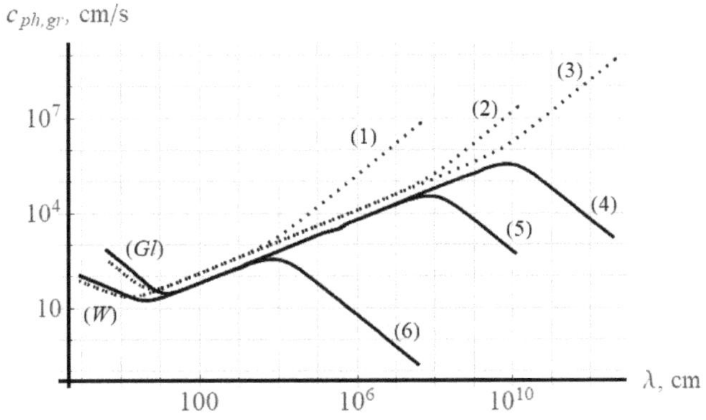

(b)

Figure 3. Dependences of the phase (dashed lines) and group (solid lines) velocities on the frequency of wave motion (**a**) and on the wavelength (**b**) in a viscous exponentially stratified fluid for a periodic solution. The curves indicated by the letter (W) are constructed for a liquid with water parameters, and by the letter (Gl)—for a liquid with glycerin parameters. The numbers indicate a different degree of stratification. Indexes (1) and (6) correspond to a weak pycnocline $N = 0.001\ \text{s}^{-1}$, indexes (2) and (5) correspond to a weakly stratified liquid $N = 0.01\ \text{s}^{-1}$ and indexes (3) and (4) correspond to a strongly stratified liquid $N = 1\,\text{s}^{-1}$.

3. Reduction to Approximation of Actually Homogeneous Fluid

The fine constituents of periodic flows—ligaments—are a consequence of the dissipative properties of the medium, which exist not only in stratified, but also in homogeneous liquids. They are described by singular solutions of dispersion equations, the appearance of which can be observed experimentally in the structure of flows in an inhomogeneous medium. If the effects associated with buoyancy are neglected, then the density of the liquid can be considered as actually homogeneous:

$$\rho = \rho_{00} \equiv \text{const} \tag{44}$$

and the basic equations of motion are simplified. For a viscous homogeneous liquid in a 2D formulation in a linear approximation equations for the stream function are shortened to:

$$\partial_t \Delta \psi - \nu \Delta\Delta\psi = 0 \tag{45}$$

Equation (45) can also be obtained by performing a limiting transition $N \to 0$ in (12). The boundary conditions that are removed to the equilibrium surface in the case of a homogeneous liquid are as follows:

$$z = 0 \begin{cases} \partial_t \zeta + \partial_x \psi = 0 \\ -g\zeta + P + 2\nu\partial_{xz}\psi + \gamma\partial_{xx}\zeta = 0 \\ \partial_{zz}\psi - \partial_{xx}\psi = 0 \end{cases} \tag{46}$$

Similarly to the basic equations of motion, the boundary conditions (46) can also be obtained from the boundary conditions (13) using a limit transition $N \to 0$. Substitution of the solution in the form of a propagating wave

$$\psi = A_+ \exp(k_z z + ik_x x - i\omega t) + A_- \exp(k_z z - ik_x x - i\omega t) \tag{47}$$

in (45) leads to dispersion relations between the components of the wave number:

$$\left(k_x^2 - k_z^2\right)\left(\nu\left(k_x^2 - k_z^2\right) - i\omega\right) = 0 \tag{48}$$

The relation (48) naturally decomposes into two solutions:

$$\begin{aligned} k_z^2 &= k_x^2 \\ k_l^2 &= k_x^2 + \tfrac{i\omega}{\nu} \end{aligned} \tag{49}$$

Here, the reassignment of one solution k_z into k_l is introduced. The component k_z corresponds to the wave solution, and the component k_l corresponds to the ligament solution. Taking into account (49), the solution of the problem in the form of a propagating wave is transformed into:

$$\psi = (A_+ \exp(ik_x x - i\omega t) + A_- \exp(-ik_x x - i\omega t))(\exp(k_z z) + B\exp(k_l z)) \tag{50}$$

Substituting (50) into the kinematic boundary condition, we obtain the relationship between the amplitudes of the velocity field and the deviation of the free surface from the equilibrium value:

$$\zeta = \frac{k_x}{\omega}(1+B)(A_- \exp(-ik_x x - i\omega t) - A_+ \exp(ik_x x - i\omega t)) \tag{51}$$

and substitution of (50) into the dynamic condition for tangential tensions leads to the expression for the amplitude of the ligament component:

$$B = -\frac{(k_x^2 + k_z^2)}{(k_x^2 + k_l^2)} = -2k_x^2\left(2k_x^2 + \frac{i\omega}{\nu}\right)^{-1} \tag{52}$$

Getting rid of the pressure in the dynamic boundary condition and taking normal components, we rewrite it as:

$$\nu\partial_{tz}\Delta\psi - \partial_{ttz}\psi + g\partial_{xx}\psi + 2\nu\partial_{txzz}\psi - \gamma\partial_{xxxx}\psi = 0 \tag{53}$$

Substituting the solution (50) into the boundary conditions (53) taking into account (49) leads to the dispersion relation for the wave constituent:

$$\gamma k_z^3 - 2i\nu\omega k_z^2 + gk_z - \omega^2 = 0 \tag{54}$$

and ligament constituent:

$$\gamma\left(k_l^2 - \frac{i\omega}{\nu}\right)^2 - 2i\nu\omega\left(k_l^2 - \frac{i\omega}{\nu}\right)k_l + g\left(k_l^2 - \frac{i\omega}{\nu}\right) + \omega k_l\left(i\nu k_l^2 - \omega\right) = 0 \tag{55}$$

It can be noticed that Equations (54) and (55) are also obtained from (16) with a limit transition $N \to 0$. Further we will transfer the dispersion relations (54) and (55) to dimensionless forms in the same way as it has been done in the previous paragraph. We will choose our own viscous scale $\delta_g^\nu = \sqrt[3]{\nu^2/g}$ as the length scale, and the ratio as the time scale will be $T_{\nu g}^\gamma = \gamma/\nu g$ and the decomposition parameter $\varepsilon = \sqrt[3]{\left(\delta_g^\nu\right)^6/\left(\delta_g^\gamma\right)^6} = \sqrt[3]{(\nu^4 g)/\gamma^3}$.

In a dimensionless form, the conventional dispersion Equation [15] for the wave component of the solution will be rewritten taking into account (54):

$$k_{*x}^3 - 2i\varepsilon^2\omega_* k_{*x}^2 + \varepsilon k_{*x} - \varepsilon^3\omega_*^2 = 0 \tag{56}$$

The solution of Equation (56) is obtained by the standard method and takes the form:

$$k_x^* = \frac{2}{3}i\varepsilon^2\omega^* - \frac{\varepsilon(3 + 4\varepsilon^3\omega^{*2})}{3\beta} + \frac{1}{3}\beta \tag{57}$$

$$
\begin{aligned}
k_x &= \tfrac{2}{3}\mathrm{i}\varepsilon^2\omega^* \pm \tfrac{1}{6}\mathrm{i}\left(\pm\mathrm{i} + \sqrt{3}\right)\beta + \frac{\left(1\pm\mathrm{i}\sqrt{3}\right)\varepsilon\left(3+4\varepsilon^3\omega^{*2}\right)}{6\cdot 2^{2/3}\beta} \\
\beta &= \left(-8\mathrm{i}\varepsilon^6\omega^{*3} + \tfrac{9}{2}\varepsilon^3\omega^*(-2\mathrm{i}+3\omega^*) + \tfrac{3}{2}\sqrt{3}\sqrt{\varepsilon^3(4 - 32\mathrm{i}\varepsilon^6\omega^{*5} + \varepsilon^3\omega^{*2}(4+9\omega^*(-4\mathrm{i}+3\omega^*)))}\right)^{1/3}
\end{aligned}
\tag{58}
$$

Decomposition (57) into a Taylor series by a small parameter ε at least up to the summands $O\left(\varepsilon^{9/2}\right)$ gives the expression:

$$
k_{*z} = \varepsilon^2\omega_*^2 + O\left(\varepsilon^{9/2}\right) \tag{59}
$$

Decomposition of the roots (58) into a Taylor series by a small parameter ε at least up to the summands $O\left(\varepsilon^{9/2}\right)$ gives the expression:

$$
k_{*z} = \pm\frac{1}{8}\mathrm{i}\sqrt{\varepsilon}\left(8 \pm 4\varepsilon^{3/2}\omega_*(2+i\omega_*) + \varepsilon^3\omega_*^2\left(4 - 4\mathrm{i}\omega_* + 3\omega_*^2\right)\right) + O\left(\varepsilon^{9/2}\right) \tag{60}
$$

As it follows from the condition of the physical realization of the roots $\mathrm{Re}(k_z) > 0$, considering the positive definite frequency of the wave motion, only one root, which is given by the expression (57), can be physically realized This solution describes the wave part of a periodic motion in a liquid.

To analyze the dispersion equation for a ligament solution (55) and carry out the nondimensionalization procedure, the characteristic scales can be selected the same as in the wave solution. Then in a dimensionless form, the dispersion equation (55) is written as:

$$
k_{*l}^4 - 2i\varepsilon\omega_* k_{*l}^2 - \varepsilon^2\omega_*^2 - 3i\varepsilon^2 k_{*l}^3\omega_* - 3\varepsilon^3 k_{*l}\omega_*^2 + \varepsilon k_{*l}^2 - i\varepsilon^2\omega_* + i\varepsilon^2\omega_* k_{*l}^3 - \varepsilon^3 k_{*l}\omega_*^2 = 0 \tag{61}
$$

Equation (61) has four roots. The analysis shows that the spatial attenuation condition is satisfied for one root only. Due to the cumbersomeness of the expression, the calculated solution is not given here.

Let us construct the dispersion dependences of the components of periodic motion in log–log scales in dimensional variables for liquids with glycerin and water parameters. Figure 4 shows the dependence of the wavelength λ on the frequency for the wave component (a) and the analog of the wavelength δ_i on the frequency for the ligament component (b). The letter (W) indicates the dependencies for water and the letter (Gl) indicates the dependencies for glycerin.

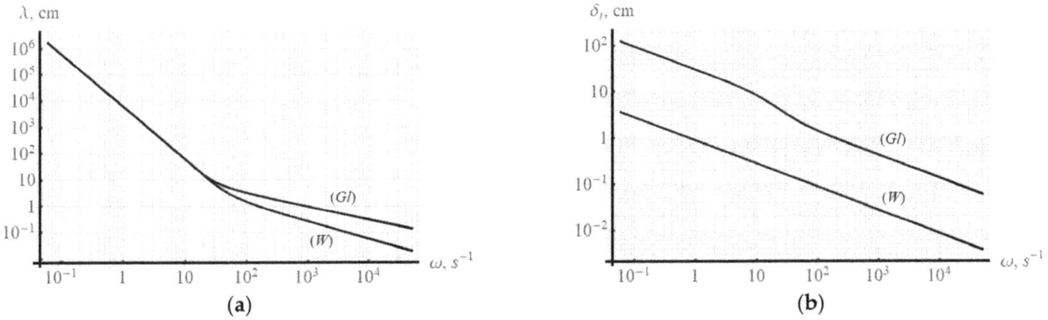

Figure 4. Wavelength dependences of the wave solution (**a**) and the ligament solution (**b**) on the frequency in a viscous homogeneous fluid. The curves indicated by the letter (W) are constructed for a liquid with water parameters, and by the letter (Gl)—for a liquid with glycerin parameters.

Figure 5 shows the dependences of the phase and group velocity on the frequency of periodic motion for the wave component of periodic motion (a) and the ligament associated with the wave component (b). The dependences are constructed for liquids with the parameters of water and glycerin. Figure 6 shows similar dependencies, but on the wavelength. There are several remarkable velocities in a viscous homogeneous liquid. The minimum group velocity and the velocity at which the group and phase velocities are compared. We show that the velocities are compared when the value of the phase velocity is minimal. The extremum condition for the phase velocity will be written as:

$$\partial_\omega c_{ph} = \partial_\omega \left(\frac{\omega}{k} \right) = \frac{1}{k^2}(k - \omega \partial_\omega k) = 0 \tag{62}$$

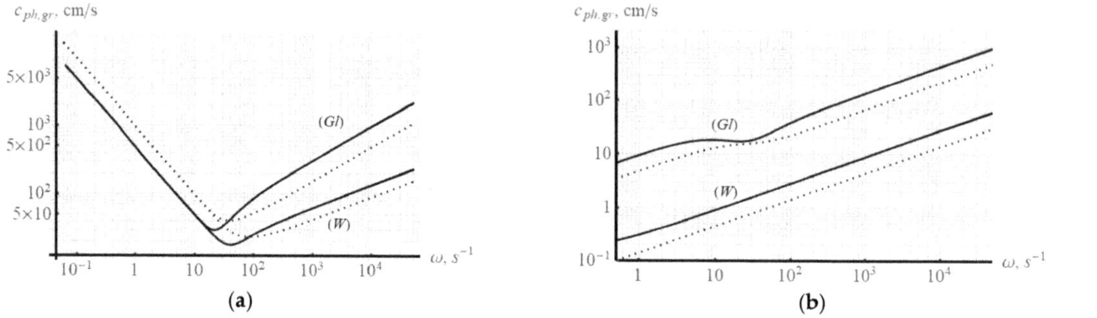

Figure 5. Dependences of the phase (dashed lines) and group (solid lines) velocities on the frequency of the wave solution (**a**) and the ligament solution (**b**) on the frequency in a viscous homogeneous fluid. The curves indicated by the letter (W) are constructed for a liquid with water parameters, and by the letter (Gl) for a liquid with glycerin parameters.

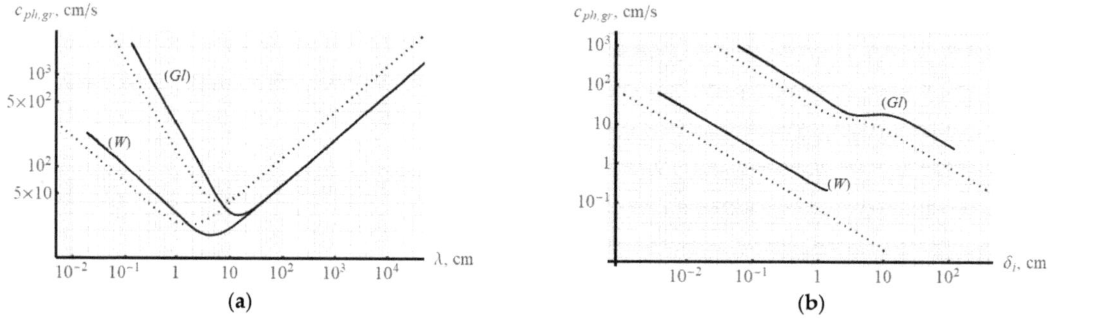

Figure 6. Dependences of the phase (dashed lines) and group (solid lines) velocities on the wavelength of the wave solution (**a**) and the ligament solution (**b**) on the frequency in a viscous homogeneous fluid. The curves indicated by the letter (W) are constructed for a liquid with water parameters, and by the letter (Gl) for a liquid with glycerin parameters.

The ratio (62) leads us to equality:

$$c_{ph} = \frac{\omega}{k} = (\partial_\omega k)^{-1} = c_{gr} \tag{63}$$

The minimum value of the group velocity for a liquid with water parameters is $c^W_{grmin} = 17.71$ cm/s and is achieved at frequency $\omega^W_{grmin} = 40.53\ s^{-1}$ and wavelength $\lambda^W_{grmin} = 4.33$ cm. For a liquid with glycerin parameters, the corresponding values are as follows: $c^{Gl}_{grmin} = 29.39$ cm/s at the frequency $\omega^W_{grmin} = 21.05\ s^{-1}$ and wavelength $\lambda^{Gl}_{grmin} = 14.88$ cm. The minimum phase velocity for a liquid with water parameters is $c^{W}_{phmin} = 23.05$ cm/s at the frequency $\omega^{W}_{phmin} = 84.82$ s^{-1} and wavelength $\lambda^{W}_{phmin} =$ 1.71 cm. For a liquid with glycerin parameters, the corresponding values are as follows: $c^{Gl}_{phmin} = 40.85$ cm/s at frequency $\omega^{W}_{phmin} = 38.96$ s^{-1} and wavelength $\lambda^{Gl}_{phmin} = 6.59$ cm.

The viscosity of the liquid affects the capillary wave motion. An increase in viscosity leads to an increase in the wavelength at a constant frequency in the region of capillary waves. As a consequence, the values of the phase and group velocity increase. The characteristic values of fluid velocities for the wave component of periodic motion increase with increasing viscosity and shift to the region of lower frequencies (longer wavelengths). Taking into account the viscosity in the model makes it possible to calculate, in addition to the wave component, the ligament component of periodic motion in a liquid.

4. Reduction to Inviscid Fluid

In an inviscid exponentially stratified fluid, the equations of motion and boundary conditions (12), (13) are reduced $\nu \to 0$:

$$z < \zeta : \begin{cases} \rho = \rho_{00}(r(z) + s(x,z,t)) \\ \rho\partial_t \mathbf{u} + \rho(\mathbf{u}\cdot\nabla)\mathbf{u} = \rho\mathbf{g} - \nabla P \\ \partial_t\rho + \mathrm{div}(\rho\mathbf{u}) = 0 \end{cases} \tag{64}$$

$$z < 0: \ \partial_{tt}\Delta\psi + N^2\exp(-z/\Lambda)\partial_{xx}\psi = 0 \tag{65}$$

$$z = 0: \ \begin{cases} \partial_t\zeta + \partial_x\psi = 0 \\ -\rho g\zeta + uprho_{00}P + \sigma\partial_{xx}\zeta = 0 \end{cases} \tag{66}$$

Dispersion relations in an inviscid liquid allow us to obtain only the wave component of periodic motion. The dispersion relations for the wave component are obtained using the limit transition $\nu \to 0$ from expressions (16) and (30) for high frequency (compared to the buoyancy frequency) and low frequency oscillations, respectively. For high-frequency wave disturbances of the free surface ($\omega > N$) we obtain:

$$\begin{cases} -N^2k_x^2 + e^{z/\Lambda}(k_x^2 - k_z^2)\omega^2 = 0 \\ gk_x^2 - k_z\omega^2 + k_x^4\gamma = 0 \end{cases} \tag{67}$$

For low frequency waves ($\omega < N$):

$$\begin{cases} -N^2k_x^2 + e^{z/\Lambda}(k_x^2 + k_z^2)\omega^2 = 0 \\ -N^2k_x^2 - e^{z/\Lambda}(k_s^2 - k_x^2)\omega^2 = 0 \end{cases} \tag{68}$$

Thus, the approximation of an ideal fluid enables us to find solutions that describe the wave constituent of periodic motion in the region of gravitational waves quite well. In the region of capillary waves, the wave constituent calculated in the model of an ideal liquid underestimates the value of the wavelength at a given frequency. Note that in the model of an ideal fluid, it is impossible to obtain a solution for the ligament constituent of periodic motion along the free surface of the fluid. This makes the solution incomplete.

5. Discussion

For the first time the theory of singular perturbations was used to describe the propagation of two-dimensional periodic perturbations over the surface of a viscous exponentially stratified incompressible fluid occupying the lower half-space. Calculations were performed in a single formulation in a wide frequency range, which includes the propagation of capillary, gravitational and internal waves. The system of equations, which includes the equations of continuity, momentum transfer and density profile with depth, replacing the equation of state, is solved together with physically justified kinematic and dynamic conditions on a free surface, taking into account the effect of surface tension. The theory of singular perturbations with respect to the compatibility condition was applied to study the propagation of infinitesimal harmonic flows with a positive definite frequency and a complex wavenumber, which considers spatial attenuation.

A complete set of solutions of the obtained dispersion relation for infinitesimal periodic perturbations includes two inseparable constituents of periodic flows. The theory of regular perturbations determines the parameters of gravitational-capillary or infra-low-frequency waves with a frequency lower than the buoyancy frequency.

The accompanying fine constituents that are ligaments are calculated by the methods of the theory of singular perturbations. The dispersion equation designed for infinitesimal periodic perturbations describes the relationship between the components of the wave vector in periodic flows in a wide frequency range $10^{-4} < \omega < 10^{3}\ \text{s}^{-1}$ which includes infra-low frequency, gravitational and capillary waves. The result, represented by length-frequency dependence, is convenient for experimental determination. The dependences of phase and group wave velocities on frequency and wavelength have also been obtained.

The dependences of phase and group velocities of regular perturbations (i.e., waves in water and glycerin) on frequency and wavelength have been represented in graphs. Wave properties are significantly modified during the transition through buoyancy frequency. The group velocity of wave propagation tends to zero and the phase one tends to infinity with the approximation of wave and buoyancy frequencies. Viscosity has a significant influence on short waves when their length becomes compared to or less than capillary scale $\lambda < \delta_g^\gamma = \sqrt{\gamma/g}$.

The wave parameters are calculated in an actual homogeneous liquid in the limit of the buoyancy frequency $N \to 0$ as well as in a constant density approximation $\rho \equiv \text{const}$, $T_b = 0$ taking into account the compatibility conditions. These calculations correspond to the given results.

Singular solutions describe ligaments (i.e., thin currents accompanying waves in a viscous stratified or homogeneous liquid). To compare them we have shown the graphs of the dependence of the wavelength and scale of the ligaments on the wave frequency in a viscous homogeneous liquid. The sizes of the ligaments differ by several orders of magnitude at low frequencies. The dependences of the phase and group velocities of waves and ligaments on the frequency and wavelength are also given.

The application of the theory of singular perturbations makes it possible to design complete solutions of the dispersion equation and the system of fundamental equations without additional hypotheses and limitations of the type of boundary layer approximation [33,34].

The evolving approach admits the extrapolation to the investigation of three-dimensional periodic perturbations. In this case, propagating surface waves accompany several types of ligaments, as well as internal waves in the thickness of a continuously stratified liquid [68,69].

In general cases, both waves and ligaments continuously interact with each other and generate new groups of flow constituents [75]. The influence of forcing together with the effects of nonlinearity and dissipation causes the evolution of the wave structure, which has different shapes at the phases of growth and attenuation.

In the low frequency range, the forcibly oscillating free surface of a stratified ocean can be a source of internal and inertial waves that transfer energy into the ocean. Internal waves, in turn, interact non-linearly with each other [76], generating new wave groups and ligaments (i.e., accompanying flows that cause the observed modulation of the surface waves) [77]).

6. Conclusions

The analysis of the linearized reduced version of the fundamental equations system by the methods of the theory of singular perturbations has been carried out with respect to the compatibility condition. It showed that the complete solutions describe waves propagating over the surface of a viscous stratified incompressible fluid, and small-scale constituents that are ligaments accompanying the waves. In extreme cases that are in limit of viscous homogeneous liquid, ideal stratified and ideal homogeneous liquid, the obtained dispersion relations for waves transfer to the widely known ones.

The experimental studies of the fine structure of surface waves in a continuously stratified liquid with high-resolution instruments that allow recording the influence of all constituents of periodic flows are of great interest. We should pay particular attention to the use of high-resolution optical methods for recording variations in the density gradient, and highly sensitive temperature or electrical conductivity sensors in a liquid with salt stratification.

Author Contributions: Conceptualization, Y.D.C. and A.A.O.; methodology, Y.D.C. and A.A.O.; software, A.A.O.; validation, Y.D.C. and A.A.O.; formal analysis, Y.D.C.; investigation, Y.D.C.; resources, Y.D.C.; data curation, Y.D.C.; writing—original draft preparation, Y.D.C. and A.A.O.; writing—review and editing, Y.D.C.; visualization, A.A.O.; supervision, Y.D.C.; project administration, Y.D.C.; funding acquisition, Y.D.C. All authors have read and agreed to the published version of the manuscript.

Funding: The work was supported by the Russian Science Foundation (project 19-19-00598" Hydrodynamics and energetics of drops and droplet jets: formation, motion, break-up, interaction with the contact surface" https://rscf.ru/en/project/19-19-00598/. (accessed on 12 August 2022).

Institutional Review Board Statement: Not applicable.

Informed Consent Statement: Not applicable.

Data Availability Statement: Not applicable.

Conflicts of Interest: The authors declare no conflict of interest.

References

1. Bernoulli, D. *Hydrodynamica, Sive de Viribus et Motibus Fluidorum Commentarii, Opus Academicum*; Sumptibus, J.R., Ed.; Argentorati: Dulseckeri, The Netherlands, 1738; 304p.
2. D'Alembert, J.-L.R. *Réflexions sur la Cause Générale des Vents*; David: Paris, France, 1747; 372p.
3. Euler, L. Principes généraux du mouvement des fluids. *Mém. L'académie Des. Sci. Berl.* **1757**, *11*, 274–315.
4. Feistel, R. Thermodynamic properties of seawater, ice and humid air: TEOS-10, before and beyond. *Ocean Sci.* **2018**, *14*, 471–502. [CrossRef]
5. Franklin, B. Behavior of oil on water. Letter to J. Pringle. In *Experiments and Observations on Electricity*; R. Cole: London, UK, 1769; pp. 142–144.
6. Lamb, H. *Treatise on the Mathematical Theory of the on the Motion of Fluids*, 6th ed.; Cambridge University Press: Cambridge, UK, 1879; 258p.
7. Rayleigh. Investigation of the character of the equilibrium of an incompressible heavy fluid of variable density. *Proc. Lond. Math. Soc.* **1882**, *1–14*, 170–177. [CrossRef]
8. Väisälä, V. Uber die Wirkung der Windschwankungen auf die Pilotbeoachtungen. *Soc. Sci. Fenn. Commentat. Phys. Math.* **1925**, *2*, 19–37.
9. Brunt, D. The period of simple vertical oscillations in the atmosphere. *Q. J. R. Meteorol. Soc.* **1927**, *53*, 30–32. [CrossRef]
10. Darrigol, O. *Worlds of Flow: A History of Hydrodynamics from the Bernoullis to Prandtl*; Oxford University Press: Oxford, UK, 2005; 356p.
11. Basile, R.; De Serio, F. Flow Field around a Vertical Cylinder in Presence of Long Waves: An Experimental Study. *Water* **2022**, *14*, 1945. [CrossRef]
12. Gerstner, F.J. Theorie der Wellen. Abhandlunger der Königlichen Böhmischen Geselschaft der Wissenschaften, Prague. *Ann. der Phys.* **1809**, *32*, 412–445. [CrossRef]
13. Rankine, W.J.M. On the exact form of waves near the surface of deep water. *Philos. Trans. R. Soc.* **1863**, *153*, 127–138.
14. Rayleigh. On waves. *Phil. Mag. Ser.* **1876**, *5*, 257–279. [CrossRef]
15. Kochin, N.E.; Kibel, I.A.; Rose, N.V. *Theoretical Hydromechanics*; Part 1; OGIZ. Gostekhizdat: Leningrad-Moscow, USSR, 1948; 728p. (In Russian)
16. Henry, D. On Gerstner's Water Wave. *J. Nonlinear Math. Phys.* **2008**, *15*, 87–95. [CrossRef]
17. Ionescu-Kruse, D. On the particle paths and the stagnation points in small-amplitude deep-water. *J. Math. Fluid Mech.* **2013**, *15*, 41–54. [CrossRef]
18. Russell, J.S. *Report on Waves. Report of the Fourteenth Meeting of the British Association for the Advancement of Science*; Murray, J., Street, A., Eds.; Scientific Research: London, UK, 1845; pp. 311–391.
19. Boussinesq, J. Theorie de I'intumescence Liquid, Appleteonde Solitaire au de Translation, se Propageantdansun Canal Rectangulaire. *Les Comptes Rendus De L'académie Des Sci.* **1871**, *72*, 755–759.

20. Korteweg, D.J.; de Vries, G. On the change of form of long waves advancing in a rectangular canal, and on a new type of long stationary waves. *Lond. Edinb. Dublin Philos. Mag. J. Sci.* **1895**, *39*, 422–443. [CrossRef]
21. Whitham, G.B. *Linear and Nonlinear Waves*; Wiley Interscience: New York, NY, USA, 1999; 660p.
22. Stokes, G.G. On the theory of oscillatory waves. *Trans. Camb. Phil. Soc.* **1847**, *8*, 441–455.
23. Stokes, G.G. Supplement to a paper on the theory of oscillatory waves. In *Mathematical and Physical Papers*; CUP: Cambridge, UK, 1880; Volume I, pp. 314–326.
24. Kistovich, A.V.; Chashechkin, Y.D. Propagating stationary surface potential waves in a deep ideal fluid. *Water Resour.* **2018**, *45*, 719–727. [CrossRef]
25. Basset, B. *A Treatise on Hydrodynamics with Numerous Examples*; Deighton, Bell and Co.: Cambridge, UK, 1888; Volume II, 368p.
26. Hough, S.S. On the Influence of Viscosity on Waves and Currents. *Proc. Lond. Math. Soc.* **1896**, *s1–28*, 264–288. [CrossRef]
27. Harrison, W.J. The Influence of Viscosity on the Oscillations of Superposed Fluids. *Proc. Lond. Math. Soc.* **1908**, *s2–6*, 96–405. [CrossRef]
28. Biesel, F. Calcul de l'amortissement d'une houle dans un liquide visqueux de profondeur finie. *La Houille Blanche* **1949**, *4*, 630–634. [CrossRef]
29. Tyvand, P.A. A note on gravity waves in a viscous liquid with surface tension. *Z. Angew. Math. Phys.* **1984**, *35*, 592–597. [CrossRef]
30. Antuono, M.; Colagross, A. The damping of viscous gravity waves. *Wave Motion* **2013**, *50*, 197–209. [CrossRef]
31. Prandtl, L. Über *Flüssigkeitsbewegung bei sehr kleiner Reibung*. In Proceedings of the Verhandlungen des Dritten Internationalen Mathematiker-Kongresses, Heidelberg, Germany, 8–13 August 1904; Teubner: Leipzig, Germany, 1905; pp. 485–491.
32. Schlichting, H. *Boundary Layer Theory*; McGraw Hill Co.: New York, NY, USA, 1955; 812p.
33. Longuet-Higgins, M.S. Mass transport in water waves. *Philos. Trans. R. Soc. Lond. Ser. A Math. Phys. Sci.* **1953**, *245*, 535–581. [CrossRef]
34. Longuet-Higgins, M.S. Mass transport in the boundary layer at a free oscillating surface. *J. Fluid Mech.* **1960**, *8*, 293–306. [CrossRef]
35. Liu, A.; Davis, S. Viscous attenuation of mean drift in water waves. *J. Fluid Mech.* **1977**, *81*, 63–84. [CrossRef]
36. Robertson, S.; Rousseaux, G. Viscous dissipation of surface waves and its relevance to analogue gravity experiments. *arXiv* **2018**, arXiv:1706.05255.
37. Dore, B. Mass transport in layered fluid systems. *J. Fluid Mech.* **1970**, *40*, 113–126. [CrossRef]
38. Thomson, W. Hydrokinetic solutions and observations. *Lond. Edinb. Dublin Philos. Mag. J. Sci.* **1871**, *42*, 362–377. [CrossRef]
39. Kistovich, A.V.; Chashechkin, Y.D. Dynamics of gravity-capillary waves on the surface of a nonuniformly heated fluid. *Izv. Atmos. Ocean. Phys.* **2007**, *43*, 95–102. [CrossRef]
40. Chen, X.; Duan, W.-Y.; Lu, D.Q. Gravity waves with effect of surface tension and fluid viscosity. *J. Hydrodyn.* **2006**, *18*, 171–176. [CrossRef]
41. Caligny, A.F.H. Expériences sur les mouvements des molécules liquides des ondes courantes, considérées dans leur mode d'action sur la marche des navires. *Compt. Rendu Acad. Sci. Paris* **1878**, *87*, 1019–1023.
42. Bagnold, R.A. Sand movement by waves: Some small-scale experiments with sand of very low density. *J. Inst. Civ. Eng.* **1947**, *27*, 447–469. [CrossRef]
43. Van Dorn, W.G. Boundary dissipation of oscillatory waves. *J. Fluid Mech.* **1966**, *24*, 769–779. [CrossRef]
44. Kalinichenko, V.A.; Chashechkin, Y.D. Structurization and restructurization of a homogeneous suspension in a standing wave field. *Fluid Dyn.* **2012**, *47*, 778–788. [CrossRef]
45. Chashechkin, Y.D. Transfer of the Substance of a Colored Drop in a Liquid Layer with Travelling Plane Gravity–Capillary Waves. *Izv. Atmos. Ocean. Phys.* **2022**, *58*, 188–197. [CrossRef]
46. Chashechkin, Y.D.; Ilinykh, A.Y. Drop decay into individual fibers at the boundary of the contact area with a target fluid. *Dokl. Phys.* **2021**, *66*, 101–105. [CrossRef]
47. Toffoli, A.; Bitner-Gregersen, E.M. Types of ocean surface waves, wave classification. In *Encyclopedia of Maritime and Offshore Engineering*; John Wiley & Sons, Ltd.: Hoboken, NJ, USA, 2017. [CrossRef]
48. Hung, L.-P.; Tsa, W.-T. The Formation of parasitic capillary ripples on gravity–capillary waves and the underlying vortical structures. *J. Phys. Oceanogr.* **2009**, *39*, 263–289. [CrossRef]
49. Kinsman, B. *Wind Waves*; Prentice-Hall: Englewood Cliffs, NJ, USA, 1965; 676p.
50. Holthuijsen, L.H. *Waves in Oceanic and Coastal Waters*; Cambridge University Press: Cambridge, UK, 2007; 388p.
51. Chashechkin, Y.D. Packets of capillary and acoustic waves of drop impact. *Her. Bauman Mosc. State Technol. Univ. Ser. Nat. Sci.* **2021**, *1*, 73–92. [CrossRef]
52. Chashechkin, Y.D.; Prohorov, V.E. Evolution of the structure of acoustic signals caused by the impact of a falling drop on a liquid. *Acoust. Phys.* **2020**, *66*, 362–374. [CrossRef]
53. Pumphrey, H.C.; Crum, L.A.; Bjorno, L. Underwater sound produced by individual drop impacts and rainfall. *J. Acoust. Soc. Am.* **1989**, *85*, 1518–1526. [CrossRef]
54. Bjorno, L. Underwater rain noise: Sources, spectra and interpretations. *J. De Phys. IV Proc. EDP Sci.* **1994**, *4*, C5/1023–C5/1030. [CrossRef]
55. Zhang, X. Enhanced dissipation of short gravity and gravity capillary waves due to parasitic capillaries. *Phys. Fluids* **2002**, *14*, L81. [CrossRef]

56. Sergievskaya, I.A.; Ermakov, S.A.; Ermoshkin, A.V.; Kapustin, I.A.; Molkov, A.A.; Danilicheva, O.A.; Shomina, O.V. Modulation of dual-polarized x-band radar backscatter due to long wind waves. *Remote Sens.* **2019**, *11*, 423. [CrossRef]
57. Slavchov, R.; Peychev, B.; Ismail, A.S. Characterization of capillary waves: A review and a new optical method. *Phys. Fluids* **2021**, *33*, 101303. [CrossRef]
58. Falcon, E.; Mordant, N. Experiments in surface gravity-capillary wave turbulence. *Annu. Rev. Fluid Mech.* **2022**, *54*, 1–27. [CrossRef]
59. Berhanu, M. Impact of the dissipation on the nonlinear interactions and turbulence of gravity-capillary waves. *Fluids* **2022**, *7*, 137. [CrossRef]
60. Thorpe, S.A. *An Introduction to Ocean Turbulence*; Cambridge University Press: Bangor, UK, 2012; 244p.
61. McWilliams, J.C.; François Colas, F.; Molemaker, M.J. Cold filamentary intensification and oceanic surface convergence lines. *Geophys. Res. Lett.* **2009**, *36*, L18602. [CrossRef]
62. Landau, L.D.; Lifshitz, E.M. *Fluid Mechanics. V.6. Course of Theoretical Physics*; Pergamon Press: Oxford, UK, 1987; 560p.
63. Müller, P. *The Equations of Oceanic Motions*; Cambridge University Press: Cambridge, UK, 2006; 291p.
64. Vallis, G.K. *Atmospheric and Oceanic Fluid Dynamics*; Cambridge University Press: Cambridge, UK, 2006; p. 745.
65. Nayfeh, A.H. *Introduction to Perturbation Technique*; John Wiley & Sons: New York, NY, USA, 1993; 536p.
66. Chashechkin, Y.D. Foundations of engineering mathematics applied for fluid flows. *Axioms* **2021**, *10*, 286. [CrossRef]
67. Lighthill, J. *Waves in Fluids*; Cambridge University Press: Cambridge, UK, 1978; 524p.
68. Bardakov, R.N.; Vasil'ev, A.Y.; Chashechkin, Y.D. Calculation and measurement of conical beams of three-dimensional periodic internal waves excited by a vertically oscillating piston. *Fluid Dyn.* **2007**, *42*, 612–626. [CrossRef]
69. Kistovich, A.V.; Chashechkin, Y.D. Fine structure of a conical beam of periodical internal waves in a stratified fluid. *Izv. Atmos. Ocean. Phys.* **2014**, *50*, 103–110. [CrossRef]
70. Zagumennyi, Y.V.; Chashechkin, Y.D. Pattern of unsteady vortex flow around plate under a zero angle of attack (two-dimensional problem). *Fluid Dyn.* **2016**, *51*, 53–70. [CrossRef]
71. Prandtl, L. *Führer Durch die Strömungslehre*; Verlagskatalog Von Friedr; Vieweg & Sohn in Braunschweig: Berlin, Germany, 1942; 337p.
72. Makarov, S.A.; Neklyudov, V.I.; Chashechkin, Y.D. Spatial structure of twodimensional monochromatic internal-wave beams in an exponentially stratified liquid. *Izv. Atmos. Ocean. Phys.* **1990**, *26*, 548–554.
73. Kistovich, Y.V.; Chashechkin, Y.D. An exact solution of a linearized problem of the radiation of monochromatic internal waves in a viscous fluid. *J. Appl. Maths Mech.* **1999**, *63*, 587–594. [CrossRef]
74. Kistovich, Y.V.; Chashechkin, Y.D. Linear theory of the propagation of internal wave beams in an arbitrarily stratified liquid. *J. Appl. Mech. Technol. Phys.* **1998**, *39*, 729–737. [CrossRef]
75. Chashechkin, Y.D. Conventional partial and new complete solutions of the fundamental equations of fluid mechanics in the problem of periodic internal waves with accompanying ligaments generation. *Mathematics* **2021**, *9*, 586. [CrossRef]
76. Veeresha, P.; Baskonus, H.M.; Gao, W. Strong Interacting internal waves in rotating ocean: Novel fractional approach. *Axioms* **2021**, *10*, 123. [CrossRef]
77. Lenain, L.; Pizzo, N. Modulation of surface gravity waves by internal waves. *J. Phys. Oceanogr.* **2021**, *51*, 2735–2748. [CrossRef]

axioms

MDPI

Article

On the Linear Quadratic Optimal Control for Systems Described by Singularly Perturbed Itô Differential Equations with Two Fast Time Scales

Vasile Drăgan †

Institute of Mathematics "Simion Stoilow" of the Romanian Academy, RO-014700 Bucharest, Romania; vasile.dragan@imar.ro

† The author is a member of Academy of the Romanian Scientists.

Received: 15 January 2019; Accepted: 28 February 2019; Published: 5 March 2019

Abstract: In this paper a stochastic optimal control problem described by a quadratic performance criterion and a linear controlled system modeled by a system of singularly perturbed Itô differential equations with two fast time scales is considered. The asymptotic structure of the stabilizing solution (satisfying a prescribed sign condition) to the corresponding stochastic algebraic Riccati equation is derived. Furthermore, a near optimal control whose gain matrices do not depend upon small parameters is discussed.

Keywords: singularly perturbed linear stochastic systems; asymptotic structure of the stabilizing solution; optimal control problem; Riccati equations of stochastic control

MSC: 93E20; 93B12; 34H15; 49N10

1. Introduction

In the last 40 years, special attention was paid to the singular perturbation techniques applied in both analysis and synthesis of control laws with prescribed performance specifications for the regulation of systems whose mathematical models are described by a system of differential equations of high dimension, and also contain a number of small parameters multiplying derivatives of a part of the state variables of the physical phenomenon under discussion.

The large number of differential equations of the mathematical model of a physical process may be caused by the presence of some "parasitic" parameters such as small time constants, resistances, inductances, capacitances, moments of inertia, small masses, etc.

The presence of such small parameters is often a source of stiffness due to the simultaneous occurrence of slow and fast phenomena. It is known that the stiffness can produce ill-conditioning of the numerical computation involved in the process of designing the optimal control. This inconvenience leads to the idea to simplify the mathematical model by neglecting the small parameters occurring in the original model. Besides the stiffness, the necessity of the simplification of the mathematical model by neglecting the small parameters is also imposed by the fact that, in many applications, the values of these parasitic quantities are not exactly known. A fundamental problem is to check if the optimal control design based on the reduced model provides a satisfactory behavior of the full system which contains fast phenomena neglected during the designing process.

Remarkable results were obtained in the problem of the design of some near optimal controllers in the case of some deterministic systems with several time scales. Such results may be found in the monographs [1–4]. A common feature of the approaches in these works is the use of the singular perturbations techniques, initially developed in connection with the study of qualitative properties of the solutions of some classes of differential equations starting with the classical work of Tichonov [5].

The interest for studying different problems regarding the singularly perturbed controlled systems is still increasing. For the reader's convenience, we refer to the recent papers [6–10].

Lately, the interest for studying optimal control problems for stochastic systems modeled by singularly perturbed Itô differential equations also increased. Unlike the deterministic case, where the reduced model is obtained by simply removing the small parameters, in the case of stochastic optimal control problems driven by systems of singularly perturbed Itô differential equations, the definition of the reduced model is not always intuitive and it is strongly dependent upon the intensity of the white noise type perturbations affecting the diffusion part of the fast equations of the mathematical model. Hence, problems related to singularly perturbed stochastic systems could not be viewed as simple extensions of there deterministic counterparts. This makes the study of this class of systems a challenging (and relatively not fully investigated) topic. The main results obtained in this field were published in [11–14].

Very few results have been reported in the literature dealing with several fast time scales. We cite here [15] for the deterministic case and [16] for the stochastic framework. Pursuing our efforts in the study of singularly perturbed stochastic systems, we consider in this paper a stochastic optimal control problem described by a quadratic performance criterion and a linear controlled system modeled by a system of singularly perturbed Itô differential equations with two fast time scales.

Unlike [17] in the deterministic case or [14] in the stochastic case, where the fast time scales have the same order of magnitude, in the present work, we consider the case in which the two fast time scales have different order of magnitude. More precisely, if $\epsilon_j > 0, j = 1,2$ are the small parameters associated with the two fast time scales, the ratio $\frac{\epsilon_2}{\epsilon_1}$ becomes the third small parameter which needs to be considered in the asymptotic analysis performed here. The most part of our study is devoted to the analysis of the asymptotic structure of the stabilizing solution of the algebraic Riccati equation involved in the computation of the optimal control of the optimization problem under consideration. The main tool in the derivation of the asymptotic structure of the stabilizing solution of the algebraic Riccati equation under consideration around the origin $(\epsilon_1, \epsilon_2, \frac{\epsilon_2}{\epsilon_1}) = (0,0,0)$ is the implicit functions theorem. To this end, we first investigate the solvability of the system of reduced equations obtained setting $\epsilon_k = 0, k = 1,2$ and $\frac{\epsilon_2}{\epsilon_1} = 0$ in the original algebraic Riccati equation. Unlike the deterministic case, in the stochastic framework considered in this paper, the system of the reduced equations is a system of strongly interconnected Riccati type algebraic equations. For this system of interconnected Riccati type equations we introduce the concept of stabilizing solution and provide a set of necessary and sufficient conditions which guarantee the existence of such a solution. Further, employing the stabilizing solution of the system of the reduced equations and the corresponding stabilizing gain matrices we show that one may apply the implicit functions theorem to obtain the existence, as well as the asymptotic structure of, the stabilizing solution of the algebraic Riccati equation associated with the optimal control problem under consideration. Based on the dominant part independent of the small parameters of the stabilizing gain matrix, we construct a near optimal control whose gain matrices can be computed without the knowledge of the precise values of the small parameters associated with the fast time scales.

The paper is organized as follows: Section 2 provides the model description and the problem formulation. In Section 3 we show how the system of reduced Riccati equations, which are not dependent upon the small parameters, can be derived. Also, we introduce the concept of the stabilizing solution for the system of reduced algebraic Riccati equations. Then, we provide conditions which guarantee the existence of this stabilizing solution. In Section 4, we obtain the existence and the asymptotic structure of the stabilizing solution for the Riccati equation associated with the original linear quadratic control problem. Finally, we show how the asymptotic structure of the stabilizing feedback gain can be used to construct a near optimal control.

2. The Problem

Let us consider the stochastic optimal control problem asking for the minimization of the quadratic functional

$$J(\mathbf{x}_0;u) = \mathbb{E}[\int_0^\infty (\sum_{j,k=0}^{2} x_j^T(t)M_{jk}x_k(t) + 2\sum_{j=0}^{2} x_j^T(t)L_ju(t) + u^T(t)Ru(t))dt \tag{1}$$

along with the trajectories of the controlled system having the state space representation described by the following system of singularly perturbed Itô differential equations

$$\begin{aligned} dx_0(t) &= (A_{00}(\epsilon)x_0(t) + A_{01}(\epsilon)x_1(t) + A_{02}(\epsilon)x_2(t) + B_0(\epsilon)u(t))dt+ \\ &+ (C_{00}(\epsilon)x_0(t) + C_{01}(\epsilon)x_1(t) + C_{02}(\epsilon)x_2(t) + D_0(\epsilon)u(t))dw(t) \\ x_0(0) &= x_0^0 \\ \epsilon_k dx_k(t) &= (A_{k0}(\epsilon)x_0(t) + A_{k1}(\epsilon)x_1(t) + A_{k2}(\epsilon)x_2(t) + B_k(\epsilon)u(t))dt+ \\ &+ \sqrt{\epsilon_k}(C_{k0}(\epsilon)x_0(t) + C_{k1}(\epsilon)x_1(t) + C_{k2}(\epsilon)x_2(t) + D_k(\epsilon)u(t))dw(t) \\ x_k(0) &= x_k^0, k = 1,2. \end{aligned} \tag{2}$$

In (1) and (2) $u(t) \in \mathbb{R}^m$ is the vector of the control parameters and $\mathbf{x}(t) = \begin{pmatrix} x_0^T(t) & x_1^T(t) & x_2^T(t) \end{pmatrix}^T \in \mathbb{R}^n$ is the vector of state parameters, $\mathbf{x}_0 = \begin{pmatrix} x_0^{0T} & x_0^{1T} & x_0^{2T} \end{pmatrix}^T$, $n = n_0 + n_1 + n_2$; $x_j(t) \in \mathbb{R}^{n_j}$, $0 \leq j \leq 2$. In (1), $M_{jk} = M_{kj}^T$, $0 \leq k,j \leq 2$, $R = R^T$. In (2), $\epsilon_k > 0$ are small parameters often not exactly known.

In order to make more intuitive the developments in this paper we assume that the small parameters $\epsilon_k, k = 1,2$ satisfy the assumption:

$\mathbf{H_1}$) $\epsilon_k = \varphi_k(\eta)$, where $\varphi_k : [0,\eta^*] \to [0,\infty)$ are nondecreasing functions with the properties:
(i) $\varphi_k(\eta) = 0$ if and only if $\eta = 0, k = 1,2$.
(ii) $\lim_{\eta\to 0_+} \varphi_k(\eta) = 0$; $\lim_{\eta\to 0_+} \frac{\varphi_2(\eta)}{\varphi_1(\eta)} = 0$.

In the sequel, the dependence of ϵ_k upon the parameter η will be suppressed.

Remark 1. *According to the terminology used in the framework of singularly perturbed differential equations, $x_0(t)$ will be called* **slow state variables** *while $x_1(t), x_2(t)$ will be named* **fast state variables***. From the condition imposed to the values of the ratio $\frac{\epsilon_2}{\epsilon_1}$ in $\mathbf{H_1}$), it follows that the states $x_2(t)$ are faster than $x_1(t)$. That is why under the assumption $\mathbf{H_1}$) system (2) is a controlled system with two fast time scales.*

In the deterministic framework, the asymptotic structure of the solutions of some systems with several time fast scales was studied in [18] while in [19] were derived uniform upper bounds of the block components of the fundamental matrix solution of the systems of linear differential equations with several fast time scales.

In (2), $\{w(t)\}_{t\geq 0}$ is a 1-dimensional standard Wiener process defined on a given probability space $(\Omega, \mathcal{F}, \mathcal{P})$. The consideration of the case with an 1-dimensional standard Wiener process is only to easy the exposition. The extension to the case of a multidimensional Wiener process can be done without difficulty.

Regarding the coefficients of system (2), we make the following assumption:

$\mathbf{H_2}$) $\epsilon = (\epsilon_1, \epsilon_2) \to (A_{jk}(\epsilon), B_j(\epsilon), C_{jk}(\epsilon), D_j(\epsilon))$ are C^1 matrix valued functions defined on a neighborhood of the origin $(\epsilon_1, \epsilon_2) = (0,0)$.

We set

$$\mathbb{A}(\epsilon) = \begin{pmatrix} A_{00}(\epsilon) & A_{01}(\epsilon) & A_{02}(\epsilon) \\ \frac{1}{\epsilon_1}A_{10}(\epsilon) & \frac{1}{\epsilon_1}A_{11}(\epsilon) & \frac{1}{\epsilon_1}A_{12}(\epsilon) \\ \frac{1}{\epsilon_2}A_{20}(\epsilon) & \frac{1}{\epsilon_2}A_{21}(\epsilon) & \frac{1}{\epsilon_2}A_{22}(\epsilon) \end{pmatrix}, \mathbb{B}(\epsilon) = \begin{pmatrix} B_0(\epsilon) \\ \frac{1}{\epsilon_1}B_1(\epsilon) \\ \frac{1}{\epsilon_2}B_2(\epsilon) \end{pmatrix},$$
$$\mathbb{C}(\epsilon) = \begin{pmatrix} C_{00}(\epsilon) & C_{01}(\epsilon) & C_{02}(\epsilon) \\ \frac{1}{\sqrt{\epsilon_1}}C_{10}(\epsilon) & \frac{1}{\sqrt{\epsilon_1}}C_{11}(\epsilon) & \frac{1}{\sqrt{\epsilon_1}}C_{12}(\epsilon) \\ \frac{1}{\sqrt{\epsilon_2}}C_{20}(\epsilon) & \frac{1}{\sqrt{\epsilon_2}}C_{21}(\epsilon) & \frac{1}{\sqrt{\epsilon_2}}C_{22}(\epsilon) \end{pmatrix}, \mathbb{D}(\epsilon) = \begin{pmatrix} D_0(\epsilon) \\ \frac{1}{\sqrt{\epsilon_1}}D_1(\epsilon) \\ \frac{1}{\sqrt{\epsilon_2}}D_2(\epsilon) \end{pmatrix} \tag{3}$$

$$\mathbb{M} = \begin{pmatrix} M_{00} & M_{01} & M_{02} \\ M_{01}^T & M_{11} & M_{12} \\ M_{02}^T & M_{12}^T & M_{22} \end{pmatrix} = \mathbb{M}^T, \mathbb{L} = \begin{pmatrix} L_0 \\ L_1 \\ L_2 \end{pmatrix}. \tag{4}$$

With these notations (1) and (2) may be written in a compact form as:

$$J(\mathbf{x}_0; u) = \mathbb{E}[\int_0^\infty (\mathbf{x}^T(t)\mathbb{M}\mathbf{x}(t) + 2\mathbf{x}^T(t)\mathbb{L}u(t) + u^T(t)Ru(t))dt] \tag{5}$$

and

$$\begin{aligned} d\mathbf{x}(t) &= (\mathbb{A}(\epsilon)\mathbf{x}(t) + \mathbb{B}(\epsilon)u(t))dt + (\mathbb{C}(\epsilon)\mathbf{x}(t) + \mathbb{D}(\epsilon)u(t))dw(t) \\ \mathbf{x}(0) &= \mathbf{x}_0. \end{aligned} \tag{6}$$

One sees that for each $\epsilon = (\epsilon_1, \epsilon_2)$ fixed, the optimal control asking for the minimization of the quadratic cost (5) in the class of controls that stabilizes system (6) is a standard stochastic linear quadratic optimal control problem, which was studied starting with [20].

In [21,22] it was shown that a stochastic linear quadratic control problem, with control dependent terms in the diffusion part of the controlled system, is still well possed even if the cost weight matrices of the states and the control are allowed to be indefinite. The optimal control is given by:

$$u_{opt}(t) = -(R + \mathbb{D}^T(\epsilon)\tilde{X}(\epsilon)\mathbb{D}(\epsilon))^{-1}(\mathbb{B}^T(\epsilon)\tilde{X}(\epsilon) + \mathbb{D}^T(\epsilon)\tilde{X}(\epsilon)\mathbb{C}(\epsilon) + \mathbb{L}^T)\mathbf{x}(t) \tag{7}$$

where $\tilde{X}(\epsilon)$ is the unique stabilizing solution of the algebraic Riccati equation of stochastic control (SARE):

$$\begin{aligned} &\mathbb{A}^T(\epsilon)X + X\mathbb{A}(\epsilon) + \mathbb{C}^T(\epsilon)X\mathbb{C}(\epsilon) - (X\mathbb{B}(\epsilon) + \mathbb{C}^T(\epsilon)X\mathbb{D}(\epsilon) + \mathbb{L})\times \\ &\times(R + \mathbb{D}^T(\epsilon)X\mathbb{D}(\epsilon))^{-1}(\mathbb{B}^T(\epsilon)X + \mathbb{D}^T(\epsilon)X\mathbb{C}(\epsilon) + \mathbb{L}^T) + \mathbb{M} = 0 \end{aligned} \tag{8}$$

satisfying the sign condition

$$R + \mathbb{D}^T(\epsilon)X\mathbb{D}(\epsilon) > 0. \tag{9}$$

The condition (9) supplies the absence of the information regarding the sign of the matrix R. In [22], necessary and sufficient conditions that guarantee the existence of the stabilizing solution of a SARE were provided as (8) satisfying the sign condition (9) and a procedure for numerical computation of this solution using the so called semidefinite programming (SDP) was proposed. Also, an iterative procedure for numerical computation of the constrained SARE of type (8) and (9) was proposed in Section 5.8 from [23]. Unfortunately, the way in which the small parameters $\epsilon_k > 0, k = 1,2$ affect the coefficients of SARE (8) and (9) may produce ill-conditioning of the numerical computation involved in obtaining the stabilizing solution $\tilde{X}(\epsilon)$ of the SARE under consideration. In order to avoid the ill-conditioning of the numerical computations generated by the high difference between the order of magnitude of the coefficients, knowledge of the asymptotic structure of the solution $\tilde{X}(\epsilon)$ in a neighborhood of the origin $\epsilon = (\epsilon_1, \epsilon_2) = (0,0)$ is useful. As a consequence of such a study, a system of Riccati type equations not depending upon the small parameters ϵ_k, $k = 1,2$, often named a system of reduced algebraic Riccati equations, which allows us to compute the dominant part of the solution $\tilde{X}(\epsilon)$ can be displayed.

In the deterministic case, see for example [1–4,24], the system of reduced algebraic Riccati equations is obtained by simply removing all of the small parameters. In the stochastic framework, when the controlled systems are modeled by singularly perturbed Itô differential equations, the definition of the system of reduced algebraic Riccati equations cannot be done by a simple neglection of the small parameters. From [11] or [12,25], one sees that the definition of the system of reduced algebraic Riccati equations is strongly dependent upon the magnitude of the white noise perturbations affecting the equations of the fast variables in the controlled system.

In order to obtain the asymptotic structure with respect to the small parameters $\epsilon_k > 0, k = 1,2$ of the stabilizing solution of SARE (8), we shall use the implicit functions theorem. To this end, we need a rigourous definition of the corresponding system of reduced algebraic Riccati equations (SRARE) and to point out a special kind of solution of this system which helps us to apply the implicit functions theorem. That is why in the next section we shall show how the system of reduced algebraic Riccati equations in the case of SARE (8) and (9) can be defined. Next, we shall introduce a concept of stabilizing solution of the obtained SRARE and we shall provide a set of conditions which guarantee the existence of this stabilizing solution of SRARE. In Section 4, using reasoning based on the implicit functions theorem, we shall obtain the asymptotic structure of the stabilizing solution of SARE (8) satisfying the sign condition (9), as well as the asymptotic structure of the corresponding stabilizing feedback gain.

3. The System of Reduced Algebraic Riccati Equations

3.1. Derivation of the System of Reduced Algebraic Riccati Equations

Setting $F = -(R + \mathbb{D}^T(\epsilon)X\mathbb{D}(\epsilon))^{-1}(\mathbb{B}^T(\epsilon)X + \mathbb{D}^T(\epsilon)X\mathbb{C}(\epsilon) + \mathbb{L}^T)$ one obtains that if X is a solution of SARE (8) satisfying the sign condition (9), then (X, F) is a solution of the system:

$$\begin{aligned} \Gamma(X,\epsilon)F + \mathbb{B}^T(\epsilon)X + \mathbb{D}^T(\epsilon)X\mathbb{C}(\epsilon) + \mathbb{L}^T &= 0 \\ \mathbb{A}^T(\epsilon)X + X\mathbb{A}(\epsilon) + \mathbb{C}^T(\epsilon)X\mathbb{C}(\epsilon) - F^T\Gamma(X,\epsilon)F + \mathbb{M} &= 0 \\ \Gamma(X,\epsilon) &= R + \mathbb{D}^T(\epsilon)X\mathbb{D}(\epsilon). \end{aligned} \tag{10}$$

Conversely, if (X, F) is a solution of system (10) satisfying $\Gamma(X,\epsilon) > 0$, then X is a solution of the constrained SARE (8) and (9). To obtain the asymptotic structure of the stabilizing solution of SARE (8) and (9), we shall analyse the asymptotic structure of the solution $(\tilde{X}(\epsilon), \tilde{F}(\epsilon))$ of system (10) with the additional property that the closed-loop system

$$d\mathbf{x}(t) = (\mathbb{A}(\epsilon) + \mathbb{B}(\epsilon)\tilde{F}(\epsilon))\mathbf{x}(t)dt + (\mathbb{C}(\epsilon) + \mathbb{D}(\epsilon)\tilde{F}(\epsilon))\mathbf{x}(t)dw(t) \tag{11}$$

is exponentially stable in mean square (ESMS).

We are looking for the solution (X, F) of (10) having the partition:

$$X = \begin{pmatrix} X_{00} & \epsilon_1 X_{01} & \epsilon_2 X_{02} \\ \epsilon_1 X_{01}^T & \epsilon_1 X_{11} & \epsilon_2 X_{12} \\ \epsilon_2 X_{02}^T & \epsilon_2 X_{12}^T & \epsilon_2 X_{22} \end{pmatrix}, \qquad F = \begin{pmatrix} F_0 & F_1 & F_2 \end{pmatrix} \tag{12}$$

where $X_{jk} \in \mathbb{R}^{n_j \times n_k}$, $F_k \in \mathbb{R}^{m \times n_k}$, $0 \le j,k \le 2$.

Employing the partitions (3) and (4) of the coefficients of SARE (8) we may obtain a partition of system (10).

To ease the exposition, let us regroup the block components from (3), (4) and (12) as:

$$\mathbb{A}(\epsilon) = \Pi^{-1}(\epsilon)\begin{pmatrix} \mathbb{A}_{11}(\epsilon) & \mathbb{A}_{12}(\epsilon) \\ \mathbb{A}_{21}(\epsilon) & A_{22}(\epsilon) \end{pmatrix}, \tag{13a}$$

$$\mathbb{B}(\epsilon) = \Pi^{-1}(\epsilon)\begin{pmatrix} \mathbb{B}_{1}(\epsilon) \\ B_{2}(\epsilon) \end{pmatrix}, \tag{13b}$$

$$\mathbb{C}(\epsilon) = \Pi^{-1}(\sqrt{\epsilon})\begin{pmatrix} \mathbb{C}_{11}(\epsilon) & \mathbb{C}_{12}(\epsilon) \\ \mathbb{C}_{21}(\epsilon) & C_{22}(\epsilon) \end{pmatrix}, \tag{13c}$$

$$\mathbb{D}(\epsilon) = \Pi^{-1}(\sqrt{\epsilon})\begin{pmatrix} \mathbb{D}_{1}(\epsilon) \\ D_{2}(\epsilon) \end{pmatrix}, \tag{13d}$$

$$\mathbb{M} = \begin{pmatrix} \mathbb{M}_{11} & \mathbb{M}_{12} \\ \mathbb{M}_{12}^T & M_{22} \end{pmatrix}, \tag{13e}$$

$$\mathbb{L} = \begin{pmatrix} \mathbb{L}_{1} \\ L_{2} \end{pmatrix}, \tag{13f}$$

where

$$\Pi(\epsilon) = diag(I_{n_0}, \epsilon_1 I_{n_1}, \epsilon_2 I_{n_2}), \tag{14a}$$

$$\mathbb{A}_{11}(\epsilon) = \begin{pmatrix} A_{00}(\epsilon) & A_{01}(\epsilon) \\ A_{10}(\epsilon) & A_{11}(\epsilon) \end{pmatrix}$$

$$\mathbb{A}_{12}(\epsilon) = \begin{pmatrix} A_{02}(\epsilon) \\ A_{12}(\epsilon) \end{pmatrix}, \mathbb{A}_{21}(\epsilon) = \begin{pmatrix} A_{20}(\epsilon) & A_{21}(\epsilon) \end{pmatrix} \tag{14b}$$

$$\mathbb{B}_{1}(\epsilon) = \begin{pmatrix} B_{0}(\epsilon) \\ B_{1}(\epsilon) \end{pmatrix}, \tag{14c}$$

$$\mathbb{C}_{11}(\epsilon) = \begin{pmatrix} C_{00}(\epsilon) & C_{01}(\epsilon) \\ C_{10}(\epsilon) & C_{11}(\epsilon) \end{pmatrix}$$

$$\mathbb{C}_{12}(\epsilon) = \begin{pmatrix} C_{02}(\epsilon) \\ C_{12}(\epsilon) \end{pmatrix}, \mathbb{C}_{21}(\epsilon) = \begin{pmatrix} C_{20}(\epsilon) & C_{21}(\epsilon) \end{pmatrix} \tag{14d}$$

$$\mathbb{D}_{1}(\epsilon) = \begin{pmatrix} D_{0}(\epsilon) \\ D_{1}(\epsilon) \end{pmatrix}, \tag{14e}$$

$$\mathbb{M}_{11} = \begin{pmatrix} M_{00} & M_{01} \\ M_{01}^T & M_{11} \end{pmatrix}, \mathbb{M}_{12} = \begin{pmatrix} M_{02} \\ M_{12} \end{pmatrix}, \mathbb{L}_{1} = \begin{pmatrix} L_{0} \\ L_{1} \end{pmatrix} \tag{14f}$$

From (12) one obtains the following structure of X and F:

$$X = \Pi(\epsilon)\begin{pmatrix} U_1(X,\epsilon) & \Delta(\epsilon)\mathbb{X}_{12} \\ \mathbb{X}_{12}^T & X_{22} \end{pmatrix} \tag{15}$$

$$F = \begin{pmatrix} \mathbb{F}_1 & F_2 \end{pmatrix}$$

where

$$U_1(X,\epsilon) = \begin{pmatrix} X_{00} & \epsilon_1 X_{01} \\ X_{01}^T & X_{11} \end{pmatrix} \quad (16a)$$
$$\mathbb{X}_{12} = \begin{pmatrix} X_{02} \\ X_{12} \end{pmatrix} \quad (16b)$$
$$\Delta(\epsilon) = diag(\epsilon_2 I_{n_0}, \epsilon_2/\epsilon_1 I_{n_1}), \quad (16c)$$
$$\mathbb{F}_1 = \begin{pmatrix} F_0 & F_1 \end{pmatrix}. \quad (16d)$$

We also have

$$\Pi^{-1}(\sqrt{\epsilon})X\Pi^{-1}(\sqrt{\epsilon}) = \begin{pmatrix} U_2(X,\epsilon) & \Delta(\sqrt{\epsilon})\mathbb{X}_{12} \\ \mathbb{X}_{12}^T\Delta(\sqrt{\epsilon}) & X_{22} \end{pmatrix} \quad (17)$$

where $U_2(X,\epsilon) = \begin{pmatrix} X_{00} & \sqrt{\epsilon_1}X_{01} \\ \sqrt{\epsilon_1}X_{01}^T & X_{11} \end{pmatrix}$, $\Delta(\sqrt{\epsilon}) = diag(\sqrt{\epsilon_2}I_{n_0}, \sqrt{\frac{\epsilon_2}{\epsilon_1}}I_{n_1})$.

With these notations we obtain the following partition of system (10)

$$\begin{aligned}
&\mathbb{B}_1^T(\epsilon)U_1(X,\epsilon) + B_2^T(\epsilon)\mathbb{X}_{12}^T + \mathbb{D}_1^T(\epsilon)U_2(X,\epsilon)\mathbb{C}_{11}(\epsilon) + D_2^T(\epsilon)\mathbb{X}_{12}^T\Delta(\sqrt{\epsilon})\mathbb{C}_{11}(\epsilon) + \quad (18a)\\
&+\mathbb{D}_1^T(\epsilon)\Delta(\sqrt{\epsilon})\mathbb{X}_{12}\mathbb{C}_{21}(\epsilon) + D_2^T(\epsilon)X_{22}\mathbb{C}_{21}(\epsilon) + \mathbb{L}_1^T + \Gamma(X,\epsilon)\mathbb{F}_1 = 0\\
&\mathbb{B}_1^T(\epsilon)\Delta(\epsilon)\mathbb{X}_{12} + B_2^T(\epsilon)X_{22} + \mathbb{D}_1^T(\epsilon)U_2(X,\epsilon)\mathbb{C}_{12}(\epsilon) + D_2^T(\epsilon)\mathbb{X}_{12}^T\Delta(\sqrt{\epsilon})\mathbb{C}_{12}(\epsilon) + \quad (18b)\\
&+\mathbb{D}_1^T(\epsilon)\Delta(\sqrt{\epsilon})X_{12}C_{22}(\epsilon) + D_2^T(\epsilon)X_{22}C_{22}(\epsilon) + L_2^T + \Gamma(X,\epsilon)F_2 = 0\\
&\mathbb{A}_{11}^T(\epsilon)U_1(X,\epsilon) + U_1^T(X,\epsilon)\mathbb{A}_{11}(\epsilon) + \mathbb{A}_{21}^T(\epsilon)\mathbb{X}_{12}^T + \mathbb{X}_{12}\mathbb{A}_{21}(\epsilon) + \\
&\mathbb{C}_{11}^T(\epsilon)U_2(X,\epsilon)\mathbb{C}_{11}(\epsilon) + \mathbb{C}_{21}^T(\epsilon)\mathbb{X}_{12}^T\Delta(\sqrt{\epsilon})\mathbb{C}_{11}(\epsilon) + \mathbb{C}_{11}^T(\epsilon)\Delta(\sqrt{\epsilon})\mathbb{X}_{12}\mathbb{C}_{21}(\epsilon) + \quad (18c)\\
&+\mathbb{C}_{21}^T(\epsilon)X_{22}\mathbb{C}_{21}(\epsilon) - \mathbb{F}_1^T\Gamma(X,\epsilon)\mathbb{F}_1 + \mathbb{M}_{11} = 0\\
&\mathbb{A}_{11}^T(\epsilon)\Delta(\epsilon)\mathbb{X}_{12} + \mathbb{A}_{21}^T(\epsilon)X_{22} + U_1^T(X,\epsilon)\mathbb{A}_{12}(\epsilon) + \mathbb{X}_{12}A_{22}(\epsilon) + \\
&\mathbb{C}_{11}^T(\epsilon)U_2(X,\epsilon)\mathbb{C}_{12}(\epsilon) + \mathbb{C}_{21}^T(\epsilon)\mathbb{X}_{12}^T\Delta(\sqrt{\epsilon})\mathbb{C}_{12}(\epsilon) + \mathbb{C}_{11}^T(\epsilon)\Delta(\sqrt{\epsilon})\mathbb{X}_{12}C_{22}(\epsilon) + \quad (18d)\\
&\mathbb{C}_{21}^T(\epsilon)X_{22}C_{22}(\epsilon) - \mathbb{F}_1^T\Gamma(X,\epsilon)F_2 + \mathbb{M}_{12} = 0\\
&\mathbb{A}_{12}^T(\epsilon)\Delta(\epsilon)\mathbb{X}_{12} + A_{22}^T(\epsilon)X_{22} + \mathbb{X}_{12}^T\Delta(\epsilon)\mathbb{A}_{12}(\epsilon) + X_{22}A_{22}(\epsilon) + \mathbb{C}_{12}^T(\epsilon)U_2(X,\epsilon)\mathbb{C}_{12}(\epsilon) + \\
&+C_{22}^T(\epsilon)\mathbb{X}_{12}^T\Delta(\sqrt{\epsilon})\mathbb{C}_{12}(\epsilon) + \mathbb{C}_{12}^T(\epsilon)\Delta(\sqrt{\epsilon})\mathbb{X}_{12}C_{22}(\epsilon) + C_{22}^T(\epsilon)X_{22}C_{22}(\epsilon) - \quad (18e)\\
&-F_2^T\Gamma(X,\epsilon)F_2 + M_{22} = 0\\
&\Gamma(X,\epsilon) = R + \mathbb{D}_1^T(\epsilon)U_2(X,\epsilon)\mathbb{D}_1(\epsilon) + D_2^T(\epsilon)\mathbb{X}_{12}^T\Delta(\sqrt{\epsilon})\mathbb{D}_1(\epsilon) + \\
&+\mathbb{D}_1^T(\epsilon)\Delta(\sqrt{\epsilon})\mathbb{X}_{12}D_2(\epsilon) + D_2^T(\epsilon)X_{22}D_2(\epsilon) \quad (18f)
\end{aligned}$$

Setting formally $\epsilon_j = 0, j = 1,2$ and $\frac{\epsilon_2}{\epsilon_1} = 0$, in (18) we obtain the equations:

$$\begin{aligned}
\mathbb{B}_1^T(0)U_1(X,0) + B_2^T(0)\mathbb{X}_{12}^T + \mathbb{D}_1^T(0)U_2(X,0)\mathbb{C}_{11}(0) + & \\
+ D_2^T(0)X_{22}C_{22} + \mathbb{L}_1^T + \Gamma(X,0)\mathbb{F}_1 &= 0 \qquad (19a)\\
B_2^T(0)X_{22} + \mathbb{D}_1^T(0)U_2(X,0)\mathbb{C}_{12}(0) + D_2^T(0)X_{22}C_{22}(0) + L_2^T + \Gamma(X,0)F_2 &= 0 \qquad (19b)\\
\mathbb{A}_{11}^T(0)U_1(X,0) + U_1^T(X,0)\mathbb{A}_{11}(0) + \mathbb{A}_{12}^T(0)\mathbb{X}_{12}^T + \mathbb{X}_{12}\mathbb{A}_{21}(0) + & \\
+\mathbb{C}_{11}^T(0)U_2(X,0)\mathbb{C}_{11}(0) + \mathbb{C}_{21}^T(0)X_{22}\mathbb{C}_{21}(0) - \mathbb{F}_1^T\Gamma(X,0)\mathbb{F}_1 + \mathbb{M}_{11} &= 0 \qquad (19c)\\
\mathbb{A}_{21}^T(0)X_{22} + U_1^T(X,0)\mathbb{A}_{12}(0) + \mathbb{X}_{12}A_{22}(0) + \mathbb{C}_{11}^T(0)U_2(X,0)\mathbb{C}_{12}(0) + & \\
+\mathbb{C}_{21}^T(0)X_{22}C_{22}(0) - \mathbb{F}_1^T\Gamma(X,0)F_2 + \mathbb{M}_{12} &= 0 \qquad (19d)\\
A_{22}^T(0)X_{22} + X_{22}A_{22}(0) + \mathbb{C}_{12}^T(0)U_2(X,0)\mathbb{C}_{12}(0) + & \\
+C_{22}^T(0)X_{22}C_{22}(0) - F_2^T\Gamma(X,0)F_2 + M_{22} &= 0 \qquad (19e)\\
\Gamma(X,0) = R + \mathbb{D}_1^T(0)U_2(X,0)\mathbb{D}_1(0) + D_2^T(0)X_{22}D_2(0). & \qquad (19f)
\end{aligned}$$

Having in mind (15) and (16), we remark that (19) is a system of nonlinear algebraic equations with the unknowns $(X_{00}, X_{01}, X_{11}, X_{02}, X_{12}, X_{22}, F_0, F_1, F_2) \in \mathcal{S}_{n_0} \times \mathbb{R}^{n_0 \times n_1} \times \mathcal{S}_{n_1} \times \mathbb{R}^{n_0 \times n_2} \times \mathbb{R}^{n_1 \times n_2} \times \mathcal{S}_{n_2} \times \mathbb{R}^{m \times n_0} \times \mathbb{R}^{m \times n_1} \times \mathbb{R}^{m \times n_2}$.

We recall that S_q denotes the linear space of symmetric matrices of size $q \times q$.

Assuming that $A_{22}(0)$ is invertible we obtain from (19d):

$$\begin{aligned}
\mathbb{X}_{12} = & -\mathbb{A}_{21}^T(0)X_{22}A_{22}^{-1}(0) - U_1^T(X,0)\mathbb{A}_{12}(0)A_{22}^{-1}(0) - \mathbb{C}_{11}^T(0)U_2(X,0)\mathbb{C}_{12}(0)A_{22}^{-1}(0) \qquad (20)\\
& -\mathbb{C}_{21}^T(0)X_{22}C_{22}(0)\mathbb{A}_{22}^{-1}(0) + \mathbb{F}_1^T\Gamma(X,0)F_2A_{22}^{-1}(0) - \mathbb{M}_{12}A_{22}^{-1}(0).
\end{aligned}$$

Substituting (20) in (19a) and (19c) we obtain after algebraic calculations:

$$\begin{aligned}
&(\mathbb{B}_1(0) - \mathbb{A}_{12}(0)A_{22}^{-1}(0)B_2(0))^T U_1(X,0) + (\mathbb{D}_1(0) - \mathbb{C}_{12}(0)A_{22}^{-1}(0)B_2(0))^T U_2(X,0)\\
&\times(\mathbb{C}_{11}(0) - \mathbb{C}_{12}(0)A_{22}^{-1}(0)\mathbb{C}_{21}(0)) + (D_2(0) - C_{22}(0)A_{22}^{-1}(0)B_2(0))^T X_{22}(\mathbb{C}_{21}(0) -\\
&-C_{22}(0)A_{22}^{-1}(0)\mathbb{A}_{21}(0)) + (I_m + F_2A_{22}^{-1}(0)B_2(0))^T\Gamma(X,0)(\mathbb{F}_1 - F_2A_{22}^{-1}(0)\mathbb{A}_{21}(0)) + \qquad (21a)\\
&+(\mathbb{L}_1 + \mathbb{M}_{12}A_{22}^{-1}(0)B_2(0))^T - (L_2 - M_{22}A_{22}^{-1}(0)B_2(0))^T A_{22}^{-1}(0)\mathbb{A}_{21}(0) = 0\\
&(\mathbb{A}_{11}(0) - \mathbb{A}_{12}(0)A_{22}^{-1}(0)\mathbb{A}_{21}(0))^T U_1(X,0) + U_1^T(X,0)(\mathbb{A}_{11}(0) - \mathbb{A}_{12}(0)A_{22}^{-1}\mathbb{A}_{21}(0)) +\\
&+(\mathbb{C}_{11}(0) - \mathbb{C}_{12}(0)A_{22}^{-1}(0)\mathbb{A}_{21}(0))^T U_2(X,0)(\mathbb{C}_{11}(0) - \mathbb{C}_{12}(0)A_{22}^{-1}(0)\mathbb{A}_{21}(0)) +\\
&+(\mathbb{C}_{21}(0) - C_{22}(0)A_{22}^{-1}(0)\mathbb{A}_{21}(0))^T X_{22}(\mathbb{C}_{21}(0) - C_{22}(0)A_{22}^{-1}(0)\mathbb{A}_{21}(0)) -\\
&-(\mathbb{F}_1 - F_2A_{22}^{-1}(0)\mathbb{A}_{21}(0))^T\Gamma(X,0)(\mathbb{F}_1 - F_2A_{22}^{-1}(0)\mathbb{A}_{21}(0)) + \mathbb{M}_{11} - \mathbb{M}_{12}A_{22}^{-1}(0)\mathbb{A}_{21}(0) - \qquad (21b)\\
&-\mathbb{A}_{21}^T(0)A_{22}^{-T}(0)\mathbb{M}_{12}^T + \mathbb{A}_{21}^T(0)A_{22}^{-T}(0)M_{22}A_{22}^{-1}(0)\mathbb{A}_{21}(0) = 0
\end{aligned}$$

Using (3) written for $(\epsilon_1, \epsilon_2) = (0,0)$ we introduce the notations

$$\begin{pmatrix} A^1_{00} & A^1_{01} \\ A^1_{10} & A^1_{11} \end{pmatrix} \triangleq \mathbb{A}_{11}(0) - \mathbb{A}_{12}(0)A^{-1}_{22}(0)\mathbb{A}_{21}(0) \tag{22a}$$

$$\begin{pmatrix} C^1_{00} & C^1_{01} \\ C^1_{10} & C^1_{11} \end{pmatrix} \triangleq \mathbb{C}_{11}(0) - \mathbb{C}_{12}(0)A^{-1}_{22}(0)\mathbb{A}_{21}(0) \tag{22b}$$

$$\begin{pmatrix} C^1_{20} & C^1_{22} \end{pmatrix} \triangleq \mathbb{C}_{21}(0) - C_{22}(0)A^{-1}_{22}(0)\mathbb{A}_{21}(0) \tag{22c}$$

$$\begin{pmatrix} B^1_0 \\ B^1_1 \end{pmatrix} \triangleq \mathbb{B}_1(0) - \mathbb{A}_{12}(0)A^{-1}_{22}(0)B_2(0) \tag{22d}$$

$$\begin{pmatrix} D^1_0 \\ D^1_1 \end{pmatrix} \triangleq \mathbb{D}_1(0) - \mathbb{C}_{12}(0)A^{-1}_{22}(0)B_2(0) \tag{22e}$$

$$D^1_2 \triangleq D_2(0) - C_{22}(0)A^{-1}_{22}(0)B_2(0) \tag{22f}$$

$$\begin{aligned}\begin{pmatrix} M^1_{00} & M^1_{01} \\ (M^1_{01})^T & M^1_{11} \end{pmatrix} \triangleq\ & \mathbb{M}_{11} - \mathbb{M}_{12}A^{-1}_{22}(0)\mathbb{A}_{21}(0) - \mathbb{A}^T_{21}(0)A^{-T}_{22}(0)M^T_{12} + \\ & +\mathbb{A}^T_{21}(0)A^{-T}_{22}(0)M_{22}A^{-1}_{22}(0)\mathbb{A}_{21}(0)\end{aligned} \tag{22g}$$

$$\begin{pmatrix} L^1_0 \\ L^1_1 \end{pmatrix} \triangleq \mathbb{L}_1 - \mathbb{A}^T_{21}(0)A^{-T}_{22}(0)L_2 - (\mathbb{M}_{12} - \mathbb{A}^T_{21}(0)A^{-T}_{22}(0)M_{22})A^{-1}_{22}(0)B_2(0) \tag{22h}$$

$$\begin{aligned}R^1 =\ & R - L^T_2 A^{-1}_{22}(0)B_2(0) - B^T_2(0)A^{-T}_{22}(0)L_2 + \\ & +B^T_2(0)A^{-T}_{22}(0)M_{22}A^{-1}_{22}(0)B_2(0).\end{aligned} \tag{22i}$$

The next result allows us to reduce the number of equations and the number of unknowns of system (19).

Lemma 1. *Assume that $A_{22}(0)$ is invertible.*

(i) If $(X_{00}, X_{01}, X_{11}, X_{02}, X_{12}, X_{22}, F_0, F_1, F_2)$ *is a solution of system (19) with the property that* $A_{22}(0) + B_2(0)F_2$ *is an invertible matrix, then* $(X_{00}, X_{01}, X_{11}, X_{22}, F_0^1, F_1^1, F_2)$ *is a solution of the following system*

$$(B_0^1)^T X_{00} + (B_1^1)^T X_{01}^T + \sum_{j=0}^{2} (D_j^1)^T X_{jj} C_{j0}^1 + (L_0^1)^T + \Gamma^1(X_{00}, X_{11}, X_{22}) F_0^1 = 0 \tag{23a}$$

$$(B_1^1)^T X_{11} + \sum_{j=0}^{2} (D_j^1)^T X_{jj} C_{j1}^1 + (L_1^1)^T + \Gamma^1(X_{00}, X_{11}, X_{22}) F_1^1 = 0 \tag{23b}$$

$$B_2^T(0) X_{22} + \sum_{j=0}^{2} D_j^T(0) X_{jj} C_{j2}(0) + L_2^T + \Gamma(X_{00}, X_{11}, X_{22}) F_2 = 0 \tag{23c}$$

$$\begin{aligned}(A_{00}^1)^T X_{00} + (A_{10}^1)^T X_{01}^T + X_{00} A_{00}^1 + X_{01} A_{10}^1 + \sum_{j=0}^{2} (C_{j0}^1)^T X_{jj} C_{j0}^1 - \\ -(F_0^1)^T \Gamma^1(X_{00}, X_{11}, X_{22}) F_0^1 + M_{00}^1 = 0\end{aligned} \tag{23d}$$

$$\begin{aligned}(A_{10}^1)^T X_{11} + X_{00} A_{01}^1 + X_{01} A_{11}^1 + \sum_{j=0}^{2} (C_{j0}^1)^T X_{jj} C_{j1}^1 - \\ (F_0^1)^T \Gamma^1(X_{00}, X_{11}, X_{22}) F_1^1 + M_{01}^1 = 0\end{aligned} \tag{23e}$$

$$(A_{11}^1)^T X_{11} + X_{11} A_{11}^1 + \sum_{j=0}^{2} (C_{j1}^1)^T X_{jj} C_{j1}^1 - (F_1^1)^T \Gamma^1(X_{00}, X_{11}, X_{22}) F_1^1 + M_1^1 = 0 \tag{23f}$$

$$A_{22}^T(0) X_{22} + X_{22} A_{22}(0) + \sum_{j=0}^{2} C_{j2}^T(0) X_{jj} C_{j2}(0) - F_2^T \Gamma(X_{00}, X_{11}, X_{22}) F_2 + M_{22} = 0 \tag{23g}$$

$$\Gamma^1(X_{00}, X_{11}, X_{22}) = R^1 + \sum_{j=0}^{2} (D_j^1)^T X_{jj} D_j^1 \tag{23h}$$

$$\Gamma(X_{00}, X_{11}, X_{22}) = R + \sum_{j=0}^{2} D_j^T(0) X_{jj} D_j(0) \tag{23i}$$

where

$$F_j^1 \triangleq (I_m + F_2 A_{22}^{-1}(0) B_2(0))^{-1} (F_j - F_2 A_{22}^{-1}(0) A_{2j}(0)), \quad j = 0, 1. \tag{24}$$

(ii) If $(X_{00}, X_{01}, X_{11}, X_{22}, F_0^1, F_1^1, F_2)$ *is a solution of system (23) with the property that* $A_{22}(0) + B_2(0)F_2$ *is an invertible matrix, then* $(X_{00}, X_{01}, X_{11}, X_{02}, X_{12}, F_0, F_1, F_2)$ *is a solution of system (19) where*

$$F_j = (I_m + F_2 A_{22}^{-1}(0) B_2(0)) F_j^1 + F_2 A_{22}^{-1}(0) A_{2j}(0), \quad j = 0, 1 \tag{25}$$

and

$$\begin{aligned} X_{02} &= -[A_{20}^T(0) X_{22} + X_{00} A_{02}(0) + X_{01} A_{12}(0) + \sum_{j=0}^{2} C_{j0}^T(0) X_{jj} C_{j2}(0) - \\ &\quad -(F_0)^T (R + \sum_{j=0}^{2} D_j^T(0) X_{jj} D_j(0)) + F_2 + M_{02}] A_{22}^{-1}(0) \end{aligned} \tag{26a}$$

$$\begin{aligned} X_{12} &= -[A_{21}^T(0) X_{22} + X_{11} A_{12}(0) + \sum_{j=0}^{2} C_{j1}^T(0) X_{jj} C_{j2}(0) - \\ &\quad -F_1^T (R + \sum_{j=0}^{2} D_j^T(0) X_{jj} D_j(0)) F_2 + M_{12}] A_{22}^{-1}(0) \end{aligned} \tag{26b}$$

Proof. The result follows directly combining (21) with (19b), (19c) and taking into account (22). It is worth noticing that if $A_{22}(0)$ and $A_{22}(0) + B_2(0)F_2$ are invertible, then $I_m + F_2A_{22}^{-1}(0)B_2(0)$ is invertible too. □

Assuming that A_{11}^1 is invertible we may compute X_{01} from (23e) as:

$$\begin{aligned} X_{01} &= -[(A_{10}^1)^T X_{11} + X_{00}A_{01}^1 + \sum_{j=0}^{2}(C_{j0}^1)^T X_{jj} C_{j1}^1 - \\ &\quad -(F_0^1)^T(R^1 + \sum_{j=0}^{2}(D_j^1)^T X_{jj} D_j^1)F_1^1 + M_{01}^1](A_{11}^1)^{-1}. \end{aligned} \tag{27}$$

Substituting (27) in (23a) and (23d) we obtain after some algebraic calculation the equations:

$$\begin{aligned} &(B_{01}^1 - A_{01}^1(A_{11}^1)^{-1}B_1^1)^T X_{00} + \sum_{j=0}^{2}(D_j^1 - C_{j1}^1(A_{11}^1)^{-1}B_1^1)^T X_{jj}\times \\ &(C_{j0}^1 - C_{j1}^1(A_{11}^1)^{-1}A_{10}^1) + (I_m + F_1^1(A_{11}^1)^{-1}B_1^1)\Gamma^1(X_{00}, X_{11}, X_{22})(F_0^1 - F_1^1(A_{11}^1)^{-1}A_{10}^1) \\ &+(L_0^1 - M_{01}^1(A_{11}^1)^{-1}B_1^1)^T - (R_1^1)^T(A_{11}^1)^{-1}A_{10}^1 + (B_1^1)^T(A_{11}^1)^{-T}M_{11}^1(A_{11}^1)^{-1}A_{10}^1 = 0 \end{aligned} \tag{28a}$$

$$\begin{aligned} &(A_{00}^1 - A_{01}^1(A_{11}^1)^{-1}A_{10}^1)^T X_{00} + X_{00}(A_{00}^1 - A_{01}^1(A_{11}^1)^{-1}A_{10}^1) + \sum_{j=0}^{2}(C_{j0}^1 - C_{j1}^1(A_{11}^1)^{-1}A_{10}^1)^T \times \\ &X_{jj}(C_{j0}^1 - C_{j1}^1(A_{11}^1)^{-1}A_{10}^1) - (F_0^1 - F_1^1(A_{11}^1)^{-1}A_{10}^1)^T\Gamma^1(X_{00}, X_{11}, X_{22})(F_0^1 - F_1^1(A_{11}^1)^{-1}A_{10}^1) \\ &+M_{00}^1 - (A_{10}^1)^T(A_{11}^1)^{-T}(M_{01}^1)^T - M_{01}^1(A_{11}^1)^{-1}A_{10}^1 + (A_{10}^1)^T(A_{11}^1)^{-T}M_{11}^1(A_{11}^1)^{-1}A_{10}^1 = 0 \end{aligned} \tag{28b}$$

We introduce the notations:

$$A_{00}^0 = A_{00}^1 - A_{01}^1(A_{11}^1)^{-1}A_{10}^1 \tag{29a}$$

$$B_0^0 = B_0^1 - A_{01}^1(A_{11}^1)^{-1}B_1^1 \tag{29b}$$

$$C_{j0}^0 = C_{j0}^1 - C_{j1}^1(A_{11}^1)^{-1}A_{10}^1 \tag{29c}$$

$$D_j^0 = D_j^1 - C_j^1(A_{11}^1)^{-1}B_1^1, 0 \leq j \leq 2 \tag{29d}$$

$$\begin{aligned} M_{00}^0 &= M_{00}^1 - (A_{10}^1)^T(A_{11}^1)^{-T}(M_{01}^1)^T - M_{01}^1(A_{11}^1)^{-1}A_{10}^1 + \\ &\quad +(A_{10}^1)^T(A_{11}^1)^{-T}M_{11}^1(A_{11}^1)^{-1}A_{10}^1 \end{aligned} \tag{29e}$$

$$L_0^0 = L_0^1 - (A_{10}^1)^T(A_{11}^1)^{-T}L_1^1 - (M_{01}^1 - (A_{10}^1)^T(A_{11}^1)^{-T}M_{11}^1)(A_{11}^1)^{-1}B_1^1 \tag{29f}$$

$$R^0 = R^1 - (B_1^1)^T(A_{11}^1)^{-T}L_1^1 - (L_1^1)^T(A_{11}^1)^{-1}B_1^1 + (B_1^1)^T(A_{11}^1)^{-T}M_{11}^1(A_{11}^1)^{-1}B_1^1. \tag{29g}$$

The next result allows us to reduce the number of unknowns and the number of the equations in system (23).

Lemma 2. *Assume that the matrices $A_{22}(0)$ and A_{11}^1 are invertible.*

(i) If $(X_{00}, X_{01}, X_{11}, X_{22}, F_0^1, F_1^1, F_2)$ *is a solution of system (23) such that* $A_{11}^1 + B_1^1 F_1^1$ *is an invertible matrix, then* $(X_{00}, X_{11}, X_{22}, F_0^1, F_1^1, F_2^2)$ *is a solution of the following system:*

$$(B_{00}^0)^T X_{00} + \sum_{j=0}^{2} (D_j^0)^T X_{jj} C_{j0}^0 + (L_0^0)^T + \Gamma^0(X_{00}, X_{11}, X_{22}) F_0^0 = 0 \tag{30a}$$

$$(B_{11}^1)^T X_{11} + \sum_{j=0}^{2} (D_j^1)^T X_{jj} C_{j1}^1 + (L_1^1)^T + \Gamma^1(X_{00}, X_{11}, X_{22}) F_1^1 = 0 \tag{30b}$$

$$(B_2^2)^T X_{22} + \sum_{j=0}^{2} (D_j^2)^T X_{jj} C_{j2}^2 + (L_2^2)^T + \Gamma^2(X_{00}, X_{11}, X_{22}) F_2^2 = 0 \tag{30c}$$

$$(A_{00}^0)^T X_{00} + X_{00} A_{00}^0 + \sum_{j=0}^{2} (C_{j0}^0)^T X_{jj} C_{j0}^0 - (F_0^0)^T \Gamma^0(X_{00}, X_{11}, X_{22}) F_0^0 + M_{00}^0 = 0 \tag{30d}$$

$$(A_{11}^1)^T X_{11} + X_{11} A_{11}^1 + \sum_{j=0}^{2} (C_{j1}^1)^T X_{jj} C_{j1}^1 - (F_1^1)^T \Gamma^1(X_{00}, X_{11}, X_{22}) F_1^1 + M_{11}^1 = 0 \tag{30e}$$

$$(A_{22}^2)^T X_{22} + X_{22} A_{22}^2 + \sum_{j=0}^{2} (C_{j2}^2)^T X_{jj} C_{j2}^2 - (F_2^2)^T \Gamma^2(X_{00}, X_{11}, X_{22}) F_2^2 + M_{22}^2 = 0 \tag{30f}$$

$$\Gamma^k(X_{00}, X_{11}, X_{22}) \triangleq R^k + \sum_{j=0}^{2} (D_j^k)^T X_{jj} D_j^k, k = 0, 1, 2 \tag{30g}$$

where

$$F_0^0 \triangleq (I_m + F_1^1 (A_{11}^1)^{-1} B_1^1)^{-1} (F_0^1 - F_1^1 (A_{11}^1)^{-1} A_{10}^1) \tag{31a}$$

$$F_2^2 \triangleq F_2 \tag{31b}$$

and

$$\begin{aligned} A_{22}^2 &\triangleq A_{22}(0),\ B_2^2 \triangleq B_2(0),\ C_{j2}^2 \triangleq C_{j2}(0),\ D_j^2 \triangleq D_j(0), \quad 0 \le j \le 2, \\ L_2^2 &= L_2, \quad M_{22}^2 = M_{22}, \quad R^2 \triangleq R \end{aligned} \tag{32}$$

(ii) If $(X_{00}, X_{11}, X_{22}, F_0^0, F_1^1, F_2^2)$ *is a solution of system (30) with the property that* $A_{11}^1 + B_{11}^1 F_1^1$ *is an invertible matrix, then* $(X_{00}, X_{01}, X_{11}, X_{22}, F_0^1, F_1^1, F_2)$ *is a solution of system (23), where*

$$\begin{aligned} F_0^1 &= (I_m + F_1^1 (A_{11}^1)^{-1} B_1^1) F_0^0 + F_1^1 (A_{11}^1)^{-1} A_{10}^1 \\ F_2 &= F_2^2 \end{aligned} \tag{33}$$

and X_{01} *is computed via (27).*

Proof. The proof may be done by direct calculation implying (23), (27), (33). The notations (32) were adopted only for the sake of symmetry of the equations (30). □

For the values of X_{jj} for which the matrices $\Gamma^k(X_{00}, X_{11}, X_{22})$ are invertible, we may eliminate the unknowns F_{kk}^k from (30) obtaining the following system of nonlinear equations with the unknown $(X_0, X_1, X_2) := (X_{00}, X_{11}, X_{22})$:

$$(A_{kk}^k)^T X_k + X_k A_{kk}^k + \sum_{j=0}^{2}(C_{jk}^k)^T X_j C_{jk}^k - (X_k B_k^k + \sum_{j=0}^{2}(C_{jk}^k)^T X_j D_j^k + L_k^k) \times$$

$$(\Gamma^k(X_0, X_1, X_2))^{-1}((B_k^k)^T X_k + \sum_{j=0}^{2}(D_j^k)^T X_j C_{jk}^k + (L_k^k)^T) + M_{kk}^k = 0, \tag{34a}$$

$$\Gamma^k(X_0, X_1, X_2) = R^k + \sum_{j=0}^{2}(D_j^k)^T X_j D_j^k, \quad k = 0, 1, 2. \tag{34b}$$

Remark 2. *(a) In the deterministic case, i.e., the special case of (2) when $C_{jk}(\epsilon) = 0$, $D_j(\epsilon) = 0$, $j, k = 0, 1, 2$, system (34) reduces to*

$$(A_{kk}^k)^T X_k + X_k A_{kk}^k - (X_k B_k^k + L_k^k)(R^k)^{-1}((B_k^k)^T X_k + (L_k^k)^T) + M_k^k = 0, k = 0, 1, 2. \tag{35}$$

System (35) is a system of three uncoupled algebraic Riccati equations of lower dimensions named the system of reduced algebraic Riccati equations (for details see e.g., [24]). That is why, in the sequel, system (34) will be named **system of reduced algebraic Riccati equations** *(SRARE), associated with SARE (8). We shall see that in this stochastic framework, system (34), plays a similar role as system (35) in the deterministic case. Unlike the deterministic case, where the system of reduced algebraic Riccati Equation (35) is obtained by simply removing the small parameters ϵ_k, $k = 1, 2$ in the controlled system, in the stochastic framework SRARE (34) cannot be obtained directly by such a procedure.*

(b) When $C_{jk}(0) = 0$, $D_j(0) = 0$, $j = 1, 2$, $k = 0, 1, 2$, system (34) becomes the system of reduced algebraic Riccati equations derived in [16]. In this special case (34) is:

$$\begin{aligned}
&(A_{00}^0)^T X_0 + X_0 A_{00}^0 + (C_{00}^0)^T X_0 C_{00}^0 - (X_0 B_0^0 + (C_{00}^0)^T X_0 D_0^0 + L_0^0) \times \\
&\quad (R^0 + (D_0^0)^T X_0 D_0^0)^{-1}((B_0^0)^T X_0 + (D_0^0)^T X_0 C_{00}^0 + (L_0^0)^T) + M_{00}^0 = 0 &(36a)\\
&(A_{11}^1)^T X_1 + X_1 A_{11}^1 + (C_{01}^1)^T X_0 C_{01}^1 - (X_1 B_1^1 + (C_{01}^1)^T X_0 D_0^1 + L_1^1) \times \\
&\quad (R^1 + (D_0^1)^T X_0 D_0^1)^{-1}((B_1^1)^T X_1 + (D_0^1)^T X_0 C_{01}^1 + (L_1^1)^T) + M_{11}^1 = 0 &(36b)\\
&(A_{22}^2)^T X_2 + X_2 A_{22}^2 + (C_{02}^2)^T X_0 C_{02}^2 - (X_2 B_2^2 + (C_{02}^2)^T X_0 D_0^2 + L_2^2) \times \\
&\quad (R^2 + (D_0^2)^T X_0 D_0^2)^{-1}((B_2^2)^T X_2 + (D_0^2)^T X_0 C_{02}^2 + (L_2^2)^T) + M_{22}^2 = 0. &(36c)
\end{aligned}$$

One sees that (36a) is the SARE of type (8) associated with the stochastic reduced linear quadratic optimal control problem described by

$$dx_0(t) = (A_{00}^0 x_0(t) + B_0^0 u(t))dt + (C_{00}^0 x_0(t) + D_0^0 u(t))dw(t), \quad x_0(0) = x_0^0$$

and

$$J_0(x_0^0; u) = \mathbb{E}[\int_0^\infty (x_0^T(t) M_{00}^0 x_0(t) + 2x_0^T(t) L_0^0 u(t) + u^T(t) R^0 u(t))dt].$$

The Equations (36b) and (36c) can be interpreted as algebraic Riccati equations associated with some deterministic reduced linear quadratic control problems described by:

$$\dot{x}_k(t) = A_{kk}^k x_k(t) + B_k^k u(t)$$

$$x_k(0) = x_k^0$$

and

$$J_k(x_k^0, u) = \int_0^{\infty} (x_k^T(t)\tilde{M}_k x_k(t) + 2x_k^T(t)\tilde{L}_k u(t) + u^T(t)\tilde{R}_k u(t)dt$$

where

$$\tilde{M}_k = M_k^k + (C_{0k}^k)^T x_0 C_{0k}^k$$
$$\tilde{L}_k = L_k^k + (C_{0k}^k)^T x_0 D_0^k$$
$$\tilde{R}_k = R^k + (D_0^k)^T x_0 D_0^k, \quad k = 1, 2.$$

The solution X_0 of SARE (36a) is involved as a parameter that affects the weights matrices from the performance criteria $J_k(x_k^0; u)$.

(c) A complete decoupling of the equations from SRARE (34) may be possible in the special case when the following conditions are simultaneously satisfied:

$$D_j(0) = 0,\ j = 0,1,2, C_{jk}(0) = 0,\ k = 1,2, C_{il}(0) = 0,\ i = 1,2,\ l = 0,1,2.$$

In the next subsection we introduce the concept of stabilizing solution of SRARE (34) and we shall provide a set of conditions equivalent to the existence of that solution.

3.2. The Stabilizing Solution of the SRARE

Let $\mathfrak{X}$ be the linear space defined by $\mathfrak{X} = \mathcal{S}_{n_0} \times \mathcal{S}_{n_1} \times \mathcal{S}_{n_2}$. An element $\mathbf{X}$ lies in $\mathfrak{X}$ if and only if $\mathbf{X} = (X_0, X_1, X_2)$, X_k being symmetric matrices of size $n_k \times n_k$.
On $\mathfrak{X}$ we introduce the inner product

$$< \mathbf{X}, \mathbf{Y} >= \sum_{j=0}^{2} Tr[X_j, Y_j] \tag{37}$$

for all $\mathbf{X} = (X_0, X_1, X_2)$, $\mathbf{Y} = (Y_0, Y_1, Y_2) \in \mathfrak{X}$. In (37) $Tr[\cdot]$ is the trace operator. Equipped with the inner product (37), $\mathfrak{X}$ becomes a finite dimensional real Hilbert space.

On $\mathfrak{X}$ we consider the order relation $\succcurlyeq$ induced by the closed, solid, convex cone

$$\mathfrak{X} = \{\mathbf{X} \in \mathfrak{X} | \mathbf{X} = (X_0, X_1, X_2), X_j \geq 0, j = 0, 1, 2\}.$$

Here, $X_j \geq 0$ means that X_j is a positive semidefinite matrix. In the sequel, we rewrite SRARE (34) as a generalized Riccati equation on $\mathfrak{X}$ as

$$\begin{aligned} \mathbf{A}^T\mathbf{X} + \mathbf{X}\mathbf{A} + \mathbf{\Pi}_1[\mathbf{X}] - (\mathbf{X}\mathbf{B} + \mathbf{\Pi}_2[\mathbf{X}] + \mathbf{L}) \times & \\ (\mathbf{R} + \mathbf{\Pi}_3[\mathbf{X}])^{-1}(\mathbf{X}\mathbf{B} + \mathbf{\Pi}_2[\mathbf{X}] + \mathbf{L})^T + \mathbf{M} &= 0 \end{aligned} \tag{38}$$

where $\mathbf{A} = (A_{00}^0, A_{11}^1, A_{22}^2) \in \mathbb{R}^{n_0 \times n_0} \times \mathbb{R}^{n_1 \times n_1} \times \mathbb{R}^{n_2 \times n_2}$, $\mathbf{B} = (B_0^0, B_1^1, B_2^2) \in \mathbb{R}^{n_0 \times m} \times \mathbb{R}^{n_1 \times m} \times \mathbb{R}^{n_2 \times m}$, $\mathbf{L} = (L_0^0, L_1^1, L_2^2) \in \mathbb{R}^{n_0 \times m} \times \mathbb{R}^{n_1 \times m} \times \mathbb{R}^{n_2 \times m}$, $\mathbf{M} = (M_0^0, M_1^1, M_2^2) \in \mathfrak{X}$, $\mathbf{R} = (R^0, R^1, R^2) \in \mathcal{S}_m \times \mathcal{S}_m \times$

$\mathcal{S}_m, \mathbf{X} \to \mathbf{\Pi}_1[\mathbf{X}] : \mathfrak{X} \to \mathfrak{X}, \mathbf{X} \to \mathbf{\Pi}_2[\mathbf{X}] : \mathfrak{X} \to \mathbb{R}^{n_0 \times m} \times \mathbb{R}^{n_1 \times m} \times \mathbb{R}^{n_2 \times n}, \mathbf{X} \to \mathbf{\Pi}_3[\mathbf{X}] : \mathfrak{X} \to \mathcal{S}_m \times \mathcal{S}_m \times \mathcal{S}_m$ are defined by

$$\begin{aligned}
\mathbf{\Pi}_1[\mathbf{X}] &= \sum_{j=0}^{2}((C_{j0}^0)^T X_j C_{j0}^0, (C_{j1}^1)^T X_j C_{j1}^1, (C_{j2}^2)^T X_j X_j C_{j2}^2) \\
\mathbf{\Pi}_2[\mathbf{X}] &= \sum_{j=0}^{2}((C_{j0}^0)^T X_j D_j^0, (C_{j1}^1)^T X_j D_j^1, (C_{j2}^2)^T X_j D_j^2) \\
\mathbf{\Pi}_3[\mathbf{X}] &= \sum_{j=0}^{2}((D_j^0)^T X_j D_j^0, (D_j^1)^T X_j D_j^1, (D_j^2)^T X_j D_j^2).
\end{aligned} \tag{39}$$

Based on the operators $\mathbf{\Pi}_k$, $k = 1,2,3$ we may define the following operator $\mathbf{X} \to \mathbf{\Pi}[\mathbf{X}] \triangleq (\mathbf{\Pi}_1[\mathbf{X}], \mathbf{\Pi}_2[\mathbf{X}], \mathbf{\Pi}_3[\mathbf{X}])$.

A feedback gain is a triple of the form $\mathbf{F} = (F_0, F_1, F_2)$ where $F_k \in \mathbb{R}^{m \times n_k}$, $k = 0,1,2$. For any feedback gain $\mathbf{F}$, we associate the following linear operator: $\mathbf{X} \to \mathcal{L}_{\mathbf{F}}[\mathbf{X}] : \mathfrak{X} \to \mathfrak{X}$ by $\mathcal{L}_{\mathbf{F}}[\mathbf{X}] = (\mathcal{L}_{\mathbf{F}0}[\mathbf{X}], \mathcal{L}_{\mathbf{F}1}[\mathbf{X}], \mathcal{L}_{\mathbf{F}2}[\mathbf{X}])$, where for each $k = 0,1,2$ we have:

$$\mathcal{L}_{\mathbf{F}k}[\mathbf{X}] = (A_{kk}^k + B_k^k F_k)X_k + X_k(A_{kk}^k + B_k^k F_k)^T + \sum_{j=0}^{2}(C_{jk}^k + D_j^k F_k)X_j(C_{jk}^k + D_j^k F_k)^T. \tag{40}$$

The next result summarizes some useful properties of the operator $\mathcal{L}_{\mathbf{F}}$.

Proposition 1. *(i) The adjoint operator $\mathcal{L}_{\mathbf{F}}^*$ of the operator $\mathcal{L}_{\mathbf{F}}$ (with respect to the inner product (37)) is given by $\mathcal{L}_{\mathbf{F}}^*[\mathbf{X}] = (\mathcal{L}_{\mathbf{F}0}^*[\mathbf{X}], \mathcal{L}_{\mathbf{F}1}^*[\mathbf{X}], \mathcal{L}_{\mathbf{F}2}^*[\mathbf{X}])$, where for each $k = 0,1,2$:*

$$\mathcal{L}_{\mathbf{F}k}^*[\mathbf{X}] = (A_{kk}^k + B_k^k F_k)^T X_k + X_k(A_{kk}^k + B_k^k F_k) + \sum_{j=0}^{2}(C_{jk}^k + D_j^k F_k)^T X_j(C_{jk}^k + D_j^k F_k). \tag{41}$$

(ii) The operator $\mathcal{L}_{\mathbf{F}}$ generates positive evolution on the space $\mathfrak{X}$ i.e., $e^{\mathcal{L}_{\mathbf{F}} t}\mathfrak{X}_+ \subset \mathfrak{X}_+$ for all $t \geq 0$.

(iii) The spectrum of the linear operator $\mathcal{L}_{\mathbf{F}}$ is located in the half plane $\mathbb{C}_- = \{\lambda \in \mathbb{C}, Re\lambda < 0\}$ if and only if there exists $\mathbf{Y} = (Y_0, Y_1, Y_2) \succ 0$ such that $\mathcal{L}_{\mathbf{F}}[\mathbf{Y}] \prec 0$.

Proof. (i) follows by direct calculation specializing the definition of the adjoint operator to the case of the operator defined in (40) and the inner product (37).

(ii) follows applying Corollary 2.2.6 from [23].

(iii) follows from the equivalence $(iv) \leftrightarrow (v)$ in the Corollary 2.3.9 from [23]. □

Now we are in the position to introduce the concept of stabilizing solution of SRARE (34).

Definition 1. *A solution $\tilde{\mathbf{X}} = (\tilde{X}_0, \tilde{X}_1, \tilde{X}_2)$ of SRARE (34) is named* **stabilizing solution** *if the spectrum of the linear operator $\mathcal{L}_{\tilde{\mathbf{F}}}$ is located in the half plane $\mathbb{C}_-$, $\mathcal{L}_{\tilde{\mathbf{F}}}$ being the linear operator of type (40) associated with the feedback gain $\tilde{\mathbf{F}} = (\tilde{F}_0, \tilde{F}_1, \tilde{F}_2)$, where for each $k = 0,1,2$*

$$\tilde{F}_k \triangleq -(R^k + \sum_{j=0}^{2}(D_j^k)^T \tilde{X}_j D_j^k)^{-1}((B_k^k)^T \tilde{X}_k + \sum_{j=0}^{2}(D_j^k)^T \tilde{X}_j C_{jk}^k + (L_k^k)^T). \tag{42}$$

Before stating the result providing the conditions which guarantee the existence of the stabilizing solution of SRARE (34), we introduce the concept of stabilizability of the triple $(\mathbf{A}, \mathbf{B}, \mathbf{\Pi})$.

Definition 2. *We say that the triple* $(\mathbf{A}, \mathbf{B}, \mathbf{\Pi})$ *is* **stabilizable** *if there exists a feedback gain* $\mathbf{F} = (F_0, F_1, F_2)$ *with the property that the spectrum of the corresponding linear operator* $\mathcal{L}_{\mathbf{F}}$ *of type (40) is inclosed in the half-plane* $\mathbb{C}_-$.

The next result provides a set of conditions equivalent to the stabilizability of the triple $(\mathbf{A}, \mathbf{B}, \mathbf{\Pi})$.

Proposition 2. *The following are equivalent:*

(i) the triple $(\mathbf{A}, \mathbf{B}, \mathbf{\Pi})$ *is stabilizable,*

(ii) there exist $\mathbf{Y} = (Y_0, Y_1, Y_2)$, $\mathbf{Z} = (Z_0, Z_1, Z_2)$, $Y_k \in \mathcal{S}_{n_k}$, $Y_k > 0$, $Z_k \in \mathbb{R}^{m \times n_k}$, $k = 0, 1, 2$, *satisfying the following system of LMIs:*

$$\begin{pmatrix} \Xi_1^k(\mathbf{Y}, \mathbf{Z}) & \Xi_2^k(\mathbf{Y}, \mathbf{Z}) \\ (\Xi_2^k(\mathbf{Y}, \mathbf{Z}))^T & \Xi_3^k(\mathbf{Y}) \end{pmatrix} < 0 \tag{43}$$

where

$$\Xi_1^k(\mathbf{Y}, \mathbf{Z}) = A_{kk}^k Y_k + Y_k (A_{kk}^k)^T + B_k^k Z_k + Z_k^T (B_k^k)^T$$

$$\Xi_2^k(\mathbf{Y}, \mathbf{Z}) = \begin{pmatrix} C_{0k}^k Y_0 + D_0^k Z_0 & C_{1k}^k Y_1 + D_1^k Z_1 & C_{2k}^k Y_2 + D_2^k Z_2 \end{pmatrix}, k = 0, 1, 2$$

$$\Xi_3^k(\mathbf{Y}) = diag(-Y_0, -Y_1, -Y_2).$$

Furthermore, if $(\mathbf{Y}, \mathbf{Z})$ *is a solution of the system of LMIs (43), then* $\mathbf{F} = (Z_0 Y_0^{-1}, Z_1 Y_1^{-1}, Z_2 Y_2^{-1})$ *is a stabilizing feedback gain.*

Proof. Following from (iii) of Proposition 3 combined with Schur complement technique. □

To obtain the asymptotic structure of the stabilizing solution of SARE (8) satisfying the sign condition (9), we shall look for conditions under which SRARE (34) has a stabilizing solution $\tilde{\mathbf{X}} = (\tilde{X}_0, \tilde{X}_1, \tilde{X}_2)$ satisfying the sign conditions

$$R^k + \sum_{j=0}^{2} (D_j^k)^T \tilde{X}_j D_j^k > 0, \quad k = 0, 1, 2. \tag{44}$$

Theorem 1. *Assume that the matrices* $A_{22}^2 \triangleq A_{22}(0)$ *and* $A_{11}^1 \triangleq A_{11}(0) - A_{12}(0) A_{22}^{-1}(0) A_{21}(0)$ *are invertible. Under these conditions the following are equivalent:*

(i) (a) the triple $(\mathbf{A}, \mathbf{B}, \mathbf{\Pi})$ *is stabilizable,*

(b) there exists $\mathbf{Y} = (Y_0, Y_1, Y_2) \in \mathfrak{X}$ *satisfying the following system of LMIs*

$$\begin{pmatrix} \Theta_{1k}(\mathbf{Y}) + M_{kk}^k & \Theta_{2k}(\mathbf{Y}) + L_k^k \\ (\Theta_{2k}(\mathbf{Y}) + L_k^k)^T & \Theta_{3k}(\mathbf{Y}) + R^k \end{pmatrix} > 0 \tag{45}$$

where

$$\Theta_{1k}(\mathbf{Y}) = (A_{kk}^k)^T Y_k + Y_k A_{kk}^k + \sum_{j=0}^{2} (C_{jk}^k)^T Y_j C_{jk}^k$$

$$\Theta_{2k}(\mathbf{Y}) = Y_k B_k^k + \sum_{j=0}^{2} (C_{jk}^k)^T Y_j D_j^k$$

$$\Theta_{3k}(\mathbf{Y}) = \sum_{j=0}^{2} (D_j^k)^T Y_j D_j^k, k = 0, 1, 2,$$

(ii) the SRARE (34) has a unique stabilizing solution $\tilde{\mathbf{X}} = (\tilde{X}_0, \tilde{X}_1, \tilde{X}_2)$ *satisfying the sign conditions (44).*

Proof. (hint) $(i) \Rightarrow (ii)$. For each $p = 0, 1, \ldots$ one computes $\mathbf{X}^{p+1} = (X_0^{p+1}, X_1^{p+1}, X_2^{p+1})$ as the unique solution of the linear equation on $\mathfrak{X}$

$$\mathcal{L}^*_{\mathbf{F}^p}[\mathbf{X}^{p+1}] + \mathbf{M} + \mathbf{L}\mathbf{F}^p + (\mathbf{F}^p)^T\mathbf{L}^T + (\mathbf{F}^p)^T\mathbf{R}\mathbf{F}^p + \frac{\gamma^2}{p+1}\mathbb{I} = 0 \tag{46}$$

where $\mathcal{L}^*_{\mathbf{F}^p}$ is the adjoint of the linear operator $\mathcal{L}_{\mathbf{F}^p}$ described by (41) with $\mathbf{F}$ replaced by $\mathbf{F}^p$, $\mathbb{I} = (I_{n_0}, I_{n_1}, I_{n_2}) \in \mathfrak{X}$. In (46), $\mathbf{F}^p = (F_0^p, F_1^p, F_2^p)$ are given by

$$F_k^p = -(R^k + \sum_{j=0}^{2}(D_j^k)^T X_j^p D_j^k)^{-1}((B_k^k)^T X_k^p + \sum_{j=0}^{2}(D_j^k)^T X_j^p C_{jk}^k + (L_k^k)^T), \ \ p \geq 1. \tag{47}$$

When $p = 0$ the feedback gain $\mathbf{F}^0 = (F_0^0, F_1^0, F_2^0)$ is obtained based on the assumption of stabilizability of the triple $(\mathbf{A}, \mathbf{B}, \mathbf{\Pi})$. It has the property that the spectrum of the corresponding linear operator $\mathcal{L}_{\mathbf{F}^0}$ is located in the half place $\mathbb{C}_-$. One shows inductively for $p = 1, 2, \ldots$ that the spectrum of each operator $\mathcal{L}_{\mathbf{F}^p}$ is located in the half plane $\mathbb{C}_-$. Hence, $\mathbf{X}^{p+1}$ is well defined as the unique solution of the linear Equation (46). Moreover, based on the assumption (i) b) from the statement, one gets that $\mathbf{X}^p \succeq \mathbf{X}^{p+1} \succeq \mathbf{Y}, \forall p \geq 0$, where $\mathbf{Y}$ is a solution of the LMIs (45). So, we have obtained that the sequence $\{\mathbf{X}^p\}_{p \geq 0}$ is convergent.

We set $\tilde{\mathbf{X}} = \lim_{p \to \infty} \mathbf{X}^p$. One proves that under the considered assumptions, $\tilde{\mathbf{X}}$ obtained in this way, is just the stabilizing solution of SRARE (34). Since, $\tilde{\mathbf{Y}} \succeq \mathbf{Y}$, where $\mathbf{Y}$ is a solution of (45), it follows that $\tilde{\mathbf{X}}$ satisfies the sign conditions (44). The uniqueness of the stabilizing solution of SRARE (34) that satisfies the sign condition (44) is a direct consequence of its maximality property.

The proof of the implication $(ii) \to (i)$ is based on the fact that if the Riccati equation of type (38) has a stabilizing solution $\tilde{\mathbf{X}}$, satisfying the sign condition (44), then the algebraic Riccati type equation obtained replacing in (38) the term $\mathbf{M}$ by $\mathbf{M} + \delta\mathbb{I}$ has a small enough solution for $\delta < 0$. The details are omitted. □

Remark 3. *The iterations described by (46) and (47) can be used for numerical computation of the stabilizing solution* $(\tilde{X}_0, \tilde{X}_1, \tilde{X}_2)$ *of SRARE (34) satisfying the sign conditions (44).*

4. The Main Results

4.1. The Asymptotic Structure of the Stabilizing Solution of SARE

In this section we shall use the stabilizing solution of SRARE (34) to derive the asymptotic structure of the stabilizing solution of SARE (8) satisfying the sign condition (9).
Let $\tilde{\mathbf{X}} = (\tilde{X}_0, \tilde{X}_1, \tilde{X}_2)$ be the stabilizing solution of SRARE (34) satisfying the sign conditions (44). Let $\tilde{\mathbf{F}} = (\tilde{F}_0, \tilde{F}_1, \tilde{F}_2)$ be the corresponding stabilizing feedback gain associated via (42). We set

$$\tilde{F}_0^1 \triangleq (I_m + \tilde{F}_1(A_{11}^1)^{-1}B_1^1)\tilde{F}_0 + \tilde{F}_1(A_{11}^1)^{-1}A_{10}^1 \tag{48a}$$

$$\tilde{F}_1^1 \triangleq \tilde{F}_1 \tag{48b}$$

$$\begin{aligned}\tilde{X}_{01} \triangleq & -[(A_{10}^1)^T\tilde{X}_1 + \tilde{X}_0 A_{01}^1 + \sum_{j=0}^{2}(C_{j0}^1)^T\tilde{X}_j C_{j1}^1 - \\ & -(\tilde{F}_0^1)^T(R^1 + \sum_{j=0}^{2}(D_j^1)^T\tilde{X}_j D_j^1)\tilde{F}_1^1 + M_{01}^1](A_{11}^1)^{-1}\end{aligned} \tag{49}$$

From (42), (48) and (49) we obtain via Lemma 2 (ii) that $(\tilde{X}_0, \tilde{X}_{01}, \tilde{X}_1, \tilde{X}_2, \tilde{F}_0^1, \tilde{F}_1^1, \tilde{F}_2)$ is a solution of system (23). To this end, we took into account that if the eigenvalues of the linear operator $\mathcal{L}_{\tilde{\mathbf{F}}}$ are inclosed in the half plane $\mathbb{C}_-$ then the matrix $A_{11}^1 + B_1^1\tilde{F}_1^1$ is a Hurwitz matrix. Hence, it is invertible.

Further, we define

$$\tilde{\tilde{F}}_j \triangleq (I_m + \tilde{F}_2 A_{22}^{-1}(0)B_2(0))\tilde{F}_j^1 + \tilde{F}_2 A_{22}^{-1}(0)A_{2j}(0), j = 0,1 \tag{50a}$$

$$\tilde{\tilde{F}}_2 \triangleq \tilde{F}_2 \tag{50b}$$

$$\begin{aligned}\tilde{X}_{02} \triangleq\ & -[A_{20}^T(0)\tilde{X}_2 + \tilde{X}_0 A_{02}(0) + \tilde{X}_{01}A_{12}(0) + \sum_{j=0}^{2} C_{j0}^T(0)\tilde{X}_j C_{j2}(0) - \\ & -\tilde{\tilde{F}}_0^T(R + \sum_{j=0}^{2} D_j^T(0)\tilde{X}_j D_j(0))\tilde{\tilde{F}}_2 + M_{02}]A_{22}^{-1}(0)\end{aligned} \tag{51a}$$

$$\begin{aligned}\tilde{X}_{12} \triangleq\ & -[A_{21}^T(0)\tilde{X}_2 + \tilde{X}_1 A_{12}(0) + \sum_{j=0}^{2} C_{j1}^T(0)\tilde{X}_j C_{j2}(0) - \\ & -\tilde{\tilde{F}}_1^T(R + \sum_{j=0}^{2} D_j^T(0)\tilde{X}_j D_j(0))\tilde{\tilde{F}}_2 + M_{12}]A_{22}^{-1}(0).\end{aligned} \tag{51b}$$

Since the eigenvalues of the linear operator $\mathcal{L}_{\tilde{\mathbf{F}}}$ are in the half plane $\mathbb{C}_-$, we deduce via (50b) that the matrix $A_{22}(0) + B_2(0)\tilde{\tilde{F}}_2$ is a Hurwitz matrix. Hence, it is invertible.

Applying Lemma 1 (ii), we deduce that $(\tilde{X}_0, \tilde{X}_{01}, \tilde{X}_1, \tilde{X}_{02}, \tilde{X}_{12}, \tilde{X}_2, \tilde{\tilde{F}}_0, \tilde{\tilde{F}}_1, \tilde{\tilde{F}}_2)$ is a solution of system (19) constructed starting from the stabilizing solution $(\tilde{X}_0, \tilde{X}_1, \tilde{X}_2)$ of SRARE (34).

Now, we are in the position to state the first main result of this paper:

Theorem 2. *Assume: (a) the assumptions* $\mathbf{H}_1$) *and* $\mathbf{H}_2$) *are fulfilled;*

(b) the matrices $A_{22}(0)$ *and* $A_{11}(0) - A_{12}(0)A_{22}^{-1}(0)A_{21}(0)$ *are invertible;*

(c) conditions from (i) of Theorem 1 are fulfilled.

Under these conditions there exists $\mu^* > 0$ *with the property that for any* $\epsilon_k > 0$, $k = 1,2$, *such that* $0 < \epsilon_1 + \epsilon_2 + \frac{\epsilon_2}{\epsilon_1} \leq (\mu^*)^2$, *the SARE (8) has a stabilizing solution* $\tilde{X}(\epsilon_1, \epsilon_2)$ *satisfying the sign condition (9). Furthermore* $\tilde{X}(\epsilon_1, \epsilon_2)$ *and the corresponding stabilizing feedback gain* $\tilde{F}(\epsilon_1, \epsilon_2)$ *have the asymptotic structure:*

$$\tilde{X}(\epsilon_1, \epsilon_2) = \begin{pmatrix} \tilde{X}_1 + O(\mu) & \epsilon_1(\tilde{X}_{01} + O(\mu)) & \epsilon_2(\tilde{X}_{02} + O(\mu)) \\ \epsilon_1(\tilde{X}_{01} + O(\mu))^T & \epsilon_1(\tilde{X}_1 + O(\mu)) & \epsilon_2(\tilde{X}_{12} + O(\mu)) \\ \epsilon_2(\tilde{X}_{02} + O(\mu))^T & \epsilon_2(\tilde{X}_{12} + O(\mu))^T & \epsilon_2(\tilde{X}_2 + O(\mu)) \end{pmatrix} \tag{52}$$

$$\tilde{F}(\epsilon_1, \epsilon_2) = \begin{pmatrix} \tilde{\tilde{F}}_0 + O(\mu) & \tilde{\tilde{F}}_1 + O(\mu) & \tilde{\tilde{F}}_2 + O(\mu) \end{pmatrix} \tag{53}$$

where $\mu = (\epsilon_1 + \epsilon_2 + \frac{\epsilon_2}{\epsilon_1})^{\frac{1}{2}}$, $(\tilde{X}_{01}, \tilde{X}_{02}, \tilde{X}_{12})$ *being computed by (49) and (51) based on the stabilizing solution* $(\tilde{X}_0, \tilde{X}_1, \tilde{X}_2)$ *of SRARE (34) satisfying the sign conditions (44) and* $\tilde{\tilde{F}}_k$ *are computed by (48) and (50) starting from the stabilizing feedback gains* $\tilde{F}_j$, $j = 0,1,2$, *associated with the stabilizing solution of SRARE (34).*

Proof. The existence of the stabilizing solution $\tilde{X}(\epsilon_1, \epsilon_2)$, as well as the asymptotic structure from (52) and (53), are obtained applying the implicit functions theorem in the case of system (18). To this end, we regard system (18) as an equation of the form:

$$\Phi(\mathbb{W}, \xi) = 0 \tag{54}$$

on the finite dimensional Banach space $\mathfrak{W} \triangleq \mathcal{S}_{n_0} \times \mathbb{R}^{n_0 \times n_1} \times \mathcal{S}_{n_1} \times \mathbb{R}^{n_0 \times n_2} \times \mathbb{R}^{n_1 \times n_2} \times \mathcal{S}_{n_2} \times \mathbb{R}^{m \times n_0} \times \mathbb{R}^{m \times n_1} \times \mathbb{R}^{m \times n_2}$. In (54), $\mathbb{W} = (X_0, X_{01}, X_1, X_{02}, X_{12}, X_2, F_0, F_1, F_2)$ and $\xi = (\sqrt{\epsilon_1}, \sqrt{\epsilon_2}, \sqrt{\frac{\epsilon_2}{\epsilon_1}})$. From (18) one sees that $\mathbb{W} \to \Phi(\mathbb{W}, \xi)$ is a C^∞-function and from the assumption $\mathbf{H}_2)$ we have that $\xi \to \Phi(\mathbb{W}, \xi)$ is a C^1-function in a neighborhood of the origin $\mathbf{0} = (0,0,0)$. We also remark that the reduced equation $\Phi(\mathbb{W}, \mathbf{0}) = 0$ coincides with system (19). So, from the developments in the first part of this section we deduce that $(\tilde{\mathbb{W}}, \mathbf{0})$ is a solution of the Equation (54) when $\tilde{\mathbb{W}} = (\tilde{X}_0, \tilde{X}_{01}, \tilde{X}_1, \tilde{X}_{02}, \tilde{X}_{12}, \tilde{X}_2, \tilde{\tilde{F}}_0, \tilde{\tilde{F}}_1, \tilde{\tilde{F}}_2)$. Let $\mathbf{\Phi}_{\mathbb{W}}(\tilde{\mathbb{W}}, \mathbf{0})$ be the partial derivative of $\mathbf{\Phi}(\mathbb{W}, \xi)$ evaluated in $(\mathbb{W}, \xi) = (\tilde{\mathbb{W}}, \mathbf{0})$.

First we show that the operator $\hat{\mathbb{W}} \to \gneqq_{\mathbb{W}}(\tilde{\mathbb{W}}, \mathbf{0})[\hat{\mathbb{W}}] : \mathfrak{W} \to \mathfrak{W}$ is injective.

To this end we consider the linear equation

$$\Phi_{\mathbb{W}}(\tilde{\mathbb{W}}, \mathbf{0})[\hat{\mathbb{W}}] = 0 \tag{55}$$

with the unknowns $\hat{\mathbb{W}} = (\hat{X}_0, \hat{X}_{01}, \hat{X}_1, \hat{X}_{02}, \hat{X}_{12}, \hat{X}_2, \hat{F}_0, \hat{F}_1, \hat{F}_2) \in \mathfrak{W}$. After some algebraic manipulations one obtains that (55) reduces to the linear equation

$$\mathcal{L}^*_{\tilde{\mathbf{F}}}[\hat{\mathbf{Y}}] = 0 \tag{56}$$

with the unknowns $\hat{\mathbf{Y}} = (\hat{X}_0, \hat{X}_1, \hat{X}_2) \in \mathfrak{X}$ and $\mathcal{L}_{\tilde{\mathbf{F}}}$ is the operator of type (40) associated with the stabilizing feedback gain $\tilde{\mathbf{F}} = (\tilde{F}_0, \tilde{F}_1, \tilde{F}_2)$. Equation (56) only has the solution $\hat{\mathbf{Y}} = (0,0,0)$ because the spectrum of the linear operator $\mathcal{L}_{\tilde{\mathbf{F}}}$ lies in the half plane $\mathbb{C}_-$. Finally, one obtains that Equation (55) has only the zero solution. This means that the kernel of the linear operator $\hat{\mathbb{W}} \to \Phi_{\mathbb{W}}(\tilde{\mathbb{W}}, \mathbf{0})[\hat{\mathbb{W}}]$ is the null subspace. Since $\mathfrak{W}$ is a finite dimensional vector space, we may conclude that $\hat{\mathbb{W}} \to \Phi_{\mathbb{W}}(\tilde{\mathbb{W}}, \mathbf{0})[\hat{\mathbb{W}}]$ is invertible. Hence, we may apply the implicit functions theorem (see [26]) in the case of Equation (54). This allows us to deduce that there exist $\mu_0 > 0$ and a C^1-function $\xi \to \mathbb{W}(\xi) : \mathcal{B}(0, \mu_0) \to \mathfrak{W}$, which satisfy $\Phi(\mathbb{W}(\xi), \xi) = 0$, for all $\xi \in \mathcal{B}(0, \mu_0) \triangleq \{\xi \in \mathbb{R}^3 | |\xi| < \mu_0\}$. Further, $\mathbb{W}(\xi) = \tilde{\mathbb{W}} + O(|\xi|)$, which yields

$$\begin{aligned} X_j(\epsilon_1, \epsilon_2) &= \tilde{X}_j + O(|\xi|), 0 \le j \le 2 \\ X_{01}(\epsilon_1, \epsilon_2) &= \tilde{X}_{01} + O(|\xi|), \\ X_{k2}(\epsilon_1, \epsilon_2) &= \tilde{X}_{k2} + O(|\xi|), k = 0,1 \\ F_l(\epsilon_1, \epsilon_2) &= \tilde{\tilde{F}}_l + O(|\xi|), 0 \le l \le 2. \end{aligned} \tag{57}$$

Plugging (57) into (12) we obtain (52), (53). We also obtain that $(\tilde{X}(\epsilon_1, \epsilon_2), \tilde{F}(\epsilon_1, \epsilon_2))$, constructed as above, satisfies (10). On the other hand, from (18f) and (52) we deduce that there exists $0 \le \mu_1 \le \mu_0$ with the property that $\tilde{X}(\epsilon_1, \epsilon_2)$ satisfies (9) for any $\epsilon_1 > 0, \epsilon_2 > 0$, such that $\epsilon_1 + \epsilon_2 + \frac{\epsilon_2}{\epsilon_1} < \mu_1^2$. Thus, we have obtained that $\tilde{X}(\epsilon_1, \epsilon_2)$ with the asymptotic structure given in (52) is a solution of SARE (8) which satisfies (9).

By a standard argument, based on singular perturbations technique, one shows that there exists $0 < \mu^* \le \mu_1$ such that the closed-loop system (11), where $\tilde{F}_1(\epsilon_1, \epsilon_2)$ has the asymptotic structure (53), is ESMS. Therefore, $\tilde{X}(\epsilon_1, \epsilon_2)$ defined by (52) is just the stabilizing solution of (8) for any $\epsilon_1 > 0, \epsilon_2 > 0$ such that $\epsilon_1 + \epsilon_2 + \frac{\epsilon_2}{\epsilon_1} \le (\mu^*)^2$. Thus the proof is complete. □

In the sequel, $(\tilde{\tilde{F}}_0 \quad \tilde{\tilde{F}}_1 \quad \tilde{\tilde{F}}_2)$ will be named dominant part of the stabilizing feedback gain.

4.2. A Near Optimal Control

In this subsection, we show that the dominant part of the optimal gain matrix $\tilde{F}(\epsilon_1, \epsilon_2)$ can be used to obtain a near optimal stabilizing feedback gain for the optimal control problem described by the quadratic functional (1) and the stochastic controlled system (2).

We consider the control

$$u_{app}(t) = \tilde{\tilde{F}}_0 x_0(t) + \tilde{\tilde{F}}_1 x_1(t) + \tilde{\tilde{F}}_2 x_2(t) \quad (58)$$

$\tilde{\tilde{F}}$ being constructed by (48) and (50) based on the stabilizing feedback gains $\tilde{F}_j$ associated with the stabilizing solution $(\tilde{X}_0, \tilde{X}_1, \tilde{X}_2)$ of SRARE (34).

Setting, $\mathbf{F}_{app} = \left(\tilde{\tilde{F}}_0, \tilde{\tilde{F}}_1, \tilde{\tilde{F}}_2\right)$ we may rewrite (58) in the following compact form:

$$u_{app}(t) = \mathbf{F}_{app}\mathbf{x}(t). \quad (59)$$

Substituting (59) in (6) we obtain the closed-loop system

$$d\mathbf{x}(t) = (\mathbb{A}(\epsilon) + \mathbb{B}(\epsilon)\mathbf{F}_{app})\mathbf{x}(t)dt + (\mathbb{C}(\epsilon) + \mathbb{D}(\epsilon)\mathbf{F}_{app})\mathbf{x}(t)dw(t). \quad (60)$$

The next result provides an upper bound of the deviation of the value $J(\mathbf{x}_0; u_{app})$ from the minimal value $J(\mathbf{x}_0, u_{opt})$.

Theorem 3. *Assume that the assumptions of Theorem 2 are fulfilled. Then there exist $\tilde{\mu} > 0$ such that the closed-loop system (60) is ESMS for any $\epsilon_k > 0$, $k = 1, 2$ which satisfy $\epsilon_1 + \epsilon_2 + \frac{\epsilon_2}{\epsilon_1} < \tilde{\mu}^2$. Moreover, the loss of the performance produced by the use of the control (59) instead of the optimal control (7) is given by*

$$0 \leq J(\mathbf{x}_0, u_{app}) - J(\mathbf{x}_0, u_{opt}) \leq \gamma(\epsilon_1 + \epsilon_2 + \frac{\epsilon_2}{\epsilon_1})|\mathbf{x}_0|^2.$$

Proof. This may be done following a similar technique as the one used in [12] in the case of a single fast time scale. The details are omitted. □

5. Conclusions

The goal of the work has been the derivation of the asymptotic structure of the stabilizing solution of an algebraic Riccati equation arising in connection with a stochastic linear quadratic optimal control problem for a controlled system described by singularly perturbed Itô differential equations with two fast time scales.

The main conclusion of our study is that, in the stochastic case when the controlled system contains state multiplicative and/or control multiplicative white noise perturbations, the reduced system of algebraic Riccati equations cannot be directly obtained by neglecting the small parameters associated with the fast time scales of the controlled system as in the deterministic framework.

In Section 3 we have shown in detail how the system of reduced algebraic Riccati equations can be defined in the considered stochastic framework. In the second part of Section 3, we have introduced the concept of a stabilizing solution of SRARE, and we have provided a set of conditions equivalent to the existence of this kind of solution of SRARE which satisfy a prescribed sign condition of type (44). Employing the stabilizing solution of SRARE, as well as the corresponding stabilizing feedback gains, we have obtained the asymptotic structure of the stabilizing solution of SARE and of the corresponding stabilizing feedback gain. The dominant part of the stabilizing feedback gain was used to construct a near optimal control whose gain matrices do not depend upon the small parameters associated with the fast time scales. The extension of the study to the case of singularly perturbed linear stochastic systems with N fast time scales, also including more complex systems such as jump Markov perturbations [27], Levy noise perturbations [28] and semi-Markov switched systems [29] remains a challenge for future research.

Funding: This research received no external funding.

Conflicts of Interest: The authors declare no conflict of interest.

References

1. Dragan, V.; Halanay, A. *Stabilization of Linear Systems*; Birkhauser: Boston, MA, USA, 1999.
2. Gajic, Z.; Lim, M.T. *Optimal Control of Singularly Perturbed Linear Systems and Applications*; Marcel Dekker Inc.: New York, NY, USA, 2001.
3. Kokotovic, P.V.; Khalil, H.; Oreilly, J. *Singular Perturbations Methods in Control: Analysis and Design*; Academic Press: London, UK, 1986.
4. Naidu, D.S. *Singular Perturbation Methodology in Control Systems*; Peter Peregrinus Limited: Stevenage, UK, 1988.
5. Tichonov, A. On the dependence of the solutions of differential equations on a small parameter. *Math. Sbornik* **1948**, *22*, 193–204.
6. Dmitriev, M.G.; Kurina, G.A. Singular perturbations in control problems. *Autom. Remote Control* **2006**, *67*, 1–43. [CrossRef]
7. Fridman, E. Robust sampled-data H_∞ control of linear singularly perturbed systems. *IEEE Trans. Autom. Control* **2006**, *51*, 470–475. [CrossRef]
8. Glizer, V.Y. Controllability conditions of linear singularly perturbed systemswith small state and input delays. *Math. Control Signals Syst.* **2016**, *28*, 1–29. [CrossRef]
9. Glizer, V.Y. Euclidean space output controllability of singularly perturbed systems with small state delays. *J. Appl. Math. Comput.* **2018**, *57*, 1–38. [CrossRef]
10. Glizer, V.Y.; Kelis, O. Asymptotic properties of an infinite horizon partial cheap control problem for linear systems with known disturbances. *Numer. Algebr. Control Optim.* **2018**, *8*, 211–235. [CrossRef]
11. Dragan, V. The linear quadratic optimization problem for a class of singularly perturbed stochastic systems. *Int. J. Innov. Comput. Inf. Control* **2005**, *1*, 53–64.
12. Dragan, V.; Mukaidani, H.; Shi, P. The Linear Quadratic Regulator Problem for a Class of Controlled Systems Modeled by Singularly Perturbed Itô Differential Equations. *SIAM J. Control Optim.* **2012**, *50*, 448–470. [CrossRef]
13. Glizer, V.Y. Stochastic singular optimal control problem with state delays: Regularization, singular perturbation and minimizing sequence. *SIAM J. Contr. Optim.* **2012**, *50*, 2862–2888. [CrossRef]
14. Sagara, M.; Mukaidani, H.; Dragan, V. Near-Optimal Control for Multiparameter Singularly Perturbed Stochastic Systems. *J. Optim. Control Appl. Methods* **2011**, *32*, 113–125. [CrossRef]
15. Pan, Z.; Basar, T. Multi-time scale zero-sum differential games with perfect states measurements. *Dyn. Control* **1995**, *5*, 7–29. [CrossRef]
16. Dragan, V. Near optimal linear quadratic regulator for controlled systems described by Itô differential equations with two fast time scales. *Ann. Acad. Rom. Sci. Ser. Math. Appl.* **2017**, *9*, 89–109.
17. Khalil, H.K.; Kokotovic, P.V. Control of linear systems with multiparameter singular erturbations. *Automatica* **1979**, *15*, 197–207. [CrossRef]
18. Hoppensteadt, F.C. On Systems of Ordinary Differential Equations with Several Parameters Multiplying the Derivatives. *J. Differ. Eq.* **1969**, *5*, 106–116. [CrossRef]
19. Dragan, V.; Halanay, A. Uniform asymptotic expansions for the fundamental matrix singularly perturbed linear systems and applications. In *Asymptotic Analysis II, Lecture Notes in Math*; Verhulst, F., Ed.; Springer: New York, NY, USA, 1983; pp. 215–347.
20. Wonham, W.M. On a matrix Riccati equation of stochastic control. *SIAM J. Control Optim.* **1968**, *6*, 312–326. [CrossRef]
21. Chen, S.; Li, X.; Zhou, X.Y. Stochastic linear quadratic regulators with indefinite control weight costs. *SIAM J. Control Optim.* **1998**, *36*, 1685–1702. [CrossRef]
22. Rami, M.A.; Zhou, X.Y. Linear matrix inequalities, Riccati equations, indefinite stochastic linear quadratic controls. *IEEE Trans. Autom. Control* **2000**, *45*, 1131–1143. [CrossRef]
23. Dragan, V.; Morozan, T.; Stoica, A.M. *Mathematical Methods in Robust Control of Linear Stochastic Systems*, 2nd ed.; Springer: New York, NY, USA, 2013.
24. Dragan, V.; Halanay, A. Suboptimal stabilization of linear systems with several time scales. *Int. J. Control* **1982**, *36*, 109–126. [CrossRef]

25. Dragan, V.; Mukaidani, H.; Shi, P. Near optimal linear quadratic regulator for a class of stochastic systems modeled by singularly perturbed Itô differential equations with state and control multiplicative white noise. *ICIC Express Lett.* **2012**, *6*, 595–602.
26. Schwartz, L. *Analyse Mathematique*; Hermna: Paris, France, 1967; Volumes I, II.
27. Wang, B.; Zhu, Q. Stability analysis of Markov switched stochastic differential equations with both stable and unstable subsystems. *Syst. Control Lett.* **2017**, *105*, 55–61. [CrossRef]
28. Zhu, Q. Stability analysis of stochastic delay differential equations with Levy noise. *Syst. Control Lett.* **2018**, *118*, 62–68. [CrossRef]
29. Wang, B.; Zhu, Q. Stability analysis of semi-Markov switched stochastic systems. *Automatica* **2018**, *94*, 72–80. [CrossRef]

axioms

MDPI

Article

Regularized Solution of Singularly Perturbed Cauchy Problem in the Presence of Rational "Simple" Turning Point in Two-Dimensional Case

Alexander Eliseev [†,‡] **and Tatjana Ratnikova** [*,‡]

Research Group on Macro-Structural Modeling of the Russian Economy, National Research University "MPEI", 111250 Moscow, Russia; predikat@bk.ru

* Correspondence: tatrat1@mail.ru; Tel.: +7-916-373-18-03

† Current address: National Research University "MPEI", Kracnokazarmennaya st., 14, 111250 Moscow, Russia.

‡ These authors contributed equally to this work.

Received: 17 September 2019; Accepted: 30 October 2019; Published: 1 November 2019

Abstract: By Lomov's S.A. regularization method, we constructed an asymptotic solution of the singularly perturbed Cauchy problem in a two-dimensional case in the case of violation of stability conditions of the limit-operator spectrum. In particular, the problem with a "simple" turning point was considered, i.e., one eigenvalue vanishes for $t = 0$ and has the form $t^{m/n}a(t)$ (limit operator is discretely irreversible). The regularization method allows us to construct an asymptotic solution that is uniform over the entire segment $[0, T]$, and under additional conditions on the parameters of the singularly perturbed problem and its right-hand side, the exact solution.

Keywords: singularly perturbed Cauchy problem; regularized asymptotic solution; rational "simple" turning point

1. Introduction

This work consists of five parts. The first part is an introduction. The second part is nomenclature. The third part presents the formulation of the Cauchy problem in the two-dimensional case if stability conditions for the spectrum of the limit operator are violated (the spectrum-stability condition means that eigenvalues of the operator $A(\tau)$ satisfy conditions $\lambda_1(\tau) \neq \lambda_2(\tau)$, $\tau \in [0, T]$ and $\lambda_i \neq 0$, $i = 1, 2$).

A "simple" pivot point of a limit operator (matrix $A(\tau)$) is understood when one eigenvalue vanishes at one point (i.e., matrix $A(\tau)$ is irreversible at this point). In [1], the case was considered of when one of the eigenvalues that had the form $\tau^n a(\tau)$, $a(\tau) \neq 0$, n was natural; in [2] the features of the solution were identified and described for a rational "simple" turning point in the one-dimensional case (when the eigenvalue had the form $\tau^{m/n}a(\tau)$, $a(\tau) \neq 0$).

In this article, we consider the case with a "simple" turning point when one of the two eigenvalues of the operator vanishes at $\tau = 0$ and has the form $\tau^{m/n}a(\tau)$, $a(\tau) \neq 0$.

The fourth part describes the formalism of the Lomov regularization method [1,3,4] that allows one to construct an asymptotic solution uniform over the entire segment $[0, T]$, under additional conditions on the parameters of a singularly perturbed problem, and its right side is the exact solution. The idea of this paper goes back to [1], in which methods were developed for solving a singularly perturbed Cauchy problem in the case of a "simple" turning point of a limit operator with a natural exponent. A lemma is given on the estimation of basic singular functions, a theorem on the point solvability of iterative problems is proved, and the leading term of the asymptotic behavior of a singularly perturbed Cauchy problem is written out.

In the fifth part of the paper, we prove a theorem on the asymptotic behavior of a regularized series and a theorem on the passage to the limit as a small parameter tends to zero. For a parabolic equation, an example of solving a singularly perturbed Cauchy problem with a fractional turning point $\lambda(\tau) = \tau^{1/2}$ is given.

The sixth part is the conclusion.

2. Problem Formulation

Consider the Cauchy problem:

$$\begin{cases} \bar{\varepsilon}\dot{u}(\tau) = A(\tau)u(\tau) + h(\tau), \\ u(0, \varepsilon) = u^0, \end{cases} \tag{1}$$

where

(1) τ is a variable, $\tau \in [0, T]$;
(2) $u(\tau)$ is a function, $u(\tau) \in C^\infty[0, T]$;
(3) $A(\tau)$ is a matrix of size (2×2), $A(\tau) \in C^\infty(0, T]$;
(4) $h(\tau)$ is a function, $h(\tau) \in C^\infty[0, T]$;
(5) $\lambda_1(\tau), \lambda_2(\tau)$ are eigenvalues of matrix $A(\tau)$; $\lambda_1(\tau) \neq \lambda_2(\tau), \tau \in [0, T]$; $\lambda_2(\tau) = \tau^{m/n}a(\tau)$, where $a(\tau) < 0, \tau \in [0, T], a(\tau) \in C^\infty[0, T]$;
(6) m, n are natural numbers;
(7) $\operatorname{Re}\lambda_1(\tau) \leq 0$;
(8) $A(t) \in C^\infty[0, T]$, where $t = \tau^{1/n}$;
(9) $\bar{\varepsilon}, \varepsilon = \bar{\varepsilon}/n \in \mathbb{R}$ there is a small parameter of the problem.

We make the change of variables in Problem (1): $t = \tau^{1/n}$. Then $\tau^{m/n} = t^m$ and

$$\frac{du}{d\tau} = \frac{du}{dt} \cdot \frac{dt}{d\tau} = \dot{u}(t)\frac{1}{n}\tau^{(1-n)/n} = \dot{u}(t)\frac{1}{n}t^{1-n}.$$

Equation (1) takes the form:

$$\frac{\bar{\varepsilon}}{n}\dot{u}(t)t^{1-n} = A(t^n)u(t) + h(t^n)$$

or

$$\frac{\bar{\varepsilon}}{n}\dot{u}(t) = t^{n-1}A(t^n)u(t) + t^{n-1}h(t^n).$$

Denote $\bar{\varepsilon}/n = \varepsilon$, $t^{n-1}A(t^n) = B(t)$. Task (1) takes the form:

$$\begin{cases} \varepsilon\dot{u}(t) = B(t)u(t) + t^{n-1}h(t^n), \\ u(0, \varepsilon) = u^0. \end{cases} \tag{2}$$

Operator $B(t)$ has eigenvalues $\bar{\lambda}_1(t) = t^{n-1}\lambda_1(t^n)$, $\bar{\lambda}_2(t) = t^p a(t^n)$, where $p = m + n - 1$, and corresponding vectors $\bar{e}_1(t) = e_1(t^n)$, $\bar{e}_2(t) = e_2(t^n)$, where $e_1(\tau)$, $e_2(\tau)$ are eigenvectors of operator $A(\tau)$, i.e.,

$$B(t)\bar{e}_1(t) = \bar{\lambda}_1(t)\bar{e}_1(t) = t^{n-1}\lambda_1(t^n)e_1(t^n);$$
$$B(t)\bar{e}_2(t) = \bar{\lambda}_2(t)\bar{e}_2(t) = t^p a(t^n)e_1(t^n).$$

Methods for solving the Cauchy problem (2) are described in [1]. Basic singularities (2) have the form:

$$e^{\varphi_i(t)/\varepsilon}, \ i = 1, 2; \ \sigma_i(t, \varepsilon) = e^{\varphi_2(t)/\varepsilon}\int_0^t e^{-\varphi_2(s)/\varepsilon}s^i ds, \ i = \overline{0, p-1}; \tag{3}$$

where $\varphi_1(t) = \int\limits_0^t s^{n-1}\lambda_1(s^n)ds, \;\; \varphi_2(t) = \int\limits_0^t s^p a(s^n)ds.$

Singularities (3) in the source variables have the form

$$e^{\varphi_1(\tau)/\varepsilon}, \;\; e^{\varphi_2(\tau)/\varepsilon}, \;\; \sigma_i(\tau,\varepsilon) = e^{\varphi_2(\tau)/\varepsilon}\int\limits_0^\tau e^{-\varphi_2(s)/\varepsilon}s^{(i+1-n)/n}ds, \;\; i=\overline{0,p-1};$$

where $\varphi_1(\tau) = \int\limits_0^\tau \lambda_1(s)ds, \;\; \varphi_2(\tau) = \int\limits_0^\tau a(s)s^{m/n}ds.$

3. Formalism of Regularization Method

Point $\varepsilon = 0$ for Problem (1) is special in the sense that classical existence theorems for the solution of the Cauchy problem do not take place. Therefore, in solving this problem, essentially singular singularities arise. When the stability condition for spectrum $A(t)$ is satisfied, singular singularities are described using exponentials of the form:

$$e^{\varphi_i(t)/\varepsilon}, \;\; \varphi_i(t) = \int\limits_0^t \lambda_i(s)ds, \;\; i=\overline{1,2}, \;\; \lambda_1(t) \neq \lambda_2(t), \;\; \lambda_i(t) \neq 0, \;\; t \in [0,T],$$

where $\varphi_i(t)$ is a smooth function (in the general case, complex) of a real variable t. To solve linear homogeneous equations, such singularities have been described by Liouville [5–8].

If stability conditions are violated for at least one point of the spectrum of operator $A(t)$, then besides exponentially essentially singularities in the solution of the inhomogeneous equation, singularities of the following form also appear:

$$\sigma_i = e^{\varphi_1(t)/\varepsilon}\int\limits_0^t e^{-\varphi_1(s)/\varepsilon}s^i ds, \;\; i=\overline{0,k-1},$$

(k is the extreme zero of $\lambda_1(t)$), which, for $\varepsilon \to 0$, has a power character of decreasing under the corresponding restrictions on $\lambda_1(t)$, while it is assumed that the remaining points of the spectrum do not vanish at $t = 0$.

Singularly perturbed problems arise in cases when the domain of definition of the initial operator, depending on ε with $\varepsilon \neq 0$, does not coincide with the domain of definition of the limit operator with $\varepsilon = 0$. When studying problems with a "simple" turning point, additional conditions arise when the domain of values of the original operator does not coincide with the domain of values of the limit operator.

Further, we need estimates of functions describing the basic singularities.

Lemma 1. *Let the conditions on the spectrum of operator $A(t)$ 5) $\div$ 7) be satisfied. Then, the estimates hold:*
(a) if $\forall t \in [0,T]$ $\mathrm{Re}\lambda_i(t) \leq 0, i = 1,2,$ then

$$|e^{\frac{1}{\varepsilon}\int\limits_0^t \bar{\lambda}_i(s)ds}| \leq C, \qquad |\sigma_k(t,\varepsilon)| \leq C,$$

where C is a constant, $k = \overline{0,p-1}$, $p = m+n-1$;
(b) if $\mathrm{Re}\,\lambda_1(t) \leq -\alpha < 0$, $\mathrm{Re}\,a(t) \leq -\alpha < 0$, then

$$|e^{\frac{1}{\varepsilon}\int\limits_0^t \bar{\lambda}_1(s)ds}| \leq e^{-\frac{\alpha t^n}{\varepsilon n}}, \;\; |e^{\frac{1}{\varepsilon}\int\limits_0^t \bar{\lambda}_2(s)ds}| \leq e^{-\frac{\alpha t^{p+1}}{\varepsilon(p+1)}}, \;\; |\sigma_k(t,\varepsilon)| \leq C\varepsilon^{\frac{k+1}{p+1}}, \; k=\overline{0,p-1}, \; p=m+n-1.$$

Proof of Lemma 1. (a) In this case, estimates are obvious.

(b) $\left|e^{\frac{1}{\varepsilon}\int\limits_0^t \bar{\lambda}_1(s)ds}\right| \leq e^{-\frac{\alpha}{\varepsilon}\int\limits_0^t s^{n-1}ds} = e^{-\frac{\alpha t^n}{\varepsilon n}}$, $\quad \left|e^{\frac{1}{\varepsilon}\int\limits_0^t \bar{\lambda}_2(s)ds}\right| \leq e^{-\frac{\alpha}{\varepsilon}\int\limits_0^t s^p ds} = e^{-\frac{\alpha t^{p+1}}{\varepsilon(p+1)}}$,

$$|\sigma_k(t,s)| = \left|\int\limits_0^t e^{\frac{1}{\varepsilon}\int\limits_s^t \bar{\lambda}_2(s)ds} s^k ds\right| \leq \int\limits_0^t e^{-\frac{\alpha}{\varepsilon}\int\limits_s^t s_1^p ds_1} s^k ds = \int\limits_0^t e^{-\frac{\alpha(t^{p+1}-s^{p+1})}{\varepsilon(p+1)}} s^k ds = \frac{\int\limits_0^t e^{\frac{\alpha s^{p+1}}{\varepsilon(p+1)}} s^k ds}{e^{\frac{\alpha t^{p+1}}{\varepsilon(p+1)}}} =$$

$$= \frac{\varepsilon^{\frac{k+1}{p+1}} \int\limits_0^{t/\varepsilon^{1/(p+1)}} e^{\frac{\alpha \xi^{p+1}}{p+1}} \xi^k d\xi}{e^{\frac{\alpha t^{p+1}}{\varepsilon(p+1)}}}.$$

Denote $\tau = \frac{t}{\varepsilon^{1/(p+1)}}$. Consider a fraction when $\tau \to \infty$; then, we have

$$\frac{\int\limits_0^{\tau} e^{\frac{\alpha \xi^{p+1}}{p+1}} \xi^k d\xi}{e^{\frac{\alpha \tau^{p+1}}{p+1}}} \sim \frac{\tau^k}{\alpha \tau^p} \underset{\tau \to \infty}{\longrightarrow} 0, \text{ as } k < p.$$

Consequently, $\sigma_k(t,\varepsilon) = \underline{O}(\varepsilon^{\frac{k+1}{p+1}})$. □

Remark 1. *Estimates in the source variables have the form:*

$$\left|e^{\frac{1}{\varepsilon}\int\limits_0^t \lambda_1(s)ds}\right| \leq e^{-\frac{\alpha t}{\varepsilon}}, \qquad \left|e^{\frac{1}{\varepsilon}\int\limits_0^t \lambda_2(s)ds}\right| \leq e^{-\frac{\alpha t^{m/n+1}}{\varepsilon(m/n+1)}}, \qquad \sigma_k(t,\varepsilon) = \underline{O}(\varepsilon^{\frac{k+1}{m/n+1}}).$$

According to the regularization method, we seek a solution of Problem (2) in the form

$$u(t,\varepsilon) = x(t,\varepsilon)e^{\varphi_1(t)/\varepsilon} + y(t,\varepsilon)e^{\varphi_2(t)/\varepsilon} + \sum_{i=0}^{p-1} z^i(t,\varepsilon)\sigma_i(t,\varepsilon) + W(t,\varepsilon), \tag{4}$$

where $x(t,\varepsilon), y(t,\varepsilon), W(t,\varepsilon), z^i(t,\varepsilon), i = \overline{0,p-1}$ are smooth with respect to t functions that depend on power on ε. Substituting Problem (4) into Problem (2), we get system

$$\begin{cases} (B(t) - \bar{\lambda}_1(t))x(t,\varepsilon) = \varepsilon\dot{x}(t,\varepsilon), \\ (B(t) - \bar{\lambda}_2(t))y(t,\varepsilon) = \varepsilon\dot{y}(t,\varepsilon), \\ (B(t) - \bar{\lambda}_2(t))z^i(t,\varepsilon) = \varepsilon\dot{z}^i(t,\varepsilon), \ i = \overline{0,p-1}, \\ B(t)W(t,\varepsilon) = \varepsilon\dot{W}(t,\varepsilon) - t^{n-1}h(t^n) + \sum\limits_{i=0}^{p-1} t^i z^i(t,\varepsilon), \\ x(0,\varepsilon) + y(0,\varepsilon) + W(0,\varepsilon) = u^0. \end{cases} \tag{5}$$

Decomposing the unknown vector functions in a series in powers of ε, we obtain a series of iterative problems:

$$\begin{cases} (B(t) - \bar{\lambda}_1(t))x_k(t) = \dot{x}_{k-1}(t), \\ (B(t) - \bar{\lambda}_2(t))y_k(t) = \dot{y}_{k-1}(t), \\ (B(t) - \bar{\lambda}_2(t))z_k^i(t,\varepsilon) = \dot{z}_{k-1}^i(t), \ i = \overline{0,p-1}, \\ B(t)W_k(t) = \dot{W}_{k-1}(t) - \delta_0^k t^{n-1}h(t^n) + \sum\limits_{i=0}^{p-1} t^i z_{k-1}^i(t), \\ x_k(0) + y_k(0) + W_k(0) = \delta_k^0 u^0. \end{cases} \tag{6}$$

To solve iterative Problems (6), we formulate a point-solvability theorem.

Theorem 1. *Let the following equation be given:*

$$B(t)u(t) \equiv t^{n-1}A(t^n) = t^{n-s}h(t^n), \;\; 0 \le s \le n-1 \tag{7}$$

and let the following conditions are met:

(1) $B(t)$ *has eigenvalues* $\bar{\lambda}_1(t) = t^{n-1}\lambda_1(t^n)$, $\bar{\lambda}_2(t) = t^p a(t^n)$ *and eigenvectors* $\bar{e}_1(t), \bar{e}_2(t)$;
(2) $h(t^n) \in C^\infty[0,T]$.

Then, Problem (7) is solvable if and only if

(a) $h_1(0) = 0, \;\; s = \overline{2, n-1}$;
(b) $h_2^{(k)} = 0, \;\; k = \overline{0, \left[\dfrac{m+s-1}{n}\right]}, \;\; s = \overline{0, n-1}$,

where $h_1(t^n), h_2(t^n)$ *are the components of decomposition* $h(t)$ *on the basis of eigenvectors of operator* $B(t)$; $u_1(t^n), u_2(t^n)$ *are the components of the expansion of* $u(t)$ *on the basis of eigenvectors of operator* $B(t)$.

Proof of Theorem 1. Let us prove the need. Let system

$$\begin{cases} t^{n-1}\lambda_1(t^n)u_1(t) = t^{n-s}h_1(t^n), \\ t^p a(t^n)u_2(t) = t^{n-s}h_2(t^n) \end{cases} \tag{8}$$

have a solution. Then,

(1) the first equation of System (8) is solvable:
 (a) if $s = 0, 1$, then $u_1(t) = t^{1-s}\dfrac{h_1(t^n)}{\lambda_1(t^n)}$,
 (b) if $s = \overline{2, n-1}$, then $h_1(0) = 0$ and $u_1(t) = t^{n+1-s}\dfrac{\bar{h}_1(t^n)}{\lambda_1(t^n)}$, where $h_1(t^n) = t^n\bar{h}_1(t^n)$;
(2) the second equation of System (8) is solvable if $(k+1)n - s \le p < (k+2)n - s$, which is equivalent to $h_2^{(k)}(0) = 0, k = \overline{0, \left[\dfrac{m+s-1}{n}\right]}$ and $u_2(t) = t^{n-j}\dfrac{\bar{h}_2(t^n)}{a(t^n)}, 0 \le j \le n-1$.

Sufficiency is obvious. □

Consider Problem (6) as $k = -1$:

$$\begin{cases} (B(t) - \bar{\lambda}_1(t))x_{-1}(t) = 0, \\ (B(t) - \bar{\lambda}_2(t))y_{-1}(t) = 0, \\ (B(t) - \bar{\lambda}_2(t))z^i_{-1}(t) = 0, \;\; i = \overline{0, p-1}, \\ B(t)W_{-1}(t) = 0, \\ x_{-1}(0) + y_{-1}(0) + W_{-1}(0) = 0. \end{cases} \tag{9}$$

Solution (9) has the form

$$x_{-1}(t) = \alpha^1_{-1}(t)\bar{e}_1(t), \;\; y_{-1}(t) = \beta^2_{-1}(t)\bar{e}_2(t), \;\; z^i_{-1}(t) = \gamma^{i,2}_{-1}(t)\bar{e}_2(t),$$
$$W_{-1}(t) \equiv 0, \;\; \alpha^1_{-1}(0) = 0, \;\; \beta^2_{-1}(0) = 0.$$

Functions $x_{-1}(t), y_{-1}(t), z^i_{-1}(t)$ are determined at the next iteration step $k = 0$ from the solvability conditions:

$$\begin{cases} (B(t) - \bar{\lambda}_1(t))x_0(t) = \dot{x}_{-1}(t), \\ (B(t) - \bar{\lambda}_2(t))y_0(t) = \dot{y}_{-1}(t), \\ (B(t) - \bar{\lambda}_2(t))z^i_0(t) = \dot{z}^i_{-1}(t), \;\; i = \overline{0, p-1}, \\ B(t)W_0(t) = -t^n h(t^n) + \sum\limits_{i=0}^{p-1} t^i z^i_{-1}(t), \\ x_0(0) + y_0(0) + W_0(0) = u^0. \end{cases} \tag{10}$$

Let be

$$\dot{\bar{e}}_i(t) = \dot{e}_i(t^n) = nt^{n-1}C_i^1(t^n)\bar{e}_1(t) + nt^{n-1}C_i^2(t^n)\bar{e}_2(t), \ \ i = 1,2.$$

Denote by $\bar{C}_i^j(t) = nt^{n-1}C_i^j(t^n), i,j = 1,2$. Then

$$\dot{\bar{e}}_i(t) = \sum_{j=1}^{2} \bar{C}_i^j(t)\bar{e}_j(t), \ \ i = 1,2.$$

System (10) takes the form:

$$\begin{cases} (B(t) - \bar{\lambda}_1(t))x_0(t) = (\dot{\alpha}_{-1}^1(t) + \bar{C}_1^1(t)\alpha_{-1}^1(t))\bar{e}_1(t) + \alpha_{-1}^1(t)\bar{C}_1^2(t)\bar{e}_2(t), \\ (B(t) - \bar{\lambda}_2(t))y_0(t) = (\dot{\beta}_{-1}^2(t) + \bar{C}_2^2(t)\beta_{-1}^2(t))\bar{e}_2(t) + \beta_{-1}^2(t)\bar{C}_2^1(t)\bar{e}_1(t), \\ (B(t) - \bar{\lambda}_2(t))z_0^i(t) = (\dot{\gamma}_{-1}^{i,2}(t) + \bar{C}_2^2(t)\gamma_{-1}^{i,2}(t))\bar{e}_2(t) + \gamma_{-1}^{i,2}(t)\bar{C}_2^1(t)\bar{e}_1(t), \ i = \overline{0, p-1}, \\ B(t)W_0(t) = -t^n h(t^n) + \sum_{i=0}^{p-1} t^i z_{-1}^i(t), \\ x_0(0) + y_0(0) + W_0(0) = u^0. \end{cases} \tag{11}$$

The conditions for the solvability of System (11) and the initial conditions at the $k = -1$ step imply that $\alpha_{-1}^1(t) \equiv 0$, $\beta_{-1}^2(t) \equiv 0$. To determine $z_{-1}^i(t)$, we wrote by coordinate the equation for the $W_0(t)$ of System (11):

$$\begin{cases} \bar{\lambda}_1(t)W_0^1(t) \equiv t^{n-1}\lambda_1(t^n)W_0^1(t) = -t^{n-1}h_1(t^n), \\ \bar{\lambda}_2(t)W_0^2(t) \equiv t^p a(t^n)W_0^2(t) = -t^{n-1}h_2(t^n) + \sum_{i=0}^{p-1} t^i \gamma_{-1}^{i,2}(t). \end{cases} \tag{12}$$

Then, $W_0^1(t) = -\dfrac{h_1(t^n)}{\lambda_1(t^n)}$. On the basis of the point-solvability theorem, we obtained:

$$\gamma_{-1}^{n-1,2}(0) = h_2(0), \ \gamma_{-1}^{2n-1,2}(0) = \dot{h}_2(0), \ \cdots, \ \gamma_{-1}^{(k+1)n-1,2}(0) = \frac{h_2^{(k)}(0)}{k!} e^{-\int_0^t \bar{C}_2^2(s)ds},$$

where $k = [m/n]$ is the integer part, so when order $\mathrm{ord}(t^i)$ is equal to order $\mathrm{ord}(t^{(j+1)n-1})$ in the expansion of $t^{n-1}h(t^n)$ in Taylor–Maclaurin series, other $\gamma_{-1}^{i,2}(0) = 0$. Thus, the solution is determined at step $k = -1$:

$$u_{-1}(t,\varepsilon) = \sum_{i=0}^{[m/n]} \gamma_{-1}^{(i+1)n-1,2}(t)\bar{e}_2(t)\sigma_{(i+1)n-1}(t,\varepsilon), \tag{13}$$

where $\gamma_{-1}^{(i+1)n-1,2}(t) = \dfrac{h_2^{(i)}(0)}{i!} e^{-\int_0^t \bar{C}_2^2(s)ds}$.

The solution at zero step $k = 0$ is written in the form

$$\begin{aligned} & x_0(t) = \alpha_0^1(t)\bar{e}_1(t), \ y_0(t) = \beta_0^2(t)\bar{e}_2(t), \\ & z_0^i(t) = \left[\begin{array}{l} \gamma_0^{i,2}(t)\bar{e}_2(t), \ \mathrm{ord}(t^i) \neq \mathrm{ord}(t^{(j+1)n-1}), \\ \gamma_0^{i,2}(t)\bar{e}_2(t) - \gamma_{-1}^{i,2}(t)\frac{\bar{C}_2^1(t)}{\bar{\lambda}_1(t) - \bar{\lambda}_2(t)}\bar{e}_1(t), \ \mathrm{ord}(t^i) = \mathrm{ord}(t^{(j+1)n-1}), \\ i = \overline{0, p-1}, \ j = \overline{0, [m/n]}, \end{array} \right. \\ & W_0(t) = -\frac{h_1(t^n)}{\lambda_1(t^n)}\bar{e}_1(t) + t^{n-s}H_0(t^n)\bar{e}_2(t), \end{aligned} \tag{14}$$

where

(a) $s = \left\{\dfrac{m}{n}\right\}n$ is the remainder of dividing m by n;

(b) $$t^{n-s}H_0(t^n) \equiv \frac{-t^{n-1}h(t^n) + \sum\limits_{i=0}^{[m/n]} t^{(i+1)n-1}\gamma_{-1}^{(i+1)n-1,2}(t)}{t^p a(t^n)}.$$

Arbitrary functions $\alpha_0^1(t)$, $\beta_0^2(t)$, $\gamma_0^{i,2}(t)$ are determined from the conditions for the solvability of the system at step $k = 1$:

$$\begin{cases} (B(t) - \bar{\lambda}_1(t))x_1(t) = (\dot{\alpha}_0^1(t) + \bar{C}_1^1(t)\alpha_0^1(t))\bar{e}_1(t) + \alpha_0^1(t)\bar{C}_1^2(t)\bar{e}_2(t), \\ (B(t) - \bar{\lambda}_2(t))y_1(t) = (\dot{\beta}_0^2(t) + \bar{C}_2^2(t)\beta_0^2(t))\bar{e}_2(t) + \beta_0^2(t)\bar{C}_2^1(t)\bar{e}_1(t), \\ (B(t) - \bar{\lambda}_1(t))z_1^i(t) = (\dot{\gamma}_0^{i,2}(t) + \bar{C}_2^2(t)\gamma_0^{i,2}(t))\bar{e}_2(t) + \gamma_0^{i,2}(t)\bar{C}_2^1(t)\bar{e}_1(t), \\ \qquad \operatorname{ord}(t^i) \neq \operatorname{ord}(t^{(j+1)n-1}), \\ (B(t) - \bar{\lambda}_2(t))z_1^i(t) = \left(\dot{\gamma}_0^{i,2}(t) + \bar{C}_2^2(t)\gamma_0^{i,2}(t) - \gamma_{-1}^{i,2}(t)\frac{\bar{C}_2^1(t)\bar{C}_1^2(t)}{\bar{\lambda}_1(t)-\bar{\lambda}_2(t)}\right)\bar{e}_2(t) + \\ \quad + \left(\dot{\gamma}_0^{i,2}(t)\bar{C}_2^1(t) - \left(\gamma_{-1}^{i,2}(t)\frac{\bar{C}_2^1(t)}{\bar{\lambda}_1(t)-\bar{\lambda}_2(t)}\right)^{\cdot} - \gamma_{-1}^{i,2}(t)\frac{\bar{C}_2^1(t)\bar{C}_1^1(t)}{\bar{\lambda}_1(t)-\bar{\lambda}_2(t)}\right)\bar{e}_1(t), \\ \qquad \operatorname{ord}(t^i) = \operatorname{ord}(t^{(j+1)n-1}), \ i = \overline{0, p-1}, \ j = \overline{0, [m/n]}, \\ B(t)W_1(t) = \dot{W}_0(t) + \sum\limits_{i=0}^{p-1} t^i z_0^i(t), \\ x_1(0) + y_1(0) + W_1(0) = 0. \end{cases} \tag{15}$$

The solvability theorem of System (15) gives

$$\alpha_0^1(t) = \left(u_1^0 + \frac{h_1(0)}{\lambda_1(0)}\right)e^{-\int\limits_0^t \bar{C}_1^1(s)ds}, \quad \beta_0^1(t) \equiv u_2^0 e^{-\int\limits_0^t \bar{C}_2^2(s)ds}.$$

Consider the equation for $W_1(t)$. Given the expression for $\bar{C}_i^j(t) = nt^{n-1}C_i^j(t^n)$, this equation can be written as follows:

$$B(t)W_1(t) = t^{n-1}(\dot{W}_0)_1(t)\bar{e}_1(t) + t^{n-1-s}(\dot{W}_0)_2(t)\bar{e}_2(t) + \sum_{i=0}^{p-1} t^i z_0^i(t). \tag{16}$$

Consider Equation (16) component-wise:

$$\begin{cases} \bar{\lambda}_1(t)W_1^1(t) = t^{n-1}(\dot{W}_0)_1(t) - \sum\limits_{i=0}^{[m/n]} \gamma_{-1}^{(i+1)n-1,2}(t)\dfrac{\bar{C}_2^1(t)t^{(i+1)n-1}}{\bar{\lambda}_1(t) - \bar{\lambda}_2(t)}, \\ \bar{\lambda}_2(t)W_1^2(t) = t^{n-1-s}(\dot{W}_0)_2(t) + \sum\limits_{i=0}^{p-1} t^i\gamma_0^{i,2}(t). \end{cases} \tag{17}$$

Solution of the first equation of System (17) is written as follows:

$$W_1^1(t) = \frac{(\dot{W}_0)_1(t)}{\lambda_1(t^n)} - \sum_{i=0}^{[m/n]} \gamma_{-1}^{(i+1)n-1,2}(t)\frac{nC_2^1(t^n)t^{in}}{\lambda_1(t^n)(\lambda_1(t^n) - t^m a(t^n))}.$$

For the solvability of the second equation of System (17), it is necessary and sufficient that

$$\gamma_0^{(i+1)n-1-s,2}(0) = -\frac{(\dot{W}_0)_2^{(i)}(0)}{i!}, \quad i = \overline{0, \left[\frac{m+s}{n}\right]},$$

here $\left[\dfrac{m+s}{n}\right] = \left[\left[\dfrac{m}{n}\right] + \dfrac{2s}{n}\right] = \left[\dfrac{m}{n}\right] + \left[\dfrac{2s}{n}\right]$.

The other $\gamma_0^{j,2}(0) = 0,\ j \neq (i+1)n - 1 - s,\ j = \overline{0, p-1}$. Defining $\gamma_0^{i,2}(0)$, we can write the expression for $z_0^i(t)$:

(a) if $j = (i+1)n - 1 - s,\ i = \overline{0, \left[\frac{m}{n}\right] + \left[\frac{2s}{n}\right]}$, then

$$\gamma_0^{j,2}(t) = -\frac{(\dot{W}_0)_2^{(i)}(0)}{i!} e^{-\int\limits_0^t \bar{C}_2^2(s)ds}, \quad z_0^j(t) = \gamma_0^{j,2}(t)\bar{C}_2(t);$$

(b) if $j \neq (i+1)n - 1 - s, j = (i+1)n - 1$, then

$$\gamma_0^{j,2}(t) = e^{-\int\limits_0^t \bar{C}_2^2(s)ds} \int\limits_0^t e^{\int\limits_0^s \bar{C}_2^2(z)dz} \gamma_{-1}^{j,2}(s) \frac{\bar{C}_2^1(s)\bar{C}_1^2(s)}{\bar{\lambda}_1(s) - \bar{\lambda}_2(s)} ds,$$

$$z_0^j(t) = \gamma_0^{j,2}(t)\bar{e}_2(t) - \gamma_{-1}^{j,2}(t)\frac{\bar{C}_2^1(t)}{\bar{\lambda}_1(t) - \bar{\lambda}_2(t)}\bar{e}_1(t);$$

(c) if $j \neq (i+1)n - 1 - s, j \neq (i+1)n - 1$, then

$$\gamma_0^{j,2}(t) \equiv 0, \quad z_0^{j,2}(t) \equiv 0.$$

The solution of the second equation of System (17) is written as follows:

$$W_1^2(t) = t^{n(1-\{\frac{2s}{n}\})} H_1(t^n),$$

where $H_1(t^n) = \dfrac{t^{n-s-1}(\dot{W}_0)_2 - \sum\limits_{i=0}^{p-1} t^i \gamma_0^{i,2}(t)}{t^p a(t^n)}$.

Thus, the solution is determined at the zero iterative step:

$$u_0(t,\varepsilon) = \alpha_0^1(t)\bar{e}_1(t)e^{-\varphi_1(t)/\varepsilon} + \beta_0^2(t)\bar{e}_2(t)e^{\varphi_2(t)/\varepsilon} + \sum_{i=0}^{[\frac{m}{n}]+[\frac{2s}{n}]} z_0^{(i+1)n-1-s}(t)\sigma_{(i+1)n-1-s}(t,\varepsilon)$$
$$+ \sum_{i=0}^{[\frac{m}{n}]} z_0^{(i+1)n-1}(t)\sigma_{(i+1)n-1}(t,\varepsilon) - \frac{h_1(t^n)}{\lambda_1(t^n)}\bar{e}_1(t) + t^{n-s}H_0(t^n)\bar{e}_2(t).$$

Similarly, according to this scheme, the solutions of subsequent iteration problems are determined. Thus, we can get an expression for any member of a regularized series.

We write the main term of the asymptotics of Problem (2):

$$u_{\mathtt{main}} = \frac{1}{\varepsilon}u_{-1}(t,\varepsilon) + u_0(t,\varepsilon).$$

4. Limit-Transition Theorem

To prove the asymptoticity of a regularized series, we prove a theorem on estimating the remainder term for $\varepsilon \to 0$.

Let be $u(t,\varepsilon) = \sum\limits_{k=-1}^{n} \varepsilon^k u_k(t,\varepsilon) + \varepsilon^{n+1}R_n(t,\varepsilon)$, where

$$u_k(t,\varepsilon) = x_k(t)e^{\varphi_1(t)/\varepsilon} + y_k(t)e^{\varphi_2(t)/\varepsilon} + \sum_{i=0}^{p-1} z_k^i(t)\sigma_i(t,\varepsilon) + W_k(t). \tag{18}$$

Substituting Problem (18) into Problem (1), we obtain the Cauchy problem for the remainder $R_n(t,\varepsilon)$:

$$\begin{cases} \varepsilon \dot{R}_n(t,\varepsilon) = B(t)R_n(t,\varepsilon) + H(t,\varepsilon), \\ R(0,\varepsilon) = 0, \end{cases} \tag{19}$$

where

$$H(t,\varepsilon) = -\left(\dot{x}_n(t)e^{\varphi_1(t)/\varepsilon} + \dot{y}_n(t)e^{\varphi_2(t)/\varepsilon} + \sum_{i=0}^{p-1} \dot{z}_n(t)\sigma_i(t,\varepsilon) + \left(\dot{W}_n(t) + \sum_{i=0}^{p-1} t^i z_n^i(t)\right)\right),$$

in this case, it is assumed that $H(t,\varepsilon)$ satisfies the conditions of the solvability theorem.

Theorem 2. *Let Cauchy Problem (1) be given and Conditions 1–9 be satisfied. Then, the estimate is correct*

$$\left\| u(t,\varepsilon) - \sum_{k=-1}^{n} \varepsilon^k u_k(t,\varepsilon) \right\| \leq C\varepsilon^{n+1},$$

where $C > 0$ in the norm $\mathbb{C}[0,T]$ for any $(t,\varepsilon) \in [0,T] \times (0,\varepsilon_0]$, $\|x(t)\|_{\mathbb{C}[0,T]} = \max_{t\in[0,T]} |x(t)|$.

Proof of Theorem 2. Solution (19) is written as follows:

$$R_n(t,\varepsilon) = \frac{1}{\varepsilon} \int_0^t U_\varepsilon(t,s)H(s,\varepsilon)ds, \tag{20}$$

where $U_\varepsilon(t,s)$ is resolving operator (fundamental solution system) satisfying system

$$\begin{cases} \varepsilon \dot{U}_\varepsilon(t,s) = B(t)U_\varepsilon(t,s), \\ U_\varepsilon(t,s)\big|_{s=t} = I. \end{cases} \tag{21}$$

Let $S(t)$ be a matrix of eigenvectors $\bar{e}_1(t)$, $\bar{e}_2(t)$ of operator $B(t)$. Then, System (21) is equivalent to system

$$\begin{cases} \varepsilon \dot{V}_\varepsilon(t,s) = \Lambda(t)V_\varepsilon(t,s) - \varepsilon S^{-1}(t)\dot{S}(t)V_\varepsilon(t,\varepsilon), \\ V_\varepsilon(t,s)\big|_{s=t} = S^{-1}(0), \end{cases} \tag{22}$$

here, $\Lambda(t) = \begin{pmatrix} \bar{\lambda}_1(t) & 0 \\ 0 & \bar{\lambda}_2(t) \end{pmatrix}$, $V_\varepsilon(t,s) = S^{-1}(t)U_\varepsilon(t,s)$. We reduce System (22) to an integral equation

$$V_\varepsilon(t,s) = e^{\frac{1}{\varepsilon}\int_s^t \Lambda(s_1)ds_1} S^{-1}(0) - \int_s^t e^{\frac{1}{\varepsilon}\int_{s_1}^t \Lambda(s_2)ds_2} S^{-1}(s_1)\dot{S}(s_1)V_\varepsilon(s_1,\varepsilon)ds_1. \tag{23}$$

Let us estimate Equation (23) at the norm $\mathbb{C}[0,T]$. Using the conditions on the spectrum of operator $B(t)$, we obtain

$$\|V_\varepsilon(t,s)\| \leq C_1\|S^{-1}(0)\| + C_2 \int_s^t \|V_\varepsilon(s_1,s)\|ds_1.$$

Using the Bellman–Gronuola inequality, we obtain $\|U_\varepsilon(t,s)\| \leq C$ on $[0,T]$.

To estimate the remaining term, it is important to take into account that operator $B(t)$ is invertible on vector functions that satisfy the conditions of the solvability theorem. Then, integrating over parts of Solution (20), we obtain chain of equalities

$$R_n(t,\varepsilon) = \frac{1}{\varepsilon}\int_0^t U_\varepsilon(t,s)H(s,\varepsilon)ds = \frac{1}{\varepsilon}\int_0^t U_\varepsilon(t,s)B(s)B^{-1}(s)H(s,\varepsilon)ds =$$

$$= -U_\varepsilon(t,s)B^{-1}(s)H(s,\varepsilon)\big|_0^t + \int_0^t U_\varepsilon(t,s)\frac{d}{ds}B^{-1}(s)H(s,\varepsilon)ds =$$

$$= -B^{-1}(t)H(t,\varepsilon) + U_\varepsilon(t,s)B^{-1}(s)H(s,\varepsilon)\big|_{s=0} + \int_0^t U_\varepsilon(t,s)\frac{d}{ds}B^{-1}(s)H(s,\varepsilon)ds.$$

Since, by virtue of Conditions 1–9, $H(t,\varepsilon)$ admits estimate $\|H(t,\varepsilon)\| \leq C_1$ in norm $\mathbb{C}[0,T]$, then remainder $R_n(t,\varepsilon)$ satisfies estimate

$$\|R_n(t,\varepsilon)\| \leq C_2 \ \forall (t,\varepsilon) \in [0,T] \times (0,\varepsilon_0].$$

Therefore, the asymptoticity of series $\sum\limits_{k=-1}^{\infty} \varepsilon^k u_k(t,\varepsilon)$ is proved. □

Theorem 3 (The limit theorem). *Let Cauchy Problem* (1) *be given and have satisfied the conditions:*

(1) *Conditions 1–9;*
(2) $h_2^{(i)}(0) = 0, i = \overline{0,[m/n]}$, *where $h_2(t)$ is the second coordinate in the expansion of $h(t) = h_1(t)e_1(t) + h_2(t)e_2(t)$ in eigenvectors of the original matrix.*

Then,

(1) *for any $\delta > 0$ $t \in [\delta, T]$, $\mathrm{Re}\lambda_i(t) \leq -\alpha < 0$*

$$\lim_{\varepsilon\to 0} u(t,\varepsilon) = -A^{-1}(t)h(t);$$

(2) *if $\mathrm{Re}\lambda_i(t) = 0$, then*

$$u(t,\varepsilon) \underset{\varepsilon\to 0}{\overset{weak}{\longrightarrow}} -A^{-1}(t)h(t) \ \text{ in a weak sense.}$$

Proof of Theorem 3. (1) Conditions $h_2^{(i)}(0) = 0, i = \overline{0,[m/n]}$ cause $u_{-1}(t,\varepsilon) = 0$. Then,

$$u_{\mathtt{main}}(t) = u_0(t,\varepsilon).$$

By virtue of the singularity estimates described in the lemma, it follows that for any $\delta > 0$ $t \in [\delta, T]$

$$\lim_{\varepsilon\to 0} u_0(t,\varepsilon) = -B^{-1}(t)t^{n-1}h(t^n),$$

equivalent in source variables $\lim\limits_{\varepsilon\to 0} u_0(t,\varepsilon) = -A^{-1}(t)h(t)$.
(2) If $\mathrm{Re}\lambda_i(t) \equiv 0$, $i = 1,2$, then singularities are rapidly oscillating exponents as $\varepsilon \to 0$. From here, according to Lebesgue's lemma, for any $\varphi(t) \in \mathbb{C}(0,T)$

$$\int_0^T (u_0(t,\varepsilon) + A^{-1}(t)h(t))\varphi(s)dt \underset{\varepsilon\to 0}{\longrightarrow} 0.$$

□

Example 1. *Consider the Cauchy problem for a parabolic equation*

$$\begin{cases} \varepsilon\dfrac{\partial u}{\partial t} - \varepsilon^2\dfrac{\partial^2 u}{\partial x^2} = -\sqrt{t}u + h(x,t), \\ u(x,0) = \varphi(x), \quad -\infty < x < \infty, \end{cases}$$

where $\varphi(x), h(x,t) \in C_0^\infty(-\infty,\infty)$ are smooth functions with compact support.
Using the technique of the regularization method outlined above, we obtain the principal term of the asymptotics of the solution:

$$u(x,t) = \frac{1}{\varepsilon}h(x,0)e^{-\frac{2}{3\varepsilon}t^{3/2}}\int_0^t e^{\frac{2s^{3/2}}{3\varepsilon}}ds + \varphi(x)e^{-\frac{2t^{3/2}}{3\varepsilon}} - \frac{\dot{h}(x,0)}{2}e^{-\frac{2t^{3/2}}{3\varepsilon}}\int_0^t e^{\frac{2s^{3/2}}{3\varepsilon}}\frac{ds}{\sqrt{s}} +$$

$$+th''(x,0)e^{-\frac{2t^{3/2}}{3\varepsilon}}\int_0^t e^{\frac{2s^{3/2}}{3\varepsilon}}ds + \frac{h(x,t)-h(x,0)}{\sqrt{t}}.$$

5. Conclusions

In this paper, the regularization method was developed into the class of singularly perturbed Cauchy problems in the case of a simple rational turning point for the limit operator (for $\varepsilon = 0$). The main singularities of the solution are highlighted:

$$e^{\varphi_1(t)/\varepsilon}, \; e^{\varphi_2(t)/\varepsilon}, \; \sigma_i(t,\varepsilon) = e^{\varphi_2(t)/\varepsilon}\int_0^t e^{-\varphi_2(s)/\varepsilon}s^{(i+1-n)/n}ds, \; i = \overline{0, p-1},$$

which allowed us to present the solution in the form:

$$u(t,\varepsilon) = x(t,\varepsilon)e^{\varphi_1(t)/\varepsilon} + y(t,\varepsilon)e^{\varphi_2(t)/\varepsilon} + \sum_{i=0}^{p-1} z^i(t,\varepsilon)\sigma_i(t,\varepsilon) + W(t,\varepsilon),$$

where $x(t,\varepsilon), y(t,\varepsilon), W(t,\varepsilon), z^i(t,\varepsilon), i = \overline{0, p-1}$ are t smooth functions that depend on power ε.

Estimates of the main singularities for $\varepsilon \to 0$ were given, and theorems on the solvability of iterative problems were proved. A theorem on the asymptotic convergence of the solution of the problem was proved, and conditions on the right-hand side of $h(t)$ were described, under which the passage to the limit theorem is valid. An example of solving the Cauchy problem for a parabolic equation with a fractional turning point $\lambda(\tau) = \tau^{1/2}$ was given.

Author Contributions: Conceptualization and writing—review and editing by A.E. and T.R.

Funding: This research received no external funding.

Conflicts of Interest: The authors declare no conflict of interest.

Nomenclature

$u(t), x(t), y(t), z^i(t), i = \overline{0, m+n-1}, w(t), h(t)$	a vector of a function of a real variable
$A(t), B(t)$	matrices of order 2×2
λ_1, λ_2	eigenvalues of matrix A
$\bar{\lambda}_1, \bar{\lambda}_2$	eigenvalues of matrix B
e_1, e_2	eigenvectors of matrix A
$\bar{e}_1, \bar{e}_2$	eigenvectors of matrix B
ε	a small task parameter
$S(t)$	a matrix of eigenvectors $\bar{e}_1, \bar{e}_2$
$\Lambda(t)$	a matrix of eigenvalues of matrix B
R_n	the remainder term of the asymptotic series

References

1. Eliseev, A.G.; Lomov, S.A. The theory of singular perturbations in the case of spectral singularities of the limit operator. *Sb. Math.* **1986**, *131*, 544–557. [CrossRef]
2. Eliseev, A.G.; Ratnikova, T.A. A singularly perturbed Cauchy problem in the presence of a rational "simple" turning point for the limit operator. *Differ. Equ. Control Process.* **2019**, *3*, 63–73.
3. Lomov, S.A. *Introduction to the General Theory of Singular Perturbations*; Nauka: Moscow, Russia, 1981.
4. Safonov, V.F.; Bobodzhanov, A.A. Course of higher mathematics. In *Singularly Perturbed Equations and the Regularization Method: Textbook*; Izdatelstvo MPEI: Moscow, Russia, 2012.
5. Liouville, J. Second memoire sur le development des fonctions en series dont divers termes sont assujettis, a une meme equation. *J. Math. Pure Appl.* **1837**, *2*, 16–35.
6. Lomov, S.A.; Eliseev, A.G. Asymptotic integration of singularly perturbed problems. *Russ. Math. Surv.* **1988**, *43*, 1–63. [CrossRef]
7. Eliseev, A.G. Singular perturbation theory for systems of differential equations in the case of multiple spectrum of the limit operator. III. *Math. USSR-Izv.* **1985**, *25*, 475–500. [CrossRef]
8. Tursunov, D.A.; Kozhbekov, K.G. Asymptotics of the solution of singularly perturbed differential equations with a fractional turning point. *News Irkutsk State Univ.* **2017**, *21*, 108–121.

axioms

MDPI

Article

Euclidean Space Controllability Conditions for Singularly Perturbed Linear Systems with Multiple State and Control Delays

Valery Y. Glizer

Department of Mathematics, ORT Braude College of Engineering, Karmiel 2161002, Israel; valery48@braude.ac.il or valgl120@gmail.com

Received: 3 February 2019; Accepted: 17 March 2019; Published: 21 March 2019

Abstract: A singularly perturbed linear time-dependent controlled system with multiple point-wise delays and distributed delays in the state and control variables is considered. The delays are small, of order of a small positive multiplier for a part of the derivatives in the system. This multiplier is a parameter of the singular perturbation. Two types of the considered singularly perturbed system, standard and nonstandard, are analyzed. For each type, two much simpler parameter-free subsystems (the slow and fast ones) are associated with the original system. It is established in the paper that proper kinds of controllability of the slow and fast subsystems yield the complete Euclidean space controllability of the original system for all sufficiently small values of the parameter of singular perturbation. Illustrative examples are presented.

Keywords: singularly perturbed system; multiple state and control delays; controllability

MSC: 34K26; 93B05; 93C23

1. Introduction

Differential systems with a small positive multiplier for a part of the highest order derivatives, called singularly perturbed differential systems, are adequate mathematical models for real-life processes with two-time-scale dynamics. In real-life problems, the small multiplier (a parameter of singular perturbation) can be a time constant, a mass, a capacitance, a geotropic reaction, and some other parameters in physics, chemistry, engineering, biology, medicine, etc (see e.g., [1–3] and references therein). An important class of singularly perturbed differential systems represents the systems with small time delays (of order the parameter of singular perturbation). Such systems arise in various real-life applications, for instance, in nuclear engineering [4], in botany [5], in physiology and medicine [6,7], in control engineering [8], and in communication engineering [9,10]. Distributed small delays appear, for instant, in stabilizing controls of singularly perturbed systems with small delays (either point-wise, or distributed, or point-wise and distributed) [11]. In such a case, a closed-loop system contains a distributed small delay. The stabilizing property of a distributed small delay also is used in the present paper (see Sections 3.3, 3.4 and 4.2).

Various topics in theory and applications of singularly perturbed controlled systems, without and with delays in state and control variables, were extensively investigated in the literature (see e.g., [1,12–14] and references therein).

Controllability of a system is one of its basic properties. This property means the ability to transfer the system from any position of a given set of initial positions to any position of a given set of terminal positions in a finite time by a proper choice of the control function. Different types of controllability for systems without or with delays were extensively studied in the literature (see e.g., [15–18] and references therein). To check whether a singularly perturbed system is controllable in a proper sense, the corresponding controllability conditions can be directly applied for any specified value of the small parameter $\varepsilon > 0$ of singular perturbation. However, the stiffness, as well as a possible high dimension of the singularly perturbed system, can considerably complicate this application. Moreover, such an application depends on the value of ε, and it should be repeated if this parameter changes. Furthermore, in most of real-life problems the current value of ε is unknown. These circumstances are crucial in the analysis of the controllability of singularly perturbed systems. They motivate the derivation of conditions, which being independent of ε, guarantee the controllability of a singularly perturbed system for all sufficiently small values of this parameter, i.e., robustly with respect to ε.

Controllability of singularly perturbed systems was analyzed in a number of works. Thus, in [19–22], the complete controllability of some linear and nonlinear undelayed systems was studied using the separation of time scales concept (see e.g., [1]). In [23] the robust complete Euclidean space controllability, as well as the controllability with respect to the slow state variable and with respect to the fast state variable, were studied for a linear standard singularly perturbed time-invariant system with a single nonsmall pointwise state delay. In [24,25], using the separation of time scales concept, parameter free conditions of complete Euclidean space controllability were obtained for linear standard singularly perturbed systems with pointwise and distributed small state delays. In [26], this result was extended to nonstandard singularly perturbed systems with multiple pointwise and distributed small delays in the state variables. In [27], parameter-free complete Euclidean space controllability conditions, which are not based on the separation of time-scales concept, were derived for a class of linear singularly perturbed systems with small state delays. In [28], a singularly perturbed linear time-dependent controlled system with a single small pointwise delay in the state and control variables was considered. Parameter-free conditions of the complete Euclidean space controllability were established for standard and nonstandard types of this system. In [29], a singularly perturbed linear time-dependent system with small state delays (multiple point-wise and distributed) was studied. Along with the set of time delay differential equations describing the dynamics of this system, a set of delay-free algebraic equations, describing the system's output, also was considered. Based on the separation of time-scales concept, different parameter-free sufficient conditions for the Euclidean space output controllability of this system were established. In [30], the complete Euclidean space controllability for one class of singularly perturbed systems with nonsmall delays (point-wise and distributed) in the state variables was studied. In [31], the defining equations method was used for analysis of the complete Euclidean space controllability of a linear singularly perturbed neutral type system with a single nonsmall pointwise delay. The particular cases of the Euclidean space output controllability, the controllability with respect to the slow state variable and the controllability with respect to the fast state variable, also were studied.

In the present paper, we consider a singularly perturbed linear time-varying system with multiple small point-wise delays and with small distributed delays in the state and control variables. The complete Euclidean space controllability of this system, robust with respect to ε, is studied. This study is based on a transformation of the complete Euclidean space controllability of the original system with delays in the state and the control to an equivalent output controllability of a new singularly perturbed system with only state delays. In the new system, the original control variable becomes an additional fast state variable. The Euclidean dimension of the slow mode equation in the new system is the same as in the original system, while the Euclidean dimension of the fast mode equation is larger than such a dimension in the original system. Further analysis is carried out based on the asymptotic decomposition of the original and transformed systems. Each system is decomposed into two much simpler ε-free subsystems, slow and fast ones. Equivalence of proper kinds of controllability of the slow subsystems, corresponding to the original and transformed systems, is established. Also, it is established the equivalence of proper kinds of controllability of the fast subsystems. Assuming the controllability of the slow and fast subsystems, associated with the transformed system, the Euclidean space output controllability of the latter is established for all sufficiently small values of $\varepsilon > 0$. Then, using the above mentioned equivalence of the controllability of the original and transformed systems, as well as of their slow and fast subsystems, the complete Euclidean space controllability of the original system, robust with respect to ε, is deduced from the assumption on proper kinds of controllability of its slow and fast subsystems. Note that the original system of the present paper is much more general than the original system of [28]. Moreover, in the present paper we propose another, more general, approach to the analysis of the nonstandard case of the original system. Also, we propose here much simpler proof of the Euclidean space output controllability of the transformed system.

The paper is organized as follows. In the next section, the rigorous problem statement, the main definitions and the objective of the paper are formulated. Some auxiliary results, including the transformation of the original system, are presented in Section 3. Section 4 is devoted to main results of the paper. An illustrative example is solved in Section 5. Conclusions are placed in Section 6.

The following main notations are applied in the paper:

1. E^n is the n-dimensional real Euclidean space.
2. The Euclidean norm of either a vector or a matrix is denoted by $\|\cdot\|$.
3. The upper index T denotes the transposition either of a vector x (x^T) or of a matrix A (A^T).
4. I_n denotes the identity matrix of dimension n.
5. The notation $O_{n_1\times n_2}$ is used for the zero matrix of the dimension $n_1 \times n_2$, excepting the cases where the dimension of zero matrix is obvious. In such cases, we use the notation 0 for the zero matrix.
6. $L^2[t_1,t_2;E^n]$ denotes the linear space of all vector-valued functions $x(\cdot):[t_1,t_2]\rightarrow E^n$ square integrable in the interval $[t_1,t_2]$; for any $x(\cdot)\in L^2[t_1,t_2;E^n]$ and $y(\cdot)\in L^2[t_1,t_2;E^n]$, the inner product in this space is defined as:

$$\langle x(\cdot),y(\cdot)\rangle_{L^2}=\int_{t_1}^{t_2}x^T(t)y(t)dt;$$

the norm of any $x(\cdot)\in L^2[t_1,t_2;E^n]$ is defined as:

$$\|x(\cdot)\|_{L^2}=\left(\int_{t_1}^{t_2}x^T(t)x(t)dt\right)^{1/2}.$$

7. $L^2_{\mathrm{loc}}[\bar{t},+\infty;E^n]$ denotes the linear space of all vector-valued functions $x(\cdot):[\bar{t},+\infty)\rightarrow E^n$ square integrable in any subinterval $[t_1,t_2]\subset[\bar{t},+\infty)$.
8. $W^{1,2}[t_1,t_2;E^n]$ denotes the corresponding Sobolev space, i.e., the linear space of all vector-valued functions $x(\cdot):[t_1,t_2]\rightarrow E^n$ square integrable in the interval $[t_1,t_2]$ with the first derivatives (generalized) square integrable in this interval.
9. $\mathrm{col}(x,y)$, where $x\in E^n$, $y\in E^m$, denotes the column block-vector of the dimension $n+m$ with the upper block x and the lower block y, i.e., $\mathrm{col}(x,y)=(x^T,y^T)^T$.
10. $\mathrm{Re}\lambda$ denotes the real part of a complex number λ.

2. Problem Formulation and Main Definitions

2.1. Original System

Consider the controlled system

$$\begin{aligned}\frac{dx(t)}{dt}&=\textstyle\sum_{j=0}^{N}\left[A_{1j}(t,\varepsilon)x(t-\varepsilon h_j)+A_{2j}(t,\varepsilon)y(t-\varepsilon h_j)\right]\\&+\textstyle\int_{-h}^{0}\left[G_1(t,\eta,\varepsilon)x(t+\varepsilon\eta)+G_2(t,\eta,\varepsilon)y(t+\varepsilon\eta)\right]d\eta\\&+\textstyle\sum_{j=0}^{N}B_{1j}(t,\varepsilon)u(t-\varepsilon h_j)+\int_{-h}^{0}H_1(t,\eta,\varepsilon)u(t+\varepsilon\eta)d\eta,\quad t\geq 0,\end{aligned}\tag{1}$$

$$\varepsilon\frac{dy(t)}{dt} = \sum_{j=0}^{N}\left[A_{3j}(t,\varepsilon)x(t-\varepsilon h_j) + A_{4j}(t,\varepsilon)y(t-\varepsilon h_j)\right] \\ + \int_{-h}^{0}\left[G_3(t,\eta,\varepsilon)x(t+\varepsilon\eta) + G_4(t,\eta,\varepsilon)y(t+\varepsilon\eta)\right]d\eta \\ + \sum_{j=0}^{N} B_{2j}(t,\varepsilon)u(t-\varepsilon h_j) + \int_{-h}^{0} H_2(t,\eta,\varepsilon)u(t+\varepsilon\eta)d\eta, \quad t \geq 0, \tag{2}$$

where $x(t) \in E^n$, $y(t) \in E^m$, $u(t) \in E^r$ ($u(t)$ is a control); $\varepsilon > 0$ is a small parameter; $N \geq 1$ is an integer; $0 = h_0 < h_1 < h_2 < \cdots < h_N = h$ are some given constants independent of ε; $A_{ij}(t,\varepsilon)$, $G_i(t,\eta,\varepsilon)$, $B_{kj}(t,\varepsilon)$, $H_k(t,\eta,\varepsilon)$, $(i = 1,\ldots,4;\ j = 0,\ldots,N;\ k = 1,2)$ are matrix-valued functions of corresponding dimensions, given for $t \geq 0$, $\eta \in [-h,0]$ and $\varepsilon \in [0,\varepsilon^0]$, ($\varepsilon^0 > 0$); the functions $A_{ij}(t,\varepsilon)$ and $B_{kj}(t,\varepsilon)$, $(i = 1,\ldots,4;\ j = 0,\ldots,N;\ k = 1,2)$ are continuous in $(t,\varepsilon) \in [0,+\infty) \times [0,\varepsilon^0]$; the functions $G_i(t,\eta,\varepsilon)$ and $H_k(t,\eta,\varepsilon)$, $(i = 1,\ldots,4;\ k = 1,2)$ are piecewise continuous in $\eta \in [-h,0]$ for any $(t,\varepsilon) \in [0,+\infty) \times [0,\varepsilon^0]$; the functions $G_i(t,\eta,\varepsilon)$ and $H_k(t,\eta,\varepsilon)$, $(i = 1,\ldots,4;\ k = 1,2)$ are continuous with respect to $(t,\varepsilon) \in [0,+\infty) \times [0,\varepsilon^0]$ uniformly in $\eta \in [-h,0]$.

For any given $\varepsilon \in (0,\varepsilon^0]$ and $u(\cdot) \in L^2_{\text{loc}}[-\varepsilon h,+\infty; E^r]$, the system (1)-(2) is a linear time-dependent nonhomogeneous functional-differential system. It is infinite-dimensional with the state variables $(x(t), x(t+\varepsilon\eta))$ and $(y(t), y(t+\varepsilon\eta))$, $\eta \in [-h,0]$. Moreover, (1)-(2) is a singularly perturbed system. The Equation (1) is the slow mode of this system, while the Equation (2) is its fast mode.

Definition 1. *For a given $\varepsilon \in (0,\varepsilon^0]$, the system (1)-(2) is said to be completely Euclidean space controllable at a given time instant $t_c > 0$ if for any $x_0 \in E^n$, $y_0 \in E^m$, $u_0 \in E^r$, $\varphi_x(\cdot) \in L^2[-\varepsilon h,0;E^n]$, $\varphi_y(\cdot) \in L^2[-\varepsilon h,0;E^m]$, $\varphi_u(\cdot) \in L^2[-\varepsilon h,0;E^r]$, $x_c \in E^n$ and $y_c \in E^m$ there exists a control function $u(\cdot) \in W^{1,2}[0,t_c;E^r]$ satisfying $u(0) = u_0$, for which the system (1)-(2) with the initial and terminal conditions*

$$x(\tau) = \varphi_x(\tau), \quad y(\tau) = \varphi_y(\tau), \quad u(\tau) = \varphi_u(\tau), \quad \tau \in [-\varepsilon h, 0), \tag{3}$$

$$x(0) = x_0, \quad y(0) = y_0, \tag{4}$$

$$x(t_c) = x_c, \quad y(t_c) = y_c, \tag{5}$$

has a solution.

2.2. Asymptotic Decomposition of the Original System

For the sake of further analysis, let us decompose asymptotically the original singularly perturbed system (1)-(2) into two much simpler ε-free subsystems, the slow and fast ones. The slow subsystem is obtained from (1)-(2) by setting formally $\varepsilon = 0$ in these controlled functional-differential equations, which yields

$$\frac{dx_s(t)}{dt} = A_{1s}(t)x_s(t) + A_{2s}(t)y_s(t) + B_{1s}(t)u_s(t), \quad t \geq 0, \tag{6}$$

$$0 = A_{3s}(t)x_s(t) + A_{4s}(t)y_s(t) + B_{2s}(t)u_s(t), \quad t \geq 0, \tag{7}$$

where $x_s(t) \in E^n$ and $y_s(t) \in E^m$ are state variables; $u_s(t) \in E^r$ is a control;

$$A_{is}(t) = \sum_{j=0}^{N} A_{ij}(t,0) + \int_{-h}^{0} G_i(t,\eta,0)d\eta, \quad i = 1,\ldots,4, \tag{8}$$

$$B_{ks}(t) = \sum_{j=0}^{N} B_{kj}(t,0) + \int_{-h}^{0} H_k(t,\eta,0)d\eta, \quad k = 1,2. \tag{9}$$

The slow subsystem (6)-(7) is a descriptor (differential-algebraic) system, and it is delay-free and ε-free.

If

$$\det A_{4s}(t) \neq 0, \quad t \geq 0, \tag{10}$$

we can eliminate the state variable $y_s(t)$ from the slow subsystem (6)-(7). Such an elimination yields the differential equation with respect to $x_s(t)$

$$\frac{dx_s(t)}{dt} = \bar{A}_s(t)x_s(t) + \bar{B}_s(t)u_s(t), \quad t \geq 0, \tag{11}$$

where

$$\bar{A}_s(t) = A_{1s}(t) - A_{2s}(t)A_{4s}^{-1}(t)A_{3s}(t), \; \bar{B}_s(t) = B_{1s}(t) - A_{2s}(t)A_{4s}^{-1}(t)B_{2s}(t). \tag{12}$$

The differential Equation (11) also is called the slow subsystem, associated with the original system (1)-(2).

The fast subsystem is derived from (2) in the following way: (a) the terms containing the state variable $(x(t), x(t+\varepsilon\eta))$, $\eta \in [-h, 0]$ are removed from (2); (b) the transformations of the variables $t = t_1 + \varepsilon\xi$, $y(t_1 + \varepsilon\xi) \stackrel{\triangle}{=} y_f(\xi)$, $u(t_1 + \varepsilon\xi) \stackrel{\triangle}{=} u_f(\xi)$ are made in the resulting system, where $t_1 \geq 0$ is any fixed time instant.

Thus, we obtain the system

$$\frac{dy_f(\xi)}{d\xi} = \sum_{j=0}^{N} A_{4j}(t_1+\varepsilon\xi,\varepsilon)y_f(\xi - h_j) + \int_{-h}^{0} G_4(t_1+\varepsilon\xi,\eta,\varepsilon)y_f(\xi+\eta)d\eta$$
$$+ \sum_{j=0}^{N} B_{2j}(t_1+\varepsilon\xi,\varepsilon)u_f(\xi - h_j) + \int_{-h}^{0} H_2(t_1+\varepsilon\xi,\eta,\varepsilon)u_f(\xi+\eta)d\eta.$$

Finally, setting formally $\varepsilon = 0$ in this system and replacing t_1 with t yield the fast subsystem

$$\begin{aligned} \tfrac{dy_f(\xi)}{d\xi} &= \textstyle\sum_{j=0}^{N} A_{4j}(t,0)y_f(\xi - h_j) + \int_{-h}^{0} G_4(t,\eta,0)y_f(\xi+\eta)d\eta \\ &+ \textstyle\sum_{j=0}^{N} B_{2j}(t,0)u_f(\xi - h_j) + \int_{-h}^{0} H_2(t,\eta,0)u_f(\xi+\eta)d\eta, \quad \xi \geq 0, \end{aligned} \tag{13}$$

where $t \geq 0$ is a parameter; $y_f(\xi) \in E^m$, $u_f(\xi) \in E^r$; $(y_f(\xi), y_f(\xi+\eta))$, $\eta \in [-h, 0]$ is a state variable, while $(u_f(\xi), u_f(\xi+\eta))$, $\eta \in [-h, 0]$ is a control variable.

The new independent variable ξ is called the stretched time, and it is expressed by the original time t in the form $\xi = (t - t_1)/\varepsilon$. Thus, for any $t > t_1$, $\xi \to +\infty$ as $\varepsilon \to +0$.

The fast subsystem (13) is a differential equation with state and control delays. It is of a lower Euclidean dimension than the original system (1)-(2), and it is ε-free.

Definition 2. *Subject to (10), the system (11) is said to be completely controllable at a given time instant $t_c > 0$ if for any $x_0 \in E^n$ and $x_c \in E^n$ there exists a control function $u_s(\cdot) \in L^2[0, t_c; E^r]$, for which (11) has a solution $x_s(t)$, $t \in [0, t_c]$, satisfying the initial and terminal conditions*

$$x_s(0) = x_0, \quad x_s(t_c) = x_c. \tag{14}$$

Definition 3. *The system (6)-(7) is said to be impulse-free controllable with respect to $x_s(t)$ at a given time instant $t_c > 0$ if for any $x_0 \in E^n$ and $x_c \in E^n$ there exists a control function $u_s(\cdot) \in L^2[0, t_c; E^r]$, for which (6)-(7) has an impulse-free solution $\mathrm{col}(x_s(t), y_s(t))$, $t \in [0, t_c]$, satisfying the initial and terminal conditions (14).*

Definition 4. *For a given $t \geq 0$, the system (13) is said to be completely Euclidean space controllable if for any $y_0 \in E^m$, $u_0 \in E^r$, $\varphi_{yf}(\cdot) \in L^2[-h, 0; E^m]$, $\varphi_{uf}(\cdot) \in L^2[-h, 0; E^r]$ and $y_c \in E^m$ there exist a number $\xi_c > 0$, independent of y_0, u_0, $\varphi_{yf}(\cdot)$, $\varphi_{uf}(\cdot)$ and y_c, and a control function $u_f(\cdot) \in W^{1,2}[0, \xi_c; E^r]$ satisfying $u_f(0) = u_0$, for which the system (13) with the initial and terminal conditions*

$$y_f(\eta) = \varphi_{yf}(\eta), \quad u_f(\eta) = \varphi_{uf}(\eta), \quad \eta \in [-h, 0); \quad y_f(0) = y_0, \tag{15}$$

$$y_f(\xi_c) = y_c, \tag{16}$$

has a solution.

2.3. Objective of the Paper

The objective of the paper is the following: using the ε-independent assumptions on the controllability of the systems (11) and (13), as well as (6)-(7) and (13), to establish the complete Euclidean space controllability of the original singularly perturbed system (1)-(2) for all sufficiently small values of $\varepsilon > 0$, i.e., robustly with respect to this parameter.

3. Auxiliary Results

In this section, some properties of systems with state and control delays are studied. Based on these results, in the next section different parameter-free conditions for the complete Euclidean space controllability of the original singularly perturbed system are derived.

3.1. Auxiliary System with Delay-Free Control

Consider the differential system, consisting of the Equations (1), (2) and the equation

$$\varepsilon\frac{du(t)}{dt} = -u(t) + v(t), \quad t \geq 0. \tag{17}$$

In this new system, $\big(x(t), x(t+\varepsilon\eta)\big)$, $\big(y(t), y(t+\varepsilon\eta)\big)$, $\big(u(t), u(t+\varepsilon\eta)\big)$, $\eta \in [-h, 0]$ are state variables, while $v(t) \in E^r$ is a control. Thus, in the system (1), (2), (17) only the state variables have delays, while the control is delay-free. Moreover, in contrast with the original system (1)-(2), the new system contains two fast modes, the Equations (2) and (17).

For the new system (1), (2), (17), we consider the algebraic output equation

$$\zeta(t) = Z\mathrm{col}\big(x(t), y(t), u(t)\big), \quad t \geq 0, \tag{18}$$

where the $(n+m) \times (n+m+r)$-matrix Z has the block form

$$Z = \Big(I_{n+m}, 0\Big). \tag{19}$$

Let us rewrite the system (1), (2), (17), (18) in a new form, more convenient for the further analysis.

For a given $\varepsilon \in (0, \varepsilon^0]$, let us introduce into the consideration the block vector $\omega(t) = \mathrm{col}\big(y(t), u(t)\big)$, $t \geq -\varepsilon h$, and the block matrices

$$\mathcal{A}_{1j}(t,\varepsilon) = A_{1j}(t,\varepsilon), \ \mathcal{A}_{2j}(t,\varepsilon) = \Big(A_{2j}(t,\varepsilon), B_{1j}(t,\varepsilon)\Big), \ j = 0, 1, \ldots, N, \ t \geq 0, \tag{20}$$

$$\mathcal{A}_{3j}(t,\varepsilon) = \begin{pmatrix} A_{3j}(t,\varepsilon) \\ O_{r\times n} \end{pmatrix}, \quad j = 0, 1, \ldots, N, \quad t \geq 0, \tag{21}$$

$$\mathcal{A}_{40}(t,\varepsilon) = \begin{pmatrix} A_{40}(t,\varepsilon) & B_{20}(t,\varepsilon) \\ O_{r\times m} & -I_r \end{pmatrix}, \quad t \geq 0, \tag{22}$$

$$\mathcal{A}_{4j}(t,\varepsilon) = \begin{pmatrix} A_{4j}(t,\varepsilon) & B_{2j}(t,\varepsilon) \\ O_{r\times m} & O_{r\times r} \end{pmatrix}, \quad j = 1, \ldots, N, \quad t \geq 0, \tag{23}$$

$$\mathcal{G}_1(t,\eta,\varepsilon) = G_1(t,\eta,\varepsilon), \ \mathcal{G}_2(t,\eta,\varepsilon) = \Big(G_2(t,\eta,\varepsilon), H_1(t,\eta,\varepsilon)\Big), \ t \geq 0, \ \eta, \in [-h, 0], \tag{24}$$

$$\mathcal{G}_3(t,\eta,\varepsilon) = \begin{pmatrix} G_3(t,\eta,\varepsilon) \\ O_{r\times n} \end{pmatrix}, \quad \mathcal{G}_4(t,\eta,\varepsilon) = \begin{pmatrix} G_4(t,\eta,\varepsilon) & H_2(t,\eta,\varepsilon) \\ O_{r\times m} & O_{r\times r} \end{pmatrix}, \tag{25}$$
$$t \geq 0, \quad \eta \in [-h, 0],$$

$$\mathcal{B}_1 = O_{n\times r}, \quad \mathcal{B}_2 = \begin{pmatrix} O_{m\times r} \\ I_r \end{pmatrix}. \tag{26}$$

Based on the above introduced vector and matrices, we can rewrite the auxiliary system (1), (2), (17), (18) in the equivalent form

$$\frac{dx(t)}{dt} = \sum_{j=0}^{N} \Big[\mathcal{A}_{1j}(t,\varepsilon)x(t-\varepsilon h_j) + \mathcal{A}_{2j}(t,\varepsilon)\omega(t-\varepsilon h_j)\Big]$$
$$+ \int_{-h}^{0} \Big[\mathcal{G}_1(t,\eta,\varepsilon)x(t+\varepsilon\eta) + \mathcal{G}_2(t,\eta,\varepsilon)\omega(t+\varepsilon\eta)\Big]d\eta, \quad t \geq 0, \tag{27}$$

$$\varepsilon\frac{d\omega(t)}{dt} = \sum_{j=0}^{N} \Big[\mathcal{A}_{3j}(t,\varepsilon)x(t-\varepsilon h_j) + \mathcal{A}_{4j}(t,\varepsilon)\omega(t-\varepsilon h_j)\Big]$$
$$+ \int_{-h}^{0} \Big[\mathcal{G}_3(t,\eta,\varepsilon)x(t+\varepsilon\eta) + \mathcal{G}_4(t,\eta,\varepsilon)\omega(t+\varepsilon\eta)\Big]d\eta + \mathcal{B}_2 v(t), \quad t \geq 0, \tag{28}$$

$$\zeta(t) = Z\text{col}\big(x(t), \omega(t)\big), \quad t \geq 0. \tag{29}$$

Definition 5. *For a given $\varepsilon \in (0, \varepsilon^0]$, the system (27)-(28), (29) is said to be Euclidean space output controllable at a given time instant $t_c > 0$ if for any $x_0 \in E^n$, $\omega_0 \in E^{m+r}$, $\varphi_x(\cdot) \in L^2[-\varepsilon h, 0; E^n]$, $\varphi_\omega(\cdot) \in L^2[-\varepsilon h, 0; E^{m+r}]$ and $\zeta_c \in E^{n+m}$ there exists a control function $v(\cdot) \in L^2[0, t_c; E^r]$, for which the solution $\text{col}\big(x(t), \omega(t)\big)$, $t \in [0, t_c]$ of the system (27)-(28) with the initial conditions*

$$x(\tau) = \varphi_x(\tau), \ \ \omega(\tau) = \varphi_\omega(\tau), \ \ \tau \in [-\varepsilon h, 0); \ \ x(0) = x_0, \ \ \omega(0) = \omega_0$$

satisfies the terminal condition $Z\text{col}\big(x(t_c), \omega(t_c)\big) = \zeta_c$.

Proposition 1. *For a given $\varepsilon \in (0, \varepsilon^0]$, the system (1)-(2) is completely Euclidean space controllable at a given time instant $t_c > 0$, if and only if the system (27)-(28), (29) is Euclidean space output controllable at this time instant.*

Proof. The proposition is proven similarly to [28] (Lemma 1). □

Now, let us decompose asymptotically the system (27)-(28), (29) into the slow and fast subsystems. We start with the slow subsystem. The dynamic part of this subsystem is obtained from (27)-(28) by setting there formally $\varepsilon = 0$. The output part of the slow subsystem is obtained from (29) by removing formally the term with the Euclidean part $\omega(t)$ of the fast state variable $\big(\omega(t), \omega(t+\varepsilon\eta)\big)$, $\eta \in [-h, 0]$. Thus, the slow subsystem has the form

$$\frac{dx_s(t)}{dt} = \mathcal{A}_{1s}(t)x_s(t) + \mathcal{A}_{2s}(t)\omega_s(t), \quad t \geq 0, \tag{30}$$

$$0 = \mathcal{A}_{3s}(t)x_s(t) + \mathcal{A}_{4s}(t)\omega_s(t) + \mathcal{B}_2 v_s(t), \quad t \geq 0, \tag{31}$$

$$\zeta_s(t) = x_s(t), \quad t \geq 0, \tag{32}$$

where $x_s(t) \in E^n$ and $\omega_s(t) \in E^{m+r}$ are state variables; $v_s(t) \in E^r$ is a control; $\zeta_s(t) \in E^n$ is an output; $\omega_s(t) = \text{col}\big(y_s(t), u_s(t)\big)$, $y_s \in E^m$, $u_s(t) \in E^r$;

$$\mathcal{A}_{is}(t) = \sum_{j=0}^{N} \mathcal{A}_{ij}(t,0) + \int_{-h}^{0} \mathcal{G}_i(t,\eta,0)d\eta, \quad i = 1,\ldots,4, \tag{33}$$

or using (8)-(9), (20)–(25)

$$\begin{gathered} \mathcal{A}_{1s}(t) = A_{1s}(t), \quad \mathcal{A}_{2s}(t) = \Big(A_{2s}(t), B_{1s}(t)\Big), \\ \mathcal{A}_{3s}(t) = \begin{pmatrix} A_{3s}(t) \\ O_{r\times n} \end{pmatrix}, \quad \mathcal{A}_{4s}(t) = \begin{pmatrix} A_{4s}(t) & B_{2s}(t) \\ O_{r\times m} & -I_r \end{pmatrix}. \end{gathered} \tag{34}$$

From the expression for $\mathcal{A}_{4s}(t)$ we have that $\det \mathcal{A}_{4s}(t) = (-1)^r \det A_{4s}(t)$. Thus, $\det \mathcal{A}_{4s}(t) \neq 0$, $t \geq 0$ if and only if $\det A_{4s}(t) \neq 0$, $t \geq 0$. Therefore, subject to (10), the differential-algebraic system (30)-(31) can be converted to the differential equation

$$\frac{dx_s(t)}{dt} = \bar{\mathcal{A}}_s(t)x_s(t) + \bar{\mathcal{B}}_s(t)v_s(t), \quad t \geq 0, \tag{35}$$

where

$$\bar{\mathcal{A}}_s(t) = \mathcal{A}_{1s}(t) - \mathcal{A}_{2s}(t)\mathcal{A}_{4s}^{-1}(t)\mathcal{A}_{3s}(t), \quad \bar{\mathcal{B}}_s(t)v_s(t) = -\mathcal{A}_{2s}(t)\mathcal{A}_{4s}^{-1}(t)\mathcal{B}_2. \tag{36}$$

Using the Equations (34) and (36), the Equation (35) can be rewritten as:

$$\frac{dx_s(t)}{dt} = \bar{A}_s(t)x_s(t) + \bar{B}_s(t)v_s(t), \quad t \geq 0, \tag{37}$$

where the matrix-valued coefficients $\bar{A}_s(t)$ and $\bar{B}_s(t)$ are given in (12). Hence, subject to (10), the slow subsystem associated with (27)-(28), (29) consists of the differential Equation (37) and the output Equation (32).

Remark 1. *Comparison of the differential Equations (37) and (11) directly yields that the former can be obtained from the latter by replacing in it $u_s(t)$ with $v_s(t)$, and vice versa. Moreover, the output in the system (37), (32) coincides with $x_s(t)$. Hence, the output controllability of this system means its controllability with respect to $x_s(t)$. Therefore, the output controllability of (37), (32) coincides with the complete controllability of (37) and, thus, it is equivalent to the complete controllability of the system (11).*

Remark 2. *Similarly to Remark 1, since the output in the system (30)-(31), (32) coincides with $x_s(t)$, then an output controllability of this system coincides with a proper controllability of its dynamic part (30)-(31) with respect to $x_s(t)$.*

Definition 6. *The system (30)-(31) is said to be impulse-free controllable with respect to $x_s(t)$ at a given time instant $t_c > 0$ if for any $x_0 \in E^n$ and $x_c \in E^n$ there exists a control function $v_s(\cdot) \in L^2[0,t_c;E^r]$, for which (30)-(31) has an impulse-free solution $\mathrm{col}(x_s(t),\omega_s(t))$, $t \in [0,t_c]$, satisfying the initial and terminal conditions $x_s(0) = x_0$ and $x_s(t_c) = x_c$.*

Proposition 2. *The system (6)-(7) is impulse-free controllable with respect to $x_s(t)$ at a given time instant $t_c > 0$ if and only if the system (30)-(31) is impulse-free controllable with respect to $x_s(t)$ at this time instant.*

Proof. Eliminating the component $u_s(t)$ of the state variable $\omega_s(t)$ from the system (30)-(31), we convert the latter to the equivalent system consisting of the equation $u_s(t) = v_s(t)$ and the system

$$\frac{dx_s(t)}{dt} = A_{1s}(t)x_s(t) + A_{2s}(t)y_s(t) + B_{1s}(t)v_s(t), \quad t \geq 0, \tag{38}$$

$$0 = A_{3s}(t)x_s(t) + A_{4s}(t)y_s(t) + B_{2s}(t)v_s(t), \quad t \geq 0, \tag{39}$$

where $A_{is}(t)$, $B_{ks}(t)$, $(i = 1, \ldots, 4;\ k = 1, 2)$ are given in (8)-(9).

Therefore, the impulse-free controllability with respect to $x_s(t)$ of the system (30)-(31) is equivalent to such a controllability of the system (38)-(39). Now, the comparison of the latter with the system (6)-(7) directly yields the statement of the proposition. □

Proceed to the fast subsystem, associated with the system (27)-(28), (29). The dynamic part of this subsystem is constructed similarly to the fast subsystem (13), associated with the original system (1)-(2). The output part of the fast subsystem is obtained from (29) by removing formally the term with the Euclidean part $x(t)$ of the state variable $\big(x(t), x(t+\varepsilon\eta)\big)$, $\eta \in [-h, 0]$. Thus the fast subsystem, associated with the auxiliary system (27)-(28), (29), consists of the differential equation

$$\frac{d\omega_f(\xi)}{d\xi} = \sum_{j=0}^{N} \mathcal{A}_{4j}(t,0)\omega_f(\xi - h_j) + \int_{-h}^{0} \mathcal{G}_4(t,\eta,0)\omega_f(\xi+\eta)d\eta + \mathcal{B}_2 v_f(\xi),\ \xi \geq 0, \tag{40}$$

and the output equation

$$\zeta_f(\xi) = \Omega_f \omega_f(\xi), \quad \xi \geq 0, \quad \Omega_f = \big(I_m, O_{m\times r}\big), \tag{41}$$

where $t \geq 0$ is a parameter; $\omega_f(\xi) \in E^{m+r}$; $\big(\omega_f(\xi), \omega_f(\xi+\eta)\big)$ is a state variable; $v_f(\xi) \in E^r$ is a control; $\zeta_f(\xi) \in E^m$ is an output.

Note that in contrast with the system (13), in the differential system (40) only the state variable has delays, while the control is undelayed.

Definition 7. *For a given $t \geq 0$, the system (40)-(41) is said to be Euclidean space output controllable if for any $\omega_0 \in E^{m+r}$, $\varphi_{\omega f}(\cdot) \in L^2[-h, 0; E^{m+r}]$ and $\zeta_{fc} \in E^m$ there exist a number $\xi_c > 0$, independent of ω_0, $\varphi_{\omega f}(\cdot)$ and ζ_{fc}, and a control function $v_f(\cdot) \in L^2[0, \xi_c; E^r]$, for which the solution $\omega_f(\xi)$, $\xi \in [0, \xi_c]$ of the differential Equation (40) with the initial conditions*

$$\omega_f(\eta) = \varphi_{\omega f}(\eta), \quad \eta \in [-h, 0); \quad \omega_f(0) = \omega_0 \tag{42}$$

satisfies the terminal condition

$$\Omega_f \omega_f(\xi_c) = \zeta_{fc}. \tag{43}$$

Lemma 1. *For a given $t \geq 0$, the system (13) is completely Euclidean space controllable if and only if the system (40)-(41) is Euclidean space output controllable.*

Proof. *Sufficiency.* Let us assume that, for some given $t \geq 0$, the system (40)-(41) is Euclidean space output controllable. Let $\omega_0 \in E^{m+r}$, $\varphi_{\omega f}(\cdot) \in L^2[-h, 0; E^{m+r}]$ and $\zeta_{fc} \in E^m$ be arbitrary given. Then, there exists a number $\xi_c > 0$, independent of ω_0, $\varphi_{\omega f}(\cdot)$ and ζ_{fc}, and a control function $v_f(\cdot) \in L^2[0, \xi_c; E^r]$, for which the differential Equation (40) with the initial (42) and terminal (43) conditions has a solution $\omega_f(\xi)$, $\xi \in [0, \xi_c]$. Let us represent the vector ω_0 and the vector-valued function $\omega_f(\xi)$ in the block form as: $\omega_0 = \mathrm{col}(y_0, u_0)$, $y_0 \in E^m$, $u_0 \in E^r$; $\omega_f(\xi) = \mathrm{col}\big(y_f(\xi), u_f(\xi)\big)$, $y_f(\xi) \in E^m$, $u_f(\xi) \in E^r$, $\xi \in [0, \xi_c]$. Also, we represent the vector-valued function $\varphi_{\omega f}(\eta)$ in the block form as: $\varphi_{\omega f}(\eta) = \mathrm{col}\big(\varphi_{yf}(\eta), \varphi_{uf}(\eta)\big)$, $\eta \in [-h, 0]$. Note, that the component $u_f(\xi)$ of the above mentioned solution $\omega_f(\xi)$ to the boundary-valued problem (40), (42), (43) satisfies the conditions $u_f(\eta) = \varphi_{uf}(\eta)$, $\eta \in [-h, 0)$ and $u_f(0) = u_0$. Moreover, since $v_f(\cdot) \in L^2[0, \xi_c; E^r]$, then $u_f(\xi) \in W^{1,2}[0, \xi_c; E^r]$. Thus, for the control function $u_f(\xi)$, the vector-valued function $y_f(\xi)$, $\xi \in [0, \xi_c]$ is a solution of the system (13) satisfying the initial condition (15) and the terminal conditions $y_f(\zeta_c) = \zeta_{fc}$. Hence, re-denoting ζ_{fc} as y_c and using Definition 4, we directly obtain that, for the given $t \geq 0$, the system (13) is completely

Euclidean space controllable. This completes the proof of the sufficiency.
Necessity. The necessity is proven similarly to the sufficiency.

Thus, the lemma is proven. □

3.2. *Output Controllability of the Auxiliary System and its Slow and Fast Subsystems: Necessary and Sufficient Conditions*

3.2.1. Output Controllability of the Auxiliary System

For a given $\varepsilon \in (0, \varepsilon^0]$, let us consider the block vector $z(t) = \mathrm{col}(x(t), \omega(t))$, $t \geq -\varepsilon h$, and the block matrices

$$\mathcal{A}_j(t,\varepsilon) = \begin{pmatrix} \mathcal{A}_{1j}(t,\varepsilon) & \mathcal{A}_{2j}(t,\varepsilon) \\ \frac{1}{\varepsilon}\mathcal{A}_{3j}(t,\varepsilon) & \frac{1}{\varepsilon}\mathcal{A}_{4j}(t,\varepsilon) \end{pmatrix}, \quad j = 0, 1, \ldots, N, \tag{44}$$

$$\mathcal{G}(t,\eta,\varepsilon) = \begin{pmatrix} \mathcal{G}_1(t,\eta,\varepsilon) & \mathcal{G}_2(t,\eta,\varepsilon) \\ \frac{1}{\varepsilon}\mathcal{G}_3(t,\eta,\varepsilon) & \frac{1}{\varepsilon}\mathcal{G}_4(t,\eta,\varepsilon) \end{pmatrix}, \quad \mathcal{B}(\varepsilon) = \begin{pmatrix} \mathcal{B}_1 \\ \frac{1}{\varepsilon}\mathcal{B}_2 \end{pmatrix} = \begin{pmatrix} O_{n\times r} \\ \frac{1}{\varepsilon}\mathcal{B}_2 \end{pmatrix}. \tag{45}$$

Thus, the auxiliary system (27)-(29), can be rewritten in the equivalent form

$$\frac{dz(t)}{dt} = \sum_{j=0}^{N} \mathcal{A}_j(t,\varepsilon)z(t-\varepsilon h_j) + \int_{-h}^{0} \mathcal{G}(t,\eta,\varepsilon)z(t+\varepsilon\eta)d\eta + \mathcal{B}(\varepsilon)v(t), \quad t \geq 0, \tag{46}$$

$$\zeta(t) = Zz(t), \quad t \geq 0. \tag{47}$$

It is clear that the system (46)-(47) is equivalent to the auxiliary system (27)-(29).

Definition 8. *For a given $\varepsilon \in (0, \varepsilon^0]$, the system (46)-(47) is said to be Euclidean space output controllable at a given time instant $t_c > 0$ if for any $z_0 \in E^{n+m+r}$, $\varphi_z(\cdot) \in L^2[-\varepsilon h, 0; E^{n+m+r}]$, and $\zeta_c \in E^{n+m}$ there exists a control function $v(\cdot) \in L^2[0, t_c; E^r]$, for which the solution $z(t)$, $t \in [0, t_c]$ of the system (46) with the initial conditions $z(\tau) = \varphi_z(\tau)$, $\tau \in [-h, 0)$, $z(0) = z_0$ satisfies the terminal condition $Zz(t_c) = \zeta_c$.*

Let, for a given $\varepsilon \in (0, \varepsilon^0]$, the $(n+m+r) \times (n+m+r)$-matrix-valued function $\Psi(\sigma, \varepsilon)$, $\sigma \in [0, t_c]$ be a solution of the terminal-value problem

$$\begin{aligned} \tfrac{d\Psi(\sigma,\varepsilon)}{d\sigma} &= -\textstyle\sum_{j=0}^{N} \left(\mathcal{A}_j(\sigma+\varepsilon h_j,\varepsilon)\right)^T \Psi(\sigma+\varepsilon h_j,\varepsilon) \\ &- \textstyle\int_{-h}^{0} \left(\mathcal{G}(t-\varepsilon\eta,\eta,\varepsilon)\right)^T \Psi(\sigma-\varepsilon\eta,\varepsilon)d\eta, \quad \sigma \in [0, t_c), \\ \Psi(t_c,\varepsilon) &= I_{n+m+r}; \quad \Psi(\sigma,\varepsilon) = 0, \quad \sigma > t_c, \end{aligned} \tag{48}$$

where it is assumed that $\mathcal{A}_{ij}(t,\varepsilon) = \mathcal{A}_{ij}(t_c,\varepsilon)$, $\mathcal{G}_i(t,\eta,\varepsilon) = \mathcal{G}_i(t_c,\eta,\varepsilon)$, $t > t_c$, $\eta \in [-h, 0]$, $\varepsilon \in [0, \varepsilon^0]$, $(i = 1, \ldots, 4;\ j = 1, \ldots, N)$. Due to the results of [32] (Section 4.3), $\Psi(\sigma, \varepsilon)$ exists and is unique for $\sigma \in [0, t_c]$, $\varepsilon \in (0, \varepsilon^0]$.

Consider the following two matrices of the dimensions $(n+m+r) \times (n+m+r)$ and $(n+m) \times (n+m)$, respectively:

$$W(t_c,\varepsilon) = \int_0^{t_c} \Psi^T(\sigma,\varepsilon)\mathcal{B}(\varepsilon)\mathcal{B}^T(\varepsilon)\Psi(\sigma,\varepsilon)d\sigma \tag{49}$$

and

$$W_Z(t_c,\varepsilon) = ZW(t_c,\varepsilon)Z^T. \tag{50}$$

Proposition 3. *For a given $\varepsilon \in (0, \varepsilon^0]$, the auxiliary system (27)-(29) is Euclidean space output controllable at a given time instant $t_c > 0$ if and only if the matrix $W_Z(t_c,\varepsilon)$ is nonsingular, i.e.,* $\det W_Z(t_c,\varepsilon) \neq 0$.

Proof. By virtue of the results of [29] (Corollary 1), the system (46)-(47) is Euclidean space output controllable at the time instant t_c if and only if $\det W_Z(t_c,\varepsilon) \neq 0$. Since this system is equivalent to the auxiliary system (27)-(29), then, due to Definitions 5 and 8, the auxiliary system also is Euclidean space output controllable at t_c if and only if $\det W_Z(t_c, \varepsilon) \neq 0$. This completes the proof of the proposition. □

3.2.2. Output Controllability of the Slow and Fast Subsystems Associated with the Auxiliary System

We start with the slow subsystem (37).

Let, for a given $t_c > 0$, the $n \times n$-matrix-valued function $\Psi_s(\sigma)$, $\sigma \in [0, t_c]$ be the unique solution of the terminal-value problem

$$\frac{d\Psi_s(\sigma)}{d\sigma} = -\big(\bar{A}_s(\sigma)\big)^T \Psi_s(\sigma), \quad \sigma \in [0, t_c), \quad \Psi_s(t_c) = I_n. \tag{51}$$

Consider the $n \times n$-matrix

$$W_s(t_c) = \int_0^{t_c} \Psi_s^T(\sigma)\bar{B}_s(\sigma)\bar{B}_s^T(\sigma)\Psi_s(\sigma)d\sigma. \tag{52}$$

By virtue of the results of [15], we have the following proposition.

Proposition 4. *Let the condition (10) be fulfilled in the interval $[0, t_c]$. Then, the slow subsystem (37), associated with the auxiliary system (27)-(29), is completely controllable at the time instant t_c, if and only if the matrix $W_s(t_c)$ is nonsingular, i.e.,* $\det W_s(t_c) \neq 0$.

Proceed to the fast subsystem (40)-(41).

Let, for any given $t \geq 0$, the $(m + r) \times (m + r)$-matrix-valued function $\Psi_f(\xi, t)$ be the unique solution of the following initial-value problem:

$$\begin{aligned} \frac{d\Psi_f(\xi)}{d\xi} &= \sum_{j=0}^{N} \big(\mathcal{A}_{4j}(t,0)\big)^T \Psi_f(\xi - h_j) \\ &+ \int_{-h}^{0} \big(\mathcal{G}_4(t,\eta,0)\big)^T \Psi_f(\xi + \eta)d\eta, \quad \xi > 0, \\ &\Psi_f(\xi) = 0, \quad \xi < 0, \quad \Psi_f(0) = I_{m+r}. \end{aligned} \tag{53}$$

Consider the $m \times m$-matrix-valued function

$$W_f(\xi, t) = \Omega_f \int_0^{\xi} \Psi_f^T(\rho, t)\mathcal{B}_2\mathcal{B}_2^T\Psi_f(\rho, t)d\rho\Omega_f^T, \quad \xi \geq 0, \quad t \geq 0. \tag{54}$$

By virtue of the results of [29] (Corollary 1), we have the following assertion.

Proposition 5. *For a given $t \geq 0$, the fast subsystem (40)-(41) of the auxiliary system (27)-(29) is Euclidean space output controllable if and only if there exists a number $\xi_c > 0$ such that the matrix $W_f(\xi_c, t)$ is nonsingular, i.e.,* $\det W_f(\xi_c, t) \neq 0$.

3.3. Linear Control Transformation in the Auxiliary System

Let us transform the control $v(t)$ in the auxiliary system (27)-(28), (29) as follows:

$$v(t) = K_1(t)\omega(t) + \int_{-h}^{0} K_2(t,\eta)\omega(t + \varepsilon\eta)d\eta + w(t), \tag{55}$$

where $w(t) \in E^r$ is a new control; $K_1(t)$ and $K_2(t, \eta)$ are any specified matrix-valued functions of the dimension $r \times (m + r)$ given for $t \geq 0$, $\eta \in [-h, 0]$; $K_1(t)$ is continuous for $t \geq 0$; $K_2(t, \eta)$ is continuous with respect to $t \geq 0$ uniformly in $\eta \in [-h, 0]$, and this function is piecewise continuous in $\eta \in [-h, 0]$ for any $t \geq 0$.

Due to this transformation, the dynamic part (27)-(28) of the system (27)-(28), (29) becomes as:

$$\frac{dx(t)}{dt} = \sum_{j=0}^{N} \Big[\mathcal{A}_{1j}(t,\varepsilon)x(t-\varepsilon h_j) + \mathcal{A}_{2j}(t,\varepsilon)\omega(t-\varepsilon h_j)\Big]$$
$$+ \int_{-h}^{0} \Big[\mathcal{G}_1(t,\eta,\varepsilon)x(t+\varepsilon\eta) + \mathcal{G}_2(t,\eta,\varepsilon)\omega(t+\varepsilon\eta)\Big]d\eta, \quad t \geq 0, \tag{56}$$

$$\varepsilon\frac{d\omega(t)}{dt} = \sum_{j=0}^{N} \Big[\mathcal{A}_{3j}(t,\varepsilon)x(t-\varepsilon h_j) + \mathcal{A}_{4j}^{K}(t,\varepsilon)\omega(t-\varepsilon h_j)\Big]$$
$$+ \int_{-h}^{0} \Big[\mathcal{G}_3(t,\eta,\varepsilon)x(t+\varepsilon\eta) + \mathcal{G}_4^{K}(t,\eta,\varepsilon)\omega(t+\varepsilon\eta)\Big]d\eta + \mathcal{B}_2 w(t), \quad t \geq 0, \tag{57}$$

where

$$\mathcal{A}_{40}^{K}(t,\varepsilon) = \mathcal{A}_{40}(t,\varepsilon) + \mathcal{B}_2 K_1(t), \quad \mathcal{A}_{4j}^{K}(t,\varepsilon) = \mathcal{A}_{4j}(t,\varepsilon), \quad j = 1, \dots, N, \tag{58}$$

$$\mathcal{G}_4^{K}(t,\eta,\varepsilon) = \mathcal{G}_4(t,\eta,\varepsilon) + \mathcal{B}_2 K_2(t,\eta). \tag{59}$$

Proposition 6. *For a given $\varepsilon \in (0, \varepsilon^0]$, the system (27)-(28), (29) is Euclidean space output controllable at a given time instant $t_c > 0$, if and only if the system (56)-(57), (29) is Euclidean space output controllable at this time instant.*

Proof. The proposition is proven similarly to [29] (Lemma 3). □

As a direct consequence of Propositions 1 and 6, we obtain the following assertion.

Corollary 1. *For a given $\varepsilon \in (0, \varepsilon^0]$, the system (1)-(2) is completely Euclidean space controllable at a given time instant $t_c > 0$, if and only if the system (56)-(57), (29) is Euclidean space output controllable at this time instant.*

Now, let us decompose asymptotically the singularly perturbed system (56)-(57), (29) into the slow and fast subsystems. This decomposition is carried out similarly to that for the system (27)-(28), (29). Thus, the slow subsystem, associated with (56)-(57), (29), consists of the differential-algebraic system

$$\frac{dx_s(t)}{dt} = \mathcal{A}_{1s}(t)x_s(t) + \mathcal{A}_{2s}(t)\omega_s(t), \quad t \geq 0, \tag{60}$$

$$0 = \mathcal{A}_{3s}(t)x_s(t) + \mathcal{A}_{4s}^{K}(t)\omega_s(t) + \mathcal{B}_2 w_s(t), \quad t \geq 0, \tag{61}$$

and the output Equation (32). In (60)-(61), (32), $x_s(t) \in E^n$ and $\omega_s(t) \in E^{m+r}$ are state variables; $w_s(t) \in E^r$ is a control; $\zeta_s(t) \in E^n$ is an output; $\mathcal{A}_{ls}(t)$, $(l = 1, 2, 3)$ are given in (33);

$$\mathcal{A}_{4s}^{K}(t) = \sum_{j=0}^{N} \mathcal{A}_{4j}^{K}(t,0) + \int_{-h}^{0} \mathcal{G}_4^{K}(t,\eta,0)d\eta. \tag{62}$$

If

$$\det \mathcal{A}_{4s}^{K}(t) \neq 0, \quad t \geq 0, \tag{63}$$

the differential-algebraic system (60)-(61) can be reduced to the differential equation with respect to $x_s(t)$

$$\frac{dx_s(t)}{dt} = \bar{\mathcal{A}}_s^{K}(t)x_s(t) + \bar{\mathcal{B}}_s^{K}(t)w_s(t), \quad t \geq 0, \tag{64}$$

where

$$\bar{\mathcal{A}}_s^K(t) = \mathcal{A}_{1s}(t) - \mathcal{A}_{2s}(t)\big(\mathcal{A}_{4s}^K(t)\big)^{-1}\mathcal{A}_{3s}(t),$$

$$\bar{\mathcal{B}}_s^K(t) = -\mathcal{A}_{2s}(t)\big(\mathcal{A}_{4s}^K(t)\big)^{-1}\mathcal{B}_2.$$

Thus, subject to (63), the slow subsystem associated with (56)-(57), (29) is (64), (32).

The fast subsystem, associated with (56)-(57), (29), consists of the differential equation with state delays

$$\frac{d\omega_f(\xi)}{d\xi} = \sum_{j=0}^{N} \mathcal{A}_{4j}^K(t,0)\omega_f(\xi - h_j) + \int_{-h}^{0} \mathcal{G}_4^K(t,\eta,0)\omega_f(\xi+\eta)d\eta \\ +\mathcal{B}_2 w_f(\xi), \quad \xi \geq 0, \tag{65}$$

and the output Equation (41). Note, that in (65), (41), $t \geq 0$ is a parameter, while ξ is an independent variable. Moreover, in this system, $\omega_f(\xi) \in E^{m+r}$; $(\omega_f(\xi), \omega_f(\xi+\eta))$, $\eta \in [-h, 0)$ is a state variable; $w_f(\xi) \in E^r$, ($w_f(\xi)$ is a control); $\zeta_f(\xi) \in E^m$, ($\zeta_f(\xi)$ is an output).

Remark 3. *Since the output in the slow subsystem in both forms, (60)-(61), (32) and (64), (32), coincides with the state variable $x_s(t)$, then an output controllability of the slow subsystem is a controllability of its dynamic part with respect to $x_s(t)$. Namely, for the slow subsystem in the form (60)-(61), (32) such a controllability is the impulse-free controllability of the system (60)-(61) with respect to $x_s(t)$. For the slow subsystem in the form (64), (32), the controllability with respect to $x_s(t)$ is the complete controllability of the system (64).*

Proposition 7. *The system (30)-(31) is impulse-free controllable with respect to $x_s(t)$ at a given time instant $t_c > 0$ if and only if the system (60)-(61) is impulse-free controllable with respect to $x_s(t)$ at this time instant.*

Proof. The proposition is proven similarly to [26] (Lemma 3). □

Based on Propositions 2 and 7, we directly obtain the following corollary.

Corollary 2. *The system (6)-(7) is impulse-free controllable with respect to $x_s(t)$ at a given time instant $t_c > 0$ if and only if the system (60)-(61) is impulse-free controllable with respect to $x_s(t)$ at this time instant.*

Proposition 8. *Let the condition (63) be satisfied. Then, the system (60)-(61) is impulse-free controllable with respect to $x_s(t)$ at a given time instant $t_c > 0$, if and only if the system (64) is completely controllable at this time instant.*

Proof. The proposition is proven similarly to [26] (Theorem 2). □

Proposition 9. *Let the conditions (10) and (63) be valid. Then, the system (37) (and therefore, the system (11)) is completely controllable at a given time instant $t_c > 0$ if and only if the system (64) is completely controllable at this time instant.*

Proof. The proposition is proven similarly to [25] (Lemma 3.6). □

By virtue of the results of [29] (Lemma 6), we have the following assertion.

Proposition 10. *For a given $t \geq 0$, the system (40)-(41) is Euclidean space output controllable if and only if the system (65), (41) is Euclidean space output controllable.*

Based on Lemma 1 and Propositions 10, we directly have the following corollary.

Corollary 3. *For a given $t \geq 0$, the system (13) is completely Euclidean space controllable if and only if the system (65), (41) is Euclidean space output controllable.*

3.4. Hybrid Set of Riccati-Type Matrix Equations

Let us denote

$$\mathcal{S}_{22} \stackrel{\triangle}{=} \mathcal{B}_2\mathcal{B}_2^T. \tag{66}$$

Consider the following set, consisting of one algebraic and two differential equations (ordinary and partial) for matrices $\mathcal{P}$, $\mathcal{Q}$, and $\mathcal{R}$:

$$\mathcal{P}(t)\mathcal{A}_{40}(t,0) + \mathcal{A}_{40}^T(t,0)\mathcal{P}(t) - \mathcal{P}(t)\mathcal{S}_{22}\mathcal{P}(t) + \mathcal{Q}(t,0) + \mathcal{Q}^T(t,0) + I_{m+r} = 0, \tag{67}$$

$$\begin{aligned} \frac{d\mathcal{Q}(t,\eta)}{d\eta} = \Big(\mathcal{A}_{40}^T(t,0) - \mathcal{P}(t)\mathcal{S}_{22}\Big)\mathcal{Q}(t,\eta) + \mathcal{P}(t)\mathcal{G}_4(t,\eta,0) \\ + \textstyle\sum_{j=1}^{N-1}\mathcal{P}(t)\mathcal{A}_{4j}(t,0)\delta(\eta+h_j) + \mathcal{R}(t,0,\eta), \end{aligned} \tag{68}$$

$$\begin{aligned} \left(\frac{\partial}{\partial\eta} + \frac{\partial}{\partial\chi}\right)\mathcal{R}(t,\eta,\chi) = \mathcal{G}_4^T(t,\eta,0)\mathcal{Q}(t,\chi) \\ + \mathcal{Q}^T(t,\eta)\mathcal{G}_4(t,\chi,0) + \sum_{j=1}^{N-1}\mathcal{A}_{4j}^T(t,0)\mathcal{Q}(t,\chi)\delta(\eta+h_j) \\ + \sum_{j=1}^{N-1}\mathcal{Q}^T(t,\eta)\mathcal{A}_{4j}(t,0)\delta(\chi+h_j) - \mathcal{Q}^T(t,\eta)\mathcal{S}_{22}(t,0)\mathcal{Q}(t,\chi), \end{aligned} \tag{69}$$

where $t \geq 0$ is a parameter; $\eta \in [-h,0]$ and $\chi \in [-h,0]$ are independent variables; $\delta(\cdot)$ is the Dirac delta-function.

The set of the Equations (67)-(69) is subject to the boundary conditions

$$\begin{aligned} \mathcal{Q}(t,-h) = \mathcal{P}(t)\mathcal{A}_{4N}(t,0), \\ \mathcal{R}(t,-h,\eta) = \mathcal{A}_{4N}^T(t,0)\mathcal{Q}(t,\eta), \quad \mathcal{R}(t,\eta,-h) = \mathcal{Q}^T(t,\eta)\mathcal{A}_{4N}(t,0). \end{aligned} \tag{70}$$

Let $t_c > 0$ be a given time instant.

In what follows of this subsection, we assume:

(I) The matrix-valued functions $\mathcal{A}_{4j}(t,0)$, $(j = 0,1,\ldots,N)$ are continuously differentiable in the interval $[0,t_c]$.

(II) The matrix-valued function $\mathcal{G}_4(t,\eta,0)$ is continuously differentiable with respect to $t \in [0,t_c]$ uniformly in $\eta \in [-h,0]$.

(III) The matrix-valued function $\mathcal{G}_4(t,\eta,0)$ is piece-wise continuous with respect to $\eta \in [-h,0]$ for each $t \in [0,t_c]$.

For the sake of the further analysis of the set (67)-(70), we introduce the following definition.

For a given $t \in [0,t_c]$, consider the state-feedback control in the fast subsystem (40)

$$\tilde{v}_f(\omega_{f,\xi}) = \widetilde{K}_{1f}(t)\omega_f(\xi) + \int_{-h}^{0}\widetilde{K}_{2f}(t,\eta)\omega_f(\xi+\eta)d\eta, \tag{71}$$

where $\omega_{f,\xi} = \{\omega_f(\xi+\eta), \eta \in [-h,0]\}$, $\widetilde{K}_{1f}(t)$ and $\widetilde{K}_{2f}(t,\eta)$ are an $r \times m$-matrix and an $r \times m$-matrix-valued function of η, respectively; $\widetilde{K}_{2f}(t,\eta)$ is piece-wise continuous in the interval $[-h,0]$.

Definition 9. *For a given $t \in [0,t_c]$, the fast subsystem (40) is called L^2-stabilizable if there exists the state-feedback control (71) such that for any given $\omega_0 \in E^{m+r}$, $\varphi_{\omega f}(\cdot) \in L^2[-h,0;E^{m+r}]$, the solution $\tilde{\omega}_f(\xi)$*

of (40) with $v_f(\xi) = \tilde{v}_f(\omega_{f,\xi})$ *and subject to the initial conditions (42) satisfies the inclusion* $\widetilde{\omega}_f(\xi) \in L^2[0, +\infty; E^{m+r}]$.

The following proposition is a direct consequence of the results of [33] (Theorems 5.9 and 6.1).

Proposition 11. *Let the assumption (III) be valid. Let, for any* $t \in [0, t_c]$*, the fast subsystem (40) be* L^2*-stabilizable. Then, for any* $t \in [0, t_c]$*, the set of the Equations (67)-(69) subject to the boundary conditions (70) has the unique solution* $\{\mathcal{P}(t),\ \mathcal{Q}(t,\eta),\ \mathcal{R}(t,\eta,\chi),\ (\eta,\chi) \in [-h,0] \times [-h,0]\}$ *such that:*

(a) $\mathcal{P}^T(t) = \mathcal{P}(t)$;
(b) *the matrix-valued function* $\mathcal{Q}(t,\eta)$ *is piece-wise absolutely continuous in* $\eta \in [-h,0]$ *with the bounded jumps at* $\eta = -h_j$, $(j = 1, \dots, N-1)$;
(c) *the matrix-valued function* $\mathcal{R}(t,\eta,\chi)$ *is piece-wise absolutely continuous in* $\eta \in [-h,0]$ *and in* $\chi \in [-h,0]$ *with the bounded jumps at* $\eta = -h_{j_1}$ *and* $\chi = -h_{j_2}$, $(j_1 = 1, \dots, N-1;\ j_2 = 1, \dots, N-1)$*, moreover,* $\mathcal{R}^T(t,\eta,\chi) = \mathcal{R}(t,\chi,\eta)$;
(d) *all roots* $\lambda(t)$ *of the equation*

$$\det\Big[\lambda I_m - \big(\mathcal{A}_{40}(t,0) - \mathcal{S}_{22}\mathcal{P}(t)\big) - \textstyle\sum_{j=1}^{N} \mathcal{A}_{4j}(t,0)\exp(-\lambda h_j) \\ - \textstyle\int_{-h}^{0} \big(\mathcal{G}_4(t,\eta,0) - \mathcal{S}_{22}\mathcal{Q}(t,\eta)\big)\exp(\lambda\eta)d\eta\Big] = 0 \tag{72}$$

satisfy the inequality

$$\mathrm{Re}\lambda(t) < -2\gamma(t), \quad t \in [0, t_c], \tag{73}$$

where $\gamma(t) > 0$ *is some function of* t.

By virtue of the results of [34] (Lemmas 4.1, 4.2 and 3.2), we directly have the following three assertions.

Proposition 12. *Let the assumptions (I)-(III) be valid. Let, for any* $t \in [0, t_c]$*, the fast subsystem (40) be* L^2*-stabilizable. Then, the matrices* $\mathcal{P}(t)$, $\mathcal{Q}(t,\eta)$, $\mathcal{R}(t,\eta,\chi)$ *are continuous functions of* $t \in [0, t_c]$ *uniformly in* $(\eta,\chi) \in [-h,0] \times [-h,0]$.

Proposition 13. *Let the assumptions (I)-(III) be valid. Let, for any* $t \in [0, t_c]$*, the fast subsystem (40) be* L^2*-stabilizable. Then, the derivatives* $d\mathcal{P}(t)/dt$, $\partial\mathcal{Q}(t,\eta)/\partial t$, $\partial\mathcal{R}(t,\eta,\chi)/\partial t$ *exist and are continuous functions of* $t \in [0, t_c]$ *uniformly in* $(\eta,\chi) \in [-h,0] \times [-h,0]$.

Proposition 14. *Let the assumptions (I)-(III) be valid. Let, for any* $t \in [0, t_c]$*, the fast subsystem (40) be* L^2*-stabilizable. Then, there exists a positive number* $\bar{\gamma}$ *such that all roots* $\lambda(t)$ *of the Equation (72) satisfy the inequality* $\lambda(t) < -2\bar{\gamma}$, $t \in [0, t_c]$.

4. Parameter-Free Controllability Conditions

In this section, we derive ε-free sufficient conditions for the Euclidean space output controllability of the auxiliary system (27)-(28), (29) and ε-free sufficient conditions for the complete Euclidean space controllability of the original system (1)-(2).

Let $t_c > 0$ be a given time instant independent of ε.

4.1. Case of the Standard System (1)-(2)

In this subsection, we assume that *the condition (10) holds for all* $t \in [0, t_c]$. In the literature, singularly perturbed systems with such a feature are called standard (see e.g., [1,12]).

In what follows, we also assume:

(AI) The matrix-valued functions $A_{ij}(t,\varepsilon)$, $B_{kj}(t,\varepsilon)$, $(i = 1,\ldots,4;\ j = 0,1,\ldots,N;\ k = 1,2)$, are continuously differentiable with respect to $(t,\varepsilon) \in [0,t_c] \times [0,\varepsilon^0]$.

(AII) The matrix-valued functions $G_i(t,\eta,\varepsilon)$, $(i = 1,\ldots,4)$ are piece-wise continuous with respect to $\eta \in [-h,0]$ for each $(t,\varepsilon) \in [0,t_c] \times [0,\varepsilon^0]$, and they are continuously differentiable with respect to $(t,\varepsilon) \in [0,t_c] \times [0,\varepsilon^0]$ uniformly in $\eta \in [-h,0]$.

(AIII) The matrix-valued functions $H_k(t,\eta,\varepsilon)$, $(k = 1,2)$ are piece-wise continuous with respect to $\eta \in [-h,0]$ for each $(t,\varepsilon) \in [0,t_c] \times [0,\varepsilon^0]$, and they are continuously differentiable with respect to $(t,\varepsilon) \in [0,t_c] \times [0,\varepsilon^0]$ uniformly in $\eta \in [-h,0]$.

(AIV) All roots $\lambda(t)$ of the equation

$$\det\left[\lambda I_m - \sum_{j=0}^{N} A_{4j}(t,0)\exp(-\lambda h_j) - \int_{-h}^{0} G_4(t,\eta,0)\exp(\lambda\eta)d\eta\right] = 0 \tag{74}$$

satisfy the inequality $\mathrm{Re}\lambda(t) < -2\beta$ for all $t \in [0,t_c]$, where $\beta > 0$ is some constant.

Lemma 2. *(Main Lemma) Let the assumptions (AI)-(AIV) be valid. Let the system (37) be completely controllable at the time instant t_c. Let, for $t = t_c$, the system (40)-(41) be Euclidean space output controllable. Then, there exists a positive number ε_1, $(\varepsilon_1 \le \varepsilon^0)$, such that for all $\varepsilon \in (0,\varepsilon_1]$, the singularly perturbed system (27)-(28), (29) is Euclidean space output controllable at the time instant t_c.*

Proof of the lemma is presented in Section 4.3.

Remark 4. *Note that the Euclidean space output controllability for singularly perturbed systems with small state delays was studied in [29]. In this paper, the case of the standard original system was treated in Theorems 1–3 where different ε-free sufficient conditions for the Euclidean space output controllability of the original system were formulated. These conditions depend considerably on relations between the Euclidean dimensions of the state and output variables of the system. However, due to the specific form (19) of the matrix of the coefficients Z in the output equation of the system (27)-(28), (29), only Theorem 1 of [29] and only in the very specific case $n \le r$ is applicable to this system. Therefore, in Section 4.3, we present the proof of Lemma 2 which is not based on the results of [29]. In particular, this proof is uniformly valid for all relations between the Euclidean dimensions of the state and output variables of the system (27)-(28), (29).*

Theorem 1. *Let the assumptions (AI)-(AIV) be valid. Let the system (11) be completely controllable at the time instant t_c. Let, for $t = t_c$, the system (13) be completely Euclidean space controllable. Then, for all $\varepsilon \in (0,\varepsilon_1]$, the singularly perturbed system (1)-(2) is completely Euclidean space controllable at the time instant t_c.*

Proof. Based on Proposition 1, Remark 1 and Lemma 1, the theorem directly follows from Lemma 2. □

4.2. Case of the Nonstandard System (1)-(2)

In this subsection, in contrast with the previous one, we consider the case where *the condition (10) does not hold at least for one value of $t \in [0,t_c]$*. In the literature, singularly perturbed systems with such a feature are called nonstandard (see e.g., [1,12]). Since the condition (10) is not satisfied for some $\bar{t} \in [0,t_c]$, then $\det A_{4s}(\bar{t}) = 0$. The latter, along with the Equation (8), means that one of the roots $\lambda(\bar{t})$ of the Equation (74) equals zero. Thus, in the case of the nonstandard system (1)-(2) the assumption (AIV) is not valid. Therefore, in this subsection, we replace this assumption as follows.

We assume:

(AV) For all $t \in [0,t_c]$ and any complex number λ with $\mathrm{Re}\lambda \ge 0$, the following equality is valid:

$$\mathrm{rank}\Big[F_A(t,\lambda) - \lambda I_m\,,\ F_B(t,\lambda)\Big] = m, \tag{75}$$

where

$$
\begin{aligned}
F_A(t,\lambda) &= \textstyle\sum_{j=0}^{N} A_{4j}(t,0)\exp(-\lambda h_j) + \int_{-h}^{0} G_4(t,\eta,0)\exp(\lambda\eta)d\eta, \\
F_B(t,\lambda) &= \textstyle\sum_{j=0}^{N} B_{2j}(t,0)\exp(-\lambda h_j) + \int_{-h}^{0} H_2(t,\eta,0)\exp(\lambda\eta)d\eta.
\end{aligned}
\tag{76}
$$

Lemma 3. *Let the assumption (AV) be valid. Then, for all $t \in [0,t_c]$ and any complex number λ with $\mathrm{Re}\lambda \geq 0$, the following equality is valid:*

$$
\begin{aligned}
\operatorname{rank}\Big[& \textstyle\sum_{j=0}^{N} \mathcal{A}_{4j}(t,0)\exp(-\lambda h_j) \\
& + \int_{-h}^{0} \mathcal{G}_4(t,\eta,0)\exp(\lambda\eta)d\eta - \lambda I_{m+r}\,,\ \mathcal{B}_2 \Big] = m + r.
\end{aligned}
\tag{77}
$$

Proof. Using the block form of the matrices $\mathcal{A}_{4j}(t,\varepsilon)$, $(j = 0,1,\ldots,N)$, $\mathcal{G}_4(t,\eta,\varepsilon)$, $\mathcal{B}_2$ (see the Equations (22), (23), (25), (26)), we can rewrite the block matrix in the left-hand side of (77) as follows:

$$
\begin{aligned}
&\left(\textstyle\sum_{j=0}^{N} \mathcal{A}_{4j}(t,0)\exp(-\lambda h_j) + \int_{-h}^{0} \mathcal{G}_4(t,\eta,0)\exp(\lambda\eta)d\eta - \lambda I_{m+r}\,,\ \mathcal{B}_2 \right) = \\
&\qquad \begin{pmatrix} F_A(t,\lambda) - \lambda I_m & F_B(t,\lambda) & O_{m\times r} \\ O_{r\times m} & -(\lambda+1)I_r & I_r \end{pmatrix}.
\end{aligned}
\tag{78}
$$

The Equation (78), along with the Equation (75), directly yields the Equation (77), which completes the proof of the lemma. □

Corollary 4. *Let the assumption (AV) be valid. Then, for any $t \in [0,t_c]$, the fast subsystem (40) is L^2-stabilizable.*

Proof. The corollary is a direct consequence of Lemma 3 and the results of [35] (Theorem 3.5). □

Theorem 2. *Let the assumptions (AI)-(AIII),(AV) be valid. Let the system (6)-(7) be impulse-free controllable with respect to $x_s(t)$ at the time instant t_c. Let, for $t = t_c$, the system (13) be completely Euclidean space controllable. Then, there exists a positive number ε_2, $(\varepsilon_2 \leq \varepsilon^0)$, such that for all $\varepsilon \in (0,\varepsilon_2]$, the singularly perturbed system (1)-(2) is completely Euclidean space controllable at the time instant t_c.*

Proof. Let us start with the auxiliary system (27)-(28), (29). Due to the assumptions (AI)-(AIII) and the Equations (20)-(26), the matrix-valued coefficients of this system satisfy the conditions similar to the assumptions (AI) and (AII) on the matrix-valued functions $A_{ij}(t,\varepsilon)$ and $G_i(t,\eta,\varepsilon)$, $(i = 1,\ldots,4;\ j = 0,1,\ldots,N)$.

For a given $\varepsilon \in (0,\varepsilon^0]$ in the auxiliary system (27)-(28), (29), let us make the control transformation (55), where

$$
K_1(t) = -\mathcal{B}_2^T\mathcal{P}(t), \quad K_2(t,\eta) = -\mathcal{B}_2^T\mathcal{Q}(t,\eta), \quad t \in [0,t_c], \ \eta \in [-h,0],
\tag{79}
$$

and $\mathcal{P}(t)$ and $\mathcal{Q}(t,\eta)$ are the components of the solution to the problem (67)-(69), (70) mentioned in Proposition 11. As a result of this transformation, we obtain the system (56)-(57), (29). By virtue of Corollary 4 and Propositions 11, 13, the matrix-valued coefficients of this system satisfy the conditions similar to the assumptions (AI) and (AII) on the matrix-valued functions $A_{ij}(t,\varepsilon)$ and $G_i(t,\eta,\varepsilon)$, $(i = 1,\ldots,4;\ j = 0,1,\ldots,N)$.

The slow and fast subsystems, associated with (56)-(57), (29), are (60)-(61) and (65), (41), respectively. Since the system (6)-(7) is impulse-free controllable with respect to $x_s(t)$ at the time instant t_c, then due to Corollary 2, the system (60)-(61) is impulse-free controllable with respect to

$x_s(t)$ at the time instant t_c. Furthermore, since, for $t = t_c$, the system (13) is completely Euclidean space controllable, then due to Corollary 3, the system (65), (41) for $t = t_c$ is Euclidean space output controllable. By virtue of Corollary 4 and Propositions 11, 14, the value $\lambda = 0$ is not a root of the Equation (72) for all $t \in [0, t_c]$. Hence, the matrix $\mathcal{A}_{4s}^K(t)$, given by (62), (79), is invertible for all $t \in [0, t_c]$. Thus, the slow subsystem (60)-(61) is reduced to the differential Equation (64). Therefore, due to Proposition 8, the above mentioned impulse-free controllability of the system (60)-(61) yields the complete controllability of the system (64) at the time instant t_c. Now, by application of Lemma 2 to the system (56)-(57), (29), we directly obtain the existence of a positive number ε_2, ($\varepsilon_2 \le \varepsilon^0$), such that for all $\varepsilon \in (0, \varepsilon_2]$, this system is Euclidean space output controllable at the time instant t_c. Finally, using Corollary 1 yields the complete Euclidean space controllability of the system (1)-(2) at the time instant t_c for all $\varepsilon \in (0, \varepsilon_2]$, which completes the proof of the theorem. □

4.3. Proof of Main Lemma (Lemma 2)

In the proof of Main Lemma, the following two auxiliary proposition are used.

4.3.1. Auxiliary Propositions

For any given $t \in [0, t_c]$ and any complex number μ, let us consider the matrix

$$\mathcal{W}(t,\mu) = \sum_{j=0}^{N} \mathcal{A}_{4j}(t,0)\exp(-\mu h_j) + \int_{-h}^{0} \mathcal{G}_4(t,\eta,0)\exp(\mu\eta)d\eta, \tag{80}$$

where $\mathcal{A}_{4j}(t,\varepsilon)$, $(j = 0, 1, \dots, N)$ and $\mathcal{G}_4(t,\eta,\varepsilon)$ are given in (22)-(23) and (25), respectively.

Proposition 15. *Let the assumption (AIV) be valid. Then, all roots $\mu(t)$ of the equation*

$$\det\left[\mu I_{m+r} - \mathcal{W}(t,\mu)\right] = 0 \tag{81}$$

satisfy the inequality $\mathrm{Re}\mu(t) < -2\nu$ for all $t \in [0, t_c]$, where $\nu = \min\{\beta, 1/4\}$.

Proof. Using (22)-(23), (25) and (80), we obtain for all $t \in [0, t_c]$:

$$\det\left[\mu I_{m+r} - \mathcal{W}(t,\mu)\right] =$$

$$\det\left[\mu I_m - \sum_{j=0}^{N} A_{41}(t,0)\exp(-\lambda h_j) - \int_{-h}^{0} G_4(t,\eta,0)\exp(\mu\eta)d\eta\right](\mu+1)^r,$$

meaning that for any $t \in [0, t_c]$ the set of all roots $\mu(t)$ of the Equation (81) consists of all roots of the Equation (74) and the root $\mu(t) \equiv -1$ of the multiplicity r. This observation, along with the assumption (AIV), directly yields the statement of the proposition. □

Let us partition the matrix-valued function $\Psi(\sigma,\varepsilon)$, given by the terminal-value problem (48), into blocks as:

$$\Psi(\sigma,\varepsilon) = \begin{pmatrix} \Psi_1(\sigma,\varepsilon) & \Psi_2(\sigma,\varepsilon) \\ \Psi_3(\sigma,\varepsilon) & \Psi_4(\sigma,\varepsilon) \end{pmatrix}, \tag{82}$$

where the blocks $\Psi_1(\sigma,\varepsilon)$, $\Psi_2(\sigma,\varepsilon)$, $\Psi_3(\sigma,\varepsilon)$ and $\Psi_4(\sigma,\varepsilon)$ are of the dimensions $n \times n$, $n \times (m+r)$, $(m+r) \times n$ and $(m+r) \times (m+r)$, respectively.

Proposition 16. *Let the assumptions (AI)-(AIV) be valid. Then, there exists a positive number ε_0, ($\varepsilon_0 \le \varepsilon^0$), such that for all $\varepsilon \in (0, \varepsilon_0]$ the matrix-valued functions $\Psi_1(\sigma,\varepsilon)$, $\Psi_2(\sigma,\varepsilon)$, $\Psi_3(\sigma,\varepsilon)$, $\Psi_4(\sigma,\varepsilon)$ satisfy the inequalities:*

$$\left\|\Psi_1(\sigma,\varepsilon) - \Psi_{1s}(\sigma)\right\| \leq a\varepsilon, \quad \left\|\Psi_2(\sigma,\varepsilon)\right\| \leq a, \ \sigma \in [0,t_c], \tag{83}$$

$$\left\|\Psi_3(\sigma,\varepsilon) - \varepsilon\Psi_{3s}(\sigma)\right\| \leq a\varepsilon\big[\varepsilon + \exp(-\nu(t_c-\sigma)/\varepsilon)\big], \quad \sigma \in [0,t_c], \tag{84}$$

$$\left\|\Psi_4(\sigma,\varepsilon) - \Psi_{4f}\big((t_c-\sigma)/\varepsilon\big)\right\| \leq a\varepsilon, \quad \sigma \in [0,t_c], \tag{85}$$

where

$$\Psi_{1s}(\sigma) = \Psi_s(\sigma), \quad \Psi_{3s}(\sigma) = -\left(\mathcal{A}_{4s}^T(\sigma)\right)^{-1}\mathcal{A}_{2s}^T(\sigma)\Psi_s(\sigma), \quad \sigma \in [0,t_c],$$
$$\Psi_{4f}(\xi) = \Psi_f(\xi,t_c), \quad \xi \geq 0;$$

the matrix-valued functions $\Psi_s(\sigma)$ and $\Psi_f(\xi,t)$ are given by the terminal-value problem (51) and the initial-value problem (53), respectively; $a > 0$ is some constant independent of ε.

Proof. Based on Proposition 15, the validity of the inequalities (83)-(85) is proven similarly to [25] (Lemma 3.2). □

Remark 5. *By virtue of Proposition 15 and the results of [36], we have the inequality*

$$\left\|\Psi_{4f}(\xi)\right\| \leq a\exp(-2\nu\xi), \quad \xi \geq 0, \tag{86}$$

where $a > 0$ is some constant.

4.3.2. Main Part of the Proof

Due to Proposition 3, in order to prove Main Lemma, it is necessary and sufficient to show the existence of a positive number ε_1 such that

$$\det W_Z(t_c,\varepsilon) \neq 0 \quad \forall \varepsilon \in (0,\varepsilon_1], \tag{87}$$

where the $(n+m)\times(n+m)$-matrix $W_Z(t_c,\varepsilon)$ is defined by the Equations (49)-(50).

Let, for a given $\varepsilon \in (0,\varepsilon_0]$, the matrix $W_1(t_c,\varepsilon)$ of the dimension $n\times n$, the matrix $W_2(t_c,\varepsilon)$ of the dimension $n\times(m+r)$ and the matrix $W_3(t_c,\varepsilon)$ of the dimension $(m+r)\times(m+r)$ be the upper left-hand, upper right-hand and lower right-hand blocks, respectively, of the symmetric matrix $W(t_c,\varepsilon)$, given by the Equation (49). Thus,

$$W(t_c,\varepsilon) = \begin{pmatrix} W_1(t_c,\varepsilon) & W_2(t_c,\varepsilon) \\ W_2^T(t_c,\varepsilon) & W_3(t_c,\varepsilon) \end{pmatrix}. \tag{88}$$

Using (49), and the block representations of the matrices $\mathcal{B}(\varepsilon)$ and $\Psi(\sigma,\varepsilon)$ (see the Equations (45) and (82)), we obtain

$$\begin{aligned} W_1(t_c,\varepsilon) = \textstyle\int_0^{t_c} \Big[&\Psi_1^T(\sigma,\varepsilon)\mathcal{S}_{11}\Psi_1(\sigma,\varepsilon) + (1/\varepsilon)\Psi_3^T(\sigma,\varepsilon)\mathcal{S}_{12}^T\Psi_1(\sigma,\varepsilon) \\ &+(1/\varepsilon)\Psi_1^T(\sigma,\varepsilon)\mathcal{S}_{12}\Psi_3(\sigma,\varepsilon) + (1/\varepsilon^2)\Psi_3^T(\sigma,\varepsilon)\mathcal{S}_{22}\Psi_3(\sigma,\varepsilon)\Big]d\sigma, \end{aligned} \tag{89}$$

$$\begin{aligned} W_2(t_c,\varepsilon) = \textstyle\int_0^{t_c} \Big[&\Psi_1^T(\sigma,\varepsilon)\mathcal{S}_{11}\Psi_2(\sigma,\varepsilon) + (1/\varepsilon)\Psi_3^T(\sigma,\varepsilon)\mathcal{S}_{12}^T\Psi_2(\sigma,\varepsilon) \\ &+(1/\varepsilon)\Psi_1^T(\sigma,\varepsilon)\mathcal{S}_{12}\Psi_4(\sigma,\varepsilon) + (1/\varepsilon^2)\Psi_3^T(\sigma,\varepsilon)\mathcal{S}_{22}\Psi_4(\sigma,\varepsilon)\Big]d\sigma, \end{aligned} \tag{90}$$

$$\begin{aligned} W_3(t_c,\varepsilon) = \textstyle\int_0^{t_c} \Big[\Psi_2^T(\sigma,\varepsilon)\mathcal{S}_{11}\Psi_2(\sigma,\varepsilon) + (1/\varepsilon)\Psi_4^T(\sigma,\varepsilon)\mathcal{S}_{12}^T\Psi_2(\sigma,\varepsilon) \\ +(1/\varepsilon)\Psi_2^T(\sigma,\varepsilon)\mathcal{S}_{12}\Psi_4(\sigma,\varepsilon) + (1/\varepsilon^2)\Psi_4^T(\sigma,\varepsilon)\mathcal{S}_{22}\Psi_4(\sigma,\varepsilon)\Big]d\sigma, \end{aligned} \tag{91}$$

where, due to (45),

$$\begin{aligned} \mathcal{S}_{11} = \mathcal{B}_1\mathcal{B}_1^T = O_{n\times n}, \quad \mathcal{S}_{12} = \mathcal{B}_1\mathcal{B}_2^T = O_{n\times(m+r)}, \\ \mathcal{S}_{22} = \mathcal{B}_2\mathcal{B}_2^T = \begin{pmatrix} O_{m\times m} & O_{m\times r} \\ O_{r\times m} & I_r \end{pmatrix}. \end{aligned} \tag{92}$$

The latter, along with (89)-(91), yields

$$W_1(t_c,\varepsilon) = (1/\varepsilon^2)\int_0^{t_c} \Psi_3^T(\sigma,\varepsilon)\mathcal{S}_{22}\Psi_3(\sigma,\varepsilon)d\sigma, \tag{93}$$

$$W_2(t_c,\varepsilon) = (1/\varepsilon^2)\int_0^{t_c} \Psi_3^T(\sigma,\varepsilon)\mathcal{S}_{22}\Psi_4(\sigma,\varepsilon)d\sigma, \tag{94}$$

$$W_3(t_c,\varepsilon) = (1/\varepsilon^2)\int_0^{t_c} \Psi_4^T(\sigma,\varepsilon)\mathcal{S}_{22}\Psi_4(\sigma,\varepsilon)d\sigma. \tag{95}$$

Let us estimate the matrices $W_1(t_c,\varepsilon)$, $W_2(t_c,\varepsilon)$ and $W_3(t_c,\varepsilon)$. We start with $W_1(t_c,\varepsilon)$. Denote

$$\Delta\Psi_3(\sigma,\varepsilon) \stackrel{\triangle}{=} \Psi_3(\sigma,\varepsilon) - \varepsilon\Psi_{3s}(\sigma). \tag{96}$$

Using this notation, we can rewrite the expression (93) for $W_1(t_c,\varepsilon)$ as:

$$\begin{aligned} W_1(t_c,\varepsilon) = (1/\varepsilon^2)\int_0^{t_c} \Big[\varepsilon^2\Psi_{3s}^T(\sigma)\mathcal{S}_{22}\Psi_{3s}(\sigma) + \varepsilon\Psi_{3s}^T(\sigma)\mathcal{S}_{22}\Delta\Psi_3(\sigma,\varepsilon) \\ +\varepsilon\big(\Delta\Psi_3(\sigma,\varepsilon)\big)^T\mathcal{S}_{22}\Psi_{3s}(\sigma) + \big(\Delta\Psi_3(\sigma,\varepsilon)\big)^T\mathcal{S}_{22}\Delta\Psi_3(\sigma,\varepsilon)\Big]d\sigma. \end{aligned} \tag{97}$$

Due to Proposition 16 (see the Equation (84)) and the Equation (96), we have $\|\Delta\Psi_3(\sigma,\varepsilon)\| \le a\varepsilon\big[\varepsilon + \exp(-\nu(t_c-\sigma)/\varepsilon)\big]$, $\sigma \in [0,t_c]$, $\varepsilon \in (0,\varepsilon_0]$. Applying this inequality to the expression (97) for the matrix $W_1(t_c,\varepsilon)$, we obtain the inequality

$$\left\|W_1(t_c,\varepsilon) - \int_0^{t_c} \Psi_{3s}^T(\sigma)\mathcal{S}_{22}\Psi_{3s}(\sigma)d\sigma\right\| \le a\varepsilon, \quad \varepsilon \in (0,\varepsilon_0], \tag{98}$$

where $a > 0$ is some constant independent of ε.

Now, let us treat the integral in the left-hand side of (98). Using the Equation (86), we have

$$\begin{aligned} W_{1s} \stackrel{\triangle}{=} \textstyle\int_0^{t_c} \Psi_{3s}^T(\sigma)\mathcal{S}_{22}\Psi_{3s}(\sigma)d\sigma \\ = \textstyle\int_0^{t_c} \Psi_s^T(\sigma)\mathcal{A}_{2s}(\sigma)\mathcal{A}_{4s}^{-1}(\sigma)\mathcal{S}_{22}\Big(\mathcal{A}_{4s}^T(\sigma)\Big)^{-1}\mathcal{A}_{2s}^T(\sigma)\Psi_s(\sigma)d\sigma. \end{aligned} \tag{99}$$

Taking into account the block form of the matrices $\mathcal{B}_2$, $\mathcal{A}_{2s}(\sigma)$ and $\mathcal{A}_{4s}(\sigma)$ (see the Equations (26), (34)) and the expression for $\bar{B}_s(\sigma)$ (see the Equation (12)), we obtain

$$
\begin{aligned}
&\mathcal{A}_{2s}(\sigma)\mathcal{A}_{4s}^{-1}(\sigma)\mathcal{B}_2 = \\
&\Big(A_{2s}(\sigma), B_{1s}(\sigma)\Big)\begin{pmatrix} A_{4s}(\sigma) & B_{2s}(\sigma) \\ O_{r\times m} & -I_r \end{pmatrix}^{-1}\begin{pmatrix} O_{m\times r} \\ I_r \end{pmatrix} = \\
&\Big(A_{2s}(\sigma), B_{1s}(\sigma)\Big)\begin{pmatrix} A_{4s}^{-1}(\sigma) & A_{4s}^{-1}(\sigma)B_{2s}(\sigma) \\ O_{r\times m} & -I_r \end{pmatrix}\begin{pmatrix} O_{m\times r} \\ I_r \end{pmatrix} = \\
&\Big(A_{2s}(\sigma)A_{4s}^{-1}(\sigma), A_{2s}(\sigma)A_{4s}^{-1}(\sigma)B_{2s}(\sigma) - B_{1s}(\sigma)\Big)\begin{pmatrix} O_{m\times r} \\ I_r \end{pmatrix} = \\
&-\big(B_{1s}(\sigma) - A_{2s}(\sigma)A_{4s}^{-1}(\sigma)B_{2s}(\sigma)\big) = -\bar{B}_s(\sigma).
\end{aligned}
\tag{100}
$$

Finally, using the expression for $\mathcal{S}_{22}$ (see the Equations (92)), as well as the Equations (52), (99) and (100), we obtain that $W_{1s} = W_s(t_c)$. The latter, along with (98), yields

$$
\|W_1(t_c,\varepsilon) - W_s(t_c)\| \le a\varepsilon, \quad \varepsilon \in (0, \varepsilon_0], \tag{101}
$$

where $a > 0$ is some constant independent of ε.

Similarly to (101), we obtain the existence of a positive number $\bar{\varepsilon}_0 \le \varepsilon_0$ such that the following inequalities are satisfied:

$$
\|W_2(t_c,\varepsilon)\| \le a, \quad \|\varepsilon W_3(t_c,\varepsilon) - W_{3f}(t_c)\| \le a\varepsilon, \quad \varepsilon \in (0, \bar{\varepsilon}_0], \tag{102}
$$

where $a > 0$ is some constant independent of ε;

$$
W_{3f}(t_c) = \int_0^{+\infty} \Psi_f^T(\rho, t_c)\mathcal{B}_2\mathcal{B}_2^T\Psi_f(\rho, t_c)d\rho. \tag{103}
$$

By virtue of the inequality (86), the integral in the expression for $W_{3f}(t_c)$ converges.

Now, let us proceed to analysis of the matrix $W_Z(t_c,\varepsilon)$. Using the Equations (19), (50) and (88), we obtains the following block representation of the matrix $W_Z(t_c,\varepsilon)$:

$$
W_Z(t_c,\varepsilon) = \begin{pmatrix} W_1(t_c,\varepsilon) & W_{21}(t_c,\varepsilon) \\ W_{21}^T(t_c,\varepsilon) & W_{31}(t_c,\varepsilon) \end{pmatrix}, \tag{104}
$$

where $W_{21}(t_c,\varepsilon)$ is the left-hand block of the dimension $n \times m$ of the matrix $W_2(t_c,\varepsilon)$, while $W_{31}(t_c,\varepsilon)$ is the upper left-hand block of the dimension $m \times m$ of the matrix $W_3(t_c,\varepsilon)$.

By virtue of (102), we immediately have that

$$
\|W_{21}(t_c,\varepsilon)\| \le a, \quad \|\varepsilon W_{31}(t_c,\varepsilon) - W_{3f,1}(t_c)\| \le a\varepsilon, \quad \varepsilon \in (0, \bar{\varepsilon}_0], \tag{105}
$$

where $W_{3f,1}(t_c)$ is the upper left-hand block of the dimension $m \times m$ of the matrix $W_{3f}(t_c)$.

Let us show that

$$
\det W_{3f,1}(t_c) \ge b, \tag{106}
$$

where $b > 0$ is some number.

Note that $W_{3f,1}(t_c)$ can be represented as:

$$
W_{3f,1}(t_c) = \Omega_f W_{3f}(t_c)\Omega_f^T, \tag{107}
$$

where Ω_f is given in (41).

Comparison of the expressions for $W_f(\xi, t)$ and $W_{3f,1}(t_c)$ (see the Equations (54) and (107)), and use of expression for $W_{3f}(t_c)$ (see the Equation (103)) yield that

$$
W_{3f,1}(t_c) = \lim_{\xi\to+\infty} W_f(\xi, t_c). \tag{108}
$$

Let us observe that, for any $\xi > 0$ and $t \in [0, t_c]$, the matrix $W_f(\xi, t)$ is positive semi-definite. Moreover, since the system (40)-(41) is Euclidean space output controllable for $t = t_c$, then by virtue of Proposition 5, $\det W_f(\xi_c, t_c) \neq 0$ with some $\xi_c > 0$. Therefore, $\det W_f(\xi_c, t_c) > 0$ and $W_f(\xi_c, t_c)$ is a positive definite matrix.

For any $\xi > \xi_c$, we have

$$W_f(\xi, t_c) = W_f(\xi_c, t_c) + \Omega_f \int_{\xi_c}^{\xi} \Psi_f^T(\rho, t)\mathcal{B}_2\mathcal{B}_2^T\Psi_f(\rho, t)d\rho\Omega_f^T,$$

and the second addend in the right-hand side of this equation is a positive semi-definite matrix. Hence, by use of the results of [37], we obtain that

$$\det W_f(\xi, t_c) \geq \det W_f(\xi_c, t_c) > 0, \quad \xi > \xi_c.$$

The latter, along with the equality (108), directly yields the inequality (106), where $b = \det W_f(\xi_c, t_c)$.

Now, we proceed to the proof of the inequality (87). Let us introduce into the consideration the matrix

$$L(\varepsilon) = \begin{pmatrix} I_n & O_{n\times m} \\ O_{m\times n} & \sqrt{\varepsilon}I_m \end{pmatrix}.$$

For any $\varepsilon > 0$, $\det L(\varepsilon) > 0$.

Using the Equation (104), we obtain

$$L(\varepsilon)W_Z(t_c, \varepsilon)L(\varepsilon) = \begin{pmatrix} W_1(t_c, \varepsilon) & \sqrt{\varepsilon}W_{21}(t_c, \varepsilon) \\ \sqrt{\varepsilon}W_{21}^T(t_c, \varepsilon) & \varepsilon W_{31}(t_c, \varepsilon) \end{pmatrix}.$$

Calculating the limit of the determinant of this matrix as $\varepsilon \to +0$, and using the inequalities (101), (105), (106) and Proposition 4, we obtain

$$\lim_{\varepsilon\to+0} \det\Big(L(\varepsilon)W_Z(t_c, \varepsilon)L(\varepsilon)\Big) = \det\begin{pmatrix} W_s(t_c) & 0 \\ 0 & W_{3f,1}(t_c) \end{pmatrix}$$
$$= \det W_s(t_c) \det W_{3f,1}(t_c) \neq 0.$$

This inequality, along with the inequality $\det L(\varepsilon) > 0$, $\varepsilon > 0$, implies the existence of a positive number ε_1 such that the inequality (87) is valid. This completes the proof of Main Lemma.

5. Examples

5.1. Example 1

Consider the following system, a particular case of (1)-(2),

$$\begin{aligned} \frac{dx(t)}{dt} &= x(t) - 4y(t) + 5y(t-\varepsilon) + \int_{-2}^{0} \eta x(t+\varepsilon\eta)d\eta \\ &\quad + (t-5)u(t) - tu(t-\varepsilon), \quad t \geq 0, \\ \varepsilon\frac{dy(t)}{dt} &= 3x(t) + (t-5)y(t) - x(t-\varepsilon) - x(t-2\varepsilon) + y(t-\varepsilon) \\ &\quad + (t-2)u(t) + tu(t-\varepsilon), \quad t \geq 0, \end{aligned} \tag{109}$$

where $x(t)$, $y(t)$ and $u(t)$ are scalars, i.e., $n = m = r = 1$; $h_1 = 1$, $h_2 = h = 2$.

We study the complete Euclidean space controllability of the system (109) at the time instant $t_c = 2$ for all sufficiently small $\varepsilon > 0$. For this purpose, let us write down the slow and fast subsystems associated with (109). Begin with the slow subsystem. For the system (109), the matrix $A_{4s}(t)$, given in (8), becomes a scalar and has the form $A_{4s}(t) = t - 4$. Thus, the condition (10) is satisfied for all $t \in [0, 2]$, meaning that the slow subsystem associated with (109) can be reduced to the differential Equation (11), i.e.,

$$\frac{dx_s(t)}{dt} = \frac{t-3}{4-t}x_s(t) + \frac{7t-22}{4-t}u_s(t), \quad t \in [0,2]. \tag{110}$$

Due to (13), the fast subsystem associated with the system (109) is

$$\frac{dy_f(\xi)}{d\xi} = (t-5)y_f(\xi) + y_f(\xi-1) + (t-2)u_f(\xi) + tu_f(\xi-1), \quad \xi \geq 0, \tag{111}$$

where $t \in [0, 2]$ is a parameter. It should be noted the following. Although the delay in the original system (109) is 2ε, the delay in the fast subsystem is 1 (but not 2), meaning that in this subsystem the coefficients for the terms with the delay 2 equal zero. Therefore, in what follows, it is sufficient to analyze the fast subsystem with the delay 1.

It is seen directly that the assumptions (AI)-(AIII) are satisfied for the system (109). Let us show the fulfillment of the assumption (AIV) for this system. Indeed, the Equation (74) becomes as:

$$\lambda - t + 5 - \exp(-\lambda) = 0. \tag{112}$$

For $\mathrm{Re}\lambda \geq -0.5$, one obtains the following:

$$\mathrm{Re}\big(\lambda - t + 5 - \exp(-\lambda)\big) \geq 2.85 - t > 0 \quad \forall t \in [0,2],$$

meaning that all roots $\lambda(t)$ of the Equation (112) satisfy the inequality $\mathrm{Re}\lambda(t) < -0.5$, $t \in [0, 2]$. Thus, for the system (109) and $t_c = 2$, the assumption (AIV) is satisfied with $\beta = 0.25$. Since the assumptions (AI)-(AIII) also are satisfied for the system (109) and $t_c = 2$, one can try to use Theorem 1 in order to find out whether the system (109) is completely Euclidean space controllable at $t_c = 2$ for all sufficiently small values of $\varepsilon > 0$. For this purpose, proper kinds of controllability of the systems (110) and (111) should be analyzed. Let us start with the system (110). Since the coefficient for $u_s(t)$ in (110) differs from zero for $t \in [0, 2]$, this system is completely controllable at the time instant $t_c = 2$.

Proceed to the system (111). Due to Lemma 1, for the given $t = t_c = 2$, this system is completely Euclidean space controllable if for this value of t the auxiliary system (40)-(41) with the scalar control $v_f(\xi)$ is Euclidean space output controllable. For $t = 2$, this system becomes

$$\begin{aligned} \frac{d\omega_f(\xi)}{d\xi} &= \widetilde{A}\omega_f(\xi) + \widetilde{H}\omega_f(\xi-1) + \widetilde{B}v_f(\xi), \\ \zeta_f(\xi) &= \widetilde{Z}\omega_f(\xi), \quad \xi \geq 0, \end{aligned} \tag{113}$$

where

$$\widetilde{A} = -\begin{pmatrix} 3 & 0 \\ 0 & 1 \end{pmatrix}, \quad \widetilde{H} = \begin{pmatrix} 1 & 2 \\ 0 & 0 \end{pmatrix}, \quad \widetilde{B} = \begin{pmatrix} 0 \\ 1 \end{pmatrix}, \quad \widetilde{Z} = (1\,,\, 0).$$

Note that the Euclidean dimension of the state variable in (113) is n_f =2, while such dimensions of the control and the output are $r_f = 1$ and $q_f = 1$, respectively. To verify the Euclidean space output controllability of the system (113), we apply the algebraic criterion for such a controllability of a time-invariant differential-difference system (see [38,39]). Using this criterion, we are going to show that the system (113) is Euclidean space output controllable at any given instant $\xi_c \in (1, 2]$ of the stretched time ξ. For this purpose, we construct the following matrices:

$$\widetilde{A}_0 = \widetilde{A} = -\begin{pmatrix} 3 & 0 \\ 0 & 1 \end{pmatrix}, \quad \widetilde{A}_1 = \begin{pmatrix} \widetilde{A}_0 & O_{2\times 2} \\ \widetilde{H} & \widetilde{A}_0 \end{pmatrix} = \begin{pmatrix} -3 & 0 & 0 & 0 \\ 0 & -1 & 0 & 0 \\ 1 & 2 & -3 & 0 \\ 0 & 0 & 0 & -1 \end{pmatrix},$$

$$\widetilde{E}_0 = I_2, \quad \widetilde{E}_1 = (O_{2\times 2}, I_2), \quad \widetilde{Z}_0 = \widetilde{Z}, \quad \widetilde{Z}_1 = \widetilde{Z}\widetilde{E}_1 = (0,0,1,0),$$

$$\widetilde{C}_0 = I_2, \quad \widetilde{B}_0 = \widetilde{B}, \quad \widetilde{B}_1 = \widetilde{C}_1\widetilde{B},$$

where

$$\widetilde{C}_1 = \begin{pmatrix} I_2 \\ \exp(\widetilde{A}_0)\widetilde{C}_0 \end{pmatrix} = \begin{pmatrix} 1 & 0 \\ 0 & 1 \\ \exp(-3) & 0 \\ 0 & \exp(-1) \end{pmatrix}.$$

Hence,

$$\widetilde{B}_1 = \begin{pmatrix} 0 \\ 1 \\ 0 \\ \exp(-1) \end{pmatrix}.$$

Due to the results of [38,39], the system (113) is Euclidean space output controllable at a given value $\xi_c \in (1,2]$ of the independent variable ξ, if and only if the rank of the following matrix equals to q_f:

$$\widetilde{D} = (\widetilde{Z}_0\widetilde{B}_0, \ldots, \widetilde{Z}_0\widetilde{A}_0^{n_f-1}\widetilde{B}_0, \widetilde{Z}_1\widetilde{B}_1, \ldots, \widetilde{Z}_1\widetilde{A}_1^{2n_f-1}\widetilde{B}_1).$$

Since each block of the matrix $\widetilde{D}$ is scalar and $q_f = 1$, then it is sufficient to show that at least one block in this matrix differs from zero. Remember that $n_f = 2$. Therefore, $\widetilde{Z}_1\widetilde{A}_1\widetilde{B}_1$ is a block of $\widetilde{D}$. Calculating this block, we obtain $\widetilde{Z}_1\widetilde{A}_1\widetilde{B}_1 = 2 \neq 0$, meaning that $\text{rank}\widetilde{D} = q_f = 1$. Thus, the system (113) is Euclidean space output controllable with any given value $\xi_c \in (1,2]$ mentioned in Definition 7. Hence, the system (111) is completely Euclidean space controllable. Therefore, by virtue of Theorem 1, the system (109) is completely Euclidean space controllable at $t_c = 2$ robustly with respect to $\varepsilon > 0$ for all its sufficiently small values.

5.2. Example 2

Consider the following particular case of the system (1)-(2):

$$\begin{aligned} \frac{dx(t)}{dt} &= 2(t-1)x(t) + 4y(t) - 2tx(t-\varepsilon) - y(t-\varepsilon) \\ &\quad + tu(t) - u(t-\varepsilon) + \int_{-1}^{0} 2t\eta u(t+\varepsilon\eta)d\eta, \quad t \geq 0, \\ \varepsilon\frac{dy(t)}{dt} &= 4x(t) - y(t) - 2x(t-\varepsilon) + y(t-\varepsilon) + 2u(t) - u(t-\varepsilon), \quad t \geq 0, \end{aligned} \tag{114}$$

where $x(t), y(t)$ and $u(t)$ are scalars, i.e., $n = m = r = 1; h = 1$.

In this example, like in the previous one, we study the complete Euclidean space controllability of the considered system. We study this controllability at the time instant $t_c = 2$ for all sufficiently small $\varepsilon > 0$.

The asymptotic decomposition of the system (114) yields the slow and fast subsystems, respectively,

$$\begin{aligned} \frac{dx_s(t)}{dt} &= -2x_s(t) + 3y_s(t) - u_s(t), \quad t \geq 0, \\ 0 &= 2x_s(t) + u_s(t), \quad t \geq 0, \end{aligned} \tag{115}$$

and

$$\frac{dy_f(\xi)}{d\xi} = -y_f(\xi) + y_f(\xi - 1) + 2u_f(\xi) - u_f(\xi - 1), \quad \xi \geq 0. \tag{116}$$

It is seen that the assumptions (AI)-(AIII) are satisfied for the system (114). The condition (10) is not satisfied for this system, meaning that (114) is a nonstandard system, and it does not satisfy the assumption (AIV). Indeed, for the system (114), the Equation (74) becomes as:

$$\lambda + 1 - \exp(-\lambda) = 0. \tag{117}$$

For this equation, $\lambda = 0$ is a single root with the nonnegative real part.

Let us show the fulfillment of the assumption (AV) for the system (114). The matrix in the Equation (75) becomes as:

$$\left[-1 + \exp(-\lambda) - \lambda\, , 2 - \exp(-\lambda) \right]. \tag{118}$$

For $\lambda = 0$, the rank of this matrix equals to the Euclidean dimension of the fast subsystem $m = 1$. Since $\lambda = 0$ is a single root with the nonnegative real part of the Equation (117), then the rank of the matrix (118) equals $m = 1$ for all complex λ with $Re\lambda \geq 0$. Thus, the assumption (AV) is fulfilled for the system (114).

Now, let us find out whether the systems (115) and (116) are controllable in the sense mentioned in Theorem 2. We start with (115). Let x_0 and x_c be any given numbers. Let $\vartheta = (x_c - x_0)/6$. One can verify immediately that for the numbers x_0 and x_c, there exists a control $u_s(t) \in L^2[0, 2; E^1]$, namely,

$$u_s(t) = -2x_0 - 3\vartheta t^2,$$

such that the system (115), subject to the initial $x_s(0) = x_0$ and terminal $x_s(2) = x_c$ conditions, has an impulse-free solution, namely,

$$x_s(t) = x_0 + 1.5\vartheta t^2, \quad y_s(t) = \vartheta t.$$

Thus, the system (115) is impulse-free controllable with respect to $x_s(t)$ at the time instant $t_c = 2$. Proceed to (116). The complete Euclidean space controllability of this system is shown similarly to such a kind of controllability of the system (111) in the previous example. Now, using Theorem 2, we obtain the complete Euclidean space controllability of the system (114) at the time instant $t_c = 2$ robustly with respect to $\varepsilon > 0$ for all its sufficiently small values.

6. Conclusions

In this paper, a singularly perturbed linear time-dependent controlled differential system with time delays (multiple point-wise and distributed) in the state and control variables was analyzed. The case where the delays are small of the order of a small positive multiplier ε for a part of the derivatives in the differential equations was treated. The complete Euclidean space controllability of the considered system, robust with respect to the small parameter ε, was studied. This study uses the asymptotic decomposition of the original system into two lower dimensions ε-free subsystems, the slow and fast ones. The slow subsystem is a differential-algebraic delay-free system. This subsystem, subject to a proper assumption, can be converted to a differential equation. The fast subsystem is a differential system with multiple point-wise delays and distributed delays in the state and the control. It was shown that proper kinds of controllability of the slow and fast subsystems yield the complete Euclidean space controllability of the original system valid for all sufficiently small values of ε.

Funding: This research received no external funding.

Conflicts of Interest: The author declares no conflict of interest.

References

1. Kokotovic, P.V.; Khalil, H.K.; O'Reilly, J. *Singular Perturbation Methods in Control: Analysis and Design*; Academic Press: London, UK, 1986.
2. Naidu, D.S.; Calise, A.J. Singular perturbations and time scales in guidance and control of aerospace systems: A survey. *J. Guid. Control Dyn.* **2001**, *24*, 1057–1078. [CrossRef]
3. O'Malley, R.E., Jr. *Historical Developments in Singular Perturbations*; Springer: New York, NY, USA, 2014.
4. Reddy, P.B.; Sannuti, P. Optimal control of a coupled-core nuclear reactor by singular perturbation method. *IEEE Trans. Autom. Control* **1975**, *20*, 766–769. [CrossRef]
5. Pena, M.L. Asymptotic expansion for the initial value problem of the sunflower equation. *J. Math. Anal. Appl.* **1989**, *143*, 471–479.
6. Lange, C.G.; Miura, R.M. Singular perturbation analysis of boundary-value problems for differential-difference equations. Part V: small shifts with layer behavior. *SIAM J. Appl. Math.* **1994**, *54*, 249–272. [CrossRef]
7. Schöll, E.; Hiller, G.; Hövel, P.; Dahlem, M.A. Time-delayed feedback in neurosystems. *Philos. Trans. R. Soc. A* **2009**, *367*, 1079–1096. [CrossRef]
8. Fridman E. Robust sampled-data H_∞ control of linear singularly perturbed systems. *IEEE Trans. Autom. Control* **2006**, *51*, 470–475. [CrossRef]
9. Stefanovic, N.; Pavel, L. A Lyapunov-Krasovskii stability analysis for game-theoretic based power control in optical links. *Telecommun. Syst.* **2011**, *47*, 19–33. [CrossRef]
10. Pavel, L. *Game Theory for Control of Optical Networks*; Birkhauser: Basel, Switzerland, 2012.
11. Glizer, V.Y. On stabilization of nonstandard singularly perturbed systems with small delays in state and control. *IEEE Trans. Autom. Control* **2004**, *49*, 1012–1016. [CrossRef]
12. Gajic, Z.; Lim, M-T. *Optimal Control of Singularly Perturbed Linear Systems and Applications. High Accuracy Techniques*; Marsel Dekker Inc.: New York, NY, USA, 2001.
13. Dmitriev, M.G.; Kurina, G.A. Singular perturbations in control problems. *Autom. Remote Control* **2006**, *67*, 1–43. [CrossRef]
14. Zhang, Y.; Naidu, D.S.; Cai, C.; Zou, Y. Singular perturbations and time scales in control theories and applications: An overview 2002–2012. *Int. J. Inf. Syst. Sci.* **2014**, *9*, 1–36.
15. Kalman, R.E. Contributions to the theory of optimal control. *Bol. Soc. Mat. Mex.* **1960**, *5*, 102–119.
16. Bensoussan, A.; Da Prato, G.; Delfour, M.C.; Mitter, S.K. *Representation and Control of Infinite Dimensional Systems*; Birkhuser: Boston, MA, USA, 2007.
17. Klamka, J. *Controllability of Dynamical Systems*; Kluwer Academic Publishers: Dordrecht, The Netherlands, 1991.
18. Klamka, J. Controllability of dynamical systems. A survey. *Bull. Pol. Acad. Sci. Tech.* **2013**, *61*, 335–342. [CrossRef]
19. Kokotovic, P.V.; Haddad, A.H. Controllability and time-optimal control of systems with slow and fast modes. *IEEE Trans. Autom. Control* **1975**, *20*, 111–113. [CrossRef]
20. Sannuti, P. On the controllability of singularly perturbed systems. *IEEE Trans. Autom. Control* **1977**, *22*, 622–624. [CrossRef]
21. Sannuti, P. On the controllability of some singularly perturbed nonlinear systems. *J. Math. Anal. Appl.* **1978**, *64*, 579–591. [CrossRef]
22. Kurina, G.A. Complete controllability of singularly perturbed systems with slow and fast modes. *Math. Notes* **1992**, *52*, 1029–1033. [CrossRef]
23. Kopeikina, T.B. Controllability of singularly perturbed linear systems with time-lag. *Differ. Equ.* **1989**, *25*, 1055–1064.
24. Glizer, V.Y. Euclidean space controllability of singularly perturbed linear systems with state delay. *Syst. Control Lett.* **2001**, *43*, 181–191. [CrossRef]
25. Glizer, V.Y. Controllability of singularly perturbed linear time-dependent systems with small state delay. *Dyn. Control* **2001**, *11*, 261–281. [CrossRef]
26. Glizer, V.Y. Controllability of nonstandard singularly perturbed systems with small state delay. *IEEE Trans. Autom. Control* **2003**, *48*, 1280–1285. [CrossRef]

27. Glizer, V.Y. Novel controllability conditions for a class of singularly perturbed systems with small state delays. *J. Optim. Theory Appl.* **2008**, *137*, 135–156. [CrossRef]
28. Glizer, V.Y. Controllability conditions of linear singularly perturbed systems with small state and input delays. *Math. Control Signals Syst.* **2016**, *28*, 1–29. [CrossRef]
29. Glizer, V.Y. Euclidean space output controllability of singularly perturbed systems with small state delays. *J. Appl. Math. Comput.* **2018**, *57*, 1–38. [CrossRef]
30. Glizer, V.Y. Euclidean space controllability conditions and minimum energy problem for time delay system with a high gain control. *J. Nonlinear Var. Anal.* **2018**, *2*, 63–90.
31. Kopeikina, T.B. Unified method of investigating controllability and observability problems of time variable differential systems. *Funct. Differ. Equ.* **2006**, *13*, 463–481.
32. Halanay, A. *Differential Equations: Stability, Oscillations, Time Lags*; Academic Press: New York, NY, USA, 1966.
33. Delfour, M.C.; McCalla, C.; Mitter, S.K. Stability and the infinite-time quadratic cost problem for linear hereditary differential systems. *SIAM J. Control* **1975**, *13*, 48–88. [CrossRef]
34. Glizer, V.Y. Dependence on parameter of the solution to an infinite horizon linear-quadratic optimal control problem for systems with state delays. *Pure Appl. Funct. Anal.* **2017**, *2*, 259–283.
35. Pritchard, A.J.; Salamon, D. The linear-quadratic control problem for retarded systems with delays in control and observation. *IMA J. Math. Control Inform.* **1985**, *2*, 335–362. [CrossRef]
36. Hale, J.K.; Verduyn Lunel, S.M. *Introduction to Functional Differential Equations*; Springer: New York, NY, USA, 1993.
37. Bellman, R. *Introduction to Matrix Analysis*; SIAM: Philadelphia, PA, USA, 1997.
38. Zmood, R.B. *On Euclidean Space and Function Space Controllability of Control Systems With Delay*; Technical report; The University of Michigan: Ann Arbor, MI, USA,1971; p. 99,
39. Zmood, R.B. The Euclidean space controllability of control systems with delay. *SIAM J. Control* **1974**, *12*, 609–623. [CrossRef]

axioms

MDPI

Article

Regularization Method for Singularly Perturbed Integro-Differential Equations with Rapidly Oscillating Coefficients and Rapidly Changing Kernels

Burkhan Kalimbetov [1,*] and Valeriy Safonov [2]

1 Department of Mathematics, Akhmed Yassawi University, B. Sattarkhanov 29, Turkestan 161200, Kazakhstan
2 Department of Higher Mathematics, National Research University «MPEI», Krasnokazarmennaya 14, 111250 Moscow, Russia; Singsaf@yandex.ru
* Correspondence: burkhan.kalimbetov@ayu.edu.kz

Received: 30 September 2020; Accepted: 10 November 2020; Published: 13 November 2020

Abstract: In this paper, we consider a system with rapidly oscillating coefficients, which includes an integral operator with an exponentially varying kernel. The main goal of the work is to develop an algorithm for the regularization method for such systems and to identify the influence of the integral term on the asymptotic behavior of the solution of the original problem.

Keywords: singular perturbation; integro-differential equation; rapidly oscillating coefficient; regularization; asymptotic convergence; resonant exhibitors

MSC: 34K26; 45J05

1. Introduction

In the study of various issues related to dynamic stability, with the properties of media with a periodic structure, in the study of other applied problems, one has to deal with differential equations with rapidly oscillating coefficients. Equations of this kind can describe some mechanical or electrical systems that are under the influence of high-frequency external forces, automatic control systems with a linear adjustable object, etc. As an example, we can cite the principle of operation of an oscillator with a small mass and a nonlinear restoring force, in which a high-frequency periodic force with a large amplitude acts. The presence of high-frequency terms creates serious problems for their direct numerical solutions. Therefore, asymptotic methods are usually applied to such equations first, the most famous of which are the Feshchenko–Shkil–Nikolenko splitting method [1–5] and the Lomov's regularization method [6–8]. It should also be noted that singularly perturbed equations are the object of study by several Russian researchers, as well as other scientists (see, for example [9–22]).

In this paper, the Lomov's regularization method is generalized to previously unexplored integro-differential equations with rapidly oscillating coefficients and with rapidly decreasing kernels of the form

$$\varepsilon\frac{dz}{dt} - a(t)z - \varepsilon g(t)\cos\frac{\beta(t)}{\varepsilon}z - \int_{t_0}^{t} e^{\frac{1}{\varepsilon}\int_s^t \mu(\theta)d\theta} K(t,s)z(s,\varepsilon)ds = h(t),\ z(t_0,\varepsilon) = z^0,\ t\in[t_0,T] \quad (1)$$

where $z = z(t,\varepsilon)$, $h(t)$, $\beta'(t) > 0$, $a(t) > 0, \mu(t) < 0$, $a(t) \neq \mu(t)$ $(\forall t \in [t_0,T])$, $g(t)$ are scalar functions, z^0 is a constant, $\varepsilon > 0$ is a small parameter. In the case $\beta(t) = 2\gamma(t)$, and of the absence of an integral term, such a system was considered in [6–8].

The limit operator $a(t)$ has a spectrum $\lambda_1(t) = a(t)$, functions $\lambda_2(t) = -i\beta'(t)$ and $\lambda_3(t) = +i\beta'(t)$ are associated with the presence in Equation (1) of a rapidly oscillating $\cos\frac{\beta(t)}{\varepsilon}$, and the function $\lambda_4(t) = \mu(t)$ characterizes the rapid change in the kernel of the integral operator.

We introduce the following notations:

$\lambda(t) = (\lambda_1(t), ..., \lambda_4(t))$,

$m = (m_1, ..., m_4)$ is multi-index with non-negative components m_j, $j = \overline{1,4}$,

$|m| = \sum_{j=1}^{4} m_j$ is multi-index height m,

$(m, \lambda(t)) = \sum_{j=1}^{4} m_j \lambda_j(t)$.

Assume that the following conditions are met:

(1) $a(t), \beta(t), \mu(t) \in C^\infty([t_0, T], \mathbb{R})$, $g(t), h(t) \in C^\infty([t_0, T], \mathbb{C})$, $K(t,s) \in C^\infty\{t_0 \le s \le t \le T, \mathbb{C}\}$;

(2) the relations $(m, \lambda(t)) = 0$, $(m, \lambda(t)) = \lambda_j(t)$, $j \in \{1, ..., 4\}$ for all multi-indices m with $|m| \ge 2$ or are not fulfilled for any $t \in [t_0, T]$, or are fulfilled identically on the whole segment $t \in [t_0, T]$.

In other words, resonant multi-indices are exhausted by the following sets

$$\Gamma_0 = \{m : (m, \lambda(t)) \equiv 0, |m| \ge 2, \forall t \in [t_0, T]\},$$
$$\Gamma_j = \{m : (m, \lambda(t)) \equiv \lambda_j(t), |m| \ge 2, \forall t \in [t_0, T]\}, \ j = \overline{1,4}.$$

Under these conditions, we will develop an algorithm for constructing a regularized [6] asymptotic solution of the problem (1).

2. Regularization of the Problem (1)

Denote by $\sigma_j = \sigma_j(\varepsilon)$ independent of the t quantities $\sigma_1 = e^{-\frac{i}{\varepsilon}\beta(t_0)}$, $\sigma_2 = e^{+\frac{i}{\varepsilon}\beta(t_0)}$, and rewrite the Equation (1) in the form

$$\begin{array}{c} \mathbf{L} z(t,\varepsilon) \equiv \varepsilon\frac{dz}{dt} - a(t)z - \varepsilon\frac{g(t)}{2}\left(e^{-\frac{i}{\varepsilon}\int_{t_0}^t \beta'(\theta)d\theta}\sigma_1 + e^{+\frac{i}{\varepsilon}\int_{t_0}^t \beta'(\theta)d\theta}\sigma_2\right) z - \\ - \int_{t_0}^t e^{\frac{1}{\varepsilon}\int_s^t \mu(\theta)d\theta} K(t,s) z(s,\varepsilon)ds = h(t), \quad z(t_0,\varepsilon) = z^0, \quad t \in [t_0, T]. \end{array} \tag{2}$$

We introduce regularizing variables

$$\tau_j = \frac{1}{\varepsilon}\int_{t_0}^t \lambda_j(\theta)d\theta \equiv \frac{\psi_j(t)}{\varepsilon}, \ j = \overline{1,4} \tag{3}$$

and instead of problem (2) we consider the problem

$$\begin{array}{c} \mathbf{L}\tilde{z}(t,\tau,\varepsilon) \equiv \varepsilon\frac{\partial \tilde{z}}{\partial t} + \sum_{j=1}^{4} \lambda_j(t)\frac{\partial \tilde{z}}{\partial \tau_j} - \lambda_1(t)\tilde{z} - \varepsilon\frac{g(t)}{2}\left(e^{\tau_2}\sigma_1 + e^{\tau_3}\sigma_2\right)\tilde{z} - \\ - \int_{t_0}^t e^{\frac{1}{\varepsilon}\int_s^t \lambda_4(\theta)d\theta} K(t,s)\tilde{z}\left(s, \frac{\psi(s)}{\varepsilon}, \varepsilon\right) ds = h(t), \ \tilde{z}(t,\tau,\varepsilon)|_{t=t_0, \tau=0} = z^0, \quad t \in [t_0, T] \end{array} \tag{4}$$

for the function $\tilde{z} = \tilde{z}(t,\tau,\varepsilon)$, where it is indicated (according to (3)): $\tau = (\tau_1, ..., \tau_4)$, $\psi = (\psi_1, ..., \psi_4)$. It is clear that if $\tilde{z} = \tilde{z}(t,\tau,\varepsilon)$ is the solution of the problem (4), then the function $z = \tilde{z}\left(t, \frac{\psi(t)}{\varepsilon}, \varepsilon\right)$ is an exact solution of the problem (2), therefore, the problem (4) is an extension of the problem (2).

However, (4) cannot be considered completely regularized, since the integral term

$$J\tilde{z} = \int_{t_0}^t e^{\frac{1}{\varepsilon}\int_s^t \lambda_4(\theta)d\theta} K(t,s)\,\tilde{z}\left(s, \frac{\psi(s)}{\varepsilon}, \varepsilon\right) ds$$

has not been regularized in it. To regularize J, we introduce a class M_ε, asymptotically invariant with respect to the operator $J\tilde{z}$ (see [6]; p. 62).

We first consider the space U of functions $z(t,\tau)$, representable by sums

$$z(t,\tau,\sigma) = z_0(t,\sigma) + \sum_{i=1}^{4} z_i(t,\sigma)e^{\tau_i} + \sum^{*}_{2\le|m|\le N_z} z^m(t,\sigma)e^{(m,\tau)},$$
$$z_0(t,\sigma), z_i(t,\sigma), z^m(t,\sigma) \in C^\infty([t_0,T],\mathbb{C}),\ i=\overline{1,4},\ 2\le|m|\le N_z \tag{5}$$

where the asterisk $*$ above the sum sign indicates that in it the summation for $|m|\ge 2$ occurs only over nonresonant multi-indices $m=(m_1,...,m_4)$, i.e., over $m \notin \bigcup_{i=0}^{4}\Gamma_i$.

Note that in (5) the degree N_z of the polynomial $z(t,\tau,\sigma)$ to exponentials e^{τ_j} depends on the element z. The elements of the space U depend on bounded in $\varepsilon>0$ constants $\sigma_1=\sigma_1(\varepsilon)$ and $\sigma_2=\sigma_2(\varepsilon)$, which do not affect the development of the algorithm described below, therefore in the notation of element (5) of this space U we omit the dependence on $\sigma=(\sigma_1,\sigma_2)$ for brevity. We show that the class $M_\varepsilon = U|_{\tau=\psi(t)/\varepsilon}$ is asymptotically invariant with respect to the operator J.

The image of the operator J on the element (5) of the space U has the form:

$$Jz(t,\tau) = \int_{t_0}^{t} e^{\frac{1}{\varepsilon}\int_s^t \lambda_4(\theta)d\theta} K(t,s)z_0(s)\,ds + \sum_{i=1}^{4}\int_{t_0}^{t} e^{\frac{1}{\varepsilon}\int_s^t \lambda_4(\theta)d\theta} K(t,s)z_i(s)e^{\frac{1}{\varepsilon}\int_{t_0}^{s}\lambda_i(\theta)d\theta}ds +$$
$$+\sum^{*}_{2\le|m|\le N_z}\int_{t_0}^{t} e^{\frac{1}{\varepsilon}\int_s^t \lambda_4(\theta)d\theta} K(t,s)z^m(s)e^{\frac{1}{\varepsilon}\int_{t_0}^{s}(m,\lambda(\theta))d\theta}ds =$$
$$= \int_{t_0}^{t} e^{\frac{1}{\varepsilon}\int_s^t \lambda_4(\theta)d\theta} K(t,s)z_0(s)\,ds + e^{\frac{1}{\varepsilon}\int_{t_0}^{t}\lambda_4(\theta)d\theta}\int_{t_0}^{t}K(t,s)z_4(s)\,ds +$$
$$+\sum_{i=1,i\ne 4}^{4} e^{\frac{1}{\varepsilon}\int_{t_0}^{t}\lambda_4(\theta)d\theta}\int_{t_0}^{t}K(t,s)z_i(s)e^{\frac{1}{\varepsilon}\int_{t_0}^{s}(\lambda_i(\theta)-\lambda_4(\theta))d\theta}ds +$$
$$+\sum^{*}_{2\le|m|\le N_z} e^{\frac{1}{\varepsilon}\int_{t_0}^{t}\lambda_4(\theta)d\theta}\int_{t_0}^{t}K(t,s)z^m(s)e^{\frac{1}{\varepsilon}\int_{t_0}^{s}(m-e_4,\lambda(\theta))d\theta}ds.$$

Integrating in parts, we have

$$J_0(t,\varepsilon) = \int_{t_0}^{t}K(t,s)z_0(s)e^{\frac{1}{\varepsilon}\int_{t_0}^{s}\lambda_4(\theta)d\theta}ds = \varepsilon\int_{t_0}^{t}\frac{K(t,s)z_0(s)}{\lambda_4(s)}de^{\frac{1}{\varepsilon}\int_{t_0}^{s}\lambda_4(\theta)d\theta} =$$
$$=\varepsilon\frac{K(t,s)z_0(s)}{\lambda_4(s)}e^{\frac{1}{\varepsilon}\int_{t_0}^{s}\lambda_4(\theta)d\theta}\Big|_{s=t_0}^{s=t} - \varepsilon\int_{t_0}^{t}\left(\frac{\partial}{\partial s}\frac{K(t,s)z_0(s)}{\lambda_4(s)}\right)e^{\frac{1}{\varepsilon}\int_{t_0}^{s}\lambda_4(\theta)d\theta}ds =$$
$$=\varepsilon\left[\frac{K(t,t)z_0(t)}{\lambda_4(t)}e^{\frac{1}{\varepsilon}\int_{t_0}^{t}\lambda_4(\theta)d\theta} - \frac{K(t,t_0)z_0(t_0)}{\lambda_4(t_0)}\right] - \varepsilon\int_{t_0}^{t}\left(\frac{\partial}{\partial s}\frac{K(t,s)z_0(s)}{\lambda_4(s)}\right)e^{\frac{1}{\varepsilon}\int_{t_0}^{s}\lambda_4(\theta)d\theta}ds.$$

Continuing this process further, we obtained the decomposition

$$J_0(t,\varepsilon) = \sum_{\nu=0}^{\infty}(-1)^\nu\varepsilon^{\nu+1}\left[\left(I_0^\nu(K(t,s)z_0(s))\right)_{s=t}e^{\frac{1}{\varepsilon}\int_{t_0}^{t}\lambda_4(\theta)d\theta} - \left(I_0^\nu(K(t,s)z_0(s))\right)_{s=t_0}\right],$$
$$\boxed{I_0^0 = \frac{1}{\lambda_4(s)}\cdot,\quad I_0^\nu = \frac{1}{\lambda_4(s)}\frac{\partial}{\partial s}I_0^{\nu-1}\ (\nu\ge 1)}.$$

Next, apply the same operation to the integrals:

$$J_{4,i}(t,\varepsilon) = e^{\frac{1}{\varepsilon}\int_{t_0}^{t}\lambda_4(\theta)d\theta}\int_{t_0}^{t}K(t,s)z_i(s)e^{\frac{1}{\varepsilon}\int_{t_0}^{s}(\lambda_i(\theta)-\lambda_4(\theta))d\theta}ds =$$
$$=\varepsilon e^{\frac{1}{\varepsilon}\int_{t_0}^{t}\lambda_4(\theta)d\theta}\int_{t_0}^{t}\frac{K(t,s)z_i(s)}{\lambda_i(s)-\lambda_4(s)}de^{\frac{1}{\varepsilon}\int_{t_0}^{s}(\lambda_i(\theta)-\lambda_4(\theta))d\theta} =$$

$$= \varepsilon e^{\frac{1}{\varepsilon}\int_{t_0}^{t}\lambda_4(\theta)d\theta}\left[\frac{K(t,s)\,z_i(s)}{\lambda_i(s)-\lambda_4(s)}e^{\frac{1}{\varepsilon}\int_{t_0}^{s}(\lambda_i(\theta)-\lambda_4(\theta))d\theta}\Big|_{s=t_0}^{s=t} -\right.$$

$$\left. -\varepsilon\int_{t_0}^{t}\left(\frac{\partial}{\partial s}\frac{K(t,s)\,z_i(s)}{\lambda_i(s)-\lambda_4(s)}\right)e^{\frac{1}{\varepsilon}\int_{t_0}^{s}(\lambda_i(\theta)-\lambda_4(\theta))d\theta}ds\right] =$$

$$= \varepsilon\left[\frac{K(t,t)z_i(t)}{\lambda_i(t)-\lambda_4(t)}e^{\frac{1}{\varepsilon}\int_{t_0}^{t}\lambda_i(\theta)d\theta} - \frac{K(t,t_0)z_i(t_0)}{\lambda_i(t_0)-\lambda_4(t_0)}e^{\frac{1}{\varepsilon}\int_{t_0}^{t}\lambda_4(\theta)d\theta}\right] -$$

$$-\varepsilon e^{\frac{1}{\varepsilon}\int_{t_0}^{t}\lambda_4(\theta)d\theta}\int_{t_0}^{t}\left(\frac{\partial}{\partial s}\frac{K(t,s)z_i(s)}{\lambda_i(s)-\lambda_4(s)}\right)e^{\frac{1}{\varepsilon}\int_{t_0}^{s}(\lambda_i(\theta)-\lambda_4(\theta))d\theta}ds =$$

$$= \Sigma_{\nu=0}^{\infty}(-1)^{\nu}\varepsilon^{\nu+1}\left[\left(I_i^{\nu}\left(K(t,s)\,z_i(s)\right)\right)_{s=t}e^{\frac{1}{\varepsilon}\int_{t_0}^{t}\lambda_i(\theta)d\theta} - \left(I_i^{\nu}\left(K(t,s)\,z_i(s)\right)\right)_{s=t_0}e^{\frac{1}{\varepsilon}\int_{t_0}^{t}\lambda_4(\theta)d\theta}\right],$$

$$\boxed{I_i^0 = \frac{1}{\lambda_i(s)-\lambda_4(s)}\cdot,\quad I_i^{\nu} = \frac{1}{\lambda_i(s)-\lambda_4(s)}\frac{\partial}{\partial s}I_i^{\nu-1},\ \nu\geq 1,\ i=\overline{1,3}}.$$

Denote bay $e_4 = (0,0,0,1)$. Then

$$J_m(t,\varepsilon) = e^{\frac{1}{\varepsilon}\int_{t_0}^{t}\lambda_4(\theta)d\theta}\int_{t_0}^{t}K(t,s)\,z^m(s)\,e^{\frac{1}{\varepsilon}\int_{t_0}^{s}(m-e_4,\lambda(\theta))d\theta}ds =$$

$$= \varepsilon e^{\frac{1}{\varepsilon}\int_{t_0}^{t}\lambda_4(\theta)d\theta}\int_{t_0}^{t}\frac{K(t,s)\,z^m(s)}{(m-e_4,\lambda(s))}de^{\frac{1}{\varepsilon}\int_{t_0}^{s}(m-e_4,\lambda(\theta))d\theta} =$$

$$= \varepsilon e^{\frac{1}{\varepsilon}\int_{t_0}^{t}\lambda_4(\theta)d\theta}\Big[\frac{K(t,s)\,z^m(s)}{(m-e_4,\lambda(s))}e^{\frac{1}{\varepsilon}\int_{t_0}^{s}(m-e_4,\lambda(\theta))d\theta}\Big|_{s=t_0}^{s=t} -$$

$$-\int_{t_0}^{t}\left(\frac{\partial}{\partial s}\frac{K(t,s)\,z^m(s)}{(m-e_4,\lambda(s))}\right)e^{\frac{1}{\varepsilon}\int_{t_0}^{s}(m-e_4,\lambda(\theta))d\theta}ds] =$$

$$= \sum_{\nu=0}^{\infty}(-1)^{\nu}\varepsilon^{\nu+1}[\left(I_{4,m}^{\nu}\left(K(t,s)\,z^m(s)\right)\right)_{s=t}e^{\frac{1}{\varepsilon}\int_{t_0}^{t}(m,\lambda(\theta))d\theta} -$$

$$-\left(I_{4,m}^{\nu}\left(K(t,s)\,z^m(s)\right)\right)_{s=t_0}e^{\frac{1}{\varepsilon}\int_{t_0}^{t}\lambda_4(\theta)d\theta}],$$

$$\boxed{I_{4,m}^0 = \frac{1}{(m-e_4,\lambda(s))}\cdot,\ I_{4,m}^{\nu} = \frac{1}{(m-e_4,\lambda(s))}\frac{\partial}{\partial s}I_{4,m}^{\nu-1},\ \nu\geq 1,\ 2\leq|m|\leq N_z.}$$

Here it is taken into account that $(m-e_4,\lambda(s)) \neq 0$, since by the definition of the space U multi-indices $m \notin \Gamma_4$. The image of the operator J on the space U element (5) is represented as a series

$$Jz(t,\tau) = e^{\frac{1}{\varepsilon}\int_{t_0}^{t}\lambda_4(\theta)d\theta}\int_{t_0}^{t}K(t,s)\,z_4(s)\,ds + \sum_{\nu=0}^{\infty}(-1)^{\nu}\varepsilon^{\nu+1}\left[\left(I_0^{\nu}\left(K(t,s)\,z_0(s)\right)\right)_{s=t}e^{\frac{1}{\varepsilon}\int_{t_0}^{t}\lambda_4(\theta)d\theta} -\right.$$

$$\left.-\left(I_0^{\nu}\left(K(t,s)\,z_0(s)\right)\right)_{s=t_0}\right] + \sum_{i=1,i\neq 4}^{4}\sum_{\nu=0}^{\infty}(-1)^{\nu}\varepsilon^{\nu+1}\left[\left(I_i^{\nu}\left(K(t,s)\,z_i(s)\right)\right)_{s=t}e^{\frac{1}{\varepsilon}\int_{t_0}^{t}\lambda_i(\theta)d\theta} -\right.$$

$$\left.-\left(I_i^{\nu}\left(K(t,s)\,z_i(s)\right)\right)_{s=t_0}e^{\frac{1}{\varepsilon}\int_{t_0}^{t}\lambda_4(\theta)d\theta}\right] +$$

$$+\sum_{2\leq|m|\leq N_z}^{*}\sum_{\nu=0}^{\infty}(-1)^{\nu}\varepsilon^{\nu+1}\left[\left(I_{4,m}^{\nu}\left(K(t,s)\,z^m(s)\right)\right)_{s=t}e^{\frac{1}{\varepsilon}\int_{t_0}^{t}(m,\lambda(\theta))d\theta} -\right.$$

$$\left.-\left(I_{4,m}^{\nu}\left(K(t,s)\,z^m(s)\right)\right)_{s=t_0}e^{\frac{1}{\varepsilon}\int_{t_0}^{t}\lambda_4(\theta)d\theta}\right],\tau = \psi(t)/\varepsilon.$$

It is easy to show (see, for example, [23], pp. 291–294) that this series converges asymptotically for $\varepsilon \to +0$ (uniformly in $t \in [t_0, T]$). This means that the class M_ε is asymptotically invariant (for $\varepsilon \to +0$) with respect to the operator J.

Let as introduce the operators $R_\nu \colon U \to U$, acting on each element $z(t,\tau) \in U$ of the form (5) according to the law:

$$R_0 z(t,\tau) = e^{\tau_4} \int_{t_0}^{t} K(t,s)\, z_4(s)\, ds, \tag{6_0}$$

$$\begin{aligned} R_1 z(t,\tau) = {} & \left[\left(I_0^0 \left(K(t,s)\, z_0(s) \right) \right)_{s=t} e^{\tau_4} - \left(I_0^0 \left(K(t,s)\, z_0(s) \right) \right)_{s=t_0} \right] + \\ & + \sum_{i=1}^{3} \left[\left(I_i^0 \left(K(t,s)\, z_i(s) \right) \right)_{s=t} e^{\tau_i} - \left(I_i^0 \left(K(t,s)\, z_i(s) \right) \right)_{s=t_0} e^{\tau_4} \right] + \\ & + \sum_{2 \le |m| \le N_z}^{*} \left[\left(I_{4,m}^0 \left(K(t,s)\, z^m(s) \right) \right)_{s=t} e^{(m,\tau)} - \left(I_{4,m}^0 \left(K(t,s)\, z^m(s) \right) \right)_{s=t_0} e^{\tau_4} \right], \end{aligned} \tag{6_1}$$

$$\begin{aligned} R_{\nu+1} z(t,\tau) = {} & \left[\left(I_0^\nu \left(K(t,s)\, z_0(s) \right) \right)_{s=t} e^{\tau_4} - \left(I_0^\nu \left(K(t,s)\, z_0(s) \right) \right)_{s=t_0} \right] + \\ & + \sum_{i=1}^{3} (-1)^\nu \left[\left(I_i^\nu \left(K(t,s)\, z_i(s) \right) \right)_{s=t} e^{\tau_i} - \left(I_i^\nu \left(K(t,s)\, z_i(s) \right) \right)_{s=t_0} e^{\tau_4} \right] + \\ & + \sum_{2 \le |m| \le N_z}^{*} \left[\left(I_{4,m}^\nu \left(K(t,s)\, z^m(s) \right) \right)_{s=t} e^{(m,\tau)} - \left(I_{4,m}^\nu \left(K(t,s)\, z^m(s) \right) \right)_{s=t_0} e^{\tau_4} \right], \nu \ge 1. \end{aligned} \tag{$6_{\nu+1}$}$$

Let now $\tilde{z}(t,\tau,\varepsilon)$ be an arbitrary continuous function in $(t,\tau) \in [t_0, T] \times \{\tau : Re\tau_j \le 0, j = \overline{1,4}\}$ with the asymptotic expansion

$$\tilde{z}(t,\tau,\varepsilon) = \sum_{k=0}^{\infty} \varepsilon^k z_k(t,\tau),\ z_k(t,\tau) \in U, \tag{7}$$

converging as $\varepsilon \to +0$ (uniformly in $(t,\tau) \in [t_0, T] \times \{\tau : Re\tau_j \le 0, j = \overline{1,4}\}$). Then the image $J\tilde{z}(t,\tau,\varepsilon)$ of this function is expanded in the asymptotic series

$$J\tilde{z}(t,\tau,\varepsilon) = \sum_{k=0}^{\infty} \varepsilon^k J z_k(t,\tau) = \sum_{r=0}^{\infty} \varepsilon^r \sum_{s=0}^{r} R_{r-s} z_s(t,\tau) \,|_{\tau=\psi(t)/\varepsilon}.$$

This equality is the basis for introducing the extension of the operator J on the series type (7):

$$\tilde{J}\tilde{z}(t,\tau,\varepsilon) \equiv \tilde{J}\left(\sum_{k=0}^{\infty} \varepsilon^k z_k(t,\tau) \right) \stackrel{def}{=} \sum_{r=0}^{\infty} \varepsilon^r \sum_{s=0}^{r} R_{r-s} z_s(t,\tau).$$

Although the operator $\tilde{J}$ is formally defined, its usefulness is obvious, since in practice they usually construct the N-th approximation of the asymptotic solution of problem (2), in which only the N-th partial sums of the series (7) will take part, which do not have a formal but true meaning. Now we can write down a problem that is completely regularized with respect to the original problem (2):

$$\begin{gathered} \mathbf{L}\,\tilde{z}(t,\tau,\varepsilon) \equiv \varepsilon \frac{\partial \tilde{z}}{\partial t} + \textstyle\sum_{j=1}^{4} \lambda_j(t) \frac{\partial \tilde{z}}{\partial \tau_j} - \lambda_1(t)\tilde{z} - \varepsilon \frac{g(t)}{2} \left(e^{\tau_2}\sigma_1 + e^{\tau_3}\sigma_2 \right) \tilde{z} - \tilde{J}\tilde{z} = h(t), \\ \tilde{z}(t,\tau,\varepsilon)|_{t=t_0,\tau=0} = z^0, \quad t \in [t_0, T]. \end{gathered} \tag{8}$$

3. Iterative Problems and Their Solvability in the Space U

Substituting series (7) into (8) and equating the coefficients for the same powers ε, we obtain the following iterative problems:

$$\mathbf{L}\, z_0\,(t,\tau) \equiv \sum_{j=1}^{4} \lambda_j\,(t)\, \frac{\partial z_0}{\partial \tau_j} - \lambda_1(t) z_0 - R_0 z_0 = h\,(t)\,,\ z_0\,(t_0,0) = z^0; \tag{9_0}$$

$$\mathbf{L}\, z_1\,(t,\tau) = -\frac{\partial z_0}{\partial t} + \frac{g(t)}{2}\,(e^{\tau_2}\sigma_1 + e^{\tau_3}\sigma_2)\, z_0 + R_1 z_0,\ z_1\,(t_0,0) = 0; \tag{9_1}$$

$$\mathbf{L}\, z_2\,(t,\tau) = -\frac{\partial z_1}{\partial t} + \frac{g(t)}{2}\,(e^{\tau_2}\sigma_1 + e^{\tau_3}\sigma_2)\, z_1 + R_1 z_1 + R_2 z_0,\ z_0\,(t_0,0) = 0; \tag{9_2}$$

$$\cdots$$

$$\mathbf{L}\, z_k\,(t,\tau) = -\frac{\partial z_{k-1}}{\partial t} + \frac{g(t)}{2}\,(e^{\tau_2}\sigma_1 + e^{\tau_3}\sigma_2)\, z_{k-1} + R_k z_0 + \ldots + R_1 z_{k-1},\ z_k\,(t_0,0) = 0,\ k \geq 1. \tag{9_k}$$

Each of the iterative problems can be written as

$$\mathbf{L}\, z\,(t,\tau) \equiv \sum_{j=1}^{4} \lambda_j\,(t)\, \frac{\partial z}{\partial \tau_j} - \lambda_1(t) z - R_0 z = H\,(t,\tau)\,,\ z\,(t_0,0) = z^*, \tag{10}$$

where $H\,(t,\tau) = H_0\,(t) + \sum_{i=1}^{4} H_i\,(t)\, e^{\tau_i} + \sum_{2\leq|m|\leq N_H}^{*} H^m\,(t)\, e^{(m,\tau)}$ is the known function of the space U, z^* is the known number of complex the space $\mathbb{C}$, and the operator R_0 has the form (see (6_0))

$$R_0 z \equiv R_0 \left(z_0\,(t) + \sum_{i=1}^{4} z_i\,(t)\, e^{\tau_i} + \sum_{2\leq|m|\leq N_z}^{*} z^m\,(t)\, e^{(m,\tau)} \right) \overset{def}{=} e^{\tau_4} \int_{t_0}^{t} K\,(t,s)\, z_4\,(s)\, ds.$$

We introduce the scalar product (for each $t \in [t_0, T]$) in the space U :

$$< z, w > \equiv < z_0\,(t) + \sum_{i=1}^{4} z_i\,(t)\, e^{\tau_i} + \sum_{2\leq|m|\leq N_z}^{*} z^m\,(t)\, e^{(m,\tau)},$$

$$w_0\,(t) + \sum_{i=1}^{4} w_i\,(t)\, e^{\tau_i} + \sum_{2\leq|m|\leq N_w}^{*} w^m\,(t)\, e^{(m,\tau)} > \overset{def}{=}$$

$$\overset{def}{=} (z_0\,(t)\,,\, w_0\,(t)) + \sum_{i=1}^{4} (z_i\,(t)\,, w_i\,(t)) + \sum_{2\leq|m|\leq\min(N_z,N_w)}^{*} (z^m\,(t)\,, w^m\,(t))\,,$$

where $(*\,,\,*)$ we denote the ordinary scalar product in the complex space $\mathbb{C}$: $(u,v) = u \cdot \bar{v}$. We prove the following statement.

Theorem 1. *Suppose that conditions (1) and (2) are satisfied and the right-hand side* $H\,(t,\tau) = H_0\,(t) + + \sum_{i=1}^{4} H_i\,(t)\, e^{\tau_i} + \sum_{2\leq|m|\leq N_H}^{*} H^m\,(t)\, e^{(m,\tau)}$ *of the Equation (10) belongs to the space* U*. Then for the solvability of the Equation (10) in* U *it is necessary and sufficient that the identities*

$$< H\,(t,\tau)\,, e^{\tau_1} > \equiv 0, \quad \forall t \in [t_0, T] \tag{11}$$

hold true.

Proof. We will determine the solution of the Equation (10) in the form of an element (5) of the space U:

$$z(t,\tau) = z_0(t) + \sum_{i=1}^{4} z_i(t) e^{\tau_i} + \sum_{2\leq|m|\leq N_H}^{*} z^m(t) e^{(m,\tau)}. \quad (12)$$

Substituting (12) into the Equation (10), we have

$$-\lambda_1(t) z_0(t) + \sum_{i=1}^{4} [\lambda_i(t) - \lambda_1(t)] z_i(t) e^{\tau_i} + \sum_{2\leq|m|\leq N_H}^{*} [(m,\lambda(t)) - \lambda_1(t)] z^m(t) e^{(m,\tau)} -$$

$$-e^{\tau_4} \int_{t_0}^{t} K(t,s) z_4(s) ds = H_0(t) + \sum_{i=1}^{4} H_i(t) e^{\tau_i} + \sum_{2\leq|m|\leq N_H}^{*} H^m(t) e^{(m,\tau)}.$$

Equating here separately the free terms and coefficients at the same exponents, we obtained the following equations:

$$-\lambda_1(t) z_0(t) = H_0(t), \quad (13_0)$$

$$[\lambda_i(t) - \lambda_1(t)] z_i(t) = H_i(t), \ i = \overline{1,3}; \quad (13_i)$$

$$[\lambda_4(t) - \lambda_1(t)] z_4(t) - \int_{t_0}^{t} K(t,s) z_4(s) ds = H_4(t); \quad (13_4)$$

$$[(m,\lambda(t)) - \lambda_1(t)] z^m(t) = H^m(t), \ m \notin \Gamma_1, \ 2 \leq |m| \leq N_H. \quad (13_m)$$

Since the function $\lambda_1(t) \neq 0 \ \forall t \in [t_0,T]$, the Equation (13_0) has a unique solution $z_0(t) = -\lambda_1^{-1}(t) H_0(t)$. Since $\lambda_4(t) - \lambda_1(t) \neq 0 \ \forall t \in [t_0,T]$, then the Equation (13_4) can be written as

$$z_4(t) = \int_{t_0}^{t} \left([\lambda_4(t) - \lambda_1(t)]^{-1} K(t,s) \right) z_4(s) ds - [\lambda_4(t) - \lambda_1(t)]^{-1} H_4(t). \quad (14)$$

Due to the smoothness of the kernel $\left([\lambda_4(t) - \lambda_1(t)]^{-1} K(t,s) \right)$ and heterogeneity $-[\lambda_4(t) - \lambda_1(t)]^{-1} H_4(t)$, this Volterra integral equation has a unique solution $z_4(t) \in C^\infty([t_0,T],\mathbb{C})$. The Equations (13_2) and (13_3) also have unique solutions

$$z_i(t) = [\lambda_i(t) - \lambda_1(t)]^{-1} H_i(t) \in C^\infty([t_0,T],\mathbb{C}), \ i = 2,3,$$

since $\lambda_i(t) \neq \lambda_1(t), i = 2,3$. The Equation (13_1) is solvable in the space $C^\infty([t_0,T],\mathbb{C})$ if and only if identities $(H_1(t), e^{\tau_1}) \equiv 0 \ \forall t \in [t_0,T]$ hold. It is easy to see that this identity coincides with identity (11).

Further, since $(m,\lambda(t)) \neq \lambda_1(t), \ 2 \leq |m| \leq N_H \ (\forall m \notin \Gamma_1)$, then the Equation (13_m) has a unique solution

$$z^m(t) = [(m,\lambda(t)) I - A(t)]^{-1} H^m(t) \in C^\infty([t_0,T],\mathbb{C}), \ 2 \leq |m| \leq N_H.$$

Thus, condition (11) is necessary and sufficient for the solvability of the Equation (10) in the space U. The Theorem 1 is proved. □

Remark 1. *If identity (11) holds, then under conditions (1) and (2), the Equation (10) has the following solution in the space* U :

$$z(t,\tau) = z_0(t) + \textstyle\sum_{i=1}^{4} z_i(t) e^{\tau_i} + \sum_{2\leq|m|\leq N_H}^{*} z^m(t) e^{(m,\tau)} \equiv z_0(t) + \alpha_1(t) e^{\tau_1} + \\ + h_{21}(t) e^{\tau_2} + h_{31}(t) e^{\tau_3} + z_4(t) e^{\tau_4} + \sum_{2\leq|m|\leq N_H}^{*} P^m(t) e^{(m,\tau)}, \quad (15)$$

where $\alpha_1(t) \in C^\infty([t_0, T], \mathbb{C})$ *are arbitrary function,* $z_0(t) = -\lambda_1^{-1}(t)H_0(t)$, $z_4(t)$ *is the solution of the integral Equation* (14)*, and introduced notations*

$$h_{21}(t) \equiv \frac{H_2(t)}{\lambda_2(t) - \lambda_1(t)}, \quad h_{31}(t) \equiv \frac{H_3(t)}{\lambda_3(t) - \lambda_1(t)}, \quad P^m(t) \equiv [(m, \lambda(t)) - \lambda_1(t)]^{-1} H^m(t).$$

4. The Remainder Term Theorem

Along with problem (10), we consider the equation

$$\mathbf{L}w(t,\tau) = -\frac{\partial z}{\partial t} + \frac{g(t)}{2}(e^{\tau_2}\sigma_1 + e^{\tau_3}\sigma_2)z + R_1 z + Q(t,\tau), \tag{16}$$

where $z = z(t,\tau)$ is the solution (15) of Equation (10), $Q(t,\tau) \in U$ is the known function of the space U (this form will have problems (9_{k+1}) after calculating the solution of the problem (9_k) in U). The right side of this equation:

$$G(t,\tau) \equiv -\frac{\partial z}{\partial t} + \frac{g(t)}{2}(e^{\tau_2}\sigma_1 + e^{\tau_3}\sigma_2)z + R_1 z + Q(t,\tau) =$$

$$= -\frac{\partial}{\partial t}\left[z_0(t) + \sum_{i=1}^{4} z_i(t)e^{\tau_i} + \sum_{2\le|m|\le N_H}^{*} z^m(t)e^{(m,\tau)}\right] +$$

$$+\frac{g(t)}{2}(e^{\tau_2}\sigma_1 + e^{\tau_3}\sigma_2)\left[z_0(t) + \sum_{i=1}^{4} z_i(t)e^{\tau_i} + \sum_{2\le|m|\le N_H}^{*} z^m(t)e^{(m,\tau)}\right] + R_1 z + Q(t,\tau),$$

may not belong to the space U, if $z = z(t,\tau) \in U$. Indeed, taking into account the form (15) of function $z = z(t,\tau) \in U$, we consider in $G(t,\tau)$, for example, the terms

$$Z(t,\tau) \equiv \frac{g(t)}{2}(e^{\tau_2}\sigma_1 + e^{\tau_3}\sigma_2)\left[z_0(t) + \sum_{i=1}^{4} z_i(t)e^{\tau_i} + \sum_{2\le|m|\le N_H}^{*} z^m(t)e^{(m,\tau)}\right] =$$

$$= \frac{g(t)}{2}z_0(t)(e^{\tau_2}\sigma_1 + e^{\tau_3}\sigma_2) + \sum_{i=1}^{4}\frac{g(t)}{2}z_i(t)(e^{\tau_i+\tau_2}\sigma_1 + e^{\tau_i+\tau_3}\sigma_2) +$$

$$+\frac{g(t)}{2}(e^{\tau_2}\sigma_1 + e^{\tau_3}\sigma_2)\sum_{2\le|m|\le N_H}^{*} P^m(t)e^{(m,\tau)}.$$

Function $Z(t,\tau) \notin U$, since it contains resonant exponentials $e^{\tau_2+\tau_3} = e^{(m,\tau)}|_{m=(0,1,1,0)}, e^{\tau_2+(m,\tau)}\,(m_2+1=m_3), e^{\tau_3+(m,\tau)}\,(m_3+1=m_2)$, and, therefore, the right-hand side $G(t,\tau) = Z(t,\tau)$ of the Equation (16) also does not belong to the U. Then, according to the well-known theory (see [6], p. 234), we need to embed $\wedge: G(t,\tau) \to \hat{G}(t,\tau)$ the right-hand side $G(t,\tau)$ of the Equation (16) into the space U. This operation is defined as follows.

Let the function $G(t,\tau) = \sum_{|m|=0}^{N} w^m(t)e^{(m,\tau)}$ contain resonant exponentials, i.e., $G(t,\tau)$, it has the form

$$G(t,\tau) = w_0(t) + \sum_{i=1}^{4} w_i(t)e^{\tau_i} + \sum_{j=0}^{4}\sum_{|m^j|=2:m^j\in\Gamma_j}^{N} w^{m^j}(t)e^{(m^j,\tau)} + \sum_{|m|=2,m\neq m^j,j=\overline{0,4}}^{N} w^m(t)e^{(m,\tau)}.$$

Then

$$\hat{G}(t,\tau) = w_0(t) + \sum_{i=1}^{4} w_i(t)e^{\tau_i} + \sum_{j=0}^{4}\sum_{|m^j|=2:\,m^j\in\Gamma_j}^{N} w^{m^j}(t)e^{\tau_j} + \sum_{|m|=2,m\neq m^j,j=\overline{0,4}}^{N} w^m(t)e^{(m,\tau)}.$$

Therefore, the embedding operation acts only on the resonant exponentials and replaces them with a unit or exponents e^{τ_j} of the first dimension according to the rule:

$$\left(e^{(m,\tau)}|_{m\in\Gamma_0}\right)^{\wedge}=e^0=1,\ \left(e^{(m,\tau)}|_{m\in\Gamma_j}\right)^{\wedge}=e^{\tau_j},\ j=\overline{1,4}.$$

Therefore, the right-hand sides of iterative problems (9_k) (if they solve sequentially) may not belong to the space U. Then, according to [6] (p. 234), the right-hand sides of these problems must be embedded in U according to the above rule. As a result, we obtained the following problems:

$$Lz_0(t,\tau)\equiv\sum_{j=1}^{4}\lambda_j(t)\frac{\partial z_0}{\partial\tau_j}-A(t)z_0-R_0z_0=h(t),z_0(t_0,0)=z^0; \qquad (\overline{9}_0)$$

$$Lz_1(t,\tau)=-\frac{\partial z_0}{\partial t}+\left[\frac{g(t)}{2}\left(e^{\tau_2}\sigma_1+e^{\tau_3}\sigma_2\right)z_0\right]^{\wedge}+R_1z_0,z_1(t_0,0)=0; \qquad (\overline{9}_1)$$

$$Lz_2(t,\tau)=-\frac{\partial z_1}{\partial t}+\left[\frac{g(t)}{2}\left(e^{\tau_2}\sigma_1+e^{\tau_3}\sigma_2\right)z_1\right]^{\wedge}+R_1z_1+R_2z_0,z_0(t_0,0)=0; \qquad (\overline{9}_2)$$

$$\cdots$$

$$Lz_k(t,\tau)=-\frac{\partial z_{k-1}}{\partial t}+\left[\frac{g(t)}{2}\left(e^{\tau_2}\sigma_1+e^{\tau_3}\sigma_2\right)z_{k-1}\right]^{\wedge}+R_kz_0+\ldots+R_1z_{k-1},$$
$$z_k(t_0,0)=0,k\geq 1 \qquad (\overline{9}_k)$$

(images of linear operators $\frac{\partial}{\partial t}$ and R_ν do not need to be embedded in the space U, since these operators act from U to U). Such a replacement will not affect the construction of an asymptotic solution to the original problem (1) (or its equivalent problem (2)), since the narrowing $\tau=\frac{\psi(t)}{\varepsilon}$ of the series of problems $(\overline{9}_k)$ will coincide with the series of problems (9_k) (see [6], pp. 234–235).

It is easy to show that applying Theorem 1 to iterative problems $(\overline{9}_k)$, we can find their solutions uniquely in the space U. As a result, we can construct series (7) with coefficients $z_k(t,\tau)\in U$. As in [23] (pp. 303–308), we proved the following statement.

Theorem 2. *Suppose that conditions (1)–(2) are satisfied for the Equation (2). Then, when* $\varepsilon\in(0,\varepsilon_0]$ *(*$\varepsilon_0>0$ *is sufficiently small) the Equation (2) has a unique solution* $z(t,\varepsilon)\in C^1([t_0,T],\mathbb{C})$*; at the same time there is the estimate*

$$||z(t,\varepsilon)-z_{\varepsilon N}(t)||_{C[t_0,T]}\leq c_N\varepsilon^{N+1},\ \forall N=0,1,2,\ldots,$$

where $z_{\varepsilon N}(t)$ *is the narrowing (for* $\tau=\frac{\psi(t)}{\varepsilon}$*) N-th partial sum of the series (7) (with coefficients* $z_k(t,\tau)\in U$ *satisfying the iterative problems* $(\overline{9}_k)$*), and the constant* $c_N>0$ *does not depend* ε *on* $\varepsilon\in(0,\varepsilon_0]$.

5. Construction of the Solution of the First Iteration Problem in the Space U

Using Theorem 1, we will tried to find a solution to the first iterative problem $(\overline{9}_0)$. Since the right-hand side $h(t)$ of the equation $(\overline{9}_0)$ satisfies condition (11), this equation has (according to (15)) a solution in the space U in the form

$$z_0(t,\tau)=z_0^{(0)}(t)+\alpha_1^{(0)}(t)e^{\tau_1}, \qquad (17)$$

where $\alpha_1^{(0)}(t)\in C^\infty([t_0,T],\mathbb{C})$ are arbitrary function, $z_0^{(0)}(t)=-\frac{h(t)}{\lambda_1(t)}$. Subordinating (17) to the initial condition $z_0(t_0,0)=z^0$, we have

$$z_0^{(0)}(t_0)+\alpha_1^{(0)}(t_0)=z^0\ \Leftrightarrow\ \alpha_1^{(0)}(t_0)=z^0+\lambda_1^{-1}(t_0)h(t_0).$$

To fully calculate the function $\alpha_1^{(0)}(t)$, we pass to the next iterative problem $(\overline{9}_1)$. Substituting the solution (17) of the equation $(\overline{9}_0)$, into it, we arrived at the following equation:

$$\mathbf{L} z_1(t,\tau) = -\frac{d}{dt}\left(z_0^{(0)}(t)\right) - \frac{d}{dt}\left(\alpha_1^{(0)}(t)\right)e^{\tau_1} + \frac{K(t,t)\, z_0^{(0)}(t)}{\lambda_4(t)}e^{\tau_4} -$$
$$-\frac{K(t,t_0)\, z_0^{(0)}(t_0)}{\lambda_4(t_0)} + \frac{g(t)}{2}\left(e^{\tau_2}\sigma_1 + e^{\tau_3}\sigma_2\right)\left(z_0^{(0)}(t) + \alpha_1^{(0)}(t)\,e^{\tau_1}\right) + \tag{18}$$
$$+\frac{K(t,t)\,\alpha_1^{(0)}(t)}{\lambda_1(t)}e^{\tau_1} - \frac{K(t,t_0)\,\alpha_j^{(0)}(t_0)}{\lambda_1(t_0)},$$

(here we used the expression (6_1) for $R_1 z(t,\tau)$ and took into account that when $z(t,\tau) = z_0(t,\tau)$ in the sum (6_1) only terms with e^{τ_1} and remain e^{τ_4}). Let us calculate

$$M = \left[\frac{g(t)}{2}\left(e^{\tau_2}\sigma_1 + e^{\tau_3}\sigma_2\right)\left(z_0^{(0)}(t) + \alpha_1^{(0)}(t)\,e^{\tau_1}\right)\right]^{\wedge} =$$
$$= \frac{1}{2}g(t)\left[\sigma_1\alpha_1^{(0)}(t)\,e^{\tau_2+\tau_1} + \sigma_2\alpha_1^{(0)}(t)\,e^{\tau_3+\tau_1} + \sigma_1 z_0^{(0)}(t)\,e^{\tau_2} + \sigma_2 z_0^{(0)}(t)\,e^{\tau_3}\right]^{\wedge}.$$

Let us analyze the exponents of the second dimension included here for their resonance:

$$e^{\tau_2+\tau_1}\big|_{\tau=\psi(t)/\varepsilon} = e^{\frac{1}{\varepsilon}\int_{t_0}^{t}(-i\beta'(\theta)+a(\theta))d\theta}, \quad e^{\tau_3+\tau_1}\big|_{\tau=\psi(t)/\varepsilon} = e^{\frac{1}{\varepsilon}\int_{t_0}^{t}(+i\beta'(\theta)+a(\theta))d\theta},$$
$$-i\beta' + a = \begin{bmatrix} 0, \\ a, \\ -i\beta', \\ +i\beta', \\ \mu, \end{bmatrix} \Leftrightarrow \varnothing; \qquad +i\beta' + a = \begin{bmatrix} 0, \\ a, \\ -i\beta', \\ +i\beta', \\ \mu, \end{bmatrix} \Leftrightarrow \varnothing.$$

Thus, exponents $e^{\tau_2+\tau_1}$ ang $e^{\tau_3+\tau_1}$ are not resonant. Then, for solvability the Equation (18) it is necessary and sufficient that the condition

$$-\frac{d}{dt}\left(\alpha_1^{(0)}(t)\right) + \frac{K(t,t)\,\alpha_1^{(0)}(t)}{\lambda_1(t)} = 0$$

is satisfied. Attaching the initial condition $\alpha_1^{(0)}(t_0) = z^0 + \lambda_1^{-1}(t_0)\,h(t_0)$, to this equation, we found uniquely the function

$$\alpha_1^{(0)}(t) = \alpha_1^{(0)}(t_0)\exp\left\{\int_{t_0}^{t}\frac{K(s,s)}{\lambda_1(s)}ds\right\},$$

and therefore, we uniquely calculate the solution (17) of the problem $(\overline{9}_0)$ in the space U. In this case, the leading term of the asymptotics of the solution to the problem (2) has the form

$$z_{\varepsilon 0}(t) = z_0^{(0)}(t) + \alpha_1^{(0)}(t_0)\exp\left\{\int_{t_0}^{t}\frac{K(s,s)}{\lambda_1(s)}ds\right\}e^{\frac{1}{\varepsilon}\int_{t_0}^{t}\lambda_1(\theta)d\theta},$$

where $\alpha_1^{(0)}(t_0) = z^0 + A^{-1}(t_0)\,h(t_0)$, $z_0^{(0)}(t) = -\lambda_1^{-1}(t_0)\,h(t)$.

Example. *Consider a model problem*

$$\varepsilon\frac{dz}{dt} = -z - \varepsilon\cos\frac{t^2+t}{\varepsilon}z - \int_{t_0}^{t} e^{\frac{-2(t-s)}{\varepsilon}}\cdot t\cdot s\cdot z(s,\varepsilon)ds + h(t),\ z(t_0,\varepsilon) = z^0,\ t\in[t_0,T](t_0\geq 0), \tag{19}$$

were $a(t) \equiv 1, \mu(t) \equiv -2, \beta(t) \equiv t^2 + t, K(t,s) \equiv t \cdot s$. *The main term of the asymptotic solution of this problem has the form*

$$z_{\varepsilon 0}\,(t) = h(t) + [z^0 - h(t_0)] \exp\,[\frac{t_0^3 - t^3}{3}] exp[\frac{t - t_0}{\varepsilon}]. \tag{20}$$

For $\varepsilon \to +0$ the function $z_{\varepsilon 0}\,(t)$ tends to the solution of the degenerate equation $-\overline{z} + h(t) = 0$ uniformly on any interval $[t_0 + \delta, T](0 < \delta \leq T - t_0)$ and at the point $t = t_0$ takes on the value $z_{\varepsilon 0}\,(t_0) = z^0$. It is seen from (20) that the leading term of the asymptotics of the solution to problem (19) does not depend on cos $\frac{t^2+t}{\varepsilon}$ and spectral value $\mu(t) \equiv -2$, but depends on the kernel $K(t,s) \equiv t \cdot s$. Further calculations show that already the asymptotic solution $z_{\varepsilon 1}\,(t) = z_{\varepsilon 0}\,(t) + \varepsilon z_1\left(t, \frac{\psi(t)}{\varepsilon}\right)$ of the first order will depend on both $\mu(t) \equiv -2$, and the frequency $\beta'(t) = 2t + 1$ of the rapidly oscillating cosine.

6. Conclusions

The function $z_{\varepsilon 0}\,(t)$ shows that when passing from a differential equation of type (1) ($K(t,s) \equiv 0$) to an integro-differential one ($K(t,s) \neq 0$), the main term of the asymptotic is influenced by the kernel $K(t,s)$ of the integral operator. However, the main term of the asymptotics is not affected by the spectral values of the integral operator $\mu(t)$ and rapidly oscillating coefficients. Their effects are detected when constructing the next approximation $z_{\varepsilon 1}\,(t)$.

Author Contributions: All authors contributed equally to this work. All authors have read and agreed to the published version of the manuscript.

Funding: This work was supported by grant No. AP05133858 of the Ministry of Education and Science of the Republic of Kazakhstan.

Conflicts of Interest: The funders had no role in the design of the study; in the collection, analyses, or interpretation of data; in the writing of the manuscript, or in the decision to publish the results.

References

1. Feschenko, S.F.; Shkil, N.I.; Nikolenko, L.D. *Asymptotic Methods in the Theory of Linear Differential Equations*; Naukova Dumka: Kiev, Ukraine, 1966.
2. Shkil, N.I. *Asymptotic Methods in Differential Equations*; Naukova Dumka: Kiev, Ukraine, 1971.
3. Daletsky, Y.L.; Krein, S.G. On differential equations in Hilbert space. *Ukr. Math. J.* **1950**, 2, 71–91.
4. Daletsky, Y.L. The asymptotic method for some differential equations with oscillating coefficients. *DAN USSR* **1962**, 143, 1026–1029.
5. Daletsky, Y.L.; Krein M.G. *Stability of Solutions of Differential Equations in a Banach Space*; Nauka: Moscow, Russia, 1970.
6. Lomov, S.A. *Introduction to General Theory of Singular Perturbations*; Vol. 112 of Translations of Math; Monographs, American Math Society: Providence, RI, USA, 1992.
7. Ryzhih, A.D. Asymptotic solution of a linear differential equation with a rapidly oscillating coefficient. *Trudy MEI* **1978**, *357*, 92–94.
8. Ryzhih, A.D. Application of the regularization method for an equation with rapidly oscillating coefficients. *Mater. All-Union. Conf. Asymptot. Methods* **1979**, *1*, 64–66.
9. Kalimbetov, B.T.; Temirbekov, M.A.; Khabibullaev, Z.O. Asymptotic solution of singular perturbed problems with an instable spectrum of the limiting operator. *Abstr. Appl. Anal.* **2012**, 120192. [CrossRef]
10. Imanbaev, N.S.; Kalimbetov, B.T.; Temirbekov, M.A. Asymptotics of solutions of singularly perturbed integro-differential equation with rapidly decreasing kernel. *Bull.-Ksu-Math.* **2013**, *72*, 55–63.
11. Bobodzhanov, A.A.; Safonov, V.F. Singularly perturbed nonlinear integro-differential systems with rapidly varying kernels. *Math. Notes.* **2002**, *72*, 605–614. [CrossRef]
12. Nefedov, N.N.; Nikitin, A.G. The Cauchy problem for a singularly perturbed integro-differential Fredholm equation. *Comput. Math. Math. Phys.* **2007**, *47*, 629–637. [CrossRef]
13. Bobodzhanov, A.A.; Safonov, V.F. Singularly perturbed integro-differential equations with diagonal degeneration of the kernel in reverse time. *J. Differ. Equ.* **2004**, *40*, 120–127. [CrossRef]

14. Bobodzhanov, A.A.; Safonov, V.F. Asymptotic analysis of integro-differential systems with an unstable spectral value of the integral operator's kernel. *Comput. Math. Math. Phys.* **2007**, *47*, 65–79. [CrossRef]
15. Bobodzhanov, A.A.; Safonov, V.F. The method of normal forms for singularly perturbed systems of Fredholm integro-differential equations with rapidly varying kernels. *Sibir. Math. J.* **2013**, *204*, 979–1002. [CrossRef]
16. Vasil'eva, A.B.; Butuzov, V.F.; Nefedov, N.N. Singularly perturbed problems with boundary and internal layers. *Proc. Steklov Inst. Math.* **2010**, *268*, 258–273. [CrossRef]
17. Imanaliev, M. *Asymptotic Methods in the Theory of Singularly Perturbed Integro-Differential Equations;* Ilim: Frunze, Kyrgyzstan, 1972.
18. Bobodzhanov, A.; Safonov, V.; Kachalov, V. Asymptotic and pseudoholomorphic solutions of singularly perturbed differential and integral equations in the lomov's regularization method. *Axioms* **2019**, *8*, 27. [CrossRef]
19. Imanaliev, M. *Methods for Solving Nonlinear Inverse Problems and Their Applications;* Ilim: Frunze, Kyrgyzstan, 1977.
20. Bobodzhanov, A.A.; Safonov, V.F. Asymptotic solutions of Fredholm integro-differential equations with rapidly changing kernels and irreversible limit operator. *Russ. Math.* **2015**, *59*, 1–15. [CrossRef]
21. Bobodzhanov, A.A.; Safonov, V.F. A problem with inverse time for a singularly perturbed integro-differential equation with diagonal degeneration of the kernel of high order. *Izv. Math.* **2016**, *80*, 285–298. [CrossRef]
22. Bobodzhanov, A.A.; Kalimbetov, B.T.; Safonov, V.F. Integro-differential problem about parametric amplification and its asymptotical integration. *Int. J. Appl. Math.* **2020**, *33*, 331–353. [CrossRef]
23. Safonov, V.F.; Bobodzhanov, A.A. *Course of Higher Mathematics. Singularly Perturbed Equations and the Regularization Method: Textbook;* Publishing House of MPEI: Moscow, Russia, 2012.

axioms

MDPI

Article

Projector Approach to Constructing Asymptotic Solution of Initial Value Problems for Singularly Perturbed Systems in Critical Case

Galina Kurina [1,2,3,*,†]

1 Voronezh State University, Voronezh 394018, Russia
2 Voronezh Institute of Law and Economics, Voronezh 394042, Russia
3 Federal Research Center "Computer Science and Control" of Russian Academy of Sciences, Moscow 119333, Russia
* Correspondence: kurina@math.vsu.ru; Tel.: +7-906-584-1966
† Voronezh State University, Universitetskaya pl.,1, Voronezh 394018, Russia.

Received: 13 February 2019; Accepted: 21 April 2019; Published: 08 May 2019

Abstract: Under some conditions, an asymptotic solution containing boundary functions was constructed in a paper by Vasil'eva and Butuzov (Differ. Uravn. 1970, 6(4), 650–664 (in Russian); English transl.: Differential Equations 1971, 6, 499–510) for an initial value problem for weakly non-linear differential equations with a small parameter standing before the derivative, in the case of a singular matrix $A(t)$ standing in front of the unknown function. In the present paper, the orthogonal projectors onto $ker A(t)$ and $ker A(t)'$ (the prime denotes the transposition) are used for asymptotics construction. This approach essentially simplifies understanding of the algorithm of asymptotics construction.

Keywords: singular perturbations; initial value problems; asymptotics; critical case; projector approach

---

1. Introduction

The bibliography of publications devoted to singularly perturbed problems is very extensive. Most of them deal with problems in which a degenerate equation, following from the original one where a small parameter is equal to zero, is resolvable with respect to a fast component of an unknown variable. If it is not so, then this more complicated case is known as critical [1], singular [2], nonstandard [3], or as a case where the unperturbed (degenerate) system is situated on the spectrum [4]. Numerous applications of singularly perturbed systems in the critical case have been listed in [5].

Vasil'eva and Butuzov were the first to study initial value problems for singularly perturbed differential and difference systems in the critical case. Asymptotic solutions of boundary value problems for such systems have been obtained in [1,2,6]. Numerical methods for singularly perturbed systems in the critical case have been researched in [7] for initial value problems, and in [8] for boundary value problems.

An asymptotic solution containing boundary functions for the initial value problem of the weakly non-linear differential equation in a real m-dimensional space X:

$$\varepsilon\frac{dx}{dt} = A(t)x + \varepsilon f(x,t,\varepsilon), \quad t \in [0,T], \tag{1}$$

$$x(0,\varepsilon) = x^0, \tag{2}$$

where $x = x(t,\varepsilon) \in X$ and the matrix $A(t)$ is singular, has been constructed in [4]. A discrete analogue of problem (1)-(2) was also considered. The results from this paper are also presented in [1,9]. In these publications, the purpose of studying equations of the last form is also explained. Here and further $\varepsilon \geq 0$ means a small parameter, and the $m \times m$ matrix $A(t)$ and the m-dimensional vector-function $f(x, t, \varepsilon)$ are sufficiently smooth with respect to their arguments.

In contrast [4], the projector approach will be used in this paper for constructing an asymptotic solution of problem (1)-(2). It allows us to represent the algorithm of the boundary functions method for constructing an asymptotic solution of initial-value singularly perturbed problems in the critical case more clearly than in [4].

Note that the projector approach has been used in [10] for constructing the zero-order asymptotic solution for a singularly perturbed linear-quadratic control problem in the critical case.

We will assume the same assumptions as in [4] that the matrix $A(t)$ has for each $t \in [0, T]$ m eigenvalues $\lambda_1(t), \lambda_2(t), ..., \lambda_m(t)$, and that they satisfy the conditions:

Assumption 1. *$\lambda_j(t) = 0$ for $j = 1, 2, ..., k$, $k < m$.*

Assumption 2. *All k eigenvectors $v_1(t), v_2(t), ..., v_k(t)$ of the matrix $A(t)$, corresponding to $\lambda_j(t) = 0$, $j = 1, 2, ..., k$, are linearly independent.*

Following [4], we will here use eigenvectors having the same smoothness as the matrix $A(t)$. The existence of such eigenvectors has been proved in [11].

Furthermore, some assumptions will be yet added.

The transposition will be denoted by the prime. By I, as usual, we mean the identity operator. For the expansion of a function $w(\varepsilon)$ into the series with respect to integer non-negative powers of ε $w(t, \varepsilon) = \sum\limits_{j \geq 0} \varepsilon^j w_j(t)$, we introduce the notation $[w(\varepsilon)]_j = w_j$.

The paper is organized as follows. In Section 2, we present the standard decomposition of the original system (1) into systems with respect to functions from the asymptotic solution, depending on t, and with respect to so-called boundary functions, depending on the argument t/ε. In the next section, we introduce orthogonal projectors of the space X onto $ker A(t)$ and $ker A(t)'$. Based on these projectors, the algorithm of constructing the zero-order asymptotic approximation of a solution of problem (1)-(2) is given in Section 4, and the algorithm of constructing the n-th order asymptotic approximation, $n \geq 1$, is developed in Section 5. Tables 1 and 2 in these two sections show the sequence of actions for finding asymptotics terms. In the sixth section, we present an example illustrating the projector approach for constructing the first-order asymptotic approximation. The last section presents our conclusions.

2. Problem Decomposition

In view of [4], we will seek the asymptotic solution of problem (1)-(2) in the form:

$$x(t, \varepsilon) = \overline{x}(t, \varepsilon) + \Pi x(\tau, \varepsilon), \tag{3}$$

where $\overline{x}(t, \varepsilon) = \sum_{j \geq 0} \varepsilon^j \overline{x}_j(t)$, $\Pi x(\tau, \varepsilon) = \sum_{j \geq 0} \varepsilon^j \Pi_j x(\tau)$, $\tau = t/\varepsilon$. Functions $\Pi_j x(\tau)$ will be found as in [4] with the help of the additional condition

$$\Pi_j x(\tau) \to 0 \text{ as } \tau \to +\infty. \tag{4}$$

Following tradition (see, for instance, [1], p. 8), a series $\sum_{j \geq 0} \varepsilon^j \overline{x}_j(t)$ with terms depending on the original argument t is called regular series, in contrast with boundary series $\sum_{j \geq 0} \varepsilon^j \Pi_j x(\tau)$ consisting of so-called boundary functions depending on the argument $\tau \geq 0$, which are essential only for arguments in some vicinities of points where additional conditions are prescribed (in a vicinity of zero in the considered case).

As usual in the theory of singular perturbations, the following representation will be used

$$f(\overline{x}(t, \varepsilon) + \Pi x(\tau, \varepsilon), t, \varepsilon) \equiv \overline{f} + \Pi f,$$

where $\overline{f} = f(\overline{x}(t,\varepsilon),t,\varepsilon) = \sum_{j\geq 0}\varepsilon^j\overline{f}_j(t)$ and $\Pi f = f(\overline{x}(\varepsilon\tau,\varepsilon)+\Pi x(\tau,\varepsilon),\varepsilon\tau,\varepsilon) - f(\overline{x}(\varepsilon\tau,\varepsilon),\varepsilon\tau,\varepsilon) = \sum_{j\geq 0}\varepsilon^j\Pi_j f(\tau)$.

Substituting expansion (3) into (1) and equating terms of the same order of ε separately depending on t and τ, we obtain the following equations for the terms of series (3):

$$\frac{d\overline{x}_{j-1}(t)}{dt} = A(t)\overline{x}_j(t) + \overline{f}_{j-1}(t), \tag{5}$$

$$\frac{d\Pi_j x(\tau)}{d\tau} = A(0)\Pi_j x(\tau) + \Pi_{j-1}f(\tau) + [(A(\varepsilon\tau)-A(0))\Pi x(\tau,\varepsilon)]_j, \tag{6}$$

where $j = 0,1,...,$

$$[(A(\varepsilon\tau)-A(0))\Pi x(\tau,\varepsilon)]_j = \sum_{k=0}^{j-1}\frac{1}{(j-k)!}\frac{d^{j-k}A}{dt^{j-k}}(0)\Pi_k x(\tau).$$

In order to write equations (5) and (6) in the same forms for the cases $j=0$ and $j>0$, we suppose that terms of expansions with negative indices are equal to zero.

Substituting expansion (3) into (2) and equating terms of the same order of ε, we obtain the equalities:

$$\overline{x}_0(0) + \Pi_0 x(0) = x^0, \tag{7}$$

$$\overline{x}_j(0) + \Pi_j x(0) = 0,\ j > 0. \tag{8}$$

3. Space Decomposition

Further, we will use the decompositions of the space X in the orthogonal sums (see, for instance, [12], p. 38)

$$X = kerA(t) \oplus imA(t)' = kerA(t)' \oplus imA(t).$$

Orthogonal projectors $P(t)$ and $Q(t)$ of the space X onto the subspaces $kerA(t)$ and $kerA(t)'$, respectively, corresponding to the decompositions of the space X into two last orthogonal sums, will be applied. We can write the explicit form of these projectors. Namely, let $V(t) = (v_1(t),...,v_k(t))$ and $S(t) = (s_1(t),...,s_k(t))$, where $s_1(t),...,s_k(t)$ are the eigenvectors of the matrix $A(t)'$ corresponding to eigenvalues $\lambda_j(t) = 0$, $j = 1,...,k$. Following [9], we believe that the eigenvectors $s_i(t)$ have been chosen in such a way that $V(t)'S(t)$ is the $k\times k$ identity matrix. We explain that this is possible. The invertibility of the matrix $V(t)'S(t)$ is proved in [1]. If $V(t)'S(t) = B(t) \neq I$, then we take the columns of the matrix $S(t)B(t)^{-1}$ as $s_1(t),...,s_k(t)$.

It easily follows from Assumption 2 that the $k\times k$ matrices $V(t)'V(t)$ and $S(t)'S(t)$ are invertible. It is not difficult to see that $P(t) = V(t)(V(t)'V(t))^{-1}V(t)'$ and $Q(t) = S(t)(S(t)'S(t))^{-1}S(t)'$ are orthogonal projectors of the space X onto the subspaces $kerA(t)$ and $kerA(t)'$, respectively, corresponding to the decompositions of the space X into the orthogonal sums.

The operator

$$A(t) = (I-Q(t))A(t)(I-P(t)) : imA(t)' \longrightarrow imA(t)$$

has the inverse operator. It will be denoted as $A(t)^+ = (I-P(t))A(t)^+(I-Q(t))$.

The following condition is assumed.

Assumption 3. *For each $t \in [0,T]$ the operator $(I-P(t))A(t)(I-P(t)) : imA(t)' \to imA(t)'$ is stable—that is, all eigenvalues of this operator have negative real parts.*

It is not difficult to prove that the operator $Q(t)P(t) : kerA(t) \to kerA(t)'$ is invertible. Let us take a vector x from $kerA(t)$. Then, $x = V(t)c(t)$, where $c(t) = (c_1(t),c_2(t),...,c_k(t))'$ and $c_i(t)$, $i = 1,2,...,k$, are some scalar functions. Consider the equation $Q(t)P(t)x = 0$. It follows from this that $V(t)'S(t)(S(t)'S(t))^{-1}S(t)'V(t)c(t) = 0$. Since $V(t)'S(t)$ is a $k \times k$ identity matrix, then $c(t) = 0$, which gives the provable invertibility.

4. Zero-Order Asymptotic Solution

From (5), we have the equation for $\overline{x}_0(t)$:

$$A(t)\overline{x}_0(t) = 0.$$

Hence,

$$(I - P(t))\overline{x}_0(t) = 0. \tag{9}$$

Using (9), we find from (7) the initial value

$$(I - P(0))\Pi_0 x(0) = (I - P(0))x^0. \tag{10}$$

From (6), we have the equation for $\Pi_0 x(\tau)$

$$\frac{d\Pi_0 x(\tau)}{d\tau} = A(0)\Pi_0 x(\tau).$$

This equation is equivalent to two ones:

$$\frac{d(I - P(0))\Pi_0 x(\tau)}{d\tau} = (I - P(0))A(0)(I - P(0))\Pi_0 x(\tau), \tag{11}$$

$$\frac{d(P(0)\Pi_0 x(\tau))}{d\tau} = P(0)A(0)(I - P(0))\Pi_0 x(\tau). \tag{12}$$

In view of Assumption 3, we obtain a unique solution of initial problem (10)-(11) satisfying the inequality

$$\| (I - P(0))\Pi_0 x(\tau) \| \leq c \exp(-\alpha\tau), \ \ \tau \geq 0,$$

with some positive constants c and α independent of τ (see, for instance, [13], p. 106). In this estimate, any norm may be used, since all norms in a finite dimensional space are equivalent. Functions satisfying the last inequality are called exponential-type boundary functions.

From (12), we get the equality $P(0)\Pi_0 x(\tau) = P(0)\Pi_0 x(0) + \int\limits_0^{\tau} P(0)A(0)(I - P(0))\Pi_0 x(s)\, ds$. Since $P(0)\Pi_0 x(\tau) \to 0$ as $\tau \to +\infty$, then $P(0)\Pi_0 x(0) = -\int_0^{+\infty} P(0)A(0)(I - P(0))\Pi_0 x(s)\, ds$. Using the exponential estimate for $(I - P(0))\Pi_0 x(s)$, we uniquely define the exponential-type boundary function $P(0)\Pi_0 x(\tau)$, namely,

$$P(0)\Pi_0 x(\tau) = -\int_{\tau}^{+\infty} P(0)A(0)(I - P(0))\Pi_0 x(s)\, ds. \tag{13}$$

Hence, the exponential-type boundary function $\Pi_0 x(\tau)$ has been found. Then, we can get the initial value from (7):

$$P(0)\overline{x}_0(0) = P(0)(x^0 - \Pi_0 x(0)). \tag{14}$$

In view of (5), the equation for $\overline{x}_1(t)$ has the form

$$A(t)\overline{x}_1(t) = -\overline{f}_0(t) + \frac{d\overline{x}_0(t)}{dt}.$$

Taking into account (9), we can write the solvability condition for the last equation in the form

$$Q(t)\frac{d(P(t)\overline{x}_0(t))}{dt} = Q(t)f(P(t)\overline{x}_0(t), t, 0).$$

Since

$$\frac{d(P(t)x(t))}{dt} = \frac{d(P(t)^2x(t))}{dt} = \frac{d(P(t))}{dt}P(t)x(t) + P(t)\frac{d(P(t)x(t))}{dt}, \tag{15}$$

we obtain the equation

$$\begin{array}{l}\frac{d(P(t)\bar{x}_0(t))}{dt} = (Q(t)P(t))^{-1}Q(t)(-\frac{dP(t)}{dt}P(t)\bar{x}_0(t) + f(P(t)\bar{x}_0(t),t,0)) \\ +(I-P(t))\frac{dP(t)}{dt}P(t)\bar{x}_0(t).\end{array} \tag{16}$$

If operator $A(t)$ is constant, then projectors $P(t) = P$ and $Q(t) = Q$ are constant too, and the last equation has the form

$$\frac{d(P\bar{x}_0(t))}{dt} = (QP)^{-1}Qf(P\bar{x}_0(t),t,0). \tag{17}$$

We will yet assume the condition.

Assumption 4. *Problem (14)–(16) has a unique solution on the segment* $[0, T]$.

A similar assumption regarding the solvability of some initial-value problem for a non-linear equation of the smaller dimension than the original one was presented in [1] (Assumption IV, p. 13).

Thus, the function $\bar{x}_0(t)$ is defined. Hence, the zero-order asymptotics for a solution of problem (1)-(2) is found.

The following Table 1 shows the sequence of finding zero-order asymptotics terms.

Table 1. The algorithm for finding the zero-order asymptotics terms.

Asymptotics Terms	Formulas
$(I-P(t))\bar{x}_0(t) = 0$	(9)
$(I-P(0))\Pi_0x(\tau)$	(10), (11)
$P(0)\Pi_0x(\tau)$	(13)
$P(t)\bar{x}_0(t)$	(14), (16)

5. Higher-Order Asymptotic Solutions

Suppose that the terms $\bar{x}_j(t)$ and $\Pi_jx(\tau)$ of expansion (3), $j = 0,1,...,n-1, n \geq 1$, have been found.

From equation (5) with $j = n$, we obtain the relation

$$A(t)\bar{x}_n(t) = \frac{d\bar{x}_{n-1}(t)}{dt} - \bar{f}_{n-1}(t),$$

where the right-hand side is known. Applying the operator I-Q(t) to this equation, we have

$$(I-Q(t))A(t)(I-P(t))\bar{x}_n(t) = (I-Q(t))(\frac{d\bar{x}_{n-1}(t)}{dt} - \bar{f}_{n-1}(t)).$$

From here, we find:

$$(I-P(t))\bar{x}_n(t) = A(t)^+(I-Q(t))(\frac{d\bar{x}_{n-1}(t)}{dt} - \bar{f}_{n-1}(t)). \tag{18}$$

Then, we can find from (8) with $j = n$ the initial value

$$(I-P(0))\Pi_nx(0) = -(I-P(0))\bar{x}_n(0). \tag{19}$$

The equation (6) with $j = n$ has the form

$$\frac{d\Pi_n x(\tau)}{d\tau} = A(0)\Pi_n x(\tau) + \Pi_{n-1} f(\tau) + [(A(\varepsilon\tau) - A(0))\Pi x(\tau, \varepsilon)]_n.$$

This equation is equivalent to two ones.

$$\frac{d(I - P(0))\Pi_n x(\tau)}{d\tau} = (I - P(0))A(0)(I - P(0))\Pi_n x(\tau) + (I - P(0))\Pi_{n-1} f(\tau) + \\ +(I - P(0))[(A(\varepsilon\tau) - A(0))\Pi x(\tau, \varepsilon)]_n, \tag{20}$$

$$\frac{d(P(0)\Pi_n x(\tau))}{d\tau} = P(0)(A(0)(I - P(0))\Pi_n x(\tau) + \Pi_{n-1} f(\tau) + \\ +[(A(\varepsilon\tau) - A(0))\Pi x(\tau, \varepsilon)]_n). \tag{21}$$

The sum of two last summands in the right-hand side in (20) is a known exponential-type boundary function. Therefore, in view of Assumption 3, we can find from (19) and (20) the exponential-type boundary function $(I - P(0))\Pi_n x(\tau)$. Note that the proof of exponential estimates for boundary functions is given in detail in monograph [14].

As the function in the braces on the right-hand side in (21) is a known exponential-type boundary function, we can get from (21) the exponential-type boundary function $P(0)\Pi_n x(\tau)$, namely

$$P(0)\Pi_n x(\tau) = -\int_\tau^{+\infty} P(0)(A(0)(I - P(0))\Pi_n x(s) + \Pi_{n-1} f(s) \\ +[(A(\varepsilon s) - A(0))\Pi x(s, \varepsilon)]_n)\, ds. \tag{22}$$

Hence, the exponential-type boundary function $\Pi_n x(\tau)$ is defined. Then, we can find from (8) with $j = n$ the initial value

$$P(0)\overline{x}_n(0) = -P(0)\Pi_n x(0). \tag{23}$$

Writing out equation (5) with $j = n + 1$, we get

$$\frac{d\overline{x}_n(t)}{dt} = A(t)\overline{x}_{n+1}(t) + \overline{f}_n(t).$$

The solvability condition for this equation has the form

$$Q(t)\frac{d(P(t)\overline{x}_n(t))}{dt} = Q(t)(\overline{f}_n(t) - \frac{d((I - P(t))\overline{x}_n(t))}{dt}).$$

In view of (15), we obtain from here the equation

$$\frac{d(P(t)\overline{x}_n(t))}{dt} = (Q(t)P(t))^{-1}Q(t)(\overline{f}_n(t) + \frac{dP(t)}{dt}P(t)\overline{x}_n(t) - \\ -\frac{d((I - P(t))\overline{x}_n(t))}{dt}) + (I - P(t))\frac{dP(t)}{dt}P(t)\overline{x}_n(t). \tag{24}$$

If operator $A(t)$ is constant, then this equation has the form:

$$\frac{d(P\overline{x}_n(t))}{dt} = (QP)^{-1}Q(\overline{f}_n(t) - \frac{d((I - P)\overline{x}_n(t))}{dt}). \tag{25}$$

It should be noted that equation (24) is linear with respect to $P(t)\overline{x}_n(t)$. As $(I - P(t))\overline{x}_n(t)$ has been found (see (18)), we can define the function $P(t)\overline{x}_n(t)$ from (23) and (24).

Hence, we have found the terms of the n-th order in expansion (3).

The following Table 2 shows the sequence of finding the n-th order terms in expansion (3).

Table 2. The algorithm for finding the n-th order asymptotics terms, $n \geq 1$.

Asymptotics Terms	Formulas
$(I - P(t))\overline{x}_n(t)$	(18)
$(I - P(0))\Pi_n x(\tau)$	(19), (20)
$P(0)\Pi_n x(\tau)$	(22)
$P(t)\overline{x}_n(t)$	(23),(24)

The previous arguments have, as a consequence, the following assertion.

Theorem 1. *Under Assumptions 1–4, the asymptotic solution of problem (1)-(2) in form (3) can be constructed with the help of orthogonal projectors onto* $\ker A(t)$ *and* $\ker A(t)'$. *The order of finding the asymptotics terms is the following:* $(I - P(t))\overline{x}_j(t)$, $(I - P(0))\Pi_j x(\tau)$, $P(0)\Pi_j x(\tau)$, $P(t)\overline{x}_j(t)$, $j \geq 0$.

6. Illustrative Example

Consider the following initial value problem of form (1)-(2) on the segment $[0, T]$:

$$\begin{aligned} \varepsilon\frac{dy}{dt} &= -y + \varepsilon z^2, \\ \varepsilon\frac{dz}{dt} &= y, \end{aligned} \tag{26}$$

$$y(0,\varepsilon) = 1, \quad z(0,\varepsilon) = 1. \tag{27}$$

Here, $t \in [0, 0.3]$, $x(t,\varepsilon) = (y(t,\varepsilon), z(t,\varepsilon))'$; $y = y(t,\varepsilon)$, $z = z(t,\varepsilon) \in \mathbf{R}$; $x^0 = (1,1)'$,

$$A(t) = A = \begin{pmatrix} -1 & 0 \\ 1 & 0 \end{pmatrix}, \; f(x,t,\varepsilon) = \begin{pmatrix} z^2 \\ 0 \end{pmatrix}.$$

Hence,

$$\lambda_1 = 0, \lambda_2 = -1, \ker A = \{(0,a)'\}, V = (0,1)', \ker A' = \{(a,a)'\},$$
$$S = (1,1)', \operatorname{im} A = \{(-a,a)'\}, \operatorname{im} A' = \{(a,0)'\},$$

$$P = \begin{pmatrix} 0 & 0 \\ 0 & 1 \end{pmatrix}, I - P = \begin{pmatrix} 1 & 0 \\ 0 & 0 \end{pmatrix}, Q = \begin{pmatrix} 1/2 & 1/2 \\ 1/2 & 1/2 \end{pmatrix},$$
$$(I-P)A(I-P) = \begin{pmatrix} -1 & 0 \\ 0 & 0 \end{pmatrix} : \operatorname{im} A' \to \operatorname{im} A',$$

$$(QP)^{-1} = \begin{pmatrix} 0 & 0 \\ 0 & 2 \end{pmatrix} : \ker A' \to \ker A, A^+ = \begin{pmatrix} -1/2 & 1/2 \\ 0 & 0 \end{pmatrix} : \operatorname{im} A \to \operatorname{im} A'.$$

We will construct the first-order approximation for the asymptotic solution of problem (26)-(27) using projectors P and Q.

Relation (9), in this case, has the form:

$$\begin{pmatrix} 1 & 0 \\ 0 & 0 \end{pmatrix} \begin{pmatrix} \overline{y}_0(t) \\ \overline{z}_0(t) \end{pmatrix} = \begin{pmatrix} \overline{y}_0(t) \\ 0 \end{pmatrix} = \begin{pmatrix} 0 \\ 0 \end{pmatrix}.$$

Therefore, $\overline{y}_0(t) = 0$.

From (10), we get $\Pi_0 y(0) = 1$.

Equation (11) has the form:

$$\frac{d}{d\tau}\begin{pmatrix} \Pi_0 y(\tau) \\ 0 \end{pmatrix} = \begin{pmatrix} -\Pi_0 y(\tau) \\ 0 \end{pmatrix}.$$

Taking into account the initial value $\Pi_0 y(0) = 1$ found from (10), we obtain from the last equation $\Pi_0 y(\tau) = e^{-\tau}$.

From (13), we find $\Pi_0 z(\tau) = -\int_\tau^{+\infty} \Pi_0 y(s) ds = -e^{-\tau}$.

From (14) and (17) we derive, respectively,

$$\begin{pmatrix} 0 & 0 \\ 0 & 1 \end{pmatrix}\begin{pmatrix} \bar{y}_0(0) \\ \bar{z}_0(0) \end{pmatrix} = \begin{pmatrix} 0 \\ \bar{z}_0(0) \end{pmatrix} = \begin{pmatrix} 0 & 0 \\ 0 & 1 \end{pmatrix}\left(\begin{pmatrix} 1 \\ 1 \end{pmatrix} - \begin{pmatrix} 1 \\ -1 \end{pmatrix}\right) = \begin{pmatrix} 0 \\ 2 \end{pmatrix}.$$

$$\frac{d}{dt}\begin{pmatrix} 0 \\ \bar{z}_0(t) \end{pmatrix} = \begin{pmatrix} 0 & 0 \\ 0 & 2 \end{pmatrix}\begin{pmatrix} 1/2 & 1/2 \\ 1/2 & 1/2 \end{pmatrix}\begin{pmatrix} (\bar{z}_0(t))^2 \\ 0 \end{pmatrix} = \begin{pmatrix} 0 \\ (\bar{z}_0(t))^2 \end{pmatrix},$$

In view of the last two relations, we have $\bar{z}_0(t) = 1/(0.5 - t)$.

It is easy to verify that conditions 1–4 are satisfied for problem (26)-(27).

Thus, we have found the zero-order asymptotic solution of form (3) $\tilde{x}_0(t,\varepsilon)$ for the solution of problems (26)-(27). Namely, we have

$$\tilde{y}_0(t,\varepsilon) = e^{-\tau},$$
$$\tilde{z}_0(t,\varepsilon) = 1/(0.5 - t) - e^{-\tau},\ \tau = t/\varepsilon.$$

Now, we will seek for the first-order asymptotics.

Equation (18) for $n = 1$ has the form:

$$\begin{pmatrix} \bar{y}_1(t) \\ 0 \end{pmatrix} = \begin{pmatrix} -1/2 & 1/2 \\ 0 & 0 \end{pmatrix}\begin{pmatrix} 1/2 & -1/2 \\ -1/2 & 1/2 \end{pmatrix}\left(\frac{d}{dt}\begin{pmatrix} 0 \\ \bar{z}_0(t) \end{pmatrix} - \begin{pmatrix} (\bar{z}_0(t))^2 \\ 0 \end{pmatrix}\right) = \begin{pmatrix} 1/(0.5-t)^2 \\ 0 \end{pmatrix}.$$

Therefore, $\bar{y}_1(t) = 1/(0.5 - t)^2$.

From (19) with $n = 1$, we get $\Pi_1 y(0) = -4$.

Equation (20) for $n = 1$ has the form:

$$\frac{d}{d\tau}\begin{pmatrix} \Pi_1 y(\tau) \\ 0 \end{pmatrix} = \begin{pmatrix} -\Pi_1 y(\tau) + 4\Pi_0 z(\tau) + (\Pi_0 z(\tau))^2 \\ 0 \end{pmatrix}.$$

From the last two relations, we obtain $\Pi_1 y(\tau) = -(3 + 4\tau + e^{-\tau})e^{-\tau}$.

From (22) with $n = 1$, we find $\Pi_1 z(\tau) = -\int_\tau^{+\infty} \Pi_1 y(s)\, ds = (7 + 4\tau + e^{-\tau}/2)e^{-\tau}$.

From (23) and (25) with $n = 1$, we derive, respectively,

$$\frac{d}{dt}\begin{pmatrix} 0 \\ \bar{z}_1(t) \end{pmatrix} = \begin{pmatrix} 0 & 0 \\ 0 & 2 \end{pmatrix}\begin{pmatrix} 1/2 & 1/2 \\ 1/2 & 1/2 \end{pmatrix}\left(\begin{pmatrix} 2\bar{z}_0(t)\bar{z}_1(t) \\ 0 \end{pmatrix} - \frac{d}{dt}\begin{pmatrix} \bar{y}_1(t) \\ 0 \end{pmatrix}\right) = \begin{pmatrix} 0 \\ 2\bar{z}_0(t)\bar{z}_1(t) - d\bar{y}_1(t)/dt \end{pmatrix},$$

$$\begin{pmatrix} 0 \\ \bar{z}_1(0) \end{pmatrix} = \begin{pmatrix} 0 \\ -\Pi_1 z(0) \end{pmatrix}.$$

In view of the last two relations, we have $\bar{z}_1(t) = (ln(0.5 - t)^2 - 15/8 + ln4)/(0.5 - t)^2$.

Thus, we have found for problems (26)-(27) the first-order asymptotic solution of form (3) $\widetilde{x}_1(t,\varepsilon)$. Namely, we have

$$\widetilde{y}_1(t,\varepsilon) = \widetilde{y}_0(t,\varepsilon) + \varepsilon(\overline{y}_1(t) + \Pi_1 y(\tau)),$$
$$\widetilde{z}_1(t,\varepsilon) = \widetilde{z}_0(t,\varepsilon) + \varepsilon(\overline{z}_1(t) + \Pi_1 z(\tau)).$$

Of course, these results can be obtained using the algorithm from [4], but we would like to demonstrate here the use of projectors for finding asymptotics terms. The results obtained by Maple 13 are given in Figures 1 and 2. They have been presented for the completeness of the paper. The solid line represents the exact solution; the dash-dotted line—the solution of the degenerate problem, the line consisting of squares represents the zero-order approximation; and the dash line represents the the first-order approximation. These graphs show that an asymptotic solution is closer to the exact one if we use higher-order asymptotics. If we use the smaller value of ε, then it will result in an asymptotic solution more similar to the exact one. The graphs of the solution of the degenerate problem and the zero-order approximation illustrate the known property of boundary functions that are essential only for arguments in some vicinities of points where additional conditions are prescribed.

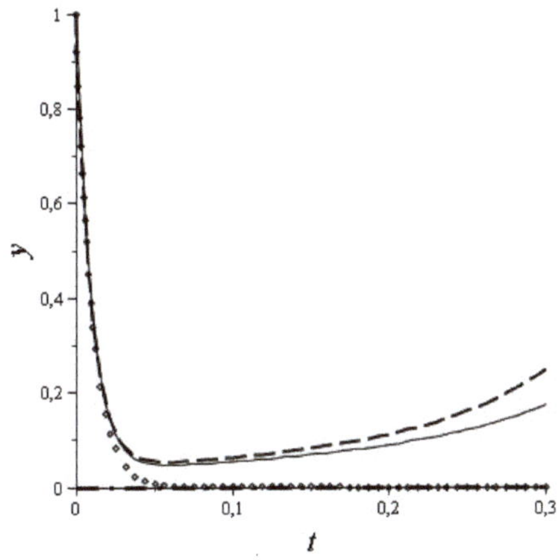

Figure 1. Trajectory $y(t,\varepsilon)$ with $\varepsilon = 0.01$ and its approximations.

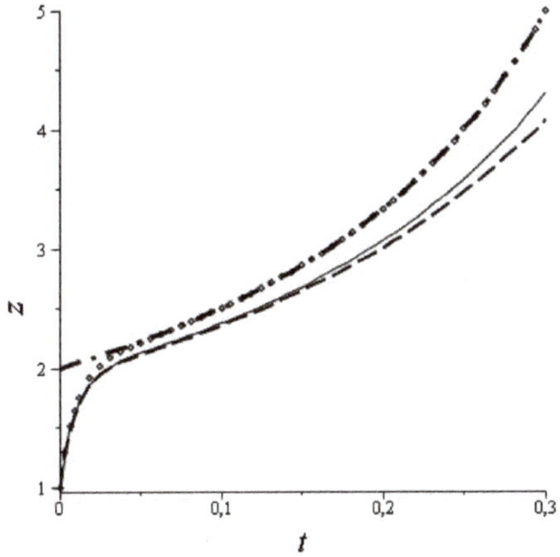

Figure 2. Trajectory $z(t,\varepsilon)$ with $\varepsilon = 0.01$ and its approximations.

7. Conclusions

This paper dealt with a new approach to the algorithm of the method of boundary functions from [4] for asymptotic solving initial value problem of form (1)-(2) in the critical case. Namely, the algorithm was formulated with the help of orthogonal projectors of the space X onto $ker A(t)$ and $ker A(t)'$. Such an approach clearly shows the structure of the algorithm for finding asymptotics terms, given in Tables 1 and 2 of the paper.

Funding: This research was funded by the RUSSIAN SCIENCE FOUNDATION, grant number 17-11-01220.

Acknowledgments: The author thank the anonymous referees for carefully reading the paper and for the numerous critical and constructive comments that have helped to improve the paper text. She thanks also N. T. Hoai for useful discussions and M. A. Kalashnikova for the help in printing the paper text.

Conflicts of Interest: The author declare no conflict of interest. The funders had no role in the design of the study; in the collection, analyses, or interpretation of data; in the writing of the manuscript, or in the decision to publish the results.

References

1. Vasil'eva, A.B.; Butuzov, V.F. *Singularly Perturbed Equations in Critical Cases*; Moscow University: Moscow, Russia, 1978; 107p; English transl.: University of Wisconsin-Madison: Wisconsin, WI, USA, 1980; 165p.
2. O'Malley, R.E., Jr. A singular singularly-perturbed linear boundary value problem. *SIAM J. Math. Anal.* **1979**, *10*, 695–708. [CrossRef]
3. Khalil, H.K. Feedback control of nonstandard singularly perturbed systems. *IEEE Trans. Autom. Control* **1989**, *34*, 1052–1060. [CrossRef]
4. Butuzov, V.F.; Vasil'eva, A.B. Differential and difference systems of equations with a small parameter in the case when the unperturbed (degenerate) system is situated on the spectrum. *Differ. Uravn.* **1970**, *6*, 650–664. (In Russian); English transl.: *Differ. Equ.* **1971**, *6*, 499–510.
5. Gu, Z.M.; Nefedov, N.N.; O'Malley, R.E., Jr. On singular singularly perturbed initial value problems. *SIAM J. Appl. Math.* **1989**, *49*, 1–25. [CrossRef]
6. Schmeiser, C.; Wess, R. Asymptotic analysis of singular singularly perturbed boundary value problems. *SIAM J. Math. Anal.* **1986**, *17*, 560–579. [CrossRef]
7. O'Malley R.E., Jr.; Flaherty, J.E. Analytical and numerical methods for nonlinear singular singularly-perturbed initial value problems. *SIAM J. Appl. Math.* **1980**, *38*, 225–248. [CrossRef]
8. Ascher, U. On some difference schemes for singular singularly-perturbed boundary value problems. *Numer. Math.* **1985**, *46*, 1–30. [CrossRef]
9. Vasil'eva, A.B.; Butuzov, V.F.; Kalachev, L.V. *The Boundary Function Method for Singular Perturbation Problems*; SIAM: Philadelphia, PA, USA, 1995.
10. Kurina, G.A.; Hoai, N.T. Projector approach for constructing the zero order asymptotic solution for the singularly perturbed linear-quadratic control problem in a critical case. *AIP Conf. Proc.* **2018**, *1997*, 020073-1–020073-7. [CrossRef]
11. Sibuya, Y. Some global properties of matrices of functions of one variable. *Math. Ann.* **1965**, *161*, 67–77. [CrossRef]
12. Kato, T. *Perturbation Theory for Linear Operators*; Mir: Moscow, Russia, 1972; 740p.
13. Daletskii, Y.L.; Krein, M.G. *Stability of Solutions of Differential Equations in Banach Space*; Nauka: Moscow, Russia, 1970; 536p.
14. Vasil'eva, A.B.; Butuzov, V.F. *Asymptotic Expansions of Solutions of Singularly Perturbed Equations*; Nauka: Moscow, Russia, 1973; 272p.

axioms

MDPI

Article

Justification of Direct Scheme for Asymptotic Solving Three-Tempo Linear-Quadratic Control Problems under Weak Nonlinear Perturbations

Galina Kurina [1,2,*] and Margarita Kalashnikova [3]

1 Faculty of Mathematics, Voronezh State University, Universitetskaya pl., 1, 394018 Voronezh , Russia
2 Federal Research Center "Computer Science and Control" of RAS, ul. Vavilova, 44/2, 119333 Moscow, Russia
3 Atos IT Solutions and Services, pr. Truda, 65, 394026 Voronezh, Russia
* Correspondence: kurina@math.vsu.ru

Abstract: The paper deals with an application of the direct scheme method, consisting of immediately substituting a postulated asymptotic solution into a problem condition and determining a series of control problems for finding asymptotics terms, for asymptotics construction of a solution of a weakly nonlinearly perturbed linear-quadratic optimal control problem with three-tempo state variables. For the first time, explicit formulas for linear-quadratic optimal control problems, from which all terms of the asymptotic expansion are found, are justified, and the estimates of the proximity between the asymptotic and exact solutions are proved for the control, state trajectory, and minimized functional. Non-increasing of the minimized functional, if a next approximation to the optimal control is used, following from the proposed algorithm of the asymptotics construction, is also established.

Keywords: optimal control problems; weak nonlinear perturbations; three-tempo variables; asymptotic solutions; the direct scheme method; estimates of asymptotic solution

MSC: 34H05; 34E13

Citation: Kurina, G.; Kalashnikova, M. Justification of Direct Scheme for Asymptotic Solving Three-Tempo Linear-Quadratic Control Problems under Weak Nonlinear Perturbations. *Axioms* **2022**, *11*, 647. https://doi.org/10.3390/axioms11110647

Academic Editor: Valery Y. Glizer

Received: 29 September 2022
Accepted: 25 October 2022
Published: 16 November 2022

1. Introduction

Systems with two-tempo variables are the main object in the study of singularly perturbed control problems (see, for instance, the reviews [1–3]). However, many practical problems contain multi-tempo fast variables. For instance, such variables arise in models of chain chemical reactions [4], fuel cells with a proton membrane [5], electrical chains [6], electromechanical processes in a synchronous machine [7], power systems [8], nuclear reactors [9], aircraft [10], ocean currents [11], rolling mills [12], two-wheeled carriages [13], forest pests [14], and epidemics [15].

Various asymptotic and numerical (see, for instance, [16]) methods are used for studying singularly perturbed systems with many small parameters standing before derivatives. Basic methods of asymptotic analysis are boundary functions method [17] and integral manifolds method ([18], ch. 7–10), which reduce the considered problem to a problem of simpler structure. The limit passage of an initial problem solution of a system with many small parameters at derivatives, when these parameters tend to zero, was studied for the first time by A.N. Tikhonov [19] and I.S. Gradstein [20]. Asymptotic solution of such problems was first constructed by A.B. Vasil'eva [21].

There are two approaches to constructing asymptotic solutions of optimal control problems. The traditional one is based on an asymptotic solution of a system following from control optimality conditions. Another approach, called the direct scheme method, consists of immediately substituting a postulated asymptotic expansion of a solution into the problem condition and receiving a series of problems for finding asymptotic terms. For two-tempo systems, it is presented, for example, in [22,23]. This method allows for establishing non-increasing of values of the minimized functional if a next optimal control approximation is used. Moreover, standard programs for solving optimal control problems can be applied for finding asymptotics terms. The direct scheme method has been, for instance, used in [24] to obtain any order asymptotic solution of a linear-quadratic optimal control problem with cheap controls of different costs.

The present paper deals with an asymptotic solution construction for the problem P_ε with weak nonlinear perturbations in a quadratic performance index and in a linear state equation. Namely, the following functional

$$J_\varepsilon(u) = \int_0^T (1/2(w(t,\varepsilon)'W(t)w(t,\varepsilon) + u(t,\varepsilon)'R(t)u(t,\varepsilon)) + \varepsilon F(w(t,\varepsilon), u(t,\varepsilon), t, \varepsilon))\, dt \quad (1)$$

is minimized on trajectories of three-tempo singularly perturbed system

$$\mathcal{E}(\varepsilon)\frac{dw(t,\varepsilon)}{dt} = A(t)w(t,\varepsilon) + B(t)u(t,\varepsilon) + \varepsilon f(w(t,\varepsilon), u(t,\varepsilon), t, \varepsilon),\ t \in [0,T], \quad (2)$$

with the initial condition

$$w(0,\varepsilon) = w^0. \quad (3)$$

Here, ε is a non-negative small parameter, $T > 0$ is fixed, the prime means transposition; $w(t,\varepsilon) = (x(t,\varepsilon)', y(t,\varepsilon)', z(t,\varepsilon)')'$, $x(t,\varepsilon) \in \mathbb{R}^{n_1}$, $y(t,\varepsilon) \in \mathbb{R}^{n_2}$, $z(t,\varepsilon) \in \mathbb{R}^{n_3}$, $u(t,\varepsilon) \in \mathbb{R}^m$; $\mathcal{E}(\varepsilon) = diag(I_{n_1}, \varepsilon I_{n_2}, \varepsilon^2 I_{n_3})$, I_{n_i} is the identity matrix of order n_i, $f = (\overset{(1)}{f}{}', \overset{(2)}{f}{}', \overset{(3)}{f}{}')'$, $\overset{(i)}{f} \in \mathbb{R}^{n_i}$, $B = (\overset{(1)}{B}{}', \overset{(2)}{B}{}', \overset{(3)}{B}{}')'$, $\overset{(i)}{B} : \mathbb{R}^m \to \mathbb{R}^{n_i}$, $i = \overline{1,3}$; all functions in (1), (2) are sufficiently smooth with respect to their arguments; for all $t \in [0,T]$ matrices $W(t)$, $R(t)$ are symmetric, moreover, $W(t)$, $R(t)$ and $S(t) = B(t)R(t)^{-1}B(t)'$ are positive definite.

It is assumed that the stability of the matrices A_{33} and $A_{22} - A_{23}A_{33}^{-1}A_{32}$ takes place. Here, and further A_{ij}, $i,j = \overline{1,3}$, mean matrices from a block representation of a matrix A with number of rows and columns n_1, n_2, n_3.

In contrast to [25], where optimal control problems for finding some zero order asymptotics terms for a solution of a nonlinear singularly perturbed problem with three-tempo state variables were formulated, here, explicit expressions of problems for receiving all asymptotic terms are obtained. Note that explicit formulas are very useful for research applying asymptotic methods for solving practical problems.

It should be noted that some results concerning the algorithm of asymptotic solving problem (1)–(3) have been presented in [26], but rigorous proofs and estimates are absent there. Note that [26] deal with matrices in (1), (2) depending on ε. However, expanding these matrices with respect to non-negative integer powers of ε and including terms depending on ε into the small nonlinearities, we obtain the problem P_ε in our statement.

It is well known that, if a linear-quadratic problem is nonsingular, then its solving is reduced to solving a system of linear differential equations resolved with respect to derivatives. Under studying nonlinear singularly perturbed optimal control problems, it is ordinarily assumed that the control problem is nonsingular, i.e., an optimal control is presented as an explicit function with respect to state and costate variables. See e.g., [27], where, apparently for the first time, singular perturbations methods were used for optimal control problems. In the present paper, unlike these cases, we do not assume the non-singularity of the considered problem for all ε and, for obtaining asymptotic estimates, we analyze a nonlinear singularly perturbed differential-algebraic system.

The essential new results obtained in this paper for problem (1)–(3) are the following:

1. The rigorous justification of explicit forms of linear-quadratic optimal control problems, solutions of which are used under constructing an asymptotic solution of nonlinear problem (1)–(3);

2. The proof of estimates of the proximity between the exact solution and asymptotic one obtained by the direct scheme method for the control, state trajectory of system (2), (3), and functional (1);
3. The proof of non-increasing values of functional (1) under using new asymptotic approximations to the optimal control and constructing minimized sequences.

Throughout the paper, the coefficient with ε^i in an expansion of a function $\omega = \omega(\varepsilon)$ in a series in powers of ε will be denoted by ω_i or $[\omega]_i$. The k-th partial sum of a series will be denoted by upper wave and the low index k or by braces with the low index k, i.e., $\widetilde{\omega}_k = \{\omega\}_k = \sum_{j=0}^{k} \varepsilon^j \omega_j$. The functions with negative indices will be considered equal to zero. Positive constants in estimates will be denoted as c and æ.

The paper is organized as follows: in Section 2, we present a formalism of asymptotics construction. Optimal control problems for finding asymptotic terms are given in Section 3. Section 4 is devoted to justification of such a choice of control problems. Namely, transformations of coefficients of expansion of minimized functional with respect to powers of ε with even and odd indices are considered. Asymptotic estimates of the proximity between the asymptotic and exact solutions are proved in Section 5. Non-increasing of the minimized functional, if a next optimal control approximation is used, is also discussed in this section. The last Section 6 contains conclusions.

2. Formalism of Asymptotics Construction

Following the boundary function method by A.B. Vasil'eva (see, for instance, [28]), we will seek a solution of problem (1)–(3) in the form

$$\vartheta(t,\varepsilon) = \overline{\vartheta}(t,\varepsilon) + \sum_{i=0}^{1}(\Pi_i\vartheta(\tau_i,\varepsilon) + Q_i\vartheta(\sigma_i,\varepsilon)). \tag{4}$$

Here, $\vartheta(t,\varepsilon) = (w(t,\varepsilon)', u(t,\varepsilon)')'$, $\overline{\vartheta}(t,\varepsilon) = \sum_{j\geq 0} \varepsilon^j \overline{\vartheta}_j(t)$, $\Pi_i\vartheta(\tau_i,\varepsilon) = \sum_{j\geq 0} \varepsilon^j \Pi_{ij}\vartheta(\tau_i)$, $Q_i\vartheta(\sigma_i,\varepsilon) = \sum_{j\geq 0} \varepsilon^j Q_{ij}\vartheta(\sigma_i)$, $\tau_i = t/\varepsilon^{i+1}$, $\sigma_i = (t-T)/\varepsilon^{i+1}$, $i = 0,1$, $\overline{\vartheta}_j(t)$ are regular functions, $\Pi_{ij}\vartheta(\tau_i)$ and $Q_{ij}\vartheta(\sigma_i)$ are boundary functions of exponential type in neighborhoods $t = 0$ and $t = T$, respectively, i.e.,

$$\|\Pi_{ij}\vartheta(\tau_i)\| \leqslant c\exp(-æ\tau_i),\ \tau_i \geqslant 0,\ \|Q_{ij}\vartheta(\sigma_i)\| \leqslant c\exp(æ\sigma_i),\ \sigma_i \leqslant 0,$$

where c and æ are positive constants independent of the arguments of functions under study.

For any sufficiently smooth function $G(w(t,\varepsilon), u(t,\varepsilon), t, \varepsilon)$, we will use the notation $G(\vartheta(t,\varepsilon), t, \varepsilon)$ and the asymptotic representation

$$G(\vartheta,t,\varepsilon) = \overline{G}(t,\varepsilon) + \sum_{i=0}^{1}(\Pi_i G(\tau_i,\varepsilon) + Q_i G(\sigma_i,\varepsilon)), \tag{5}$$

$\overline{G}(t,\varepsilon) = G(\overline{\vartheta}(t,\varepsilon),t,\varepsilon) = \sum_{j\geq 0}\varepsilon^j \overline{G}_j(t)$, $\Pi_0 G(\tau_0,\varepsilon) = G(\overline{\vartheta}(\varepsilon\tau_0,\varepsilon) + \Pi_0\vartheta(\tau_0,\varepsilon), \varepsilon\tau_0, \varepsilon)$
$- G(\overline{\vartheta}(\varepsilon\tau_0,\varepsilon), \varepsilon\tau_0,\varepsilon) = \sum_{j\geq 0}\varepsilon^j \Pi_{0j}G(\tau_0)$, $\Pi_1 G(\tau_1,\varepsilon) = G(\overline{\vartheta}(\varepsilon^2\tau_1,\varepsilon) + \Pi_0\vartheta(\varepsilon\tau_1,\varepsilon)$
$+ \Pi_1\vartheta(\tau_1,\varepsilon), \varepsilon^2\tau_1,\varepsilon) - G(\overline{\vartheta}(\varepsilon^2\tau_1,\varepsilon) + \Pi_0\vartheta(\varepsilon\tau_1,\varepsilon), \varepsilon^2\tau_1,\varepsilon) = \sum_{j\geq 0}\varepsilon^j \Pi_{1j}G(\tau_1)$,
$Q_0 G(\sigma_0,\varepsilon) = G(\overline{\vartheta}(T+\varepsilon\sigma_0,\varepsilon) + Q_0\vartheta(\sigma_0,\varepsilon), T+\varepsilon\sigma_0,\varepsilon) - G(\overline{\vartheta}(T+\varepsilon\sigma_0,\varepsilon), T+\varepsilon\sigma_0,\varepsilon)$
$= \sum_{j\geq 0}\varepsilon^j Q_{0j}G(\sigma_0)$, $Q_1 G(\sigma_1,\varepsilon) = G(\overline{\vartheta}(T+\varepsilon^2\sigma_1,\varepsilon) + Q_0\vartheta(\varepsilon\sigma_1,\varepsilon) + Q_1\vartheta(\sigma_1,\varepsilon), T+\varepsilon^2\sigma_1,\varepsilon)$
$- G(\overline{\vartheta}(T+\varepsilon^2\sigma_1,\varepsilon) + Q_0\vartheta(\varepsilon\sigma_1,\varepsilon), T+\varepsilon^2\sigma_1,\varepsilon) = \sum_{j\geq 0}\varepsilon^j Q_{1j}G(\sigma_1)$.

Substitute (4) in (1) and present the integrand in the form of sum (4). Passing in the integrals from the expressions depending on τ_i, σ_i, $i = 0,1$, to integrals over the corresponding intervals $[0,+\infty)$ and $(-\infty,0]$, we obtain the following expansion of the functional (1)

$$J_\varepsilon(u) = \sum_{j\geq 0}\varepsilon^j J_j. \tag{6}$$

Substituting expansion (4) into system (2) and initial value (3), using (5), then equating terms of the same powers of ε, separately depending on regular and different boundary functions, we obtain relations for defining asymptotics terms.

Introducing the notation $E_1 = diag(I_{n_1}, 0, 0)$, $E_2 = diag(0, I_{n_2}, 0)$, $E_3 = diag(0, 0, I_{n_3})$, and $\phi(\vartheta, t, \varepsilon) = A(t)w(t,\varepsilon) + B(t)u(t,\varepsilon) + \varepsilon f(w(t,\varepsilon), u(t,\varepsilon), t, \varepsilon)$, we obtain the following equations:

$$E_1 \frac{d\overline{w}_j(t)}{dt} + E_2 \frac{d\overline{w}_{j-1}(t)}{dt} + E_3 \frac{d\overline{w}_{j-2}(t)}{dt} = [\overline{\phi}(t,\varepsilon)]_j, \tag{7}$$

$$\begin{aligned}(E_1+E_2)\frac{d\Pi_{0j}w(\tau_0)}{d\tau_0} + E_3\frac{d\Pi_{0(j-1)}w(\tau_0)}{d\tau_0} &= E_1[\Pi_0\phi(\tau_0,\varepsilon)]_{j-1} \\ &+(E_2+E_3)[\Pi_0\phi(\tau_0,\varepsilon)]_j,\end{aligned} \tag{8}$$

$$\begin{aligned}(E_1+E_2)\frac{dQ_{0j}w(\sigma_0)}{d\sigma_0} + E_3\frac{dQ_{0(j-1)}w(\sigma_0)}{d\sigma_0} &= E_1[Q_0\phi(\sigma_0,\varepsilon)]_{j-1} \\ &+(E_2+E_3)[Q_0\phi(\sigma_0,\varepsilon)]_j,\end{aligned} \tag{9}$$

$$\frac{d\Pi_{1j}w(\tau_1)}{d\tau_1} = E_1[\Pi_1\phi(\tau_1,\varepsilon)]_{j-2} + E_2[\Pi_1\phi(\tau_1,\varepsilon)]_{j-1} + E_3[\Pi_1\phi(\tau_1,\varepsilon)]_j, \tag{10}$$

$$\frac{dQ_{1j}w(\sigma_1)}{d\sigma_1} = E_1[Q_1\phi(\sigma_1,\varepsilon)]_{j-2} + E_2[Q_1\phi(\sigma_1,\varepsilon)]_{j-1} + E_3[Q_1\phi(\sigma_1,\varepsilon)]_j. \tag{11}$$

From Equations (8)–(11) at $j = 0$, (10) and (11) at $j = 1$, we found the corresponding boundary functions

$$\begin{aligned}&E_1\Pi_{00}w(\tau_0) = 0,\ E_1\Pi_{10}w(\tau_1) = E_1\Pi_{11}w(\tau_1) = 0,\ E_1Q_{00}w(\sigma_0) = 0, \\ &E_1Q_{10}w(\sigma_1) = E_1Q_{11}w(\sigma_1) = 0,\ E_2\Pi_{10}w(\tau_1) = 0,\ E_2Q_{10}w(\sigma_1) = 0.\end{aligned} \tag{12}$$

In view of the last equalities, from (3), we obtain relations for initial values

$$E_1\overline{w}_0(0) = E_1w^0,\ E_1(\overline{w}_1(0) + \Pi_{01}w(0)) = 0, \tag{13}$$

$$E_1(\overline{w}_j(0) + \Pi_{0j}w(0) + \Pi_{1j}w(0)) = 0,\ j \geq 2, \tag{14}$$

$$E_2(\overline{w}_0(0) + \Pi_{00}w(0)) = E_2w^0, \tag{15}$$

$$E_2(\overline{w}_j(0) + \Pi_{0j}w(0) + \Pi_{1j}w(0)) = 0,\ j \geq 1, \tag{16}$$

$$E_3(\overline{w}_j(0) + \Pi_{0j}w(0) + \Pi_{1j}w(0)) = \begin{cases} E_3w^0,\ j = 0, \\ 0,\ j \geq 1. \end{cases} \tag{17}$$

Remark 1. *If boundary functions $\Pi_{ij}w$, $Q_{ij}w$, $i = 0,1$, $j = \overline{0, n-1}$ have been found, then, from Equations (8)–(11), it follows the corollary that functions $E_1\Pi_{in}w(\tau_i)$, $E_1Q_{in}w(\sigma_i)$, $i = 0,1$, $E_2\Pi_{1n}w(\tau_1)$, $E_2Q_{1n}w(\sigma_1)$, and $E_1\Pi_{1(n+1)}w(\tau_1)$, $E_1Q_{1(n+1)}w(\sigma_1)$ are known.*

3. Optimal Control Problems for Finding Asymptotics Terms

In this section, forms of control problems for finding asymptotics terms will be given. In contrast to [26], the justification of these relations will be presented.

With the help of the notations,

$$\begin{aligned}\rho(\vartheta,\psi,t,\varepsilon) &= W(t)w(t,\varepsilon) - A(t)'\psi(t,\varepsilon) + \varepsilon(F_w(\vartheta,t,\varepsilon)' - f_w(\vartheta,t,\varepsilon)'\psi(t,\varepsilon)), \\ \chi(\vartheta,\psi,t,\varepsilon) &= R(t)u(t,\varepsilon) - B(t)'\psi(t,\varepsilon) + \varepsilon(F_u(\vartheta,t,\varepsilon)' - f_u(\vartheta,t,\varepsilon)'\psi(t,\varepsilon)),\end{aligned}$$

five optimal control problems $\overline{P}_j$, $\Pi_{ij}P$, $Q_{ij}P$, $i = 0,1$, for determining asymptotics terms in expansion (4) will be written. Costate variables in these problems will be denoted as $\overline{\psi}_j(t)$, $\Pi_{ij}\psi(\tau_i)$, $Q_{ij}\psi(\sigma_i)$, $i = 0,1$, respectively.

Furthermore, the hat and the low index k in a function notation will mean that the function is calculated with the functional argument equal to the k-th partial sum of the corresponding expansion, e.g., $\widehat{f}_k(t,\varepsilon) = f(\widetilde{\vartheta}_k(t,\varepsilon),t,\varepsilon)$.

In the following expressions with ρ and χ in the performance indices of the formulated optimal control problems, we take $\psi(t,\varepsilon) = \sum_{j=0}^{\infty}\varepsilon^j(\overline{\psi}_j(t) + (\varepsilon E_1 + E_2 + E_3)(\Pi_{0j}\psi(\tau_0) + Q_{0j}\psi(\sigma_0)) + (\varepsilon^2 E_1 + \varepsilon E_2 + E_3)(\Pi_{1j}\psi(\tau_1) + Q_{1j}\psi(\sigma_1)))$.

Regular functions $\overline{\vartheta}_j(t)$, $t \in [0,T]$, are determined as solutions of problems $\overline{P}_j$, which consist of minimizing the functional

$$\overline{J}_j(\overline{u}_j) = \overline{w}_j(T)'E_1(Q_{0(j-1)}\psi(0) + Q_{1(j-2)}\psi(0)) + \int_0^T (\overline{w}_j(t)'(\frac{1}{2}W(t)\overline{w}_j(t)$$
$$+[\widehat{\overline{\rho}}_{j-1}(t,\varepsilon)]_j - E_2\frac{d\overline{\psi}_{j-1}(t)}{dt} - E_3\frac{d\overline{\psi}_{j-2}(t)}{dt}) + \overline{u}_j(t)'(\frac{1}{2}R(t)\overline{u}_j(t) + [\widehat{\overline{\chi}}_{j-1}(t,\varepsilon)]_j))\,dt$$

on trajectories of system (7) with initial conditions from (13) or (14) in dependence on j.

The boundary functions $\Pi_{0j}\vartheta(\tau_0)$, $\tau_0 \in [0,+\infty)$ are determined from optimal control problems $\Pi_{0j}P$ consisting of minimizing the functional

$$\Pi_{0j}J(\Pi_{0j}u) = \int_0^{+\infty} (\Pi_{0j}w(\tau_0)'(\frac{1}{2}W(0)\Pi_{0j}w(\tau_0) + [\widehat{\Pi}_{0(j-1)}\rho(\tau_0,\varepsilon)]_j - E_3\frac{d\Pi_{0(j-1)}\psi(\tau_0)}{d\tau_0})$$
$$+\Pi_{0j}u(\tau_0)'(\frac{1}{2}R(0)\Pi_{0j}u(\tau_0) + [\widehat{\Pi}_{0(j-1)}\chi(\tau_0,\varepsilon)]_j))\,d\tau_0$$

on trajectories of system (8) with the conditions $\Pi_{0j}x(+\infty) = 0$ and (15) or (16) in dependence on j.

The boundary functions $Q_{0j}\vartheta(\sigma_0)$, $\sigma_0 \in (-\infty,0]$, are determined from optimal control problems $Q_{0j}P$ consisting of minimizing the functional

$$Q_{0j}J(Q_{0j}u) = Q_{0j}w(0)'E_2(\overline{\psi}_j(T) + Q_{1(j-1)}\psi(0))$$
$$+\int_{-\infty}^0 (Q_{0j}w(\sigma_0)'(\frac{1}{2}W(T)Q_{0j}w(\sigma_0) + [\widehat{Q}_{0(j-1)}\rho(\sigma_0,\varepsilon)]_j - E_3\frac{dQ_{0(j-1)}\psi(\sigma_0)}{d\sigma_0})$$
$$+Q_{0j}u(\sigma_0)'(\frac{1}{2}R(T)Q_{0j}u(\sigma_0) + [\widehat{Q}_{0(j-1)}\chi(\sigma_0,\varepsilon)]_j))\,d\sigma_0$$

on trajectories of system (9) with the condition $(E_1 + E_2)Q_{0j}w(-\infty) = 0$.

The boundary functions $\Pi_{1j}\vartheta(\tau_1)$, $\tau_1 \in [0,+\infty)$, are determined from optimal control problems $\Pi_{1j}P$ consisting of minimizing the functional

$$\Pi_{1j}J(\Pi_{1j}u) = \int_0^{+\infty} (\Pi_{1j}w(\tau_1)'(\frac{1}{2}W(0)\Pi_{1j}w(\tau_1) + [\widehat{\Pi}_{1(j-1)}\rho(\tau_1,\varepsilon)]_j)$$
$$+\Pi_{1j}u(\tau_1)'(\frac{1}{2}R(0)\Pi_{1j}u(\tau_1) + [\widehat{\Pi}_{1(j-1)}\chi(\tau_1,\varepsilon)]_j))\,d\tau_1$$

on trajectories of system (10) with the conditions $(E_1 + E_2)\Pi_{1j}w(+\infty) = 0$ and (17).

The boundary functions $Q_{1j}\vartheta(\sigma_1)$, $\sigma_1 \in (-\infty,0]$, are determined from optimal control problems $Q_{1j}P$ consisting of minimizing the functional

$$Q_{1j}J(Q_{1j}u) = Q_{1j}w(0)'E_3(\overline{\psi}_j(T) + Q_{0j}\psi(0)) + \int_{-\infty}^0 (Q_{1j}w(\sigma_1)'(\frac{1}{2}W(T)Q_{1j}w(\sigma_1)$$

$$+[\widehat{Q}_{1(j-1)}\rho(\sigma_1,\varepsilon)]_j) + Q_{1j}u(\sigma_1)'(\frac{1}{2}R(T)Q_{1j}u(\sigma_1) + [\widehat{Q}_{1(j-1)}\chi(\sigma_1,\varepsilon)]_j))\,d\sigma_1$$

on trajectories of system (11) with the condition $Q_{1j}w(-\infty) = 0$.

Remark 2. *Though the original problem (1)–(3) is nonlinear, the considered optimal control problems $\overline{P}_j$, $\Pi_{ij}P$, $Q_{ij}P$, $i = 0,1$, are linear-quadratic.*

Solutions of the formulated optimal control problems can be found from the control optimality conditions in the Pontryagin maximum principle form. Namely, a solution of the problem $\overline{P}_j$ can be found from (7), (13), or (14) in dependence on j, and the relations

$$B(t)'\overline{\psi}_j(t) - R(t)\overline{u}_j(t) - [\widehat{\overline{\chi}}_{j-1}(t,\varepsilon)]_j = 0, \tag{18}$$

$$E_1\frac{d\overline{\psi}_j(t)}{dt} = W(t)\overline{w}_j(t) - A(t)'\overline{\psi}_j(t) + [\widehat{\overline{\rho}}_{j-1}(t,\varepsilon)]_j - E_2\frac{d\overline{\psi}_{j-1}(t)}{dt} - E_3\frac{d\overline{\psi}_{j-2}(t)}{dt}, \tag{19}$$

$$E_1\overline{\psi}_j(T) = -E_1(Q_{0(j-1)}\psi(0) + Q_{1(j-2)}\psi(0)). \tag{20}$$

A solution of the problem $\Pi_{0j}P$ with $E_1\Pi_{0j}w(+\infty) = 0$ can be found from (8), (12) and (15) or (16) in dependence on j, and the relations

$$B(0)'(E_2+E_3)\Pi_{0j}\psi - R(0)\Pi_{0j}u - [\widehat{\Pi}_{0(j-1)}\chi(\tau_0,\varepsilon)]_j = 0, \tag{21}$$

$$\begin{aligned}(E_1+E_2)\frac{d\Pi_{0j}\psi}{d\tau_0} &= W(0)\Pi_{0j}w - A(0)'(E_2+E_3)\Pi_{0j}\psi\\ &+[\widehat{\Pi}_{0(j-1)}\rho(\tau_0,\varepsilon)]_j - E_3\frac{d\Pi_{0(j-1)}\psi}{d\tau_0},\end{aligned} \tag{22}$$

$$(E_1+E_2)\Pi_{0j}\psi(+\infty) = 0.$$

A solution of the problem $Q_{0j}P$ with $(E_1+E_2)Q_{0j}w(-\infty) = 0$ can be found from (9), (12) and the relations

$$B(T)'(E_2+E_3)Q_{0j}\psi - R(T)Q_{0j}u - [\widehat{Q}_{0(j-1)}\chi(\sigma_0,\varepsilon)]_j = 0,$$

$$\begin{aligned}(E_1+E_2)\frac{dQ_{0j}\psi}{d\sigma_0} &= W(T)Q_{0j}w - A(T)'(E_2+E_3)Q_{0j}\psi\\ &+[\widehat{Q}_{0(j-1)}\rho(\sigma_0,\varepsilon)]_j - E_3\frac{dQ_{0(j-1)}\psi}{d\sigma_0},\end{aligned}$$

$$E_1Q_{0j}\psi(-\infty) = 0,\ E_2Q_{0j}\psi(0) = -E_2(\overline{\psi}_j(T) + Q_{1(j-1)}\psi(0)). \tag{23}$$

A solution of the problem $\Pi_{1j}P$ with $(E_1+E_2)\Pi_{1j}w(+\infty) = 0$ can be found from (10), (12), (17) in dependence on j, and the relations

$$B(0)'E_3\Pi_{1j}\psi - R(0)\Pi_{1j}u - [\widehat{\Pi}_{1(j-1)}\chi(\tau_1,\varepsilon)]_j = 0, \tag{24}$$

$$\frac{d\Pi_{1j}\psi}{d\tau_1} = W(0)\Pi_{1j}w - A(0)'E_3\Pi_{1j}\psi + [\widehat{\Pi}_{1(j-1)}\rho(\tau_1,\varepsilon)]_j, \tag{25}$$

$$\Pi_{1j}\psi(+\infty) = 0.$$

A solution of the problem $Q_{1j}P$ with $Q_{1j}w(-\infty) = 0$ can be found from (11), (12) in dependence on j, and the relations

$$B(T)'E_3Q_{1j}\psi - R(T)Q_{1j}u - [\widehat{Q}_{1(j-1)}\chi(\sigma_1,\varepsilon)]_j = 0,$$

$$\frac{dQ_{1j}\psi}{d\sigma_1} = W(T)Q_{1j}w - A(T)'E_3Q_{1j}\psi + [\widehat{Q}_{1(j-1)}\rho(\sigma_1,\varepsilon)]_j,$$

$$(E_1+E_2)Q_{1j}\psi(-\infty)=0,\ E_3Q_{1j}\psi(0)=-E_3(\overline{\psi}_j(T)+Q_{0j}\psi(0)). \tag{26}$$

In view of the control optimality condition in the Pontryagin maximum principle, a solution of the problem (1)–(3) satisfies (2), (3) and the following relations, including the costate variable $\varphi(t,\varepsilon)=(\zeta(t,\varepsilon)',\eta(t,\varepsilon)',\theta(t,\varepsilon)')'$,

$$B(t)'\varphi-R(t)u-\varepsilon(F_u(\vartheta,t,\varepsilon)'-f_u(\vartheta,t,\varepsilon)'\varphi)=0, \tag{27}$$

$$\mathcal{E}(\varepsilon)\frac{d\varphi}{dt}=W(t)w-A(t)'\varphi+\varepsilon(F_w(\vartheta,t,\varepsilon)'-f_w(\vartheta,t,\varepsilon)'\varphi), \tag{28}$$

$$\varphi(T,\varepsilon)=0. \tag{29}$$

An asymptotic solution of problems (2), (3), (27)–(29) can be constructed in the form (4), i.e., in addition, we set

$$\varphi(t,\varepsilon)=\overline{\varphi}(t,\varepsilon)+\sum_{i=0}^{1}(\Pi_i\varphi(\tau_i,\varepsilon)+Q_i\varphi(\sigma_i,\varepsilon)), \tag{30}$$

where all terms have the properties of the corresponding terms in (4).

Substitute asymptotic expansions (4), (30) into (27)–(29) and use presentation (5). Introducing the notation $g(\vartheta,\varphi,t,\varepsilon)=\rho(\vartheta,\varphi,t,\varepsilon)$, $h(\vartheta,\varphi,t,\varepsilon)=\chi(\vartheta,\varphi,t,\varepsilon)$ and equating terms of the same power of ε separately depending on t, τ_i, σ_i, $i=0,1$, we obtain the relations

$$B(t)'\overline{\varphi}_j-R(t)\overline{u}_j-[\widehat{\overline{h}}_{j-1}(t,\varepsilon)]_j=0,$$

$$E_1\frac{d\overline{\varphi}_j}{dt}+E_2\frac{d\overline{\varphi}_{j-1}}{dt}+E_3\frac{d\overline{\varphi}_{j-2}}{dt}=W(t)\overline{w}_j-A(t)'\overline{\varphi}_j+[\widehat{\overline{g}}_{j-1}(t,\varepsilon)]_j,$$

$$B(0)'\Pi_{ij}\varphi-R(0)\Pi_{ij}u-[\widehat{\Pi}_{i(j-1)}h(\tau_i,\varepsilon)]_j=0,$$

$$\begin{aligned}E_1\frac{d\Pi_{ij}\varphi}{d\tau_i}+E_2\frac{d\Pi_{i(j-1)}\varphi}{d\tau_i}+E_3\frac{d\Pi_{i(j-2)}\varphi}{d\tau_i}&=W(0)\Pi_{i(j-i-1)}w\\ -A(0)'\Pi_{i(j-i-1)}\varphi+[\widehat{\Pi}_{i(j-i-2)}g(\tau_i,\varepsilon)]_{j-i-1},&\end{aligned} \tag{31}$$

$$B(T)'Q_{ij}\varphi-R(T)Q_{ij}u-[\widehat{Q}_{i(j-1)}h(\sigma_i,\varepsilon)]_j=0,$$

$$\begin{aligned}E_1\frac{dQ_{ij}\varphi}{d\sigma_i}+E_2\frac{dQ_{i(j-1)}\varphi}{d\sigma_i}+E_3\frac{dQ_{i(j-2)}\varphi}{d\sigma_i}&=W(T)Q_{i(j-i-1)}w\\ -A(T)'Q_{i(j-i-1)}\varphi+[\widehat{Q}_{i(j-i-2)}g(\sigma_i,\varepsilon)]_{j-i-1},&\end{aligned} \tag{32}$$

$$\overline{\varphi}_j(T)+Q_{0j}\varphi(0)+Q_{1j}\varphi(0)=0. \tag{33}$$

It follows from (31), (32) with $j=0$ and $i=j=1$ that

$$E_1\Pi_{00}\varphi(\tau_0)=0,\ E_1\Pi_{10}\varphi(\tau_1)=E_1\Pi_{11}\varphi(\tau_1)=0,\ E_1Q_{00}\varphi(\sigma_0)=0,$$
$$E_1Q_{10}\varphi(\sigma_1)=E_1Q_{11}\varphi(\sigma_1)=0,\ E_2\Pi_{10}\varphi(\tau_1)=0,\ E_2Q_{10}\varphi(\sigma_1)=0.$$

4. Justification of Formalism of Asymptotics Construction

This section deals with the establishment of a relation between the forms of coefficients in the expansion (6) of the minimized functional with respect to powers of ε and the expressions of the performance indices in optimal control problems formulated in the previous section. The following theorem, which was given in [26] without any rigorous proof, will be further justified.

Theorem 1. *The sum $\overline{J}_j+\Pi_{1(j-1)}J+Q_{1(j-1)}J$ of the performance indices in problems $\overline{P}_j$, $\Pi_{1(j-1)}P$, $Q_{1(j-1)}P$ is obtained by transforming the coefficient J_{2j} in expansion (6) and dropping terms, which* *are known after solving problems $\overline{P}_k$, $\Pi_{0k}P$, $Q_{0k}P$, $k = \overline{0, j-1}$, $\Pi_{1k}P$, $Q_{1k}P$, $k = \overline{0, k-2}$. The sum $\Pi_{0j}J + Q_{0j}J$ of the performance indices in problems $\Pi_{0j}P$, $Q_{0j}P$ is obtained by transforming the coefficient J_{2j+1} in expansion (6) and dropping terms, which are known after solving problems $\overline{P}_k$, $k = \overline{0, j}$, $\Pi_{ik}P$, $Q_{ik}P$, $i = 0, 1$, $k = \overline{0, j-1}$.*

Proof. Denote the integrand in (1) by means $\mathbb{F}(\vartheta, t, \varepsilon)$. In view of (5), we can present J_k in the form

$$J_k = \int_0^T \overline{\mathbb{F}}_k(t)\,dt + \int_0^{+\infty} \Pi_{0(k-1)}\mathbb{F}(\tau_0)\,d\tau_0$$
$$+ \int_{-\infty}^0 Q_{0(k-1)}\mathbb{F}(\sigma_0)d\sigma_0 + \int_0^{+\infty} \Pi_{1(k-2)}\mathbb{F}(\tau_1)d\tau_1 + \int_{-\infty}^0 Q_{1(k-2)}\mathbb{F}(\sigma_1)\,d\sigma_1. \tag{34}$$

It is clear that the last expression contains the asymptotics terms with numbers more than it is necessary in this theorem, for instance, $\overline{\mathbb{F}}_{2n}(t) = [\mathbb{F}(\tilde{\overline{\vartheta}}_{2n}(t,\varepsilon), t, \varepsilon)]_{2n}$. In order to prove the theorem, we will use control optimality conditions for formulated previously control problems.

It is evident that the coefficient J_0 in (6) is the performance index in problem $\overline{P}_0$.

We will analyze the coefficient J_1. In view of (34) with $k = 1$, we have

$$J_1 = \int_0^T \overline{\mathbb{F}}_1(t)\,dt + \int_0^{+\infty} \Pi_{00}\mathbb{F}(\tau_0)\,d\tau_0 + \int_{-\infty}^0 Q_{00}\mathbb{F}(\sigma_0)\,d\sigma_0$$
$$= \int_0^T (\overline{w}_1(t)'W(t)\overline{w}_0(t) + \overline{u}_1(t)'R(t)\overline{u}_0(t) + [\hat{\overline{\mathbb{F}}}_0(t,\varepsilon)]_1)\,dt$$
$$+ \int_0^{+\infty} (\frac{1}{2}(\Pi_{00}w(\tau_0)'W(0)\Pi_{00}w(\tau_0) + \Pi_{00}u(\tau_0)'R(0)\Pi_{00}u(\tau_0))$$
$$+\Pi_{00}w(\tau_0)'W(0)\overline{w}_0(0) + \Pi_{00}u(\tau_0)'R(0)\overline{u}_0(0))\,d\tau_0$$
$$+ \int_{-\infty}^0 (\frac{1}{2}(Q_{00}w(\sigma_0)'W(T)Q_{00}w(\sigma_0) + Q_{00}u(\sigma_0)'R(T)Q_{00}u(\sigma_0))$$
$$+Q_{00}w(\sigma_0)'W(T)\overline{w}_0(T) + Q_{00}u(\sigma_0)'R(T)\overline{u}_0(T))\,d\sigma_0.$$

Transforming the following expression from J_1 with the help of control optimality conditions for the problem $\overline{P}_0$ (see (18)–(20) with $j = 0$), the integration by parts, and also (12), (7) with $j = 1$, (8), (9) with $j = 0$ and $j = 1$, and (15), we have

$$\int_0^T (\overline{w}_1(t)'W(t)\overline{w}_0(t) + \overline{u}_1(t)'R(t)\overline{u}_0(t))\,dt + \int_0^{+\infty} (\Pi_{00}w(\tau_0)'W(0)\overline{w}_0(0)$$
$$+\Pi_{00}u(\tau_0)'R(0)\overline{u}_0(0))\,d\tau_0 + \int_{-\infty}^0 (Q_{00}w(\sigma_0)'W(T)\overline{w}_0(T) + Q_{00}u(\sigma_0)'R(T)\overline{u}_0(T))\,d\sigma_0$$
$$= \int_0^T (\overline{w}_1(t)'(E_1\frac{d\overline{\psi}_0(t)}{dt} + A(t)'\overline{\psi}_0(t)) + \overline{u}_1(t)'B(t)'\overline{\psi}_0(t))\,dt$$
$$+ \int_0^{+\infty} (\Pi_{00}w(\tau_0)'(E_1\frac{d\overline{\psi}_0}{dt}(0) + A(0)'\overline{\psi}_0(0)) + \Pi_{00}u(\tau_0)'B(0)'\overline{\psi}_0(0))\,d\tau_0$$
$$+ \int_{-\infty}^0 (Q_{00}w(\sigma_0)'(E_1\frac{d\overline{\psi}_0}{dt}(T) + A(T)'\overline{\psi}_0(T)) + Q_{00}u(\sigma_0)'B(T)'\overline{\psi}_0(T))\,d\sigma_0$$
$$= \overline{w}_1(t)'E_1\overline{\psi}_0(t)|_0^T + \int_0^T \overline{\psi}_0(t)'(-E_1\frac{d\overline{w}_1}{dt}(t) + A(t)\overline{w}_1(t) + B(t)\overline{u}_1(t))\,dt$$
$$+ \int_0^{+\infty} \overline{\psi}_0(0)'(A(0)\Pi_{00}w(\tau_0) + B(0)\Pi_{00}u(\tau_0))\,d\tau_0$$
$$+ \int_{-\infty}^0 \overline{\psi}_0(T)'(A(T)Q_{00}w(\sigma_0) + B(T)Q_{00}u(\sigma_0))\,d\sigma_0$$
$$= \Pi_{01}w(0)'E_1\overline{\psi}_0(0) + \int_0^T \overline{\psi}_0(t)'(E_2\frac{d\overline{w}_0}{dt}(t) - [\hat{\overline{\phi}}_0(t,\varepsilon)]_1)\,dt$$

$$+\int_0^{+\infty}\overline{\psi}_0(0)'(E_1\frac{d\Pi_{01}w(\tau_0)}{d\tau_0}+E_2\frac{d\Pi_{00}w(\tau_0)}{d\tau_0})\,d\tau_0$$
$$+\int_{-\infty}^{0}\overline{\psi}_0(T)'(E_1\frac{dQ_{01}w(\sigma_0)}{d\sigma_0}+E_2\frac{dQ_{00}w(\sigma_0)}{d\sigma_0})\,d\sigma_0$$
$$=\Pi_{01}w(0)'E_1\overline{\psi}_0(0)+\int_0^T\overline{\psi}_0(t)'(E_2\frac{d\overline{w}_0}{dt}(t)-[\widehat{\overline{\phi}}_0(t,\varepsilon)]_1)\,dt$$
$$-\overline{\psi}_0(0)'(E_1\Pi_{01}w(0)+E_2\Pi_{00}w(0))+\overline{\psi}_0(T)'E_2Q_{00}w(0)$$
$$=\overline{\psi}_0(T)'E_2Q_{00}w(0)+\int_0^T\overline{\psi}_0(t)'(E_2\frac{d\overline{w}_0}{dt}(t)-[\widehat{\overline{\phi}}_0(t,\varepsilon)]_1)\,dt-\overline{\psi}_0(0)'E_2(w^0-\overline{w}_0(0)).$$

Taking into account this relation and the previous expression for J_1, and also dropping terms, which are known after solving the problem $\overline{P}_0$, we see that the transformed expression for J_1 is the sum $\Pi_{00}J+Q_{00}J$.

Assuming that the problems $\overline{P}_0,\Pi_{00}P,Q_{00}P$ have been solved, we transform by similar way the coefficient J_2 in (6). According to (34), J_2 has the form:

$$\int_0^T\overline{\mathbb{F}}_2(t)\,dt+\int_0^{+\infty}\Pi_{01}\mathbb{F}(\tau_0)\,d\tau_0+\int_{-\infty}^{0}Q_{01}\mathbb{F}(\sigma_0)\,d\sigma_0$$
$$+\int_0^{+\infty}\Pi_{10}\mathbb{F}(\tau_1)\,d\tau_1+\int_{-\infty}^{0}Q_{10}\mathbb{F}(\sigma_1)\,d\sigma_1.$$

Write down the unknown terms in $\overline{\mathbb{F}}_2(t)$

$$\overline{w}_2(t)'W(t)\overline{w}_0(t)+\overline{u}_2(t)'R(t)\overline{u}_0(t)+\overline{w}_1(t)'(1/2W(t)\overline{w}_1(t)+\overline{F}_{w_0}(t)')$$
$$+\overline{u}_1(t)'(1/2R(t)\overline{u}_1(t)+\overline{F}_{u_0}(t)').$$

Transforming $\int_0^T(\overline{w}_2'W\overline{w}_0+\overline{u}_2'R\overline{u}_0)\,dt$ with the help of optimality conditions (18), (19) at $j=0$, integrating by parts, (7) at $j=2$, (20) at $j=0$, and dropping known terms, we obtain $-\overline{\psi}_0(0)'(E_1\overline{w}_2(0)+E_2\overline{w}_1(0))+\overline{\psi}_0(T)'E_2\overline{w}_1(T)-\int_0^T(\overline{w}_1'E_2d\overline{\psi}_0/dt+\overline{\psi}_0'(\overline{f}_{w_0}\overline{w}_1+\overline{f}_{u_0}\overline{u}_1))\,dt$.

The unknown expression in $\Pi_{01}\mathbb{F}(\tau_0)$ is

$$\Pi_{01}w(\tau_0)'W(0)(\overline{w}_0(0)+\Pi_{00}w(\tau_0))+\Pi_{00}w(\tau_0)'W(0)\overline{w}_1(0)$$
$$+\Pi_{01}u(\tau_0)'R(0)(\overline{u}_0(0)+\Pi_{01}u(\tau_0))+\Pi_{00}u(\tau_0)'R(0)\overline{u}_1(0).$$

The integral of this expression will be transformed using control optimality conditions for problems $\overline{P}_0$ and $\Pi_{00}P$, Equations (7) at $j=1$, (8) at $j=1,2$, the formula of integration by parts and Remark 1. Dropping known terms, we have $-\Pi_{00}\psi(0)'((E_1+E_2)\overline{w}_1(0)+E_2\Pi_{01}w(0))-\overline{\psi}_0(0)'(E_2\Pi_{01}w(0)+E_1\Pi_{02}w(0))$.

Similarly, we transform the third integral in J_2, depending on an unknown expression

$$\int_{-\infty}^{0}(Q_{01}w(\sigma_0)'W(T)(\overline{w}_0(T)+Q_{00}w(\sigma_0))+Q_{00}w(\sigma_0)'W(T)\overline{w}_1(T)$$
$$+Q_{01}u(\sigma_0)'R(T)(\overline{u}_0(T)+Q_{00}u(\sigma_0))+Q_{00}u(\sigma_0)'R_0(T)\overline{u}_1(T))\,d\sigma_0$$
$$=Q_{00}\psi(0)'((E_1+E_2)\overline{w}_1(T)+E_2Q_{01}w(0))+\overline{\psi}_0(T)'E_2Q_{01}w(0).$$

The unknown expression in $\Pi_{10}\mathbb{F}(\tau_1)$ is

$$\Pi_{10}w(\tau_1)'W(0)(\overline{w}_0(0)+\Pi_{00}w(0))+\Pi_{10}u(\tau_1)'R(0)(\overline{u}_0(0)+\Pi_{00}u(0))$$
$$+1/2(\Pi_{10}w(\tau_1)'W(0)\Pi_{10}w(\tau_1)+\Pi_{10}u(\tau_1)'R(0)\Pi_{10}u(\tau_1)).$$

Transform the integral

$$\int_0^{+\infty}(\Pi_{10}w(\tau_1)'W(0)(\overline{w}_0(0)+\Pi_{00}w(0))+\Pi_{10}u(\tau_1)'R(0)(\overline{u}_0(0)+\Pi_{00}u(0)))\,d\tau_1$$

with the help of optimality conditions for problems $\overline{P}_0$, $\Pi_{00}P$, (10) at $j=0,1,2$, (12) and integration by parts. Dropping known terms, we have $-\overline{\psi}_0(0)'(E_1\Pi_{12}w(0)+E_2\Pi_{11}w(0)+E_3\Pi_{10}w(0))-\Pi_{00}\psi(0)'(E_2\Pi_{11}w(0)+E_3\Pi_{10}w(0))$.

Transforming in a similar way the fifth integral in J_2, depending on unknown terms, we obtain

$$\begin{aligned}
&\int_{-\infty}^{0}(Q_{10}w(\sigma_1)'W(T)(\overline{w}_0(T)+Q_{00}w(0))+Q_{10}u(\sigma_1)'R(T)(\overline{u}_0(T)+Q_{00}u(0))\\
&\quad+1/2(Q_{10}w(\sigma_1)'W(T)Q_{10}w(\sigma_1)+Q_{10}u(\sigma_1)'R(T)Q_{10}u(\sigma_1)))\,d\sigma_1\\
&=\overline{\psi}_0(T)'(E_2Q_{11}w(0)+E_3Q_{10}w(0))+Q_{00}\psi(0)'(E_2Q_{11}w(0)+E_3Q_{10}w(0))\\
&\quad+1/2\int_{-\infty}^{0}(Q_{10}w(\sigma_1)'W(T)Q_{10}w(\sigma_1)+Q_{10}u(\sigma_1)'R(T)Q_{10}u(\sigma_1))\,d\sigma_1.
\end{aligned}$$

Substituting the transformed relations into J_2, taking into account the second equality in (13), (14) at $j=2$, (16) at $j=1$, (17) and (23) at $j=0$, and also Remark 1, and finally dropping known terms, we obtain the theorem statement for the coefficient J_2.

Introduce the notation

$$\vartheta(t,\varepsilon)-\widetilde{\vartheta}_{j-1}(t,\varepsilon)=\Delta_j\overline{\vartheta}(t,\varepsilon)+\sum_{i=0}^{1}(\Delta_j\Pi_i\vartheta(\tau_i,\varepsilon)+\Delta_jQ_i\vartheta(\sigma_i,\varepsilon)), \tag{35}$$

where $\Delta_j\overline{\vartheta}(t,\varepsilon)=\overline{\vartheta}(t,\varepsilon)-\widetilde{\overline{\vartheta}}_{j-1}(t,\varepsilon)=\varepsilon^j\overline{\vartheta}_j(t)+\alpha(\varepsilon^{j+1})$, $\Delta_j\Pi_i\vartheta(\tau_i,\varepsilon)=\Pi_i\vartheta(\tau_i,\varepsilon)-\widetilde{\Pi}_{i(j-1)}\vartheta(\tau_i,\varepsilon)=\varepsilon^j\Pi_{ij}\vartheta(\tau_i)+\alpha(\varepsilon^{j+1})$, $\Delta_jQ_i\vartheta(\sigma_i,\varepsilon)=Q_i\vartheta(\tau_i,\varepsilon)-\widetilde{Q}_{i(j-1)}\vartheta(\sigma_i,\varepsilon)=\varepsilon^jQ_{ij}\vartheta(\sigma_i)+\alpha(\varepsilon^{j+1})$, $i=0,1$, $\alpha(\varepsilon^{j+1})$ is a sum of the expansion terms of order ε^{j+1} and higher.

Assuming that the problems $\overline{P}_j$, $\Pi_{0j}P$, $Q_{0j}P$ and $\Pi_{1(j-1)}P$, $Q_{1(j-1)}P$, $j=\overline{0,n-1}$ have been solved, we will transform each term in the coefficient J_{2n}, having the presentation (34) with $k=2n$.

Using the notation (35), we can see that the unknown terms in $\overline{\mathbb{F}}_{2n}(t)$ are the following:

$$\begin{aligned}
&\overline{w}_n(t)'(1/2W(t)\overline{w}_n(t)+[\widehat{\overline{F}}_{w(n-1)}(t,\varepsilon)']_{n-1})\\
&+\overline{u}_n(t)'(1/2R(t)\overline{u}_n(t)+[\widehat{\overline{F}}_{u(n-1)}(t,\varepsilon)']_{n-1})\\
&+[\Delta_{n+1}\overline{w}(t,\varepsilon)'(W(t)\widetilde{\overline{w}}_{n-1}(t,\varepsilon)+\{\varepsilon\widehat{\overline{F}}_{w(n-1)}(t,\varepsilon)'\}_{n-1})]_{2n}\\
&+[\Delta_{n+1}\overline{u}(t,\varepsilon)'(R(t)\widetilde{\overline{u}}_{n-1}(t,\varepsilon)+\{\varepsilon\widehat{\overline{F}}_{u(n-1)}(t,\varepsilon)'\}_{n-1})]_{2n}.
\end{aligned}$$

Multiplying the Equations (18), (19) by ε^j, $j=\overline{0,k}$, and summing up the obtained equations, we obtain the following relations

$$\begin{aligned}
&\{R(t)\widetilde{\overline{u}}_k(t,\varepsilon)+\varepsilon\widehat{\overline{F}}_{u(k-1)}(t,\varepsilon)'\}_k=\{B(t)'\widetilde{\overline{\psi}}_k(t,\varepsilon)+\varepsilon\widehat{\overline{f}}_{u(k-1)}(t,\varepsilon)'\widetilde{\overline{\psi}}_{k-1}(t,\varepsilon)\}_k,\\
&\{W(t)\widetilde{\overline{w}}_k(t,\varepsilon)+\varepsilon\widehat{\overline{F}}_{w(k-1)}(t,\varepsilon)'\}_k=\{\mathcal{E}(\varepsilon)\frac{d\widetilde{\overline{\psi}}_k(t,\varepsilon)}{dt}\}_k\\
&\quad+\{A(t)'\widetilde{\overline{\psi}}_k(t,\varepsilon)+\varepsilon\widehat{\overline{f}}_{w(k-1)}(t,\varepsilon)'\widetilde{\overline{\psi}}_{k-1}(t,\varepsilon)\}_k.
\end{aligned} \tag{36}$$

Substituting $\vartheta(t,\varepsilon)$ from (35) with $j = n+1$ into (2) and equating terms depending on t, we obtain the equation

$$\mathcal{E}(\varepsilon)(\frac{d\widetilde{\overline{w}}_n(t,\varepsilon)}{dt} + \frac{d\Delta_{n+1}\overline{w}(t,\varepsilon)}{dt}) = A(t)(\widetilde{\overline{w}}_n(t,\varepsilon) + \Delta_{n+1}\overline{w}(t,\varepsilon))$$
$$+B(t)(\widetilde{\overline{u}}_n(t,\varepsilon) + \Delta_{n+1}\overline{u}(t,\varepsilon)) + \varepsilon f(\widetilde{\overline{\vartheta}}_n(t,\varepsilon) + \Delta_{n+1}\overline{\vartheta}(t,\varepsilon), t, \varepsilon). \tag{37}$$

We will use the next easily proved formula from [29], which is valid for any sufficiently smooth vector functions $a(t,\varepsilon)$, $b(t,\varepsilon)$ and a matrix $\mathbb{D}(t,\varepsilon)$ of the corresponding size,

$$_k = [\{b(t,\varepsilon)\}_l'\mathbb{D}(t,\varepsilon)a(t,\varepsilon)]_k$$
$$-[\{b(t,\varepsilon)\}_l'\mathbb{D}(t,\varepsilon)\{a(t,\varepsilon)\}_{k-l-1}]_k,\ k,l \in I\!N,\ k \geqslant l. \tag{38}$$

Using (36) with $k = n-1$, (37), (38) with $l = n-1$, $k = 2n$, we can rewrite

$$\int_0^T ([\Delta_{n+1}\overline{w}(t,\varepsilon)'(W(t)\widetilde{\overline{w}}_{n-1}(t,\varepsilon) + \{\varepsilon\widehat{\overline{F}}_{w(n-1)}(t,\varepsilon)'\}_{n-1})]_{2n}$$
$$+[\Delta_{n+1}\overline{u}(t,\varepsilon)'(R(t)\widetilde{\overline{u}}_{n-1}(t,\varepsilon) + \{\varepsilon\widehat{\overline{F}}_{u(n-1)}(t,\varepsilon)'\}_{n-1})]_{2n})\,dt$$

in the following way

$$\int_0^T ([\Delta_{n+1}\overline{w}(t,\varepsilon)'\{\mathcal{E}(\varepsilon)\frac{d\widetilde{\overline{\psi}}_{n-1}(t,\varepsilon)}{dt}\}_{n-1}]_{2n} + [\widetilde{\overline{\psi}}_{n-1}(t,\varepsilon)'(A(t)\Delta_{n+1}\overline{w}(t,\varepsilon)$$
$$+B(t)\Delta_{n+1}\overline{u}(t,\varepsilon)]_{2n} + [\Delta_{n+1}\overline{w}(t,\varepsilon)'\{\varepsilon\widehat{\overline{f}}_{w(n-1)}(t,\varepsilon)'\widetilde{\overline{\psi}}_{n-1}(t,\varepsilon)\}_{n-1}]_{2n}$$
$$+[\Delta_{n+1}\overline{u}(t,\varepsilon)'\{\varepsilon\widehat{\overline{f}}_{u(n-1)}(t,\varepsilon)'\widetilde{\overline{\psi}}_{n-1}(t,\varepsilon)\}_{n-1}\rangle]_{2n})\,dt$$
$$= \int_0^T ([\Delta_{n+1}\overline{w}(t,\varepsilon)'\{\mathcal{E}(\varepsilon)\frac{d\widetilde{\overline{\psi}}_{n-1}(t,\varepsilon)}{dt}\}_{n-1}]_{2n} + [\widetilde{\overline{\psi}}_{n-1}(t,\varepsilon)'(\mathcal{E}(\varepsilon)(\frac{d\Delta_{n+1}\overline{w}(t,\varepsilon)}{dt}$$
$$+\frac{d\widetilde{\overline{w}}_n(t,\varepsilon)}{dt}) - A(t)\widetilde{\overline{w}}_n(t,\varepsilon) - B(t)\widetilde{\overline{u}}_n(t,\varepsilon) - \varepsilon f(\widetilde{\overline{\vartheta}}_n(t,\varepsilon) + \Delta_{n+1}\overline{\vartheta}(t,\varepsilon), t, \varepsilon))]_{2n}$$
$$+[\widetilde{\overline{\psi}}_{n-1}(t,\varepsilon)'(\{\varepsilon\widehat{\overline{f}}_{w(n-1)}(t,\varepsilon)\}_{n-1}\Delta_{n+1}\overline{w}(t,\varepsilon) + \{\varepsilon\widehat{\overline{f}}_{u(n-1)}(t,\varepsilon)\}_{n-1}\Delta_{n+1}\overline{u}(t,\varepsilon))]_{2n})\,dt.$$

Integrating by parts in the first term of the last expression, taking into account the equality $\Delta_{n+1}\overline{\vartheta}(t,\varepsilon) = \Delta_n\overline{\vartheta}(t,\varepsilon) - \varepsilon^n\overline{\vartheta}_n(t)$, decomposing $f(\widetilde{\overline{\vartheta}}_n(t,\varepsilon) + \Delta_{n+1}\overline{\vartheta}(t,\varepsilon), t, \varepsilon)$ in the neighborhood of $\widetilde{\overline{\vartheta}}_{n-1}(t,\varepsilon)$, and omitting known terms, we obtain

$$[\Delta_{n+1}\overline{w}(t,\varepsilon)'\mathcal{E}(\varepsilon)\widetilde{\overline{\psi}}_{n-1}(t,\varepsilon)]_{2n}|_0^T + (\overline{\psi}_{n-1}(t)'E_2\overline{w}_n(t) + \overline{\psi}_{n-2}(t)'E_3\overline{w}_n(t))|_0^T$$
$$+\int_0^T (\overline{w}_n'(-E_2\frac{d\overline{\psi}_{n-1}}{dt} - E_3\frac{d\overline{\psi}_{n-2}}{dt} - [\{\widetilde{\overline{\psi}}_{n-1}(t,\varepsilon)'\varepsilon\widehat{\overline{f}}_{\vartheta(n-1)}(t,\varepsilon)\}_n(\varepsilon^n\overline{\vartheta}_n + \Delta_{n+1}\vartheta(t,\varepsilon))]_{2n}$$
$$+[\widetilde{\overline{\psi}}_{n-1}(t,\varepsilon)'(\{\varepsilon\widehat{\overline{f}}_{w(n-1)}(t,\varepsilon)\}_{n-1}\Delta_{n+1}\overline{w}(t,\varepsilon) + \{\varepsilon\widehat{\overline{f}}_{u(n-1)}(t,\varepsilon)\}_{n-1}\Delta_{n+1}\overline{u}(t,\varepsilon))]_{2n})\,dt$$
$$= [\Delta_n\overline{w}(t,\varepsilon)'(\frac{1}{\varepsilon}E_1 + E_2 + \varepsilon E_3)\widetilde{\overline{\psi}}_{n-1}(t,\varepsilon)]_{2n-1}|_0^T$$
$$-\int_0^T (\overline{w}_n(t)'(E_2\frac{d\overline{\psi}_{n-1}}{dt} + E_3\frac{d\overline{\psi}_{n-2}}{dt} + [\varepsilon\widehat{\overline{f}}_{w(n-1)}(t,\varepsilon)'\widetilde{\overline{\psi}}_{n-1}(t,\varepsilon)]_n)$$
$$+[\overline{u}_n(t)'([\varepsilon\widehat{\overline{f}}_{u(n-1)}(t,\varepsilon)'\widetilde{\overline{\psi}}_{n-1}(t,\varepsilon)]_n)\,dt.$$

Taking into account the last relation, omitting known terms, we obtain the following expression for the first term of J_{2n}:

$$\int_0^T \overline{\mathbb{F}}_{2n}(t)\,dt = [\Delta_n\overline{w}(t,\varepsilon)'(\frac{1}{\varepsilon}E_1 + E_2 + \varepsilon E_3)\widetilde{\overline{\psi}}_{n-1}(t,\varepsilon)]_{2n-1}|_0^T + \overline{J}_n$$

$$-\overline{w}_n(T)'E_1(Q_{0(n-1)}\psi(0) + Q_{1(n-2)}\psi(0)).$$

The next step is the transformation of the unknown parts of $\Pi_{0(2n-1)}\mathbb{F}(\tau_0)$, which, after substituting (35) and some transformations, is given below

$$[\Delta_n\Pi_0 w(\tau_0,\varepsilon)'(\{W(\varepsilon\tau_0)\widetilde{\overline{w}}_{n-1}(\varepsilon\tau_0,\varepsilon)\}_{n-1} + \{\varepsilon\widehat{\overline{F}}_{w(n-1)}(\varepsilon\tau_0,\varepsilon)'\}_{n-1})]_{2n-1}$$
$$+[\Delta_n\Pi_0 u(\tau_0,\varepsilon)'(\{R(\varepsilon\tau_0)\widetilde{\overline{u}}_{n-1}(\varepsilon\tau_0,\varepsilon)\}_{n-1} + \{\varepsilon\widehat{\overline{F}}_{u(n-1)}(\varepsilon\tau_0,\varepsilon)'\}_{n-1})]_{2n-1}$$
$$+[(\Delta_n\overline{w}(\varepsilon\tau_0,\varepsilon) + \Delta_n\Pi_0 w(\tau_0,\varepsilon))'(\{W(\varepsilon\tau_0)\widetilde{\Pi}_{0(n-1)}w(\tau_0,\varepsilon)\}_{n-1}$$
$$+\{\varepsilon\widehat{\Pi}_{0(n-1)}F_w(\tau_0,\varepsilon)'\}_{n-1})]_{2n-1} + [(\Delta_n\overline{u}(\varepsilon\tau_0,\varepsilon)$$
$$+\Delta_n\Pi_0 u(\tau_0,\varepsilon))'(\{R(\varepsilon\tau_0)\widetilde{\Pi}_{0(n-1)}u(\tau_0,\varepsilon)\}_{n-1} + \{\varepsilon\widehat{\Pi}_{0(n-1)}F_u(\tau_0,\varepsilon)'\}_{n-1})]_{2n-1}.$$

Substituting $\vartheta(t,\varepsilon)$ from (35) into (2) and considering terms depending on τ_0, we obtain the equation

$$\begin{aligned}
&(\frac{1}{\varepsilon}E_1 + E_2 + \varepsilon E_3)(\frac{d\widetilde{\Pi}_{0(n-1)}w(\tau_0,\varepsilon)}{d\tau_0} + \frac{d\Delta_n\Pi_0 w(\tau_0,\varepsilon)}{d\tau_0}) \\
&= A(\varepsilon\tau_0)(\widetilde{\Pi}_{0(n-1)}w(\tau_0,\varepsilon) + \Delta_n\Pi_0 w(\tau_0,\varepsilon)) + B(\varepsilon\tau_0)(\widetilde{\Pi}_{0(n-1)}u(\tau_0,\varepsilon) \\
&+\Delta_n\Pi_0 u(\tau_0,\varepsilon)) + \varepsilon(f(\widetilde{\overline{\vartheta}}_{n-1}(\varepsilon\tau_0,\varepsilon) + \widetilde{\Pi}_{0(n-1)}\vartheta(\tau_0,\varepsilon) \\
&+\Delta_n\overline{\vartheta}(\varepsilon\tau_0,\varepsilon) + \Delta_n\Pi_0\vartheta(\tau_0,\varepsilon),\varepsilon\tau_0,\varepsilon) - f(\widetilde{\overline{\vartheta}}_{n-1}(\varepsilon\tau_0,\varepsilon) + \Delta_n\overline{\vartheta}(\varepsilon\tau_0,\varepsilon),\varepsilon\tau_0,\varepsilon)).
\end{aligned} \tag{39}$$

Using (36) with $k = n-1$, (39) and (38), we transform the following expression:

$$\int_0^{+\infty}([\Delta_n\Pi_0 w(\tau_0,\varepsilon)'(\{W(\varepsilon\tau_0)\widetilde{\overline{w}}_{n-1}(\varepsilon\tau_0,\varepsilon)\}_{n-1} + \{\varepsilon\widehat{\overline{F}}_{w(n-1)}(\varepsilon\tau_0,\varepsilon)'\}_{n-1})]_{2n-1}$$
$$+[\Delta_n\Pi_0 u(\tau_0,\varepsilon)'(\{R(\varepsilon\tau_0)\widetilde{\overline{u}}_{n-1}(\varepsilon\tau_0,\varepsilon)\}_{n-1} + \{\varepsilon\widehat{\overline{F}}_{u(n-1)}(\varepsilon\tau_0,\varepsilon)'\}_{n-1})]_{2n-1})\,d\tau_0.$$

Omitting known terms, we have

$$\int_0^{+\infty}([\Delta_n\Pi_0 w(\tau_0,\varepsilon)'(\{\mathcal{E}(\varepsilon)\frac{d\widetilde{\overline{\psi}}_{n-1}}{dt}(\varepsilon\tau_0,\varepsilon)\}_{n-1} + \{A(\varepsilon\tau_0)'\widetilde{\overline{\psi}}_{n-1}(\varepsilon\tau_0,\varepsilon)\}_{n-1}$$
$$+\{\varepsilon\widehat{\overline{f}}_{w(n-1)}(\varepsilon\tau_0,\varepsilon)'\widetilde{\overline{\psi}}_{n-1}(\varepsilon\tau_0,\varepsilon)\}_{n-1})]_{2n-1} + [\Delta_n\Pi_0 u(\tau_0,\varepsilon)'(\{B(\varepsilon\tau_0)'\widetilde{\overline{\psi}}_{n-1}(\varepsilon\tau_0,\varepsilon)\}_{n-1}$$
$$+\{\varepsilon\widehat{\overline{f}}_{u(n-1)}(\varepsilon\tau_0,\varepsilon)'\widetilde{\overline{\psi}}_{n-1}(\varepsilon\tau_0,\varepsilon)\}_{n-1})]_{2n-1})\,d\tau_0 = \int_0^{+\infty}(\Pi_{0n}w'E_1\frac{d\overline{\psi}_{n-1}}{dt}(0)$$
$$+[\Delta_n\Pi_0 w(\tau_0,\varepsilon)'\{(\frac{1}{\varepsilon}E_1 + E_2 + \varepsilon E_3)\frac{d\widetilde{\overline{\psi}}_{n-2}}{dt}(\varepsilon\tau_0,\varepsilon)\}_{n-2}]_{2n-2}$$
$$+[\widetilde{\overline{\psi}}_{n-1}(\varepsilon\tau_0,\varepsilon)'(\frac{1}{\varepsilon}E_1 + E_2 + \varepsilon E_3)\frac{d\Delta_n\Pi_0 w}{d\tau_0}(\tau_0,\varepsilon)]_{2n-1}$$
$$-[\widetilde{\overline{\psi}}_{n-1}(\varepsilon\tau_0,\varepsilon)'(\varepsilon\widehat{\Pi}_{0(n-1)}f_\vartheta(\tau_0,\varepsilon)\Delta_n\overline{\vartheta}(\varepsilon\tau_0,\varepsilon) + \varepsilon f_\vartheta(\widetilde{\overline{\vartheta}}_{n-1}(\varepsilon\tau_0,\varepsilon)$$
$$+\widetilde{\Pi}_{0(n-1)}\vartheta(\tau_0,\varepsilon),\varepsilon\tau_0,\varepsilon)\Delta_n\Pi_0\vartheta(\tau_0,\varepsilon))]_{2n-1} + [\widetilde{\overline{\psi}}_{n-1}(\varepsilon\tau_0,\varepsilon)'(\varepsilon\widehat{\overline{f}}_{w(n-1)}(\varepsilon\tau_0,\varepsilon)\Delta_n\Pi_0 w(\tau_0,\varepsilon)$$
$$+\varepsilon\widehat{\overline{f}}_{u(n-1)}(\varepsilon\tau_0,\varepsilon)\Delta_n\Pi_0 u(\tau_0,\varepsilon))]_{2n-1})\,d\tau_0.$$

From here, applying the formula of integrating by parts and Remark 1, omitting known terms, we obtain

$$-[\Delta_n\Pi_0 w(0,\varepsilon)'(\frac{1}{\varepsilon}E_1 + E_2 + \varepsilon E_3)\widetilde{\overline{\psi}}_{n-1}(0,\varepsilon)]_{2n-1}$$
$$-\int_0^{+\infty}([\widetilde{\overline{\psi}}_{n-1}(\varepsilon\tau_0,\varepsilon)'(\varepsilon\widehat{\Pi}_{0(n-1)}f_\vartheta(\Delta_n\overline{\vartheta}(\varepsilon\tau_0,\varepsilon) + \Delta_n\Pi_0\vartheta(\tau_0,\varepsilon))]_{2n-1})\,d\tau_0.$$

Multiplying the Equations (21), (22) by ε^j, $j = \overline{0,k}$, and summing up the obtained equations, we obtain the equalities

$$\begin{aligned}
&\{R(\varepsilon\tau_0)\widetilde{\Pi}_{0k}u\}_k + \{\varepsilon\widehat{\Pi}_{0(k-1)}F_u(\tau_0,\varepsilon)'\}_k \\
&= \{B(\varepsilon\tau_0)'(\varepsilon E_1 + E_2 + E_3)\widetilde{\Pi}_{0k}\psi\}_k + \{\varepsilon\widehat{\Pi}_{0(k-1)}f_u(\tau_0,\varepsilon)'\overline{\widetilde{\psi}}_{k-1}(\varepsilon\tau_0,\varepsilon)\}_k \\
&+\{\varepsilon f_u(\overline{\widetilde{\vartheta}}_{k-1}(\varepsilon\tau_0,\varepsilon) + \widetilde{\Pi}_{0(k-1)}\vartheta(\tau_0,\varepsilon),\varepsilon\tau_0,\varepsilon)'(\varepsilon E_1 + E_2 + E_3)\widetilde{\Pi}_{0(k-1)}\psi\}_k, \\
&\{W(\varepsilon\tau_0)\widetilde{\Pi}_{0k}w\}_k + \{\varepsilon\widehat{\Pi}_{0(k-1)}F_w(\tau_0,\varepsilon)'\}_k = \{(E_1 + E_2 + \varepsilon E_3)\frac{d\widetilde{\Pi}_{0k}\psi}{d\tau_0}\}_k \\
&+\{A(\varepsilon\tau_0)'(\varepsilon E_1 + E_2 + E_3)\widetilde{\Pi}_{0k}\psi\}_k + \{\varepsilon\widehat{\Pi}_{0(k-1)}f_w(\tau_0,\varepsilon)'\overline{\widetilde{\psi}}_{k-1}(\varepsilon\tau_0,\varepsilon)\}_k \\
&+\{\varepsilon f_w(\overline{\widetilde{\vartheta}}_{k-1}(\varepsilon\tau_0,\varepsilon) + \widetilde{\Pi}_{0(k-1)}\vartheta(\tau_0,\varepsilon),\varepsilon\tau_0,\varepsilon)'(\varepsilon E_1 + E_2 + E_3)\widetilde{\Pi}_{0(k-1)}\psi\}_k.
\end{aligned} \tag{40}$$

We transform

$$\begin{aligned}
&\int_0^{+\infty}([(\Delta_n\overline{w}(\varepsilon\tau_0,\varepsilon) + \Delta_n\Pi_0 w(\tau_0,\varepsilon))'(\{W(\varepsilon\tau_0)\widetilde{\Pi}_{0(n-1)}w(\tau_0,\varepsilon)\}_{n-1} \\
&+\{\varepsilon\widehat{\Pi}_{0(n-1)}F_w(\tau_0,\varepsilon)'\}_{n-1})]_{2n-1} + [(\Delta_n\overline{u}(\varepsilon\tau_0,\varepsilon) \\
&+\Delta_n\Pi_0 u(\tau_0,\varepsilon))'(\{R(\varepsilon\tau_0)\widetilde{\Pi}_{0(n-1)}u(\tau_0,\varepsilon)\}_{n-1} + \{\varepsilon\widehat{\Pi}_{0(n-1)}F_u(\tau_0,\varepsilon)'\}_{n-1})]_{2n-1})\, d\tau_0.
\end{aligned}$$

Using (40) and (38), as a result, we obtain

$$\begin{aligned}
&\int_0^{+\infty}([(\Delta_n\overline{w}(\varepsilon\tau_0,\varepsilon) + \Delta_n\Pi_0 w(\tau_0,\varepsilon))'\{(E_1 + E_2 + \varepsilon E_3)\frac{d\widetilde{\Pi}_{0(n-1)}\psi}{d\tau_0}\}_{n-1}]_{2n-1} \\
&+[\widetilde{\Pi}_{0(n-1)}\psi'(\varepsilon E_1 + E_2 + E_3)(A(\varepsilon\tau_0)\Delta_n\overline{w}(\varepsilon\tau_0,\varepsilon) + B(\varepsilon\tau_0)\Delta_n\overline{u}(\varepsilon\tau_0,\varepsilon))]_{2n-1} \\
&+[\widetilde{\Pi}_{0(n-1)}\psi'(\varepsilon E_1 + E_2 + E_3)(A(\varepsilon\tau_0)\Delta_n\Pi_0 w(\tau_0,\varepsilon) + B(\varepsilon\tau_0)\Delta_n\Pi_0 u(\tau_0,\varepsilon))]_{2n-1} \\
&+[\overline{\widetilde{\psi}}_{n-1}(\varepsilon\tau_0,\varepsilon)'(\{\varepsilon\widehat{\Pi}_{0(n-1)}f_w(\tau_0,\varepsilon)\}_{n-1}(\Delta_n\overline{w}(\varepsilon\tau_0,\varepsilon) + \Delta_n\Pi_0 w(\tau_0,\varepsilon)) \\
&+\{\varepsilon\widehat{\Pi}_{0(n-1)}f_u(\tau_0,\varepsilon)\}_{n-1}(\Delta_n\overline{u}(\varepsilon\tau_0,\varepsilon) + \Delta_n\Pi_0 u(\tau_0,\varepsilon)))]_{2n-1} \\
&+[\widetilde{\Pi}_{0(n-1)}\psi'(\varepsilon E_1 + E_2 + E_3)(\{\varepsilon f_w(\overline{\widetilde{\vartheta}}_{n-1}(\varepsilon\tau_0,\varepsilon) + \widetilde{\Pi}_{0(n-1)}\vartheta,\varepsilon\tau_0,\varepsilon)\}_{n-1}(\Delta_n\overline{w}(\varepsilon\tau_0,\varepsilon) \\
&+\Delta_n\Pi_0 w(\tau_0,\varepsilon)) + \{\varepsilon f_u(\overline{\widetilde{\vartheta}}_{n-1}(\varepsilon\tau_0,\varepsilon) + \widetilde{\Pi}_{0(n-1)}\vartheta,\varepsilon\tau_0,\varepsilon)\}_{n-1}(\Delta_n\overline{u}(\varepsilon\tau_0,\varepsilon) \\
&+\Delta_n\Pi_0 u(\tau_0,\varepsilon)))]_{2n-1})\, d\tau_0.
\end{aligned}$$

In view of (37) and (39), we obtain from the last expression, omitting known terms, the following:

$$\begin{aligned}
&\int_0^{+\infty}([(\Delta_n\overline{w}(\varepsilon\tau_0,\varepsilon) + \Delta_n\Pi_0 w(\tau_0,\varepsilon))'\{(E_1 + E_2 + \varepsilon E_3)\frac{d\widetilde{\Pi}_{0(n-1)}\psi(\tau_0,\varepsilon)}{d\tau_0}\}_{n-1}]_{2n-1} \\
&+[\widetilde{\Pi}_{0(n-1)}\psi(\tau_0,\varepsilon)'(\varepsilon E_1 + E_2 + E_3)((E_1 + \varepsilon E_2 + \varepsilon^2 E_3)(\frac{d\overline{\widetilde{w}}_{n-1}(\varepsilon\tau_0,\varepsilon)}{dt} \\
&+\frac{d\Delta_n\overline{w}(\varepsilon\tau_0,\varepsilon)}{dt}) - A(\varepsilon\tau_0)\overline{\widetilde{w}}_{n-1}(\varepsilon\tau_0,\varepsilon) - B(\varepsilon\tau_0)\overline{\widetilde{u}}_{n-1}(\varepsilon\tau_0,\varepsilon) \\
&-\{\varepsilon\overline{\widehat{f}}_{\vartheta(n-1)}(\varepsilon\tau_0,\varepsilon)\}_{n-1}\Delta_n\overline{\vartheta}(\varepsilon\tau_0,\varepsilon))]_{2n-1} \\
&+[\widetilde{\Pi}_{0(n-1)}\psi(\tau_0,\varepsilon)'(\varepsilon E_1 + E_2 + E_3)((\frac{1}{\varepsilon}E_1 + E_2 + \varepsilon E_3)(\frac{d\widetilde{\Pi}_{0(n-1)}w(\tau_0,\varepsilon)}{d\tau_0} + \frac{d\Delta_n\Pi_0 w(\tau_0,\varepsilon)}{d\tau_0}) \\
&-A(\varepsilon\tau_0)\widetilde{\Pi}_{0(n-1)}w(\tau_0,\varepsilon) - B(\varepsilon\tau_0)\widetilde{\Pi}_{0(n-1)}u(\tau_0,\varepsilon) - \{\varepsilon\widehat{\Pi}_{0(n-1)}f_\vartheta(\tau_0,\varepsilon)\}_{n-1}\Delta_n\overline{\vartheta}(\varepsilon\tau_0,\varepsilon) \\
&-\{\varepsilon f_\vartheta(\overline{\widetilde{\vartheta}}_{n-1}(\varepsilon\tau_0,\varepsilon) + \widetilde{\Pi}_{0(n-1)}\vartheta(\tau_0,\varepsilon),\varepsilon\tau_0,\varepsilon)\}_{n-1}\Delta_n\Pi_0\vartheta(\tau_0,\varepsilon))]_{2n-1} \\
&+[\overline{\widetilde{\psi}}_{n-1}(\varepsilon\tau_0,\varepsilon)'(\{\varepsilon\widehat{\Pi}_{0(n-1)}f_w(\tau_0,\varepsilon)\}_{n-1}(\Delta_n\overline{w}(\varepsilon\tau_0,\varepsilon) + \Delta_n\Pi_0 w(\tau_0,\varepsilon))
\end{aligned}$$

$$+\{\varepsilon\widehat{\Pi}_{0(n-1)}f_u(\tau_0,\varepsilon)\}_{n-1}(\Delta_n\overline{u}(\varepsilon\tau_0,\varepsilon)+\Delta_n\Pi_0 u(\tau_0,\varepsilon)))]_{2n-1}$$
$$+[\widetilde{\Pi}_{0(n-1)}\psi(\tau_0,\varepsilon)'(\varepsilon E_1+E_2+E_3)(\{\varepsilon f_w(\widetilde{\overline{\vartheta}}_{n-1}(\varepsilon\tau_0,\varepsilon)$$
$$+\widetilde{\Pi}_{0(n-1)}\vartheta(\tau_0,\varepsilon),\varepsilon\tau_0,\varepsilon)\}_{n-1}(\Delta_n\overline{w}(\varepsilon\tau_0,\varepsilon)+\Delta_n\Pi_0 w(\tau_0,\varepsilon))+\{\varepsilon f_u(\widetilde{\overline{\vartheta}}_{n-1}(\varepsilon\tau_0,\varepsilon)$$
$$+\widetilde{\Pi}_{0(n-1)}\vartheta(\tau_0,\varepsilon),\varepsilon\tau_0,\varepsilon)\}_{n-1}(\Delta_n\overline{u}(\varepsilon\tau_0,\varepsilon)+\Delta_n\Pi_0 u(\tau_0,\varepsilon)))]_{2n-1})\,d\tau_0.$$

Integrating by parts in the last expression and dropping known terms, we obtain

$$-[(\Delta_n\overline{w}(0,\varepsilon)+\Delta_n\Pi_0 w(0,\varepsilon))'(E_1+E_2+\varepsilon E_3)\widetilde{\Pi}_{0(n-1)}\psi(0,\varepsilon)]_{2n-1}$$
$$+\int_0^{+\infty}[\widetilde{\overline{\psi}}_{n-1}(\varepsilon\tau_0,\varepsilon)'\{\varepsilon\widehat{\Pi}_{0(n-1)}f_\vartheta(\tau_0,\varepsilon)\}_{n-1}(\Delta_n\overline{\vartheta}(\varepsilon\tau_0,\varepsilon)+\Delta_n\Pi_0\vartheta(\tau_0,\varepsilon))]_{2n-1}\,d\tau_0.$$

Summing up the results, obtained from the transformed terms of the integral $\int_0^{+\infty}\Pi_{0(2n-1)}\mathbb{F}\,d\tau_0$, after dropping known terms, we have

$$-[\Delta_n\Pi_0 w(0,\varepsilon)'(\frac{1}{\varepsilon}E_1+E_2+\varepsilon E_3)\widetilde{\overline{\psi}}_{n-1}(0,\varepsilon)]_{2n-1}$$
$$-[(\Delta_n\overline{w}(0,\varepsilon)+\Delta_n\Pi_0 w(0,\varepsilon))'(E_1+E_2+\varepsilon E_3)\widetilde{\Pi}_{0(n-1)}\psi(0,\varepsilon)]_{2n-1}.$$

Performing similar transformations for $\int_{-\infty}^0 Q_{0(2n-1)}\mathbb{F}\,d\sigma_0$, we obtain the following result:

$$[\Delta_n Q_0 w(0,\varepsilon)'(\frac{1}{\varepsilon}E_1+E_2+\varepsilon E_3)\widetilde{\overline{\psi}}_{n-1}(T,\varepsilon)]_{2n-1}$$
$$+[(\Delta_n\overline{w}(T,\varepsilon)+\Delta_n Q_0 w(0,\varepsilon))'(E_1+E_2+\varepsilon E_3)\widetilde{Q}_{0(n-1)}\psi(0,\varepsilon)]_{2n-1}.$$

Furthermore, we apply the analogous transformations for the forth term of J_{2n}. The integral over the interval $[0,+\infty)$ of unknown terms of $\Pi_{1(2n-2)}\mathbb{F}(\tau_1)$ is presented as the sum

$$\begin{aligned}&\sum_{i=1}^{4}s_i+\int_0^{+\infty}(\Pi_{1(n-1)}w(\tau_1)'([W(\varepsilon^2\tau_1)\widetilde{\overline{w}}_{n-1}(\varepsilon^2\tau_1,\varepsilon)]_{n-1}+[\widehat{\overline{F}}_{w(n-2)}(\varepsilon^2\tau_1,\varepsilon)']_{n-2})\\&+\Pi_{1(n-1)}u(\tau_1)'([R(\varepsilon^2\tau_1)\widetilde{\overline{u}}_{n-1}(\varepsilon^2\tau_1,\varepsilon)]_{n-1}+[\widehat{\overline{F}}_{u(n-2)}(\varepsilon^2\tau_1,\varepsilon)']_{n-2}))\,d\tau_1\\&+\int_0^{+\infty}(\Pi_{1(n-1)}w(\tau_1)'([W(\varepsilon^2\tau_1)\widetilde{\Pi}_{0(n-1)}w(\varepsilon\tau_1,\varepsilon)]_{n-1}+[\widehat{\Pi}_{0(n-2)}F_w(\varepsilon\tau_1,\varepsilon)']_{n-2})\\&+\Pi_{1(n-1)}u(\tau_1)'([R(\varepsilon^2\tau_1)\widetilde{\Pi}_{0(n-1)}u(\varepsilon\tau_1,\varepsilon)]_{n-1}+[\widehat{\Pi}_{0(n-2)}F_u(\varepsilon\tau_1,\varepsilon)']_{n-2}))\,d\tau_1\\&+\int_0^{+\infty}(\Pi_{1(n-1)}w(\tau_1)'(1/2W(0)\Pi_{1(n-1)}w(\tau_1)+[W(\varepsilon^2\tau_1)\widetilde{\Pi}_{1(n-2)}w(\tau_1,\varepsilon)]_{n-1}\\&+[\widehat{\Pi}_{1(n-2)}F_w(\tau_1,\varepsilon)']_{n-2})+\Pi_{1(n-1)}u(\tau_1)'(1/2R(0)\Pi_{1(n-1)}u(\tau_1)\\&+[R(\varepsilon^2\tau_1)\widetilde{\Pi}_{1(n-2)}u(\tau_1,\varepsilon)]_{n-1}+[\widehat{\Pi}_{1(n-2)}F_u(\tau_1,\varepsilon)']_{n-2}))\,d\tau_1,\end{aligned}\tag{41}$$

where the expressions for s_i, $i=\overline{1,4}$, will be written later when they will be transformed.

Substituting $\vartheta(t,\varepsilon)$ from (35) into (2) and considering terms depending on τ_1, we obtain the equation

$$\begin{aligned}&(\frac{1}{\varepsilon^2}E_1+\frac{1}{\varepsilon}E_2+E_3)(\frac{d\widetilde{\Pi}_{1(n-1)}w(\tau_1,\varepsilon)}{d\tau_1}+\frac{d\Delta_n\Pi_1 w(\tau_1,\varepsilon)}{d\tau_1})\\&=A(\varepsilon^2\tau_1)(\widetilde{\Pi}_{1(n-1)}w(\tau_1,\varepsilon)+\Delta_n\Pi_1 w(\tau_1,\varepsilon))+B(\varepsilon^2\tau_1)(\widetilde{\Pi}_{1(n-1)}u(\tau_1,\varepsilon)\\&+\Delta_n\Pi_1 u(\tau_1,\varepsilon))+\varepsilon(f(\widetilde{\overline{\vartheta}}_{n-1}(\varepsilon^2\tau_1,\varepsilon)+\widetilde{\Pi}_{0(n-1)}\vartheta(\varepsilon\tau_1,\varepsilon)+\widetilde{\Pi}_{1(n-1)}\vartheta(\tau_1,\varepsilon)\\&+\Delta_n\overline{\vartheta}(\varepsilon^2\tau_1,\varepsilon)+\Delta_n\Pi_0\vartheta(\varepsilon\tau_1,\varepsilon)+\Delta_n\Pi_1\vartheta(\tau_1,\varepsilon),\varepsilon^2\tau_1,\varepsilon)\\&-f(\widetilde{\overline{\vartheta}}_{n-1}(\varepsilon^2\tau_1,\varepsilon)+\widetilde{\Pi}_{0(n-1)}\vartheta(\varepsilon\tau_1,\varepsilon)+\Delta_n\overline{\vartheta}(\varepsilon^2\tau_1,\varepsilon)+\Delta_n\Pi_0\vartheta(\varepsilon\tau_1,\varepsilon),\varepsilon^2\tau_1,\varepsilon)).\end{aligned}\tag{42}$$

Using (36) with $k = n-2$ and $t = \varepsilon^2\tau_1$ in the expression

$$s_1 = \int_0^{+\infty} ([\Delta_n\Pi_1 w(\tau_1,\varepsilon)'(\{W(\varepsilon^2\tau_1)\widetilde{\overline{w}}_{n-2}(\varepsilon^2\tau_1,\varepsilon)\}_{n-2} + \{\varepsilon\widehat{\overline{F}}_{w(n-2)}(\varepsilon^2\tau_1,\varepsilon)'\}_{n-2})]_{2n-2}$$
$$+[\Delta_n\Pi_1 u(\tau_1,\varepsilon)'(\{R(\varepsilon^2\tau_1)\widetilde{\overline{u}}_{n-2}(\varepsilon^2\tau_1,\varepsilon)\}_{n-2} + \{\varepsilon\widehat{\overline{F}}_{u(n-2)}(\varepsilon^2\tau_1,\varepsilon)'\}_{n-2})]_{2n-2})\,d\tau_1,$$

we obtain

$$\int_0^{+\infty} ([\Delta_n\Pi_1 w(\tau_1,\varepsilon)'(E_1+\varepsilon E_2+\varepsilon^2 E_3)\frac{d\widetilde{\overline{\psi}}_{n-2}}{dt}(\varepsilon^2\tau_1,\varepsilon))]_{2n-2}$$
$$+[\Delta_n\Pi_1 w(\tau_1,\varepsilon)'(\{A(\varepsilon^2\tau_1)'\widetilde{\overline{\psi}}_{n-2}(\varepsilon^2\tau_1,\varepsilon)\}_{n-2} + \{\varepsilon\widehat{\overline{f}}_{w(n-2)}(\varepsilon^2\tau_1,\varepsilon)'\widetilde{\overline{\psi}}_{n-2}(\varepsilon^2\tau_1,\varepsilon)\}_{n-2})]_{2n-2}$$
$$+[\Delta_n\Pi_1 u(\tau_1,\varepsilon)'(\{B(\varepsilon^2\tau_1)'\widetilde{\overline{\psi}}_{n-2}(\varepsilon^2\tau_1,\varepsilon)\}_{n-2} + \{\varepsilon\widehat{\overline{f}}_{u(n-2)}(\varepsilon^2\tau_1,\varepsilon)'\widetilde{\overline{\psi}}_{n-2}(\varepsilon^2\tau_1,\varepsilon)\}_{n-2})]_{2n-2})\,d\tau_1.$$

Then, applying (38) with $k = 2n-2$, $l = n-2$, and (42), we have

$$\int_0^{+\infty} ([\Delta_n\Pi_1 w(\tau_1,\varepsilon)'(E_1+\varepsilon E_2+\varepsilon^2 E_3)\frac{d\widetilde{\overline{\psi}}_{n-2}}{dt}(\varepsilon^2\tau_1,\varepsilon)]_{2n-2}$$
$$+[\,\widetilde{\overline{\psi}}_{n-2}(\varepsilon^2\tau_1,\varepsilon)'(A(\varepsilon^2\tau_1)\Delta_n\Pi_1 w(\tau_1,\varepsilon) + B(\varepsilon^2\tau_1)\Delta_n\Pi_1 u(\tau_1,\varepsilon))]_{2n-2}$$
$$+[\widetilde{\overline{\psi}}_{n-2}(\varepsilon^2\tau_1,\varepsilon)'(\varepsilon\widehat{\overline{f}}_{w(n-2)}(\varepsilon^2\tau_1,\varepsilon)\Delta_n\Pi_1 w(\tau_1,\varepsilon) + \varepsilon\widehat{\overline{f}}_{u(n-2)}(\varepsilon^2\tau_1,\varepsilon)\Delta_n\Pi_1 u(\tau_1,\varepsilon))]_{2n-2})\,d\tau_1$$
$$= \int_0^{+\infty} ([\Delta_n\Pi_1 w(\tau_1,\varepsilon)'(E_1+\varepsilon E_2+\varepsilon^2 E_3)\frac{d\widetilde{\overline{\psi}}_{n-2}}{dt}(\varepsilon^2\tau_1,\varepsilon)]_{2n-2}$$
$$+[\widetilde{\overline{\psi}}_{n-2}(\varepsilon^2\tau_1,\varepsilon)'((\frac{1}{\varepsilon^2}E_1 + \frac{1}{\varepsilon}E_2 + E_3)(\frac{d\widetilde{\Pi}_{1(n-1)}w(\tau_1,\varepsilon)}{d\tau_1} + \frac{d\Delta_n\Pi_1 w(\tau_1,\varepsilon)}{d\tau_1})$$
$$-A(\varepsilon^2\tau_1)\widetilde{\Pi}_{1(n-1)}w(\tau_1,\varepsilon) - B(\varepsilon^2\tau_1)\widetilde{\Pi}_{1(n-1)}u(\tau_1,\varepsilon) - \varepsilon(f(\widetilde{\overline{\vartheta}}_{n-1}(\varepsilon^2\tau_1,\varepsilon) + \widetilde{\Pi}_{0(n-1)}\vartheta(\varepsilon\tau_1,\varepsilon)$$
$$+\widetilde{\Pi}_{1(n-1)}\vartheta(\tau_1,\varepsilon) + \Delta_n\overline{\vartheta}(\varepsilon^2\tau_1,\varepsilon) + \Delta_n\Pi_0\vartheta(\varepsilon\tau_1,\varepsilon) + \Delta_n\Pi_1\vartheta(\tau_1,\varepsilon),\varepsilon^2\tau_1,\varepsilon)$$
$$-f(\widetilde{\overline{\vartheta}}_{n-1}(\varepsilon^2\tau_1,\varepsilon) + \widetilde{\Pi}_{0(n-1)}\vartheta(\varepsilon\tau_1,\varepsilon) + \Delta_n\overline{\vartheta}(\varepsilon^2\tau_1,\varepsilon) + \Delta_n\Pi_0\vartheta(\varepsilon\tau_1,\varepsilon),\varepsilon^2\tau_1,\varepsilon)))]_{2n-2}$$
$$+[\widetilde{\overline{\psi}}_{n-2}(\varepsilon^2\tau_1,\varepsilon)'\varepsilon\widehat{\overline{f}}_{\vartheta(n-2)}(\varepsilon^2\tau_1,\varepsilon)\Delta_n\Pi_1\vartheta(\tau_1,\varepsilon)]_{2n-2})\,d\tau_1.$$

Integrating by parts in the last relation, using Remark 1, dropping known terms, we obtain

$$-[\Delta_n\Pi_1 w(0,\varepsilon)'(\frac{1}{\varepsilon^2}E_1 + \frac{1}{\varepsilon}E_2 + E_3)\widetilde{\overline{\psi}}_{n-2}(0,\varepsilon)]_{2n-2}$$
$$+\int_0^{+\infty}(\underbrace{[\widetilde{\overline{\psi}}_{n-2}(\varepsilon^2\tau_1,\varepsilon)'(\frac{1}{\varepsilon^2}E_1 + \frac{1}{\varepsilon}E_2 + E_3)\frac{d\widetilde{\Pi}_{1(n-1)}w(\tau_1,\varepsilon)}{d\tau_1}]_{2n-2}}_{\{1\}}$$
$$+[\widetilde{\overline{\psi}}_{n-2}(\varepsilon^2\tau_1,\varepsilon)'\varepsilon\widehat{\overline{f}}_{\vartheta(n-2)}(\varepsilon^2\tau_1,\varepsilon)\Delta_n\Pi_1\vartheta(\tau_1,\varepsilon)]_{2n-2}$$
$$-[\widetilde{\overline{\psi}}_{n-2}(\varepsilon^2\tau_1,\varepsilon)'(\varepsilon\widetilde{\Pi}_{1(n-2)}f_\vartheta(\tau_1,\varepsilon)(\Delta_n\overline{\vartheta}(\varepsilon^2\tau_1,\varepsilon) + \Delta_n\Pi_0\vartheta(\varepsilon\tau_1,\varepsilon))$$
$$+\varepsilon(\underbrace{\widehat{\overline{f}}_{\vartheta(n-2)}(\varepsilon^2\tau_1,\varepsilon)}_{\{1\}} + \underbrace{\widetilde{\Pi}_{0(n-2)}f_\vartheta(\varepsilon\tau_1,\varepsilon)}_{\{2\}} + \widetilde{\Pi}_{1(n-2)}f_\vartheta(\tau_1,\varepsilon))\varepsilon^{n-1}\Pi_{1(n-1)}\vartheta(\tau_1) \qquad (43)$$
$$+\varepsilon f_\vartheta(\widetilde{\overline{\vartheta}}_{n-2}(\varepsilon^2\tau_1,\varepsilon) + \widetilde{\Pi}_{0(n-2)}\vartheta(\varepsilon\tau_1,\varepsilon) + \widetilde{\Pi}_{1(n-2)}\vartheta,\varepsilon^2\tau_1,\varepsilon)\Delta_n\Pi_1\vartheta(\tau_1,\varepsilon))]_{2n-2}$$
$$-\underbrace{\Pi_{1(n-1)}w(\tau_1)'[A(\varepsilon^2\tau_1)'\widetilde{\overline{\psi}}_{n-2}(\varepsilon^2\tau_1,\varepsilon)]_{n-1}}_{\{1\}}$$
$$-\underbrace{\Pi_{1(n-1)}u(\tau_1)'[B(\varepsilon^2\tau_1)'\widetilde{\overline{\psi}}_{n-2}(\varepsilon^2\tau_1,\varepsilon)]_{n-1})}_{\{1\}}\,d\tau_1.$$

Consider together the first integral in (41) and some terms with $\Pi_{1(n-1)}\vartheta(\tau_1,\varepsilon)$ in the transformed last expression for s_1, marked by $\{1\}$, namely, the expression of the form

$$\int_0^{+\infty}(\Pi_{1(n-1)}w(\tau_1)'([W(\varepsilon^2\tau_1)\overline{\widetilde{w}}_{n-1}(\varepsilon^2\tau_1,\varepsilon)]_{n-1}+[\widehat{\overline{F}}_{w(n-2)}(\varepsilon^2\tau_1,\varepsilon)']_{n-2}$$
$$-[A(\varepsilon^2\tau_1)'\overline{\widetilde{\psi}}_{n-2}(\varepsilon^2\tau_1,\varepsilon)]_{n-1}-[\widehat{\overline{f}}_{w(n-2)}(\varepsilon^2\tau_1,\varepsilon)'\overline{\widetilde{\psi}}_{n-2}(\varepsilon^2\tau_1,\varepsilon)]_{n-2})$$
$$+\Pi_{1(n-1)}u(\tau_1)'([R(\varepsilon^2\tau_1)\overline{\widetilde{u}}_{n-1}(\varepsilon^2\tau_1,\varepsilon)]_{n-1}+[\widehat{\overline{F}}_{u(n-2)}(\varepsilon^2\tau_1,\varepsilon)']_{n-2}$$
$$-[B(\varepsilon^2\tau_1)'\overline{\widetilde{\psi}}_{n-2}(\varepsilon^2\tau_1,\varepsilon)]_{n-1}-[\widehat{\overline{f}}_{u(n-2)}(\varepsilon^2\tau_1,\varepsilon)'\overline{\widetilde{\psi}}_{n-2}(\varepsilon^2\tau_1,\varepsilon)]_{n-2})$$
$$+[\overline{\widetilde{\psi}}_{n-2}(\varepsilon^2\tau_1,\varepsilon)'(\frac{1}{\varepsilon^2}E_1+\frac{1}{\varepsilon}E_2+E_3)\frac{d\widetilde{\Pi}_{1(n-1)}w(\tau_1,\varepsilon)}{d\tau_1}]_{2n-2})\,d\tau_1.$$

Transforming this expression with the help of (36) with $k=n-1$ and (10) at $j=n-1,n,n+1$ and omitting some known terms, we have

$$\int_0^{+\infty}(\Pi_{1(n-1)}w(\tau_1)'[E_1\frac{d\overline{\widetilde{\psi}}_{n-1}}{dt}(\varepsilon^2\tau_1,\varepsilon)+E_2\frac{d\overline{\widetilde{\psi}}_{n-2}}{dt}(\varepsilon^2\tau_1,\varepsilon)+E_3\frac{d\overline{\widetilde{\psi}}_{n-3}}{dt}(\varepsilon^2\tau_1,\varepsilon)]_{n-1}$$
$$+\overline{\psi}_{n-1}(0)'(E_1\frac{d\Pi_{1(n+1)}w(\tau_1)}{d\tau_1}+E_2\frac{d\Pi_{1n}w(\tau_1)}{d\tau_1}+E_3\frac{d\Pi_{1(n-1)}w(\tau_1)}{d\tau_1}$$
$$+[\overline{\widetilde{\psi}}_{n-2}(\varepsilon^2\tau_1,\varepsilon)'(\frac{1}{\varepsilon^2}E_1+\frac{1}{\varepsilon}E_2+E_3)\frac{d\widetilde{\Pi}_{1(n-1)}w(\tau_1,\varepsilon)}{d\tau_1})]_{2n-2})\,d\tau_1.$$

From here, using Remark 1, integrating by parts and omitting known terms, we obtain

$$-\overline{\psi}_{n-1}(0)'(E_1\Pi_{1(n+1)}w(0)+E_2\Pi_{1n}w(0)+E_3\Pi_{1(n-1)}w(0)).$$

Further changes concern the expression

$$s_2=\int_0^{+\infty}([\Delta_n\Pi_1w(\tau_1,\varepsilon)'(\{W(\varepsilon^2\tau_1)\widetilde{\Pi}_{0(n-2)}w(\varepsilon\tau_1,\varepsilon)\}_{n-2}$$
$$+\{\varepsilon\widehat{\Pi}_{0(n-2)}F_w(\varepsilon\tau_1,\varepsilon)\}_{n-2})]_{2n-2}+[\Delta_n\Pi_1u(\tau_1,\varepsilon)'(\{R(\varepsilon^2\tau_1,\varepsilon)\widetilde{\Pi}_{0(n-2)}u(\varepsilon\tau_1,\varepsilon)\}_{n-2}$$
$$+\{\varepsilon\widehat{\Pi}_{0(n-2)}F_u(\varepsilon\tau_1,\varepsilon)\}_{n-2})]_{2n-2})\,d\tau_1.$$

It will be transformed using (40) with $k=n-2$ and (38) in the following way:

$$\int_0^{+\infty}([\Delta_n\Pi_1w(\tau_1,\varepsilon)'(E_1+E_2+\varepsilon E_3)\frac{d\widetilde{\Pi}_{0(n-2)}\psi(\varepsilon\tau_1,\varepsilon)}{d\tau_0}]_{2n-2}$$
$$+[\widetilde{\Pi}_{0(n-2)}\psi(\varepsilon\tau_1,\varepsilon)'(\varepsilon E_1+E_2+E_3)(A(\varepsilon^2\tau_1)\Delta_n\Pi_1w(\tau_1,\varepsilon)+B(\varepsilon^2\tau_1)\Delta_n\Pi_1u(\tau_1,\varepsilon))]_{2n-2}$$
$$+[\Delta_n\Pi_1w(\tau_1,\varepsilon)'(\{\varepsilon\widehat{\Pi}_{0(n-2)}f_w(\varepsilon\tau_1,\varepsilon)'\overline{\widetilde{\psi}}_{n-2}(\varepsilon^2\tau_1,\varepsilon)\}_{n-2}$$
$$+\{\varepsilon f_w(\overline{\widetilde{\vartheta}}_{n-2}(\varepsilon^2\tau_1,\varepsilon)+\widetilde{\Pi}_{0(n-2)}\vartheta(\varepsilon\tau_1,\varepsilon),\varepsilon^2\tau_1,\varepsilon)'(\varepsilon E_1+E_2+E_3)\widetilde{\Pi}_{0(n-2)}\psi(\varepsilon\tau_1,\varepsilon)\}_{n-2})]_{2n-2}$$
$$+[\Delta_n\Pi_1u(\tau_1,\varepsilon)'(\{\varepsilon\widehat{\Pi}_{0(n-2)}f_u(\varepsilon\tau_1,\varepsilon)'\overline{\widetilde{\psi}}_{n-2}(\varepsilon^2\tau_1,\varepsilon)\}_{n-2}+\{\varepsilon f_u(\overline{\widetilde{\vartheta}}_{n-2}(\varepsilon^2\tau_1,\varepsilon)$$
$$+\widetilde{\Pi}_{0(n-2)}\vartheta(\varepsilon\tau_1,\varepsilon),\varepsilon^2\tau_1,\varepsilon)'(\varepsilon E_1+E_2+E_3)\widetilde{\Pi}_{0(n-2)}\psi(\varepsilon\tau_1,\varepsilon)\}_{n-2})]_{2n-2})\,d\tau_1.$$

From here, using (42) and omitting some known terms, we have

$$\int_0^{+\infty}([\Delta_n\Pi_1w(\tau_1,\varepsilon)'(E_1+E_2+\varepsilon E_3)\frac{d\widetilde{\Pi}_{0(n-2)}\psi(\varepsilon\tau_1,\varepsilon)}{d\tau_0}]_{2n-2}+[\widetilde{\Pi}_{0(n-2)}\psi(\varepsilon\tau_1,\varepsilon)'(\varepsilon E_1$$
$$+E_2+E_3)((\frac{1}{\varepsilon^2}E_1+\frac{1}{\varepsilon}E_2+E_3)(\frac{d\widetilde{\Pi}_{1(n-1)}w(\tau_1,\varepsilon)}{d\tau_1}+\frac{d\Delta_n\Pi_1w(\tau_1,\varepsilon)}{d\tau_1})$$

$$-A(\varepsilon^2\tau_1)\widetilde{\Pi}_{1(n-1)}w(\tau_1,\varepsilon) - B(\varepsilon^2\tau_1)\widetilde{\Pi}_{1(n-1)}u(\tau_1,\varepsilon) - \varepsilon\widehat{\Pi}_{1(n-2)}f_\vartheta(\tau_1,\varepsilon)(\varepsilon^{n-1}\overline{\vartheta}_{n-1}(\varepsilon^2\tau_1,\varepsilon)$$
$$+\varepsilon^{n-1}\Pi_{0(n-1)}\vartheta(\varepsilon\tau_1,\varepsilon) + \Delta_n\overline{\vartheta}(\varepsilon^2\tau_1,\varepsilon) + \Delta_n\Pi_0\vartheta(\varepsilon\tau_1,\varepsilon)) - \varepsilon f_\vartheta(\overline{\overline{\vartheta}}_{n-2}(\varepsilon^2\tau_1,\varepsilon)$$
$$+\widetilde{\Pi}_{0(n-2)}\vartheta(\varepsilon\tau_1,\varepsilon) + \widetilde{\Pi}_{1(n-2)}\vartheta(\tau_1,\varepsilon),\varepsilon^2\tau_1,\varepsilon)\}_{n-2})(\varepsilon^{n-1}\Pi_{1(n-1)}\vartheta(\tau_1)$$
$$+\Delta_n\Pi_1\vartheta(\tau_1,\varepsilon)))]_{2n-2} + [\Delta_n\Pi_1 w(\tau_1,\varepsilon)'(\{\varepsilon\widehat{\Pi}_{0(n-2)}f_w(\varepsilon\tau_1,\varepsilon)'\overline{\overline{\psi}}_{n-2}(\varepsilon^2\tau_1,\varepsilon)\}_{n-2}$$
$$+\{\varepsilon f_w(\overline{\overline{\vartheta}}_{n-2}(\varepsilon^2\tau_1,\varepsilon) + \widetilde{\Pi}_{0(n-2)}\vartheta(\varepsilon\tau_1,\varepsilon),\varepsilon^2\tau_1,\varepsilon)'(\varepsilon E_1 + E_2 + E_3)\widetilde{\Pi}_{0(n-2)}\psi(\varepsilon\tau_1,\varepsilon)\}_{n-2})]_{2n-2}$$
$$+[\Delta_n\Pi_1 u(\tau_1,\varepsilon)'(\{\varepsilon\widehat{\Pi}_{0(n-2)}f_u(\varepsilon\tau_1,\varepsilon)'\overline{\overline{\psi}}_{n-2}(\varepsilon^2\tau_1,\varepsilon)\}_{n-2} + \{\varepsilon f_u(\overline{\overline{\vartheta}}_{n-2}(\varepsilon^2\tau_1,\varepsilon)$$
$$+\widetilde{\Pi}_{0(n-2)}\vartheta(\varepsilon\tau_1,\varepsilon),\varepsilon^2\tau_1,\varepsilon)'(\varepsilon E_1 + E_2 + E_3)\widetilde{\Pi}_{0(n-2)}\psi(\varepsilon\tau_1,\varepsilon)\}_{n-2})]_{2n-2})\,d\tau_1.$$

Integrating by parts the first term in the last expression and dropping known terms, we obtain

$$\begin{aligned}
&-[\widetilde{\Pi}_{0(n-2)}\psi(0,\varepsilon)'(\frac{1}{\varepsilon}E_1 + \frac{1}{\varepsilon}E_2 + E_3)\Delta_n\Pi_1 w(0,\varepsilon)]_{2n-2} \\
&+\int_0^{+\infty}(\underbrace{[\widetilde{\Pi}_{0(n-2)}\psi(\varepsilon\tau_1,\varepsilon)'(\frac{1}{\varepsilon}E_1 + \frac{1}{\varepsilon}E_2 + E_3)\frac{d\widetilde{\Pi}_{1(n-1)}w(\tau_1,\varepsilon)}{d\tau_1}]_{2n-2}}_{\{2\}} \\
&-[\widetilde{\Pi}_{0(n-2)}\psi(\varepsilon\tau_1,\varepsilon)'(\varepsilon E_1 + E_2 + E_3)(\varepsilon\widehat{\Pi}_{1(n-2)}f_\vartheta(\tau_1,\varepsilon)(\Delta_n\overline{\vartheta}(\varepsilon^2\tau_1,\varepsilon) + \Delta_n\Pi_0\vartheta(\varepsilon\tau_1,\varepsilon)) \\
&+\varepsilon(\widehat{\Pi}_{1(n-2)}f_\vartheta(\tau_1,\varepsilon) + \underbrace{f_\vartheta(\overline{\overline{\vartheta}}_{n-2}(\varepsilon^2\tau_1,\varepsilon) + \widetilde{\Pi}_{0(n-2)}\vartheta(\varepsilon\tau_1,\varepsilon),\varepsilon^2\tau_1,\varepsilon))}_{\{2\}}\varepsilon^{n-1}\Pi_{1(n-1)}\vartheta(\tau_1) \\
&+\varepsilon f_\vartheta(\overline{\overline{\vartheta}}_{n-2}(\varepsilon^2\tau_1,\varepsilon) + \widetilde{\Pi}_{0(n-2)}\vartheta(\varepsilon\tau_1,\varepsilon) + \widetilde{\Pi}_{1(n-2)}\vartheta(\tau_1,\varepsilon),\varepsilon^2\tau_1,\varepsilon)\Delta_n\Pi_1\vartheta(\tau_1,\varepsilon))]_{2n-2} \\
&\underbrace{-\Pi_{1(n-1)}w(\tau_1)'[A(\varepsilon^2\tau_1)'(\varepsilon E_1 + E_2 + E_3)\widetilde{\Pi}_{0(n-2)}\psi(\varepsilon\tau_1,\varepsilon)]_{n-1}}_{\{2\}} \\
&\underbrace{-\Pi_{1(n-1)}u(\tau_1)'[B(\varepsilon^2\tau_1)'(\varepsilon E_1 + E_2 + E_3)\widetilde{\Pi}_{0(n-2)}\psi(\varepsilon\tau_1,\varepsilon)]_{n-1}}_{\{2\}} \\
&+[\Delta_n\Pi_1\vartheta(\tau_1,\varepsilon)'(\{\varepsilon\widehat{\Pi}_{0(n-2)}f_\vartheta(\varepsilon\tau_1,\varepsilon)'\overline{\overline{\psi}}_{n-2}(\varepsilon^2\tau_1,\varepsilon)\}_{n-2} + \{\varepsilon f_\vartheta(\overline{\overline{\vartheta}}_{n-2}(\varepsilon^2\tau_1,\varepsilon) \\
&+\widetilde{\Pi}_{0(n-2)}\vartheta(\varepsilon\tau_1,\varepsilon),\varepsilon^2\tau_1,\varepsilon)'(\varepsilon E_1 + E_2 + E_3)\widetilde{\Pi}_{0(n-2)}\psi(\varepsilon\tau_1,\varepsilon)\}_{n-2})]_{2n-2})\,d\tau_1.
\end{aligned} \tag{44}$$

Consider together the second integral in (41) and some terms with $\Pi_{1(n-1)}\vartheta(\tau_1,\varepsilon)$ from (43) and (44), marked by $\{2\}$, namely the expression of the form

$$\int_0^{+\infty}(\Pi_{1(n-1)}w(\tau_1)'([W(\varepsilon^2\tau_1)\widetilde{\Pi}_{0(n-1)}w(\varepsilon\tau_1,\varepsilon)]_{n-1} + [\widehat{\Pi}_{0(n-2)}F_w(\varepsilon\tau_1,\varepsilon)']_{n-2}$$
$$-[A(\varepsilon^2\tau_1)'(\varepsilon E_1 + E_2 + E_3)\widetilde{\Pi}_{0(n-2)}\psi(\varepsilon\tau_1,\varepsilon)]_{n-1} - [\varepsilon\widehat{\Pi}_{0(n-2)}f_w(\varepsilon\tau_1,\varepsilon)'\overline{\overline{\psi}}_{n-2}(\varepsilon^2\tau_1,\varepsilon)]_{n-1}$$
$$-[\varepsilon f_w(\overline{\overline{\vartheta}}_{n-2}(\varepsilon^2\tau_1,\varepsilon) + \widetilde{\Pi}_{0(n-2)}\vartheta(\varepsilon\tau_1,\varepsilon),\varepsilon^2\tau_1,\varepsilon)'(\varepsilon E_1 + E_2 + E_3)\widetilde{\Pi}_{0(n-2)}\psi(\varepsilon\tau_1,\varepsilon)]_{n-1})$$
$$+\Pi_{1(n-1)}u(\tau_1)'([R(\varepsilon^2\tau_1)\widetilde{\Pi}_{0(n-1)}u(\varepsilon\tau_1,\varepsilon)]_{n-1} + [\widehat{\Pi}_{0(n-2)}F_u(\varepsilon\tau_1,\varepsilon)']_{n-2}$$
$$-[B(\varepsilon^2\tau_1)'(\varepsilon E_1 + E_2 + E_3)\widetilde{\Pi}_{0(n-2)}\psi(\varepsilon\tau_1,\varepsilon)]_{n-1} - [\varepsilon\widehat{\Pi}_{0(n-2)}f_u(\varepsilon\tau_1,\varepsilon)'\overline{\overline{\psi}}_{n-2}(\varepsilon^2\tau_1,\varepsilon)]_{n-1}$$
$$-[\varepsilon f_u(\overline{\overline{\vartheta}}_{n-2}(\varepsilon^2\tau_1,\varepsilon) + \widetilde{\Pi}_{0(n-2)}\vartheta(\varepsilon\tau_1,\varepsilon),\varepsilon^2\tau_1,\varepsilon)'(\varepsilon E_1 + E_2 + E_3)\widetilde{\Pi}_{0(n-2)}\psi(\varepsilon\tau_1,\varepsilon)]_{n-1})$$
$$+[\widetilde{\Pi}_{0(n-2)}\psi(\varepsilon\tau_1,\varepsilon)'(\frac{1}{\varepsilon}E_1 + \frac{1}{\varepsilon}E_2 + E_3)\frac{d\widetilde{\Pi}_{1(n-1)}w(\tau_1,\varepsilon)}{d\tau_1}]_{2n-2})\,d\tau_1.$$

We will transform this expression with the help of (40) with $k = n-1$ and (10) with $j = n-1, n$. Omitting known terms, we obtain

$$\int_0^{+\infty} ([\widetilde{\Pi}_{0(n-2)}\psi(\varepsilon\tau_1,\varepsilon)'(\frac{1}{\varepsilon}E_1 + \frac{1}{\varepsilon}E_2 + E_3)\frac{d\widetilde{\Pi}_{1(n-1)}w(\tau_1,\varepsilon)}{d\tau_1}]_{2n-2}$$
$$+\Pi_{1(n-1)}w(\tau_1)'[(E_1+E_2)\frac{d\widetilde{\Pi}_{0(n-1)}\psi}{d\tau_0}(\varepsilon\tau_1,\varepsilon) + E_3\frac{d\widetilde{\Pi}_{0(n-2)}\psi}{d\tau_0}(\varepsilon\tau_1,\varepsilon)]_{n-1}$$
$$+\Pi_{0(n-1)}\psi(0)'(E_2\frac{d\Pi_{1n}w(\tau_1)}{d\tau_1} + E_3\frac{d\Pi_{1(n-1)}w(\tau_1)}{d\tau_1}))\,d\tau_1.$$

Integrating by parts, using Remark 1 and omitting known terms, we have

$$-\Pi_{0(n-1)}\psi(0)'(E_2\Pi_{1n}w(0) + E_3\Pi_{1(n-1)}w(0)).$$

Multiplying Equations (24), (25) by ε^j, $j = \overline{0, n-2}$ and summing up the obtained results, we obtain the equalities

$$\begin{aligned}
&\{R(\varepsilon^2\tau_1)\widetilde{\Pi}_{1(n-2)}u(\tau_1,\varepsilon)\}_{n-2} + \{\varepsilon\widehat{\Pi}_{1(n-3)}F_u(\tau_1,\varepsilon)'\}_{n-2}\\
&= \{B(\varepsilon^2\tau_1)'(\varepsilon^2E_1 + \varepsilon E_2 + E_3)\widetilde{\Pi}_{1(n-2)}\psi(\tau_1,\varepsilon)\}_{n-2}\\
&+\{\varepsilon\widehat{\Pi}_{1(n-3)}f_u(\tau_1,\varepsilon)'\widetilde{\overline{\psi}}_{n-3}(\varepsilon^2\tau_1,\varepsilon)\}_{n-2}\\
&+\{\varepsilon\widehat{\Pi}_{1(n-3)}f_u(\tau_1,\varepsilon)'(\varepsilon E_1 + E_2 + E_3)\widetilde{\Pi}_{0(n-3)}\psi(\varepsilon\tau_1,\varepsilon)\}_{n-2}\\
&+\{\varepsilon f_u(\widetilde{\overline{\vartheta}}_{n-3}(\varepsilon^2\tau_1,\varepsilon) + \widetilde{\Pi}_{0(n-3)}\vartheta(\varepsilon\tau_1,\varepsilon) + \widetilde{\Pi}_{1(n-3)}\vartheta(\tau_1,\varepsilon), \varepsilon^2\tau_1,\varepsilon)'(\varepsilon^2E_1\\
&+\varepsilon E_2 + E_3)\widetilde{\Pi}_{1(n-3)}\psi(\tau_1,\varepsilon)\}_{n-2},
\end{aligned} \tag{45}$$

$$\begin{aligned}
&\{W(\varepsilon^2\tau_1)\widetilde{\Pi}_{1(n-2)}w(\tau_1,\varepsilon)\}_{n-2} + \{\varepsilon\widehat{\Pi}_{1(n-3)}F_w(\tau_1,\varepsilon)'\}_{n-2} = \frac{d\widetilde{\Pi}_{1(n-2)}\psi(\tau_1,\varepsilon)}{d\tau_1}\\
&+\{A(\varepsilon^2\tau_1)'(\varepsilon^2E_1 + \varepsilon E_2 + E_3)\widetilde{\Pi}_{1(n-2)}\psi(\tau_1,\varepsilon)\}_{n-2}\\
&+\{\varepsilon\widehat{\Pi}_{1(n-3)}f_w(\tau_1,\varepsilon)'\widetilde{\overline{\psi}}_{n-3}(\varepsilon^2\tau_1,\varepsilon)\}_{n-2}\\
&+\{\varepsilon\widehat{\Pi}_{1(n-3)}f_w(\tau_1,\varepsilon)'(\varepsilon E_1 + E_2 + E_3)\widetilde{\Pi}_{0(n-3)}\psi(\varepsilon\tau_1,\varepsilon)\}_{n-2}\\
&+\{\varepsilon f_w(\widetilde{\overline{\vartheta}}_{n-3}(\varepsilon^2\tau_1,\varepsilon) + \widetilde{\Pi}_{0(n-3)}\vartheta(\varepsilon\tau_1,\varepsilon) + \widetilde{\Pi}_{1(n-3)}\vartheta(\tau_1,\varepsilon), \varepsilon^2\tau_1,\varepsilon)'(\varepsilon^2E_1\\
&+\varepsilon E_2 + E_3)\widetilde{\Pi}_{1(n-3)}\psi(\tau_1,\varepsilon)\}_{n-2}.
\end{aligned} \tag{46}$$

We will transform the expression

$$s_3 = \int_0^{+\infty} ([(\Delta_n\overline{w}(\varepsilon^2\tau_1,\varepsilon) + \Delta_n\Pi_0w(\varepsilon\tau_1,\varepsilon))'(\{W(\varepsilon^2\tau_1)\widetilde{\Pi}_{1(n-2)}w(\tau_1,\varepsilon)\}_{n-2}$$
$$+\{\varepsilon\widehat{\Pi}_{1(n-2)}F_w(\tau_1,\varepsilon)'\}_{n-2})]_{2n-2} + [(\Delta_n\overline{u}(\varepsilon^2\tau_1,\varepsilon)$$
$$+\Delta_n\Pi_0u(\varepsilon\tau_1,\varepsilon))'(\{R(\varepsilon^2\tau_1)\widetilde{\Pi}_{1(n-2)}u(\tau_1,\varepsilon)\}_{n-2} + \{\varepsilon\widehat{\Pi}_{1(n-2)}F_u(\tau_1,\varepsilon)'\}_{n-2})]_{2n-2})\,d\tau_1.$$

Using (45), (46), (38) with $k = 2n-2$, $l = n-2$, (37), (39) and omitting known terms, we have

$$\int_0^{+\infty} ([(\Delta_n\overline{w}(\varepsilon^2\tau_1,\varepsilon) + \Delta_n\Pi_0w(\varepsilon\tau_1,\varepsilon))'\frac{d\widetilde{\Pi}_{1(n-2)}\psi(\tau_1,\varepsilon)}{d\tau_1}]_{2n-2}$$
$$+[\widetilde{\Pi}_{1(n-2)}\psi(\tau_1,\varepsilon)'(\varepsilon^2E_1 + \varepsilon E_2 + E_3)((E_1 + \varepsilon E_2 + \varepsilon^2E_3)(\frac{d\widetilde{\overline{w}}_{n-1}(\varepsilon^2\tau_1,\varepsilon)}{dt}$$
$$+\frac{d\Delta_n\overline{w}(\varepsilon^2\tau_1,\varepsilon)}{dt}) - A(\varepsilon^2\tau_1)\widetilde{\overline{w}}_{n-1}(\varepsilon^2\tau_1,\varepsilon) - B(\varepsilon^2\tau_1)\widetilde{\overline{u}}_{n-1}(\varepsilon^2\tau_1,\varepsilon)$$
$$-\varepsilon\widehat{\overline{f}}_{\vartheta(n-2)}(\varepsilon^2\tau_1,\varepsilon)(\varepsilon^{n-1}\overline{\vartheta}_{n-1}(\varepsilon^2\tau_1,\varepsilon) + \Delta_n\overline{\vartheta}(\varepsilon^2\tau_1,\varepsilon)))]_{2n-2}$$

$$
\begin{aligned}
&+[\widetilde{\Pi}_{1(n-2)}\psi(\tau_1,\varepsilon)'(\varepsilon^2E_1+\varepsilon E_2+E_3)((\frac{1}{\varepsilon}E_1+E_2+\varepsilon E_3)(\frac{d\widetilde{\Pi}_{0(n-1)}w(\varepsilon\tau_1,\varepsilon)}{d\tau_0}\\
&+\frac{d\Delta_n\Pi_0 w(\varepsilon\tau_1,\varepsilon)}{d\tau_0})-A(\varepsilon^2\tau_1)\widetilde{\Pi}_{0(n-1)}w(\varepsilon\tau_1,\varepsilon)-B(\varepsilon^2\tau_1)\widetilde{\Pi}_{0(n-1)}u(\varepsilon\tau_1,\varepsilon)\\
&-\varepsilon\widehat{\Pi}_{0(n-2)}f_\vartheta(\varepsilon\tau_1,\varepsilon)(\varepsilon^{n-1}\overline{\vartheta}_{n-1}(\varepsilon^2\tau_1,\varepsilon)+\Delta_n\overline{\vartheta}(\varepsilon^2\tau_1,\varepsilon))\\
&-\varepsilon f_\vartheta(\widetilde{\overline{\vartheta}}_{n-2}(\varepsilon^2\tau_1,\varepsilon)+\widetilde{\Pi}_{0(n-2)}\vartheta(\varepsilon\tau_1,\varepsilon),\varepsilon^2\tau_1,\varepsilon)(\varepsilon^{n-1}\Pi_{0(n-1)}\vartheta(\varepsilon\tau_1,\varepsilon)+\Delta_n\Pi_0\vartheta(\varepsilon\tau_1,\varepsilon)))]_{2n-2}\\
&+[(\Delta_n\overline{\vartheta}(\varepsilon^2\tau_1,\varepsilon)+\Delta_n\Pi_0\vartheta(\varepsilon\tau_1,\varepsilon))'(\{\varepsilon\widehat{\Pi}_{1(n-2)}f_\vartheta(\tau_1,\varepsilon)'\widetilde{\overline{\psi}}_{n-2}(\varepsilon^2\tau_1,\varepsilon)\}_{n-2}\\
&+\{\varepsilon\widehat{\Pi}_{1(n-2)}f_\vartheta(\tau_1,\varepsilon)'(\varepsilon E_1+E_2+E_3)\widetilde{\Pi}_{0(n-2)}\psi(\varepsilon\tau_1,\varepsilon)\}_{n-2}+\{\varepsilon f_\vartheta(\widetilde{\overline{\vartheta}}_{n-2}(\varepsilon^2\tau_1,\varepsilon)\\
&+\widetilde{\Pi}_{0(n-2)}\vartheta(\varepsilon\tau_1,\varepsilon)+\widetilde{\Pi}_{1(n-2)}\vartheta(\tau_1,\varepsilon),\varepsilon^2\tau_1,\varepsilon)'(\varepsilon^2E_1+\varepsilon E_2\\
&+E_3)\widetilde{\Pi}_{1(n-2)}\psi(\tau_1,\varepsilon)\}_{n-2})]_{2n-2})\,d\tau_1.
\end{aligned}
$$

From here, applying the formula of integrating by parts, and omitting known terms, we obtain the unknown part from s_3

$$
\begin{aligned}
&-[(\Delta_n\overline{w}(0,\varepsilon)+\Delta_n\Pi_0 w(0,\varepsilon))'\widetilde{\Pi}_{1(n-2)}\psi(0,\varepsilon)]_{2n-2}\\
&-\int_0^{+\infty}([\widetilde{\Pi}_{1(n-2)}\psi(\tau_1,\varepsilon)'(\varepsilon^2E_1+\varepsilon E_2+E_3)(\varepsilon\widehat{\overline{f}}_{\vartheta(n-2)}(\varepsilon^2\tau_1,\varepsilon)\Delta_n\overline{\vartheta}(\varepsilon^2\tau_1,\varepsilon)\\
&+\varepsilon\widehat{\Pi}_{0(n-2)}f_\vartheta(\varepsilon\tau_1,\varepsilon)\Delta_n\overline{\vartheta}(\varepsilon^2\tau_1,\varepsilon)\\
&+\varepsilon f_\vartheta(\widetilde{\overline{\vartheta}}_{n-2}(\varepsilon^2\tau_1,\varepsilon)+\widetilde{\Pi}_{0(n-2)}\vartheta(\varepsilon\tau_1,\varepsilon),\varepsilon^2\tau_1,\varepsilon)\Delta_n\Pi_0\vartheta(\varepsilon\tau_1,\varepsilon))]_{2n-2}\\
&-[(\Delta_n\overline{\vartheta}(\varepsilon^2\tau_1,\varepsilon)+\Delta_n\Pi_0\vartheta(\varepsilon\tau_1,\varepsilon))'(\{\varepsilon\widehat{\Pi}_{1(n-2)}f_\vartheta(\tau_1,\varepsilon)'\widetilde{\overline{\psi}}_{n-2}(\varepsilon^2\tau_1,\varepsilon)\}_{n-2}\\
&+\{\varepsilon\widehat{\Pi}_{1(n-2)}f_\vartheta(\tau_1,\varepsilon)'(\varepsilon E_1+E_2+E_3)\widetilde{\Pi}_{0(n-2)}\psi(\varepsilon\tau_1,\varepsilon)\}_{n-2}\\
&+\{\varepsilon f_\vartheta(\widetilde{\overline{\vartheta}}_{n-2}(\varepsilon^2\tau_1,\varepsilon)+\widetilde{\Pi}_{0(n-2)}\vartheta(\varepsilon\tau_1,\varepsilon)\\
&+\widetilde{\Pi}_{1(n-2)}\vartheta(\tau_1,\varepsilon),\varepsilon^2\tau_1,\varepsilon)'(\varepsilon^2E_1+\varepsilon E_2+E_3)\widetilde{\Pi}_{1(n-2)}\psi(\tau_1,\varepsilon)\}_{n-2})]_{2n-2})\,d\tau_1.
\end{aligned}
$$

Furthermore, applying the same algorithm, we will transform the expression

$$
\begin{aligned}
s_4=&\int_0^{+\infty}([\Delta_n\Pi_1 w(\tau_1,\varepsilon)'(\{W(\varepsilon^2\tau_1)\widetilde{\Pi}_{1(n-2)}w(\tau_1,\varepsilon)\}_{n-2}+\{\varepsilon\widehat{\Pi}_{1(n-2)}F_w(\tau_1,\varepsilon)'\}_{n-2})]_{2n-2}\\
&+[\Delta_n\Pi_1 u(\tau_1,\varepsilon)'(\{R(\varepsilon^2\tau_1)\widetilde{\Pi}_{1(n-2)}u(\tau_1,\varepsilon)\}_{n-2}+\{\varepsilon\widehat{\Pi}_{1(n-2)}F_u(\tau_1,\varepsilon)'\}_{n-2})]_{2n-2})\,d\tau_1.
\end{aligned}
$$

Using (45), (46), (38) with $k=2n-2$, $l=n-2$, (42) and omitting known terms, we obtain

$$
\begin{aligned}
&\int_0^{+\infty}([\Delta_n\Pi_1 w(\tau_1,\varepsilon)'\frac{d\widetilde{\Pi}_{1(n-2)}\psi(\tau_1,\varepsilon)}{d\tau_1}]_{2n-2}+[\widetilde{\Pi}_{1(n-2)}\psi(\tau_1,\varepsilon)'(\varepsilon^2E_1+\varepsilon E_2+E_3)((\frac{1}{\varepsilon^2}E_1\\
&+\frac{1}{\varepsilon}E_2+E_3)(\frac{d\widetilde{\Pi}_{1(n-1)}w(\tau_1,\varepsilon)}{d\tau_1}+\frac{d\Delta_n\Pi_1 w(\tau_1,\varepsilon)}{d\tau_1})-A(\varepsilon^2\tau_1)\widetilde{\Pi}_{1(n-1)}w(\tau_1,\varepsilon)\\
&-B(\varepsilon^2\tau_1)\widetilde{\Pi}_{1(n-1)}u(\tau_1,\varepsilon)-\varepsilon\widehat{\Pi}_{1(n-2)}f_\vartheta(\tau_1,\varepsilon)(\Delta_n\overline{\vartheta}(\varepsilon^2\tau_1,\varepsilon)+\Delta_n\Pi_0\vartheta(\varepsilon\tau_1,\varepsilon))\\
&-\varepsilon f_\vartheta(\widetilde{\overline{\vartheta}}_{n-2}(\varepsilon^2\tau_1,\varepsilon)+\widetilde{\Pi}_{0(n-2)}\vartheta(\varepsilon\tau_1,\varepsilon)+\widetilde{\Pi}_{1(n-2)}(\tau_1,\varepsilon),\varepsilon^2\tau_1,\varepsilon)(\varepsilon^{n-1}\Pi_{1(n-1)}\vartheta(\tau_1)\\
&+\Delta_n\Pi_1\vartheta(\tau_1,\varepsilon)))]_{2n-2}+[\,\Delta_n\Pi_1\vartheta(\tau_1,\varepsilon)'(\{\varepsilon\widehat{\Pi}_{1(n-2)}f_\vartheta(\tau_1,\varepsilon)'\widetilde{\overline{\psi}}_{n-2}(\varepsilon^2\tau_1,\varepsilon)\}_{n-2}\\
&+\{\varepsilon\widehat{\Pi}_{1(n-2)}f_\vartheta(\tau_1,\varepsilon)'(\varepsilon E_1+E_2+E_3)\widetilde{\Pi}_{0(n-2)}\psi(\varepsilon\tau_1,\varepsilon)\}_{n-2}+\{\varepsilon f_\vartheta(\widetilde{\overline{\vartheta}}_{n-2}(\varepsilon^2\tau_1,\varepsilon)\\
&+\widetilde{\Pi}_{0(n-2)}\vartheta(\varepsilon\tau_1,\varepsilon)+\widetilde{\Pi}_{1(n-2)}\vartheta(\tau_1,\varepsilon),\varepsilon^2\tau_1,\varepsilon)'(\varepsilon^2E_1+\varepsilon E_2\\
&+E_3)\widetilde{\Pi}_{1(n-2)}\psi(\tau_1,\varepsilon)\}_{n-2})]_{2n-2})\,d\tau_1.
\end{aligned}
$$

Due to formula of integrating by parts, after omitting known terms, we obtain the following:

$$-[\Delta_n\Pi_1 w(0,\varepsilon)'\widetilde{\Pi}_{1(n-2)}\psi(0,\varepsilon)]_{2n-2}$$
$$+\int_0^{+\infty}(\Pi_{1(n-1)}w(\tau_1))'(-[A(\varepsilon^2\tau_1)'(\varepsilon^2E_1+\varepsilon E_2+E_3)\widetilde{\Pi}_{1(n-2)}\psi(\tau_1,\varepsilon)]_{n-1}$$
$$-[\varepsilon f_w(\widetilde{\overline{\vartheta}}_{n-2}(\varepsilon^2\tau_1,\varepsilon)+\widetilde{\Pi}_{0(n-2)}\vartheta(\varepsilon\tau_1,\varepsilon)+\widetilde{\Pi}_{1(n-2)}\vartheta(\tau_1,\varepsilon),\varepsilon^2\tau_1,\varepsilon)'(\varepsilon^2E_1+\varepsilon E_2$$
$$+E_3)\widetilde{\Pi}_{1(n-2)}\psi(\tau_1,\varepsilon)]_{n-1})-\Pi_{1(n-1)}u(\tau_1)'([B(\varepsilon^2\tau_1)'(\varepsilon^2E_1+\varepsilon E_2+E_3)\widetilde{\Pi}_{1(n-2)}\psi(\tau_1,\varepsilon)]_{n-1}$$
$$-[\varepsilon f_u(\widetilde{\overline{\vartheta}}_{n-2}(\varepsilon^2\tau_1,\varepsilon)+\widetilde{\Pi}_{0(n-2)}\vartheta(\varepsilon\tau_1,\varepsilon)+\widetilde{\Pi}_{1(n-2)}\vartheta(\tau_1,\varepsilon),\varepsilon^2\tau_1,\varepsilon)'(\varepsilon^2E_1$$
$$+\varepsilon E_2+E_3)\widetilde{\Pi}_{1(n-2)}\psi(\tau_1,\varepsilon)]_{n-1})-[\widetilde{\Pi}_{1(n-2)}\psi(\tau_1,\varepsilon)'(\varepsilon^2E_1+\varepsilon E_2$$
$$+E_3)(\varepsilon\widehat{\Pi}_{1(n-2)}f_\vartheta(\tau_1,\varepsilon)(\Delta_n\overline{\vartheta}(\varepsilon^2\tau_1,\varepsilon)+\Delta_n\Pi_0\vartheta(\varepsilon\tau_1,\varepsilon))$$
$$+\varepsilon f_\vartheta(\widetilde{\overline{\vartheta}}_{n-2}(\varepsilon^2\tau_1,\varepsilon)+\widetilde{\Pi}_{0(n-2)}\vartheta(\varepsilon\tau_1,\varepsilon)+\widetilde{\Pi}_{1(n-2)}\vartheta(\tau_1,\varepsilon),\varepsilon^2\tau_1,\varepsilon)\Delta_n\Pi_1\vartheta(\tau_1,\varepsilon))]_{2n-2}$$
$$+[\Delta_n\Pi_1\vartheta(\tau_1,\varepsilon)'(\{\varepsilon\widehat{\Pi}_{1(n-2)}f_\vartheta(\tau_1,\varepsilon)'\widetilde{\overline{\psi}}_{n-2}(\varepsilon^2\tau_1,\varepsilon)\}_{n-2}$$
$$+\{\varepsilon\widehat{\Pi}_{1(n-2)}f_\vartheta(\tau_1,\varepsilon)'(\varepsilon E_1+E_2+E_3)\widetilde{\Pi}_{0(n-2)}\psi(\varepsilon\tau_1,\varepsilon)\}_{n-2}+\{\varepsilon f_\vartheta(\widetilde{\overline{\vartheta}}_{n-2}(\varepsilon^2\tau_1,\varepsilon)$$
$$+\widetilde{\Pi}_{0(n-2)}\vartheta(\varepsilon\tau_1,\varepsilon)+\widetilde{\Pi}_{1(n-2)}\vartheta(\tau_1,\varepsilon),\varepsilon^2\tau_1,\varepsilon)'(\varepsilon^2E_1+\varepsilon E_2$$
$$+E_3)\widetilde{\Pi}_{1(n-2)}\psi(\tau_1,\varepsilon)\}_{n-2})]_{2n-2})\,d\tau_1.$$

Summing up the obtained terms of transformed expressions and considering separately four groups of terms, depending on $\Delta_n\overline{\vartheta}$, $\Delta_n\Pi_0\vartheta$, $\Delta_n\Pi_1\vartheta$, and without these variables, we can write out the transformed forth term of J_{2n} in the following form:

$$-[\Delta_n\Pi_1 w(0,\varepsilon)'(\frac{1}{\varepsilon}E_1+E_2+\varepsilon E_3)\widetilde{\overline{\psi}}_{n-1}(0,\varepsilon)]_{2n-1}-\Pi_{1(n-1)}w(0)'E_3(\overline{\psi}_{n-1}(0)$$
$$+\Pi_{0(n-1)}\psi(0))-[\Delta_n\Pi_1 w(0,\varepsilon)'(E_1+E_2+\varepsilon E_3)\widetilde{\Pi}_{0(n-1)}\psi(0,\varepsilon)]_{2n-1}$$
$$-[(\Delta_n\overline{w}(0,\varepsilon)+\Delta_n\Pi_0 w(0,\varepsilon)+\Delta_n\Pi_1 w(0,\varepsilon))'\widetilde{\Pi}_{1(n-2)}\psi(0,\varepsilon)]_{2n-2}+\Pi_{1(n-1)}J.$$

Transforming the fifth term in J_{2n} in the same way, we write the final result as

$$[\Delta_n Q_1 w(0,\varepsilon)'(\frac{1}{\varepsilon}E_1+E_2+\varepsilon E_3)\widetilde{\overline{\psi}}_{n-1}(T,\varepsilon)]_{2n-1}$$
$$+[\Delta_n Q_1 w(0,\varepsilon)'(E_1+E_2+\varepsilon E_3)\widetilde{Q}_{0(n-1)}\psi(0,\varepsilon)]_{2n-1}$$
$$+[(\Delta_n\overline{w}(T,\varepsilon)+\Delta_n Q_0 w(0,\varepsilon)+\Delta_n Q_1 w(0,\varepsilon))'\widetilde{Q}_{1(n-2)}\psi(0,\varepsilon)]_{2n-2}+Q_{1(n-1)}J.$$

Substituting $w(t,\varepsilon)$ from (35) in (3), we obtain the relation

$$\widetilde{\overline{w}}_{j-1}(0,\varepsilon)+\Delta_j\overline{w}(0,\varepsilon)+\sum_{i=0}^{1}(\widetilde{\Pi}_{i(j-1)}w(0,\varepsilon)+\Delta_j\Pi_i w(0,\varepsilon)$$
$$+\widetilde{Q}_{i(j-1)}w(-T/\varepsilon^{i+1},\varepsilon)+\Delta_j Q_i w(-T/\varepsilon^{i+1},\varepsilon))=w^0. \tag{47}$$

Summing up (20) and the second relations in (23), (26), we obtain the equality

$$\overline{\psi}_j(T)+E_1(Q_{0(j-1)}\psi(0)+Q_{1(j-2)}\psi(0))$$
$$+E_2(Q_{0j}\psi(0)+Q_{1(j-1)}\psi(0))+E_3(Q_{0j}\psi(0)+Q_{1j}\psi(0))=0.$$

Multiplying this equation by ε^j, $j=\overline{0,n-1}$, and summing up the obtained results, we have

$$\widetilde{\overline{\psi}}_{n-1}(T)+\varepsilon E_1(\widetilde{Q}_{0(n-2)}\psi(0)+\varepsilon\widetilde{Q}_{1(n-3)}\psi(0))$$
$$+E_2(\widetilde{Q}_{0(n-1)}\psi(0)+\varepsilon\widetilde{Q}_{1(n-2)}\psi(0))+E_3(\widetilde{Q}_{0(n-1)}\psi(0)+\widetilde{Q}_{1(n-1)}\psi(0))=0. \tag{48}$$

Summing up the remaining parts of the transformed terms for J_{2n}, applying (47), (48) to the non-integrand terms, taking into account Remark 1 and omitting known terms, we finally have $\overline{J}_n + \Pi_{1(n-1)}J + Q_{1(n-1)}J$, which proves Theorem 1 for J_{2n}.

In addition to the previous assumption on solvability of the problems $\overline{P}_j$, $\Pi_{0j}P$, $Q_{0j}P$ and $\Pi_{1(j-1)}P$, $Q_{1(j-1)}P$, $j = \overline{0, n-1}$, we will assume that the problems $\overline{P}_n$, $\Pi_{1(n-1)}P$, $Q_{1(n-1)}P$ have been solved.

Let us consider the coefficient J_{2n+1} having the form (34) with $k = 2n+1$.

We transform separately the terms of J_{2n+1} using the previous algorithm for transforming similar terms in J_{2n}. Summing up the obtained expressions for five terms and dropping known terms, we obtain the sum of the performance indices $\Pi_{0n}J + Q_{0n}J$.

Thus, Theorem 1 is completely proved. □

5. Asymptotic Estimates

Suppose that the problems $\overline{P}_j$, $\Pi_{ij}P$, $Q_{ij}P$, $i = 0, 1$, $j = \overline{0, n}$, have been solved. We will prove asymptotic estimates of the proximity between the asymptotic solution obtained by the direct scheme method $\widetilde{\vartheta}_n(t,\varepsilon) = \sum_{j=0}^{n} \varepsilon^j(\overline{\vartheta}_j(t) + \sum_{i=0}^{1} \Pi_{ij}\vartheta(\tau_i) + Q_{ij}\vartheta(\sigma_i))$ and the exact solution of the problem P_ε.

We will use here the notation for asymptotics remainder terms

$$
\begin{gathered}
r_n w = w - \widetilde{w}_n = (r_n x', r_n y', r_n z')', r_n u = u - \widetilde{u}_n, r_n\vartheta = \vartheta - \widetilde{\vartheta}_n = (r_n w', r_n u')', \\
r_n\varphi = \varphi - \widetilde{\varphi}_n = (r_n\zeta', r_n\eta', r_n\theta')', \ \widetilde{X}_n = \begin{pmatrix} \widetilde{x}_n \\ \widetilde{\zeta}_n \end{pmatrix}, \\
\widetilde{Y}_n = \begin{pmatrix} \widetilde{y}_n \\ \widetilde{\eta}_n \end{pmatrix}, \ \widetilde{Z}_n = \begin{pmatrix} \widetilde{z}_n \\ \widetilde{\theta}_n \end{pmatrix}, \ r_n X = \begin{pmatrix} r_n x \\ r_n\zeta \end{pmatrix}, \ r_n Y = \begin{pmatrix} r_n y \\ r_n\eta \end{pmatrix}, \ r_n Z = \begin{pmatrix} r_n z \\ r_n\theta \end{pmatrix}.
\end{gathered}
\tag{49}
$$

In comparison with the notation in the previous section, we have, e.g., $r_n u = \Delta_{n+1}u$ and so on.

Since the matrix $R(t)$ is positive definite, we obtain from (27) the following relation

$$
u(t,\varepsilon) = R(t)^{-1}B(t)'\varphi + \varepsilon R(t)^{-1}(f_u(\vartheta,t,\varepsilon)'\varphi - F_u(\vartheta,t,\varepsilon)').
$$

Taking into account this equality and substituting the expressions for $\vartheta(t,\varepsilon)$, $\varphi(t,\varepsilon)$ from (49) into (2), (27) and (28), we obtain the equations for the remainders

$$
r_n u = \overset{(1)}{\mathcal{A}}(t) r_n X + \overset{(1)}{\mathcal{B}}(t) r_n Y + \overset{(1)}{\mathcal{C}}(t) r_n Z + \overset{(1)}{g}(r_n\vartheta, r_n\varphi, t, \varepsilon), \tag{50}
$$

$$
\frac{d r_n X}{dt} = \overset{(2)}{\mathcal{A}}(t) r_n X + \overset{(2)}{\mathcal{B}}(t) r_n Y + \overset{(2)}{\mathcal{C}}(t) r_n Z + \overset{(2)}{g}(r_n\vartheta, r_n\varphi, t, \varepsilon), \tag{51}
$$

$$
\varepsilon\frac{d r_n Y}{dt} = \overset{(3)}{\mathcal{A}}(t) r_n X + \overset{(3)}{\mathcal{B}}(t) r_n Y + \overset{(3)}{\mathcal{C}}(t) r_n Z + \overset{(3)}{g}(r_n\vartheta, r_n\varphi, t, \varepsilon), \tag{52}
$$

$$
\varepsilon^2\frac{d r_n Z}{dt} = \overset{(4)}{\mathcal{A}}(t) r_n X + \overset{(4)}{\mathcal{B}}(t) r_n Y + \overset{(4)}{\mathcal{C}}(t) r_n Z + \overset{(4)}{g}(r_n\vartheta, r_n\varphi, t, \varepsilon), \tag{53}
$$

where

$$
\overset{(1)}{\mathcal{A}} = (0 \ \ R^{-1}\overset{(1)}{B'}), \ \overset{(1)}{\mathcal{B}} = (0 \ \ R^{-1}\overset{(2)}{B'}), \ \overset{(1)}{\mathcal{C}} = (0 \ \ R^{-1}\overset{(3)}{B'}),
$$

$$
\overset{(2)}{\mathcal{A}} = \begin{pmatrix} A_{11} & S_{11} \\ W_{11} & -A'_{11} \end{pmatrix}, \overset{(2)}{\mathcal{B}} = \begin{pmatrix} A_{12} & S_{12} \\ W_{12} & -A'_{21} \end{pmatrix}, \overset{(2)}{\mathcal{C}} = \begin{pmatrix} A_{13} & S_{13} \\ W_{13} & -A'_{31} \end{pmatrix},
$$

$$
\overset{(3)}{\mathcal{A}} = \begin{pmatrix} A_{21} & S'_{12} \\ W'_{12} & -A'_{12} \end{pmatrix}, \overset{(3)}{\mathcal{B}} = \begin{pmatrix} A_{22} & S_{22} \\ W_{22} & -A'_{22} \end{pmatrix}, \overset{(3)}{\mathcal{C}} = \begin{pmatrix} A_{23} & S_{23} \\ W_{23} & -A'_{32} \end{pmatrix},
$$

$$\overset{(4)}{\mathcal{A}} = \begin{pmatrix} A_{31} & S'_{13} \\ W'_{13} & -A'_{13} \end{pmatrix}, \overset{(4)}{\mathcal{B}} = \begin{pmatrix} A_{32} & S'_{23} \\ W'_{23} & -A'_{23} \end{pmatrix}, \overset{(4)}{\mathcal{C}} = \begin{pmatrix} A_{33} & S_{33} \\ W_{33} & -A'_{33} \end{pmatrix},$$

$$\overset{(1)}{g}(r_n\vartheta, r_n\varphi, t, \varepsilon) = R(t)^{-1}B(t)'\widetilde{\varphi}_n + \widetilde{u}_n$$
$$+\varepsilon R(t)^{-1}(f_u(\widetilde{\vartheta}_n + r_n\vartheta, t, \varepsilon)'(\widetilde{\varphi}_n + r_n\varphi) - F_u(\widetilde{\vartheta}_n + r_n\vartheta, t, \varepsilon)'),$$

$$\overset{(2)}{g}(r_n\vartheta, r_n\varphi, t, \varepsilon) = \overset{(2)}{\mathcal{A}}(t)\widetilde{X}_n + \overset{(2)}{\mathcal{B}}(t)\widetilde{Y}_n + \overset{(2)}{\mathcal{C}}(t)\widetilde{Z}_n$$
$$-d\widetilde{X}_n/dt + \varepsilon \overset{(2)}{h}(\widetilde{\vartheta}_n + r_n\vartheta, \widetilde{\varphi}_n + r_n\varphi, t, \varepsilon),$$

$$\overset{(3)}{g}(r_n\vartheta, r_n\varphi, t, \varepsilon) = \overset{(3)}{\mathcal{A}}(t)\widetilde{X}_n + \overset{(3)}{\mathcal{B}}(t)\widetilde{Y}_n + \overset{(3)}{\mathcal{C}}(t)\widetilde{Z}_n$$
$$-\varepsilon d\widetilde{Y}_n/dt + \varepsilon \overset{(3)}{h}(\widetilde{\vartheta}_n + r_n\vartheta, \widetilde{\varphi}_n + r_n\varphi, t, \varepsilon),$$

$$\overset{(4)}{g}(r_n\vartheta, r_n\varphi, t, \varepsilon) = \overset{(4)}{\mathcal{A}}(t)\widetilde{X}_n + \overset{(4)}{\mathcal{B}}(t)\widetilde{Y}_n + \overset{(4)}{\mathcal{C}}(t)\widetilde{Z}_n$$
$$-\varepsilon^2 d\widetilde{Z}_n/dt + \varepsilon \overset{(4)}{h}(\widetilde{\vartheta}_n + r_n\vartheta, \widetilde{\varphi}_n + r_n\varphi, t, \varepsilon),$$

$$\overset{(2)}{h}(\widetilde{\vartheta}_n + r_n\vartheta, \widetilde{\varphi}_n + r_n\varphi, t, \varepsilon) = ((\overset{(1)}{B}(t)R(t)^{-1}(f_u(\widetilde{\vartheta}_n + r_n\vartheta, t, \varepsilon)'(\widetilde{\varphi}_n + r_n\varphi) - F_u(\widetilde{\vartheta}_n$$
$$+r_n\vartheta, t, \varepsilon)') + \overset{(1)}{f}(\widetilde{\vartheta}_n + r_n\vartheta, t, \varepsilon))', (F_x(\widetilde{\vartheta}_n + r_n\vartheta, t, \varepsilon)' - \overset{(1)}{f_x}(\widetilde{\vartheta}_n + r_n\vartheta, t, \varepsilon)'(\widetilde{\varphi}_n + r_n\varphi))')',$$

$$\overset{(3)}{h}(\widetilde{\vartheta}_n + r_n\vartheta, \widetilde{\varphi}_n + r_n\varphi, t, \varepsilon) = ((\overset{(2)}{B}(t)R(t)^{-1}(f_u(\widetilde{\vartheta}_n + r_n\vartheta, t, \varepsilon)'(\widetilde{\varphi}_n + r_n\varphi) - F_u(\widetilde{\vartheta}_n$$
$$+r_n\vartheta, t, \varepsilon)') + \overset{(2)}{f}(\widetilde{\vartheta}_n + r_n\vartheta, t, \varepsilon))', (F_y(\widetilde{\vartheta}_n + r_n\vartheta, t, \varepsilon)' - \overset{(2)}{f_y}(\widetilde{\vartheta}_n + r_n\vartheta, t, \varepsilon)'(\widetilde{\varphi}_n + r_n\varphi))')',$$

$$\overset{(4)}{h}(\widetilde{\vartheta}_n + r_n\vartheta, \widetilde{\varphi}_n + r_n\varphi, t, \varepsilon) = ((\overset{(3)}{B}(t)R(t)^{-1}(f_u(\widetilde{\vartheta}_n + r_n\vartheta, t, \varepsilon)'(\widetilde{\varphi}_n + r_n\varphi) - F_u(\widetilde{\vartheta}_n$$
$$+r_n\vartheta, t, \varepsilon)') + \overset{(3)}{f}(\widetilde{\vartheta}_n + r_n\vartheta, t, \varepsilon))', (F_z(\widetilde{\vartheta}_n + r_n\vartheta, t, \varepsilon)' - \overset{(3)}{f_z}(\widetilde{\vartheta}_n + r_n\vartheta, t, \varepsilon)'(\widetilde{\varphi}_n + r_n\varphi))')'.$$

For brevity, the arguments t, ε are dropped in some of the last relations.

In view of the algorithm of asymptotics construction, namely Equalities (12)–(17) and (33), we obtain the boundary conditions

$$\begin{aligned} r_n w(0,\varepsilon) &= -\widetilde{Q}_{0n} w(-T/\varepsilon,\varepsilon) - \widetilde{Q}_{1n} w(-T/\varepsilon^2,\varepsilon), \\ r_n\varphi(T,\varepsilon) &= -\widetilde{\Pi}_{0n}\varphi(T/\varepsilon,\varepsilon) - \widetilde{\Pi}_{1n}\varphi(T/\varepsilon^2,\varepsilon). \end{aligned} \tag{54}$$

Using variables' changes,

$$\rho_n w(t,\varepsilon) = r_n w(t,\varepsilon) - r_n w(0,\varepsilon), \ \rho_n\varphi(t,\varepsilon) = r_n\varphi(t,\varepsilon) - r_n\varphi(T,\varepsilon) \tag{55}$$

and the notation $\rho_n v(t,\varepsilon) = (r_n u', \rho_n w', \rho_n\varphi')'$, system (50)–(54) can be written as

$$r_n u = \overset{(1)}{\mathcal{A}}(t)\rho_n X + \overset{(1)}{\mathcal{B}}(t)\rho_n Y + \overset{(1)}{\mathcal{C}}(t)\rho_n Z + \overset{(1)}{\chi}(\rho_n v, t, \varepsilon), \tag{56}$$

$$\frac{d\rho_n X}{dt} = \overset{(2)}{\mathcal{A}}(t)\rho_n X + \overset{(2)}{\mathcal{B}}(t)\rho_n Y + \overset{(2)}{\mathcal{C}}(t)\rho_n Z + \overset{(2)}{\chi}(\rho_n v, t, \varepsilon), \tag{57}$$

$$\varepsilon\frac{d\rho_n Y}{dt} = \overset{(3)}{\mathcal{A}}(t)\rho_n X + \overset{(3)}{\mathcal{B}}(t)\rho_n Y + \overset{(3)}{\mathcal{C}}(t)\rho_n Z + \overset{(3)}{\chi}(\rho_n v, t, \varepsilon), \tag{58}$$

$$\varepsilon^2\frac{d\rho_n Z}{dt} = \overset{(4)}{\mathcal{A}}(t)\rho_n X + \overset{(4)}{\mathcal{B}}(t)\rho_n Y + \overset{(4)}{\mathcal{C}}(t)\rho_n Z + \overset{(4)}{\chi}(\rho_n v, t, \varepsilon), \tag{59}$$

$$\rho_n w(0,\varepsilon) = 0,\ \rho_n \varphi(T,\varepsilon) = 0, \tag{60}$$

where

$$\rho_n X = \begin{pmatrix} \rho_n x \\ \rho_n \zeta \end{pmatrix},\ \rho_n Y = \begin{pmatrix} \rho_n y \\ \rho_n \eta \end{pmatrix},\ \rho_n Z = \begin{pmatrix} \rho_n z \\ \rho_n \theta \end{pmatrix},$$

$$\overset{(i)}{\chi}(\rho_n v, t, \varepsilon) = \overset{(i)}{g}(r_n u(t,\varepsilon), \rho_n w(t,\varepsilon) + r_n w(0,\varepsilon), \rho_n \varphi(t,\varepsilon) + r_n \varphi(T,\varepsilon), t, \varepsilon) +$$
$$+ \overset{(i)}{\mathcal{A}}(t)\begin{pmatrix} r_n x(0,\varepsilon) \\ r_n \zeta(T,\varepsilon) \end{pmatrix} + \overset{(i)}{\mathcal{B}}(t)\begin{pmatrix} r_n y(0,\varepsilon) \\ r_n \eta(T,\varepsilon) \end{pmatrix} + \overset{(i)}{\mathcal{C}}(t)\begin{pmatrix} r_n z(0,\varepsilon) \\ r_n \theta(T,\varepsilon) \end{pmatrix},\ i = \overline{1,4}.$$

Taking into account the algorithm of the asymptotics construction and the form of the functions $\overset{(i)}{\chi}$, $i = \overline{1,4}$, we obtain two important properties, namely:

(1) for $t \in [0,T]$, $0 < \varepsilon \leqslant \varepsilon_0$, the following inequalities take place

$$\begin{gathered} \|\overset{(i)}{\chi}(0,t,\varepsilon)\| \le c\varepsilon^{n+1},\ i = 1,4,\ \|\overset{(3)}{\chi}(0,t,\varepsilon)\| \le c(\varepsilon^{n+1} + \varepsilon^n \exp(-æt/\varepsilon^2) \\ +\varepsilon^n \exp(æ(t-T)/\varepsilon^2)),\ \|\overset{(2)}{\chi}(0,t,\varepsilon)\| \le c(\varepsilon^{n+1} + \varepsilon^n \exp(-æt/\varepsilon) \\ +\varepsilon^n \exp(æ(t-T)/\varepsilon) + \varepsilon^{n-1}\exp(-æt/\varepsilon^2) + \varepsilon^{n-1}\exp(æ(t-T)/\varepsilon^2)), \end{gathered} \tag{61}$$

where c and æ are positive constants independent of t, ε,

(2) for any $q > 0$, there exist such constants $\delta = \delta(q)$ and $\varepsilon_0 = \varepsilon_0(q)$ that, for $\|v_i\|_{C_{[0,T]}} \leqslant \delta$, $i = 1,2$, $0 < \varepsilon \leqslant \varepsilon_0$

$$\|\overset{(i)}{\chi}(v_1,t,\varepsilon) - \overset{(i)}{\chi}(v_2,t,\varepsilon)\|_{C_{[0,T]}} \leqslant q\|v_1 - v_2\|_{C_{[0,T]}},\ i = \overline{1,4}. \tag{62}$$

It follows from the form of the matrix $\overset{(4)}{\mathcal{C}}(t)$ that the boundary value problem

$$\varepsilon^2\frac{dZ}{dt} = \overset{(4)}{\mathcal{C}}(t)Z,\ Z = (Z_1', Z_2')',\ Z_1(0) = 0,\ Z_2(T) = 0, \tag{63}$$

is uniquely solvable [30]. Therefore, there exists a matrix Green function $\overset{(4)}{G}(t,s,\varepsilon)$ for this problem.

For eigenvalues of the matrix $\overset{(4)}{\mathcal{C}}(t)$, we suppose the condition:
I. $\lambda_i(t) \neq \lambda_j(t)$ for $i \neq j$, $t \in [0,T]$.

Then, in the matrix $\mathfrak{B} = \begin{pmatrix} \mathfrak{B}_{11} & \mathfrak{B}_{12} \\ \mathfrak{B}_{21} & \mathfrak{B}_{22} \end{pmatrix}$, consisting of eigenvectors of the matrix $\overset{(4)}{\mathcal{C}}(t)$, the matrices $\mathfrak{B}_{ii}$, $i = 1,2$, are nondegenerate. Hence, the condition 4^0 from ([28], c.125) is valid and therefore due to ([28], n. 9) for sufficiently small $\varepsilon > 0$ the matrix Green function $\overset{(4)}{G}(t,s,\varepsilon)$ satisfies the inequality

$$\|\overset{(4)}{G}(t,s,\varepsilon)\| \le c\exp(-æ|t-s|/\varepsilon^2),\ t,s \in [0,T]. \tag{64}$$

Furthermore, we need the following three lemmas.

Lemma 1. *If $G(t,s)$ is a matrix Green function of the boundary value problem*

$$\frac{dx}{dt} = A(t)x + f(t), t \in [0,T],\ Px(0) = 0,\ (I-P)x(T) = 0,$$

where the matrix $A(t)$ is continuous with respect to t and invertible for all $t \in [0,T]$, and P is a projector, then

$$\frac{\partial G(t,s)}{\partial t} = -A(t)\frac{\partial G(t,s)}{\partial s}A(s)^{-1},\ t \neq s.$$

The proof of this lemma is given in [24]. It follows from the explicit form for the matrix Green function

$$G(t,s) = \begin{cases} -V(t,0)((I-P)V(T,0)(I-P))^{-1}(I-P)V(T,s),\ t \leqslant s, \\ V(t,s) - V(t,0)((I-P)V(T,0)(I-P))^{-1}(I-P)V(T,s),\ t \geqslant s, \end{cases} \tag{65}$$

where $V(t,s) = V(t)V(s)^{-1}$, $dV(t,s)/dt = A(t)V(t,s)$, $V(s,s) = I$.

Lemma 2. *The boundary value problem*

$$\begin{aligned} \mathcal{E}(0)\frac{d\overline{w}}{dt} &= A(t)\overline{w} + S(t)\overline{\varphi}, \\ \mathcal{E}(0)\frac{d\overline{\varphi}}{dt} &= W(t)\overline{w} - A(t)'\overline{\varphi}, \\ \mathcal{E}(0)\overline{w}(0) = 0,&\ \mathcal{E}(0)\overline{\varphi}(T) = 0, \end{aligned}$$

where $\overline{w} = (\overline{x}', \overline{y}', \overline{z}')'$, $\overline{\varphi} = (\overline{\xi}', \overline{\eta}', \overline{\theta}')'$, is uniquely solvable.

Proof. Multiply scalarly the first equation of the considered system by $\overline{\varphi}$ and the second equation in this system by $\overline{x}$. Adding the obtained results, we have $d/dt(\overline{\varphi}'\mathcal{E}(0)\overline{w}) = \overline{\varphi}'S(t)\overline{\varphi} + \overline{w}'W(t)\overline{w}$. Integrating this equality over the interval $[0,T]$, in view of the boundary values, we obtain $\int_0^T(\overline{\varphi}'S(t)\overline{\varphi} + \overline{w}'W(t)\overline{w})\,dt = 0$. Taking into account the positive definiteness of $S(t)$ and $W(t)$, we obtain $\overline{w}(t) = \overline{\varphi}(t) = 0$, i.e., the unique solvability is proved. □

Lemma 3. *If $\mathcal{G}$ is a contractive mapping in a Banach space X, $x_0 = 0$, $x_k = \mathcal{G}(x_{k-1})$, $k = 1,2,...,$ and $\|x_1\| \leqslant a$, then $\|x_k\| \leqslant a/(1-q)$.*

See the proof of this lemma in [24].

Theorem 2. *Solution $\vartheta_*(t,\varepsilon)$ of problem P_ε for sufficiently small $\varepsilon > 0$, $t \in [0,T]$, satisfy the inequality*

$$\|\vartheta_*(t,\varepsilon) - \widetilde{\vartheta}_n(t,\varepsilon)\| \leq c\varepsilon^{n+1}.$$

Proof. The proof of this theorem is based on transforming systems (56)–(59) with boundary values (60) to a system of integral equations, using estimates for matrix Green functions and applying to the obtained system the principle of contractive mappings.

Using Green function $\overset{(4)}{G}(t,s,\varepsilon)$, we have from (59) the integral equation

$$\rho_n Z(t,\varepsilon) = \frac{1}{\varepsilon^2}\int_0^T \overset{(4)}{G}(t,s,\varepsilon)(\overset{(4)}{\mathcal{A}}(s)\rho_n X + \overset{(4)}{\mathcal{B}}(s)\rho_n Y)\,ds + \overset{(4)}{\mathcal{G}}(\rho_n\vartheta, t, \varepsilon), \tag{66}$$

where $\overset{(4)}{\mathcal{G}}(\rho_n v, t, \varepsilon) = 1/\varepsilon^2 \int_0^T \overset{(4)}{G}(t,s,\varepsilon)\overset{(4)}{\chi}(\rho_n v, s, \varepsilon)\, ds$.

In view of (61), (62), and (64), the function $\overset{(4)}{\mathcal{G}}(\rho_n \vartheta, t, \varepsilon)$ satisfies the properties (2) and (3) $\overset{(4)}{\mathcal{G}}(0,t,\varepsilon) \leqslant c\varepsilon^{n+1}$.

Furthermore, we will denote functions, appearing under transformations of the problems (56)–(60) with the properties (2) and (3), by $\overset{(j)}{\mathcal{G}}(\vartheta, t, \varepsilon)$, $j = \overline{1,4}$. Specific forms of these functions are omitted since they are insignificant for the proof.

In transforming (66), we will use the formula following from (63) and Lemma 1

$$\overset{(4)}{G}(t,s,\varepsilon) = -\varepsilon^2 \frac{\partial}{\partial s}(\overset{(4)}{G}(t,s,\varepsilon)\overset{(4)}{\mathcal{C}}(s)^{-1}) + \varepsilon^2 \overset{(4)}{G}(t,s,\varepsilon)\frac{d}{ds}(\overset{(4)}{\mathcal{C}}(s)^{-1}),\ t \neq s. \tag{67}$$

We present the integral in (66), containing the first term on the right side in (67) as the sum of integrals over the intervals $[0,t]$ and $[t,T]$ and integrate by parts. Taking into account the jump of function $\overset{(4)}{G}(t,s,\varepsilon)$ at $s = t$, i.e., the equality $\overset{(4)}{G}(t,t+0,\varepsilon) - \overset{(4)}{G}(t,t-0,\varepsilon) \equiv -I_{2n_3}$, following from (65), and estimate (64), we obtain

$$\begin{aligned}\rho_n Z(t,\varepsilon) = &-\overset{(4)}{\mathcal{C}}(t)^{-1}(\overset{(4)}{\mathcal{A}}(t)\rho_n X(t,\varepsilon) + \overset{(4)}{\mathcal{B}}(t)\rho_n Y(t,\varepsilon))\\ &+\overset{(4)}{G}(t,0,\varepsilon)\overset{(4)}{\mathcal{C}}(0)^{-1}(\overset{(4)}{\mathcal{A}}(0)\rho_n X(0,\varepsilon) + \overset{(4)}{\mathcal{B}}(0)\rho_n Y(0,\varepsilon))\\ &-\overset{(4)}{G}(t,T,\varepsilon)\overset{(4)}{\mathcal{C}}(T)^{-1}(\overset{(4)}{\mathcal{A}}(T)\rho_n X(T,\varepsilon) + \overset{(4)}{\mathcal{B}}(T)\rho_n Y(T,\varepsilon)) + \overset{(4)}{\mathcal{G}}(\rho_n v, t, \varepsilon).\end{aligned} \tag{68}$$

Substitute (68) into (58). Introducing the notation $\Lambda(t) = \overset{(3)}{\mathcal{B}}(t) - \overset{(3)}{\mathcal{C}}(t)\overset{(4)}{\mathcal{C}}(t)^{-1}\overset{(4)}{\mathcal{B}}(t)$, we write the obtained equation in the following way:

$$\begin{aligned}\varepsilon\frac{d\rho_n Y}{dt} = &(\overset{(3)}{\mathcal{A}}(t) - \overset{(3)}{\mathcal{C}}(t)\overset{(4)}{\mathcal{C}}(t)^{-1}\overset{(4)}{\mathcal{A}}(t))\rho_n X(t,\varepsilon) + \Lambda(t)\rho_n Y(t,\varepsilon)\\ &+\overset{(3)}{\mathcal{C}}(t)(\overset{(4)}{G}(t,0,\varepsilon)\overset{(4)}{\mathcal{C}}(0)^{-1}(\overset{(4)}{\mathcal{A}}(0)\rho_n X(0,\varepsilon) + \overset{(4)}{\mathcal{B}}(0)\rho_n Y(0,\varepsilon))\\ &-\overset{(4)}{G}(t,T,\varepsilon)\overset{(4)}{\mathcal{C}}(T)^{-1}(\overset{(4)}{\mathcal{A}}(T)\rho_n X(T,\varepsilon) + \overset{(4)}{\mathcal{B}}(T)\rho_n Y(T,\varepsilon))) + \overset{(3)}{\mathcal{G}}(\rho_n \vartheta, t, \varepsilon).\end{aligned} \tag{69}$$

Let us study the structure of the matrix $\Lambda = \Lambda(t)$.

For brevity, we will sometimes omit the argument t. Due to our assumption, it follows from [30] that the Hamiltonian matrix $\overset{(4)}{\mathcal{C}}(t)$ is invertible and its inverse has the form $\overset{(4)}{\mathcal{C}}{}^{-1} = \begin{pmatrix} D_1 & D_2 \\ D_3 & -D_1' \end{pmatrix}$, where D_2 and D_3 are symmetric. Similarly to the proof in [30] of the non-negative definiteness of the matrices D_2 and D_3, it is proved that, in view of the positive definiteness of S_{33} and W_{33}, the matrices D_2 and D_3 are also positive definite.

Let the matrix $\Lambda(t)$ have the block presentation $\begin{pmatrix} \Lambda_1(t) & \Lambda_2(t) \\ \Lambda_3(t) & \Lambda_4(t) \end{pmatrix}$. Write out the explicit expressions for $\Lambda_i(t)$, $i = \overline{1,4}$:

$$\begin{aligned}\Lambda_1 &= A_{22} - A_{23}(D_1 A_{32} + D_2 W_{23}') - S_{23}(D_3 A_{32} - D_1' W_{23}'),\\ \Lambda_2 &= S_{22} - A_{23}(D_1 S_{23}' - D_2 A_{23}') - S_{23}(D_3 S_{23}' + D_1' A_{23}'),\\ \Lambda_3 &= W_{22} - W_{23}(D_1 A_{32} + D_2 W_{23}') + A_{32}'(D_3 A_{32} - D_1' W_{23}'),\\ \Lambda_4 &= -A_{22}' - W_{23}(D_1 S_{23}' - D_2 A_{23}') + A_{32}'(D_3 S_{23}' + D_1' A_{23}').\end{aligned}$$

Comparing $\Lambda_1(t)$ with $\Lambda_4(t)$, it is not difficult to see that $\Lambda_4(t) = -\Lambda_1(t)'$. It also follows from the form of the matrices $\Lambda_2(t)$ and $\Lambda_3(t)$ that these matrices are symmetric.

Introducing for an arbitrary $b \in \mathbb{R}^{n_2}$ the notation

$$b_1 = A'_{23}b,\ b_2 = S'_{23}b,\ b_3 = \begin{pmatrix} b \\ -(D'_1 b_1 + D_3 b_2) \end{pmatrix},$$
$$b_4 = A_{32}b,\ b_5 = W'_{23}b,\ b_6 = \begin{pmatrix} b \\ -(D_1 b_4 + D_2 b_5) \end{pmatrix},$$

we obtain

$$b'\Lambda_2 b = b'_3 \begin{pmatrix} S_{22} & S_{23} \\ S'_{23} & S_{33} \end{pmatrix} b_3 + (D_1 b_2 - D_2 b_1)' W_{33} (D_1 b_2 - D_2 b_1)$$
$$+ b'_1 (D_2 - D_2 W_{33} D_2 - D_1 S_{33} D'_1) b_1 + 2 b'_1 (D_2 W_{33} D_1 - D_1 S_{33} D_3) b_2$$
$$+ b'_2 (D_3 - D'_1 W_{33} D_1 - D_3 S_{33} D_3) b_2,$$
$$b'\Lambda_3 b = b'_6 \begin{pmatrix} W_{22} & W_{23} \\ W'_{23} & W_{33} \end{pmatrix} b_6 + (D_3 b_4 - D'_1 b_5)' S_{33} (D_3 b_4 - D'_1 b_5)$$
$$+ b'_4 (D_3 - D_3 S_{33} D_3 - D'_1 W_{33} D_1) b_4 + 2 b'_4 (D_3 S_{33} D'_1 - D'_1 W_{33} D_2) b_5$$
$$+ b'_5 (D_2 - D_1 S_{33} D'_1 - D_2 W_{33} D_2) b_5.$$

Taking into account the equalities

$$A_{33} D_1 + S_{33} D_3 = I_{n_3},\ A_{33} D_2 - S_{33} D'_1 = 0,$$
$$W_{33} D_1 - A'_{33} D_3 = 0,\ W_{33} D_2 + A'_{33} D'_1 = I_{n_3},$$

we obtain that three last summands in the expressions for $b'\Lambda_2 b$ and $b'\Lambda_3 b$ are equal to zero.

In view of positive definiteness of matrices $S(t)$ and $W(t)$, the matrices $\begin{pmatrix} S_{22} & S_{23} \\ S'_{23} & S_{33} \end{pmatrix}$, S_{33}, $\begin{pmatrix} W_{22} & W_{23} \\ W'_{23} & S_{33} \end{pmatrix}$, W_{33} are positive definite too. Then, the positive definiteness of matrices $\Lambda_2(t)$ and $\Lambda_3(t)$ follows from the obtained forms for $b'\Lambda_2(t)b$ and $b'\Lambda_3 b$.

Thus, the matrix $\Lambda(t)$ has the form $\begin{pmatrix} \Lambda_1(t) & \Lambda_2(t) \\ \Lambda_3(t) & -\Lambda_1(t)' \end{pmatrix}$, where $\Lambda_2(t)$ and $\Lambda_3(t)$ are positive definite.

We will suppose yet one condition

II. Eigenvalues of the matrix $\Lambda(t)$ satisfy the condition *I*.

Then, the boundary value problem

$$\varepsilon \frac{dY}{dt} = \Lambda(t)Y,\ Y = (Y'_1, Y'_2)',\ Y_1(0) = 0,\ Y_2(T) = 0 \tag{70}$$

has a unique solution and, for the corresponding matrix Green function $\overset{(3)}{G}(t,s,\varepsilon)$, the following inequality is valid

$$\|\overset{(3)}{G}(t,s,\varepsilon)\| \leqslant c \exp(-æ|t-s|/\varepsilon), t, s \in [0,T]. \tag{71}$$

With the help of the Green function $\overset{(3)}{G}(t,s,\varepsilon)$, using (64), (71), we obtain from (69) the following

$$\rho_n Y(t,\varepsilon) = \frac{1}{\varepsilon} \int_0^T \overset{(3)}{G}(t,s,\varepsilon)(\overset{(3)}{\mathcal{A}}(s) - \overset{(3)}{C}(s)\overset{(4)}{C}(s)^{-1}\overset{(4)}{\mathcal{A}}(s))\rho_n X(s,\varepsilon)\, ds + \overset{(3)}{\mathcal{G}}(\rho_n \vartheta, t, \varepsilon). \tag{72}$$

The following formula follows from Lemma 1 and (70)

$$\overset{(3)}{G}(t,s,\varepsilon) = -\varepsilon\frac{\partial}{\partial s}(\overset{(3)}{G}(t,s,\varepsilon)\Lambda(s)^{-1}) + \varepsilon\overset{(3)}{G}(t,s,\varepsilon)\frac{d}{ds}(\Lambda(s)^{-1}),\ t \neq s. \tag{73}$$

Using this formula, we present the integral in (72), containing the first term from the right side (73), as a sum of integrals over the intervals $[0,t]$ and $[t,T]$ and integrate by parts. In view of the jump of function $\overset{(3)}{G}(t,s,\varepsilon)$ at $s = t$ and estimate (71), we have

$$\begin{aligned} \rho_n Y(t,\varepsilon) &= -\Lambda(t)^{-1}(\overset{(3)}{\mathcal{A}}(t) - \overset{(3)}{\mathcal{C}}(t)\overset{(4)}{\mathcal{C}}(t)^{-1}\overset{(4)}{\mathcal{A}}(t))\rho_n X(t,\varepsilon) \\ &+ \overset{(3)}{G}(t,0,\varepsilon)\Lambda(0)^{-1}(\overset{(3)}{\mathcal{A}}(0) - \overset{(3)}{\mathcal{C}}(0)\overset{(4)}{\mathcal{C}}(0)^{-1}\overset{(4)}{\mathcal{A}}(0))\rho_n X(0,\varepsilon) \\ &- \overset{(3)}{G}(t,T,\varepsilon)\Lambda(T)^{-1}(\overset{(3)}{\mathcal{A}}(T) - \overset{(3)}{\mathcal{C}}(T)\overset{(4)}{\mathcal{C}}(T)^{-1}\overset{(4)}{\mathcal{A}}(T))\rho_n X(T,\varepsilon) + \overset{(3)}{\mathcal{G}}(\rho_n\vartheta,t,\varepsilon). \end{aligned} \tag{74}$$

Taking into account (68), (74), we obtain from (57) the following equation:

$$\begin{aligned} \frac{d\rho_n X(t,\varepsilon)}{dt} &= \Omega(t)\rho_n X(t,\varepsilon) + (\overset{(2)}{\mathcal{B}}(t) - \overset{(2)}{\mathcal{C}}(t)\overset{(4)}{\mathcal{C}}(t)^{-1}\overset{(4)}{\mathcal{B}}(t)) \\ &\times(\overset{(3)}{G}(t,0,\varepsilon)\Lambda(0)^{-1}(\overset{(3)}{\mathcal{A}}(0) - \overset{(3)}{\mathcal{C}}(0)\overset{(4)}{\mathcal{C}}(0)^{-1}\overset{(4)}{\mathcal{A}}(0))\rho_n X(0,0) \\ &- \overset{(3)}{G}(t,T,\varepsilon)\Lambda(T)^{-1}(\overset{(3)}{\mathcal{A}}(T) - \overset{(3)}{\mathcal{C}}(T)\overset{(4)}{\mathcal{C}}(T)^{-1}\overset{(4)}{\mathcal{A}}(T))\rho_n X(T,\varepsilon) \\ &+ \overset{(2)}{\mathcal{C}}(t)(\overset{(4)}{G}(t,0,\varepsilon)\overset{(4)}{\mathcal{C}}(0)^{-1}(\overset{(4)}{\mathcal{A}}(0)\rho_n X(0,\varepsilon) + \overset{(4)}{\mathcal{B}}(0)\rho_n Y(0,\varepsilon)) \\ &- \overset{(4)}{G}(t,T,\varepsilon)\overset{(4)}{\mathcal{C}}(T)^{-1}(\overset{(4)}{\mathcal{A}}(T)\rho_n X(T,\varepsilon) + \overset{(4)}{\mathcal{B}}(T)\rho_n Y(T,\varepsilon)) \\ &+ \overset{(2)}{\chi}(\rho_n\vartheta,t,\varepsilon) + \overset{(2)}{\mathcal{G}}(\rho_n\vartheta,t,\varepsilon), \end{aligned} \tag{75}$$

where

$$\begin{aligned} \Omega(t) = \overset{(2)}{\mathcal{A}}(t) - \overset{(2)}{\mathcal{C}}(t)\overset{(4)}{\mathcal{C}}(t)^{-1}\overset{(4)}{\mathcal{A}}(t) - (\overset{(2)}{\mathcal{B}}(t) - \overset{(2)}{\mathcal{C}}(t)\overset{(4)}{\mathcal{C}}(t)^{-1}\overset{(4)}{\mathcal{B}}(t))\Lambda(t)^{-1}(\overset{(3)}{\mathcal{A}}(t) \\ - \overset{(3)}{\mathcal{C}}(t)\overset{(4)}{\mathcal{C}}(t)^{-1}\overset{(4)}{\mathcal{A}}(t)). \end{aligned}$$

It follows from Lemma 2 that the boundary value problem

$$\frac{d}{dt}\begin{pmatrix} \overline{x} \\ \overline{\zeta} \end{pmatrix} = \Omega(t)\begin{pmatrix} \overline{x} \\ \overline{\zeta} \end{pmatrix},\ \overline{x}(0) = 0,\ \overline{\zeta}(T) = 0$$

is uniquely solvable. Hence, there exists the matrix Green function $\overset{(2)}{G}(t,s)$ of the last boundary value problem, which is bounded, i.e.,

$$\|\overset{(2)}{G}(t,s)\| \leq c,\ t,s \in [0,T]. \tag{76}$$

Due to (61), (62), (64), (71), and (76) from the expression of the solution $\rho_n X(t,\varepsilon)$ of Equation (75) with the boundary values from (60), written by the help of the matrix Green function $\overset{(2)}{G}(t,s)$, it follows that $\rho_n X(t,\varepsilon) = \overset{(2)}{\mathcal{G}}(\rho_n\vartheta,t,\varepsilon)$. Furthermore, from (74), (68), and (56), and properties (1), (2), we successively obtain $\rho_n Y(t,\varepsilon) = \overset{(3)}{\mathcal{G}}(\rho_n\vartheta,t,\varepsilon)$,

$\rho_n Z(t,\varepsilon) = \overset{(4)}{\mathcal{G}}(\rho_n \vartheta, t, \varepsilon), r_n u(t,\varepsilon) = \overset{(1)}{\mathcal{G}}(\rho_n \vartheta, t, \varepsilon)$. Thus, for determining $\rho_n \vartheta$, we have in the space of continuous functions with values in $\mathbb{R}^m \times \mathbb{R}^{n_1} \times \mathbb{R}^{n_2} \times \mathbb{R}^{n_3}$ the equation

$$\rho_n \vartheta = \mathcal{G}(\rho_n \vartheta, t, \varepsilon), \tag{77}$$

where the function $\mathcal{G}$ for sufficiently small $\varepsilon > 0$, $t \in [0,T]$ satisfies properties (2), (3). If we will take in condition (2) $q < 1$, then $\mathcal{G}$ is a contraction mapping in $C_{[0,T]}$. According to the contractive mappings principle, Equation (77) has a unique solution, and this solution can be found by the method of successive approximations.

According to (3) and Lemma 3, all successive approximations are not more than $c\varepsilon^{n+1}$. Hence, a solution of the Equation (77) will have the same estimate. Due to (49) and (55), it proves the theorem statement. □

Theorem 3. *Under conditions of Theorem 2, for sufficiently small $\varepsilon > 0$, the following inequality for the performance index is valid*

$$J_\varepsilon(\tilde{u}_n) - J_\varepsilon(u_*) \leq c\varepsilon^{2n+2}.$$

Proof. Denoting by $\tilde{s}$ a solution of the problem (2)–(3) at $u = \tilde{u}_n$, we present the solution of the problem P_ε in the form $w_* = \tilde{s} + \delta w$, $u_* = \tilde{u}_n + \Delta u$, then δw satisfies the system

$$\mathcal{E}(\varepsilon)\frac{d\delta w}{dt} = A(t)\delta w + B(t)\Delta u + \varepsilon(f(w_*, u_*, t, \varepsilon) - f(w_* - \delta w, u_* - \Delta u, t, \varepsilon)),$$
$$\delta w(0,\varepsilon) = 0.$$

In view of Theorem 2,

$$\|\Delta u\| = \|r_n u\| \leqslant c\varepsilon^{n+1}. \tag{78}$$

Using this estimate and the condition of stability of the matrices A_{33} and $A_{22} - A_{23}A_{33}^{-1}A_{32}$, we can prove the estimate

$$\|\delta w(t,\varepsilon)\| \leqslant c\varepsilon^{n+1}. \tag{79}$$

Introducing the notation $\Delta J = J_\varepsilon(\tilde{u}_n) - J_\varepsilon(u_*)$, we present it in the form

$$\Delta J = \frac{1}{2}\int_0^T (\delta w' W(t)\delta w + \Delta u' R(t)\Delta u)\,dt$$
$$+ \int_0^T (-\delta w' W(t) w_* - \Delta u' R(t) u_* + \varepsilon(F(\tilde{w}, \tilde{u}_n, t, \varepsilon) - F(w_*, u_*, t, \varepsilon))\,dt.$$

Using control optimality condition (27), (28) for the problem (1)–(3), after integrating by parts and taking into account the equation for δw and the boundary values $\delta w(0,\varepsilon)$, $\varphi(T,\varepsilon)$, we have

$$\Delta J = \frac{1}{2}\int_0^T (\delta w' W(t)\delta w + \Delta u' R(t)\Delta u)\,dt + \varepsilon\int_0^T (\varphi'(f(w_*, u_*, t, \varepsilon)$$
$$- f(w_* - \delta w, u_* - \Delta u, t, \varepsilon) - f_w(w_*, u_*, t, \varepsilon)\delta w - f_u(w_*, u_*, t, \varepsilon)\Delta u)$$
$$+ F(w_* - \delta w, u_* - \Delta u, t, \varepsilon) - F(w_*, u_*, t, \varepsilon) + F_w(w_*, u_*, t, \varepsilon)\delta w + F_u(w_*, u_*, t, \varepsilon)\Delta u)\,dt.$$

From here, in view of (78) and (79), we obtain the theorem assertion. □

Denote by $\tilde{u}_{i*}$ the i-th order approximation for an optimal control constructed by the direct scheme method.

Theorem 4. *For sufficiently small $\varepsilon > 0$, the following inequalities are valid*

$$\begin{aligned} &J_\varepsilon(\widetilde{u}_{*(n-1)}) \geqslant J_\varepsilon(\widetilde{u}_{*(n-1)} + \varepsilon^n \overline{u}_{*n}) \\ &\geqslant J_\varepsilon(\widetilde{u}_{*(n-1)} + \varepsilon^n(\overline{u}_{*n} + \Pi_{0n}u_* + Q_{0n}u_*)) \geqslant J_\varepsilon(\widetilde{u}_{*n}), \ n \geqslant 1. \end{aligned} \tag{80}$$

If an addition to $\widetilde{u}_{*(n-1)}$ is non-zero, then the corresponding inequality is strict.

Proof. The proof of this theorem is based on Theorem 1.

Asymptotic solution of the form (4) can be constructed for a solution of problem (2), (3) at control $u = \widetilde{u}_{*n}$ for each n. Moreover, the terms of these asymptotic solution coincide with corresponding terms of asymptotic expansion of optimal trajectory to n-th order inclusively.

Substitute expansions for $\widetilde{u}_{*(n-1)}(t,\varepsilon)$, $\widetilde{u}_{*n}(t,\varepsilon)$ and corresponding trajectories into $J_\varepsilon(u)$. After applying the expansion (6) and Theorem 1, we see that the first $2n$ coefficients from (6) coincided and a difference between $J_\varepsilon(\widetilde{u}_{*(n-1)})$ and $J_\varepsilon(\widetilde{u}_{*n})$ appears for the first time in the coefficient J_{2n}.

Taking into account the equality $\widetilde{u}_{*n} = \widetilde{u}_{*(n-1)} + \varepsilon^n(\overline{u}_{*n} + \sum_{i=0}^{1}(\Pi_{in}u_* + Q_{in}u_*))$, consider the expressions for the coefficients J_{2n}, J_{2n+1} and J_{2n+2} separately.

If we omit the terms known after solving problems $\overline{P}_j$, $\Pi_{0j}P$, $Q_{0j}P$, $j = \overline{0, n-1}$, $\Pi_{1j}P$, $Q_{1j}P$, $j = \overline{0, n-2}$, we obtain from J_{2n} the sum of the performance indices $\overline{J}_n + \Pi_{1(n-1)}J + Q_{1(n-1)}J$. For $J_\varepsilon(\widetilde{u}_{*n})$, we have $\overline{J}_n = \overline{J}_n(\overline{u}_n)$ and, for $J_\varepsilon(\widetilde{u}_{*(n-1)})$, we have $\overline{J}_n = \overline{J}_n(0)$. Since $\overline{u}_{*n}$ is found from the problem of minimizing the functional $\overline{J}_n$, we obtain the first inequality in (80).

By a similar way, we obtain from the form J_{2n+1} the sum $\Pi_{0n}J + Q_{0n}J$. For $J_\varepsilon(\widetilde{u}_{*n})$, we have $\Pi_{0n}J = \Pi_{0n}J(\Pi_{0n}u_*)$, $Q_{0n}J = Q_{0n}J(Q_{0n}u_*)$. For $J_\varepsilon(\widetilde{u}_{*(n-1)})$, we have $\Pi_{0n}J = \Pi_{0n}J(0)$, $Q_{0n}J = Q_{0n}J(0)$. Since $\Pi_{0n}u_*$ and $Q_{0n}u_*$ are found by means of minimizing the functionals $\Pi_{0n}J$ and $Q_{0n}J$, respectively, we obtain the second inequality in (80).

Analogously, the third inequality in (80) follows from the form J_{2n+2}.

The assertion concerning non-zero additions to $\widetilde{u}_{*(n-1)}$ follows from the unique solvability of linear-quadratic control problems. □

Remark 3. *From Theorems 3 and 4, it follows that the sequences $\{\widetilde{u}_{*(n-1)}\}$, $\{\widetilde{u}_{*(n-1)} + \varepsilon^n\overline{u}_{*n}\}$, $\{\widetilde{u}_{*(n-1)} + \varepsilon^n(\overline{u}_{*n} + \Pi_{0n}u_* + Q_{0n}u_*)\}$ are minimizing.*

6. Conclusions

In this paper, unlike the previous one [26], devoted to a similar problem, detailed proofs of linear-quadratic optimal control problems forms, from which terms of asymptotic solution of given nonlinear optimal control problem are found, are presented in Theorem 1. Note that all problems for finding asymptotic terms are obtained in an explicit form. It is very comfortable for research applying asymptotic methods for solving practical problems.

For the first time, asymptotic estimates of the proximity between the exact and asymptotic solutions are established for the control, state trajectory in Theorem 2 and for the minimized functional in Theorem 3.

It should be noted that, in view of Theorem 4, values of the minimized functional with a control, which is an asymptotic approximation to the optimal control u_* respectively of the form $\widetilde{u}_{*(n-1)}$, $\widetilde{u}_{*(n-1)} + \varepsilon^n\overline{u}_{*n}$, $\widetilde{u}_{*(n-1)} + \varepsilon^n(\overline{u}_{*n} + \Pi_{0n}u_* + Q_{0n}u_*)$, $\widetilde{u}_{*n}$, do not increase. It follows from Theorem 3 that the corresponding sequences of the controls are minimizing.

In the future, it is useful to give a program realization of applying the direct scheme method for problems of type (1)–(3). The results obtained in the paper can be used for constructing asymptotic solutions of practical optimal control problems with three-tempo state variables and weak nonlinear perturbations in a linear state equation and a quadratic performance index.

The advantage of applying a direct scheme method is the possibility to use standard software packages for solving optimal control problems in order to find terms of asymptotic solution. As it is proved in this paper, for problems with three-tempo state variables, the found sequence of approximations to the optimal control $\{\widetilde{u}_{*n}\}$ is minimizing.

Author Contributions: Problem statement and methodology, G.K.; algorithm of asymptotics construction, G.K. and M.K.; asymptotic estimates, G.K.; writing—original draft preparation, M.K.; writing—review and editing G.K. All authors have read and agreed to the published version of the manuscript.

Funding: The work of the first author was supported by the Russian Science Foundation (Project No. 21-11-00202).

Institutional Review Board Statement: Not applicable.

Informed Consent Statement: Not applicable.

Data Availability Statement: Not applicable.

Acknowledgments: The authors thank the three Reviewers for helpful comments and N.T. Hoai and A.S. Kostenko for useful discussions.

Conflicts of Interest: The authors declare no conflict of interest. The funders had no role in the design of the study; in the collection, analysis, or interpretation of data; in the writing of the manuscript, or in the decision to publish the results.

References

1. Kokotovic, P.V.; O'Malley, R.E., Jr.; Sannuti, P. Singular perturbations and order reduction in control theory—An overview. *Automatica* **1976**, *12*, 123–132. [CrossRef]
2. Dmitriev, M.G.; Kurina, G.A. Singular perturbations in control problems. *Autom. Remote Control* **2006**, *67*, 1–43. [CrossRef]
3. Zhang, Y.; Naidu, D.S.; Cai, C.; Zou, Y. Singular perturbations and time scales in control theories and applications: An overview 2002–2012. *Int. J. Inf. Syst. Sci.* **2014**, *9*, 1–36.
4. Sayasov, Y.S.; Vasil'eva, A.B. Justification and conditions for the applicability of the Semenov-Bodenstein method of quasistationary concentrations. *Zh. Fiz. Khim.* **1955**, *29*, 802–810. (In Russian)
5. Radisavljević-Gajić, V.; Milanović, M.; Rose, P. *Multi-Stage and Multi-Time Scale Feedback Control of Linear Systems with Applications to Fuel Cells*; Mechanical Engineering Series; Springer: Cham, Switzerland, 2019; 214p.
6. Meng, X.; Wang, Q.; Zhou, N.; Xiao, S.; Chi, Y. Multi-time scale model order reduction and stability consistency certification of inverter-interfaced DG system in AC microgrid. *Energies* **2018**, *11*, 254. [CrossRef]
7. Voropaeva, N.V.; Sobolev, V.A. *Decomposition of Multi-Tempo Systems*; SMS: Samara, Russia, 2000; 290p. (In Russian)
8. Kishor Babu, G.; Krishnarayalu, M.S. Application of singular perturbation method to two parameter discrete power system model. *J. Control Instrument. Eng.* **2017**, *3*, 1–13.
9. Shimjith, S.R.; Tiwari, A.P.; Bandyopadhyay, B. *Lecture Notes in Control and Information Sciences, 431. Modeling and Control of a Large Nuclear Reactor. A Three-Time-Scale Approach*; Springer: Berlin/Heidelberg, Germany; New York, NY, USA; Dordrecht, The Netherlands; London, UK, 2013; 138p.
10. Naidu, D.S.; Calise, A.J. Singular perturbations and time scales in guidance and control of aerospace systems: A survey. *J. Guid. Control Dyn.* **2001**, *24*, 1057–1078. [CrossRef]
11. Il'in, A.M.; Kamenkovich, V.M. On structure of boundary layer in two-dimensional theory of ocean currents. *Okeanologiya* **1964**, *4*, 756–769. (In Russian)
12. Jamshidi, M. Three-stage near-optimum design of nonlinear-control processes. *Proc. IEE* **1974**, *121*, 886–892. [CrossRef]
13. Vlakhova, A.V.; Novozhilov, I.V. On skidding of a wheeled vehicle when one of the wheels locks or slips. *J. Math. Sci.* **2007**, *146*, 5803–5810. [CrossRef]
14. Brøns, M.; Desroches, M.; Krupa, M. Mixed-mode oscillations due to a singular Hopf bifurcation in a forest pest model. *Math. Popul. Stud. Int. J. Math. Demogr.* **2015**, *22*, 71–79. [CrossRef]
15. Archibasov, A.A.; Korobeinikov, A.; Sobolev, V.A. Asymptotic expansions of solutions in a singularly perturbed model of virus evolution. *Comput. Math. Math. Phys.* **2015**, *55*, 240–250. [CrossRef]
16. O'Riordan, E.; Pickett, M.L.; Shishkin, G.I. Singularly perturbed problems. Modeling reaction-convection-diffusion processes. *Comput. Methods Appl. Math.* **2003**, *3*, 424–442. [CrossRef]
17. Vasil'eva, A.B. Asymptotic methods in the theory of ordinary differential equations containing small parameters in front of the higher derivatives. *Comput. Math. Math. Phys.* **1963**, *3*, 823–863. [CrossRef]
18. Voropaeva, N.V.; Sobolev, V.A. *Geometric Decomposition of Singularly Perturbed Systems*; Fizmatlit: Moscow, Russia, 2009; 256p. (In Russian)
19. Tikhonov, A.N. On systems of differential equations containing parameters. *Matem. Sb. N. Ser.* **1950**, *27*, 147–156. (In Russian)
20. Gradshtein, I.S. Differential equations in with factors at derivatives are various powers of a small parameter. *Dokl. AN SSSR* **1952**, *LXXXII*, 5–8. (In Russian)

21. Vasil'eva, A.B. Asymptotic formulas for solutions of systems of ordinary differential equations containing parameters of various orders of smallness at derivatives. *Dokl. AN SSSR* **1959**, *128*, 1110–1113. (In Russian)
22. Belokopytov, S.V.; Dmitriev, M.G. Solution of classical optimal control problems with a boundary layer. *Autom. Remote Control* **1989**, *50*, 907–917.
23. Dmitriev, M.G.; Kurina, G.A. Direct scheme for constructing asymptotics of classical optimal control problems solution. In *Programmnye Sistemy*; Nauka Fizmatlit: Moscow, Russia, 1999; pp. 44–55. (In Russian)
24. Kalashnikova, M.A.; Kurina, G.A. Direct scheme for the asymptotic solution of linear-quadratic problems with cheap control of different costs. *Differ. Equ.* **2019**, *55*, 84–104. [CrossRef]
25. Kalashnikova, M.A.; Kurina, G.A. Asymptotics of a solution of three-tempo optimal control problem. In *Trudy XII Vserossijskogo Soveshchanija po Problemam Upravlenija VSPU-2014*; IPU RAN: Moscow, Russia, 2014; pp. 1560–1570. (In Russian)
26. Kalashnikova, M.; Kurina, G. Direct scheme of constructing asymptotic solution of three-tempo linear-quadratic control problems with weak nonlinear perturbations. In Proceedings of the 2022 16 International Conference on Stability and Oscillations of Nonlinear Control Systems (Pyatnitsky's Conference), Moscow, Russia, 1–3 June 2022.
27. Bagirova, N.H.; Vasil'eva, A.B.; Imanaliev, M.I. A question on the asymptotic solution of an optimal control problem. *Differ. Uravn.* **1967**, *3*, 1895–1902. (In Russian)
28. Vasil'eva, A.B.; Butuzov, V.F. *Asymptotic Expansions of Solutions of Singularly Perturbed Equations*; Nauka: Moscow, Russia, 1973; 272p. (In Russian)
29. Nguen, T.H. Asymptotic Solution of Singularly Perturbed Linear-quadratic Optimal Control Problems with Discontinuous Coefficients. Ph.D. Dissertation, Voronezh State University, Voronezh, Russia, 2010. (In Russian)
30. Kurina, G.A. On linear Hamiltonian systems not resolved with respect to derivative. *Differ. Uravn.* **1986**, *22*, 193–198. (In Russian)

axioms

MDPI

Article

Singularly Perturbed Cauchy Problem for a Parabolic Equation with a Rational "Simple" Turning Point

Tatiana Ratnikova †

Moscow Power Engineering Institute, National Research University, 111250 Moscow, Russia; tatrat1@mail.ru; Tel.: +7-916-373-18-03

† The results of the work are obtained in the framework of the state contract of the Ministry of Education and Science of the Russian Federation (project no. FSWF-2020-0022).

Received: 8 October 2020; Accepted: 25 November 2020; Published: 27 November 2020

Abstract: The aim of the research is to develop the regularization method. By Lomov's regularization method, we constructed a uniform asymptotic solution of the singularly perturbed Cauchy problem for a parabolic equation in the case of violation of stability conditions of the limit-operator spectrum. The problem with a "simple" turning point is considered in the case, when the eigenvalue vanishes at $t = 0$ and has the form $t^{m/n}a(t)$. The asymptotic convergence of the regularized series is proved.

Keywords: singularly perturbed Cauchy problem; parabolic equation; asymptotic solution; rational "simple" turning point

1. Introduction

Singularly perturbed problems with an unstable spectrum of the limit operator for ordinary partial differential equations and partial differential equations have been studied by many authors [1–8]. The most difficult of them are problems with point instability, namely turning points.

For the first time, these problems arose, in particular, in quantum mechanics. One of the first methods of solution was the WKB method, a method of semiclassical calculation, which was developed in 1926 by G. Wentzel, H.A. Kramers, and L. Brillouin. At the same time, H. Jeffreys generalized the method of approximate solution of linear differential equations of the second order, including the solution of the Schrodinger equation. Methods for solving problems with spectral features were and are being developed by: the school of V.P. Maslov, the school of A.B. Vasilieva – V.F. Butuzov – N.N. Nefedov, and the school of S.A. Lomov, among others.

The regularization method defines three groups of turning points.

1. "Simple" turning point: The eigenvalues of the limit operator are isolated from each other, and one eigenvalue at separate points of t vanishes.

2. "Weak" turning point: At least one pair of eigenvalues intersect at separate points of t, but the limit operator preserves the diagonal structure up to the intersection points. The eigenvector basis remains smooth in t.

3. "Strong" turning point: At least one pair of eigenvalues intersect at separate points of t, but the limit operator changes the diagonal structure to Jordan at the intersection points. The basis at the intersection points loses its smoothness in t.

Classic turning points are of the third type. A feature of the problem with a "simple" turning point presented in the article is the pointwise irreversibility of the limit operator $t^{m/n}$.

In the present paper, using the Lomov's regularization method [1], a regularized asymptotic solution of the singularly perturbed Cauchy problem on the entire segment $[0, T]$ for a parabolic equation is constructed in the presence of a "simple" rational turning point of the limit operator.

The point $\varepsilon = 0$ for the singularly perturbed Cauchy problem is singular in the sense that the classical existence theorems for the solution of the Cauchy problem do not hold at this point. Therefore, in the solution of singularly perturbed problems, essentially special singularities arise that describe the irregular dependence of the solution on ε [9].

One of the advantages of the regularization method is that it makes it possible to construct a global asymptotic solution over the entire domain of integration, and, under certain conditions on the coefficients of the equation, it gives an exact solution.

The fractional turning point of order $1/2$ for an ordinary differential equation of the first order was considered by the method of boundary functions by [6].

In this article, we continue on the "simple" rational turning point work carried out in [10,11].

2. Formulation of the Problem and Construction of an Asymptotic Solution

Consider the Cauchy problem

$$\begin{cases} \varepsilon\dfrac{\partial u}{\partial t} = \varepsilon^2\left(\dfrac{\partial}{\partial x}\left(k(x)\dfrac{\partial u}{\partial x}\right)\right) - t^{m/n}a(t)u + h(x,t), \\ u(x,0) = f(x), \quad -\infty < x < +\infty, \end{cases} \tag{1}$$

where $u = u(x,t,\varepsilon)$ is a function depending on variables x and t and real parameters ε, $\varepsilon > 0$ and $\varepsilon << 1$.

Let the following conditions be satisfied:

$$\begin{aligned} &(1) \quad h(x,t) \in C^{\infty}(R \times [0,T]), \text{ the function } h(x,t) \text{ and all its derivatives are bounded} \\ &\qquad \text{on } \{R \times [0,T]\}; \\ &(2) \quad k(x) \in C^{\infty}(R), \ k(x) \geq k_0 > 0; \\ &(3) \quad f(x) \in C^{\infty}(R), \text{ the function } f(x) \text{ and all its derivatives are bounded to } R; \\ &(4) \quad a(t) \in C^{\infty}([0,T]), \ a(t) > 0; and \\ &(5) \quad m,n \in N, \ p = m+n-1, \ m/n = r \text{ is a fractional number.} \end{aligned} \tag{2}$$

These conditions ensure the existence and uniqueness of the bounded solution and the possibility of constructing the asymptotic series of the problem (1) [12].

Singularly perturbed problems arise when the domain of definition of the initial operator depending on ε at $\varepsilon \neq 0$ does not coincide with the domain of definition of the limit operator at $\varepsilon = 0$.

Under the stability condition for the spectrum of the limiting operator, essentially singular singularities are described using exponentials of the form $e^{\varphi_i(t)/\varepsilon}$, $\varphi_i(t) = \int\limits_0^t \lambda_i(s)ds$, $i = \overline{1,n}$, where $\lambda_i(t)$ are the eigenvalues of the limit operator and $\varphi_i(t)$ are smooth (in general, complex) functions of a real variable t.

If the stability conditions for the limit operator are violated for at least one point of the spectrum of the limit operator, then new singularities arise in the solution of the inhomogeneous equation. When studying problems with a "simple" turning point, we are faced with a problem when the range of values of the original operator does not coincide with the range of values of the limit operator.

Special singularities of the problem (1) have the form:

$$e^{\varphi_i(t)/\varepsilon}, \quad \varphi(t) = \int\limits_0^t s^{m/n}a(s)ds, \quad \sigma_i = e^{-\varphi(t)/\varepsilon}\int\limits_0^t e^{\varphi(s)/\varepsilon}s^{(i+1-n)/n}ds, \ i = \overline{0,(p-1)}.$$

We look for a solution $u(x,t,\varepsilon)$ in the form [1,9]:

$$u(x,t,\varepsilon) = e^{-\varphi(t)/\varepsilon}X(x,t,\varepsilon) + \sum_{i=0}^{p-1} Z^i(x,t,\varepsilon)\sigma_i + W(x,t,\varepsilon). \tag{3}$$

We denote differentiation with respect to t by " $\cdot$ " and differentiation with respect to x by " $'$ " or $\frac{\partial}{\partial x}$. Substituting (3) into (1) we get the equation

$$\varepsilon\left[e^{-\varphi(t)/\varepsilon}\dot{X}+\sum_{i=0}^{p-1}\left(t^{(i+1-n)/n}Z^i+\sigma_i\dot{Z}^i\right)+\dot{W}\right]=$$

$$=\varepsilon^2\left[e^{-\varphi(t)/\varepsilon}\frac{\partial}{\partial x}\left(k(x)\frac{\partial X}{\partial x}\right)+\sum_{i=0}^{p-1}\frac{\partial}{\partial x}\left(k(x)\frac{\partial Z^i}{\partial x}\right)\sigma_i+\frac{\partial}{\partial x}\left(k(x)\frac{\partial W}{\partial x}\right)\right]-t^{m/n}a(t)W+h(x,t).$$

By identifying the coefficients in the linear combination of $e^{-\varphi(t)/\varepsilon}$, σ_i and 1, we obtain the system

$$\begin{cases}\dot{X}=\varepsilon\frac{\partial}{\partial x}\left(k(x)\frac{\partial X}{\partial x}\right),\\ \dot{Z}^i=\varepsilon\frac{\partial}{\partial x}\left(k(x)\frac{\partial Z^i}{\partial x}\right),\ i=\overline{0,(p-1)},\\ t^{m/n}a(t)W=-\varepsilon\dot{W}+\varepsilon^2\frac{\partial}{\partial x}\left(k(x)\frac{\partial W}{\partial x}\right)-\varepsilon\sum_{i=0}^{p-1}t^{(i+1-n)/n}Z^i+h(x,t),\\ X(x,0)+W(x,0)=f(x).\end{cases} \quad (4)$$

We look for a solution (4) in the form of power series in ε:

$$\begin{cases}X(x,t,\varepsilon)=\sum_{k=-1}^{\infty}\varepsilon^k X_k(x,t),\\ Z^i(x,t,\varepsilon)=\sum_{k=-1}^{\infty}\varepsilon^k Z^i_k(x,t),\ i=\overline{0,(p-1)},\\ W(x,t,\varepsilon)=\sum_{k=-1}^{\infty}\varepsilon^k W_k(x,t).\end{cases} \quad (5)$$

Substituting (5) into (4) we get a series of iterative tasks:

$$\begin{cases}\dot{X}_k=\frac{\partial}{\partial x}\left(k(x)\frac{\partial X_{k-1}}{\partial x}\right),\\ \dot{Z}^i_k=\frac{\partial}{\partial x}\left(k(x)\frac{\partial Z^i_{k-1}}{\partial x}\right),\ i=\overline{0,(p-1)},\\ t^{m/n}a(t)W_k=-\dot{W}_{k-1}+\frac{\partial}{\partial x}\left(k(x)\frac{\partial W_{k-2}}{\partial x}\right)-\sum_{i=0}^{p-1}t^{(i+1-n)/n}Z^i_{k-1}+\delta_0^k h(x,t),\\ X_k(x,0)+W_k(x,0)=\delta_0^k f(x),\ k=\overline{-1,\infty};\ \text{if the index } (k-1)\le -2,\\ \qquad (k-2)\le -2, \text{ then the term is by definition } 0.\end{cases} \quad (6)$$

To solve iterative problems, the solvability theorem is used (see Section 4).

Consider the system (6) with $k=-1$:

$$\begin{cases}\dot{X}_{-1}=0,\\ \dot{Z}^i_{-1}=0,\ i=\overline{0,(p-1)},\\ t^{m/n}a(t)W_{-1}\equiv 0,\\ X_{-1}(x,0)+W_{-1}(x,0)=0.\end{cases} \quad (7)$$

From (7) it follows that

$$\begin{cases}X_{-1}(x,t)=X_{-1}(x,0)=0,\\ Z^i_{-1}(x,t)=Z^i_{-1}(x,0),\ i=\overline{0,(p-1)},\\ W_{-1}(x,t)\equiv 0.\end{cases} \quad (8)$$

The functions $Z^i_{-1}(x,0)$ are found from the solvability condition for the system (6) for $k = 0$:

$$\begin{cases} \dot{X}_0 = 0, \\ \dot{Z}^i_0 = \dfrac{\partial}{\partial x}\left(k(x)\dfrac{\partial Z^i_{-1}(x,0)}{\partial x}\right), \quad i = \overline{0,(p-1)}, \\ t^{m/n}a(t)W_0 = -\sum\limits_{i=0}^{p-1} t^{(i+1-n)/n} Z^i_{-1}(x,0) + h(x,t), \\ X_0(x,0) + W_0(x,0) = f(x). \end{cases} \tag{9}$$

Let us expand $h(x,t)$ by Maclaurin's formula in t:

$$h(x,t) = h(x,0) + t\dot{h}(x,0) + \ldots + t^{[m/n]}h^{[m/n]}(x,0) + t^{[m/n]+1}h_0(x,t).$$

From the condition of solvability of the equation for W_0 at $t = 0$, it follows that, if $\frac{i+1-n}{n} = j$, $i = \overline{0,(p-1)}, 0 \le j \le \left[\frac{m}{n}\right]$, then

$$Z^{(j+1)n-1}_{-1}(x,0) = \frac{h^{(j)}(x,0)}{j!}, \tag{10}$$

where $\left[\frac{m}{n}\right]$ is the whole part of m/n; if $\frac{i+1-n}{n} \neq j$, $i = \overline{0,(p-1)}, 0 \le j \le \left[\frac{m}{n}\right]$, then

$$Z^i_{-1}(x,0) = 0.$$

As a result, we get the solution at the (-1) iteration step:

$$u_{-1}(x,t,\varepsilon) = \frac{1}{\varepsilon}\sum_{j=0}^{[m/n]} Z^{(j+1)n-1}_{-1}(x,0)\sigma_{(j+1)n-1}(t,\varepsilon) = \frac{1}{\varepsilon}\sum_{j=0}^{[m/n]} \frac{h^{(j)}(x,0)}{j!}\sigma_{(j+1)n-1}(t,\varepsilon). \tag{11}$$

Considering (10) we can write the system (9) as:

$$\begin{cases} \dot{X}_0 = 0, \\ \dot{Z}^i_0 = 0, \ \text{when } i \neq (j+1)n-1, \ 0 \le j \le [m/n], \\ \dot{Z}^i_0 = \dfrac{\partial}{\partial x}\left(k(x)\dfrac{\partial Z^i_{-1}(x,0)}{\partial x}\right), \quad i = (j+1)n-1, \ 0 \le j \le [m/n], \\ t^{m/n}a(t)W_0 = -\sum\limits_{j=0}^{[m/n]} t^j Z^{(j+1)n-1}_{-1}(x,0) + h(x,t), \\ X_0(x,0) + W_0(x,0) = f(x). \end{cases} \tag{12}$$

The right-hand side of the equation with respect to W_0 of the system (12), due to the choice of $Z^i_{-1}(x,0)$, satisfies the solvability theorem (see Section 4):

$$W_0(x,t) = \frac{h(x,t) - \sum\limits_{j=0}^{[m/n]} t^j Z^{(j+1)n-1}_{-1}(x,0)}{t^{m/n}a(t)} = t^{s/n}h_0(x,t), \ 1 \le s \le n-1 \tag{13}$$

(s is an integer). Notice, that $W_0(x,0) = 0$.

Solving (12) we get

$$\begin{cases} X_0(x,t) = X_0(x,0) = f(x), \\ Z_0^i(x,t) = Z_0^i(x,0), \ i \neq (j+1)n-1, \ 0 \leq j \leq [m/n], \\ Z_0^i(x,t) = t\dfrac{\partial}{\partial x}\left(k(x)\dfrac{\partial Z_{-1}^i(x,0)}{\partial x}\right) + Z_0^i(x,0), \ i = (j+1)n-1, \ i = \overline{0,(p-1)}, \\ W_0(x,t) = t^{s/n}h_0(x,t), \end{cases} \tag{14}$$

where $Z_{-1}^i(x,0)$ are determined from (10).

$Z_0^i(x,0)$ at the iteration step $k = 0$ are unknown. The functions $Z_0^i(x,0)$ are found from the condition of solvability for $t = 0$ of the iterative problem for $k = 1$:

$$\begin{cases} \dot{X}_1(x,t) = f(x), \\ \dot{Z}_1^i(x,t) = \dfrac{\partial}{\partial x}\left(k(x)\dfrac{\partial Z_0^i(x,t)}{\partial x}\right), \ i = \overline{0,(p-1)}, \\ t^{m/n}a(t)W_1(x,t) = -\dot{W}_0 - \sum\limits_{i=0}^{p-1} t^{(i+1-n)/n}Z_0^i(x,t), \\ X_1(x,0) + W_1(x,0) = 0; \end{cases} \tag{15}$$

$$\dot{W}_0(x,t) = \frac{s}{n}t^{\frac{s}{n}-1}h_0(x,t) + t^{\frac{s}{n}}\dot{h}_0(x,t) = t^{\frac{s}{n}-1}\left[\frac{s}{n}h_0(x,t) + \dot{h}(x,t)t\right] =$$

$$= t^{\frac{s}{n}-1}h_1(x,t) = t^{-\{\frac{m}{n}\}}h_1(x,t),$$

where $\{\frac{m}{n}\}$ is the fractional part of m/n.

To determine $W_1(x,t)$, we subordinate the right side of the equation to the conditions of point solvability. To do this, we expand $h_1(x,t)$ by Maclaurin's formula in t:

$$\dot{W}_0(x,t) = t^{-\{\frac{m}{n}\}}h_1(x,0) + t^{-\{\frac{m}{n}\}+1}h_1(x,0) + \ldots + t^{-\{\frac{m}{n}\}+k}\frac{h^k(x,0)}{k!} + t^{-\{\frac{m}{n}\}+k+1}h_2(x,t), \tag{16}$$

where $k = \left[\frac{m}{n} + \{\frac{m}{n}\}\right]$, $\{\frac{m}{n}\} = \frac{s}{n}$, $1 \leq s \leq n-1$. Then,

$$\begin{aligned} &\text{(a) if } j - \{\tfrac{m}{n}\} = \tfrac{i+1-n}{n}, \text{ i.e } i = n(j+1)-s-1, \ j = \overline{0,k}, \text{ then } Z_0^i(x,0) = -\frac{h_1^{(j)}(x,0)}{j!}; \\ &\text{(b) if } i \neq (j+1)n-s-1, \ j = \overline{0,k}, \text{ then } Z_0^i(x,0) = 0. \end{aligned} \tag{17}$$

Using this scheme, you can find a solution at any iteration step.

For $u_0(x,t,\varepsilon)$, we get

$$u_0(x,t,\varepsilon) = e^{-\varphi(t)/\varepsilon}f(x) + \sum_{i=0}^{p-1}\sigma_i Z_0^i(x,t) + \frac{h(x,t) - \sum\limits_{j=0}^{[m/n]}\frac{t^j h^{(j)}(x,0)}{j!}}{t^{m/n}a(t)}, \tag{18}$$

where $Z_0^i(x,t)$ are determined from (14) and (17).

Thus, the main term of the asymptotics of the solution is written in the form

$$u_{\mathtt{main}}(x,t,\varepsilon) = \frac{1}{\varepsilon}\sum_{j=0}^{[m/n]}\frac{h^{(j)}(x,0)}{j!}\sigma_{(j+1)n-1}(t,\varepsilon) + e^{-\varphi(t)/\varepsilon}f(x) + \sum_{i=0}^{p-1}\sigma_i(t,\varepsilon)Z_0^i(x,t) +$$

$$+\frac{h(x,t)-\sum\limits_{j=0}^{[m/n]}\frac{h^{(j)}(x,0)}{j!}t^j}{t^{m/n}a(t)}.$$

3. Remainder Estimate

Let $(N+1)$ iterative problems be solved. Then, the solution to the Cauchy problem can be represented in the form

$$u(x,t,\varepsilon)=\sum_{k=-1}^{N}u_k(x,t,\varepsilon)\varepsilon^k+\varepsilon^{N+1}R_N(x,t,\varepsilon). \tag{19}$$

Substituting (19) into (1) and taking into account that $u_k(x,t,\varepsilon)$ are solutions to iterative problems, we obtain the Cauchy problem for determining the remainder $R_N(x,t,\varepsilon)$:

$$\begin{cases} L(R_N)=\varepsilon\dot{R}_N-\varepsilon^2\dfrac{\partial}{\partial x}\left(k(x)\dfrac{\partial R_N}{\partial x}\right)+t^{m/n}a(t)R_N=H(x,t,\varepsilon),\\ R_N(x,0,\varepsilon)=0,\ \ -\infty<x<+\infty, \end{cases} \tag{20}$$

where

$$H(x,t,\varepsilon)=H_1(x,t)+\varepsilon H_2(x,t,\varepsilon),$$

$$H_1(x,t)=-\dot{W}_N+\frac{\partial}{\partial x}\left(k(x)\frac{\partial W_{N-1}}{\partial x}\right)-\sum_{i=0}^{p-1}t^{\frac{i+1}{n}-1}Z_N^i(x,t),$$

$$H_2(x,t,\varepsilon)=-e^{-\varphi(t)/\varepsilon}\frac{\partial}{\partial x}\left(k(x)\frac{\partial X_N}{\partial x}\right)-\frac{\partial}{\partial x}\left(k(x)\frac{\partial W_N}{\partial x}\right)-\sum_{i=0}^{p-1}\frac{\partial}{\partial x}\left(k(x)\frac{\partial Z_N^i}{\partial x}\right)\sigma_i(t,\varepsilon).$$

The remainder estimate is based on the maximum principle for parabolic problems [12]. This principle is used in the form of generality that we need to estimate the remainder. The classical solution to the problem (1) is a function $R(x,t,\varepsilon)$, continuous in $Q_T=(-\infty,+\infty)\times[0,T]\times(0,\varepsilon]$, having continuous $\frac{\partial R}{\partial t}$, $\frac{\partial R}{\partial x}$, $\frac{\partial^2 R}{\partial x^2}$ in Q_T, and satisfies Equation (1) and the initial conditions at $t=0$ at all points of Q_T.

Theorem 1 (remainder estimate). *Let the conditions be satisfied:*

(1) *Conditions (1)–(5) of the Cauchy problem (1);*
(2) $|H(x,t,\varepsilon)|\le M_1\ \forall(x,t)\in(-\infty,+\infty)\times[0,T]\ \forall\varepsilon\in(0,\varepsilon_0]$, $M_1>0$;
(3) $k(x)<M(x^2+1)$, $|k'(x)|<M\sqrt{x^2+1}$, $M>0$; *and*
(4) $R_N>-m$, $m>0$.

Then, $|R_N(x,t,\varepsilon)|\le M_2\ \forall(x,t)\in(-\infty,+\infty)\times[0,T]\ \forall\varepsilon\in(0,\varepsilon_0]$, $M_2>0$.

Proof of Theorem 1. We denote $t^{m/n}a(t)=q(t)$. We prove the theorem in two stages.

Stage 1. Consider the homogeneous Cauchy problem in the domain $D_L=\{[-L,L]\times[0,T]\}$:

$$\begin{cases} L(u)=\varepsilon\dfrac{\partial u}{\partial t}-\varepsilon^2\dfrac{\partial}{\partial x}\left(k(x)\dfrac{\partial u}{\partial x}\right)+q(t)u=0,\\ u(x,0,\varepsilon)=0. \end{cases} \tag{21}$$

In the proof of this theorem, ideas from [12] are used to estimate the solution. We introduce the function $w = u + e^{\alpha t/\varepsilon}\frac{m}{L^2}(x^2 + pt)$. Then,

$$L(w) = L(u) + L\left(e^{\alpha t/\varepsilon}\frac{m}{L^2}(x^2 + pt)\right) =$$
$$= [(\alpha + q)(x^2 + pt) + \varepsilon p - \varepsilon^2 2xk'(x) - \varepsilon^2 2k(x)]e^{\alpha t/\varepsilon}\frac{m}{L^2} \geq$$
$$\geq e^{\alpha t/\varepsilon}\frac{m}{L^2}[(\alpha + q)(x^2 + pt) + \varepsilon p - \varepsilon^2 2Mx\sqrt{x^2+1} - \varepsilon^2 2M(x^2+1)].$$

(1) For $|x| \geq 1$,

$$L(w) \geq e^{\alpha t/\varepsilon}\frac{m}{L^2}[(\alpha + q)(x^2 + pt) + \varepsilon p - \varepsilon^2 8Mx^2] \geq e^{\alpha t/\varepsilon}\frac{m}{L^2}[\alpha - \varepsilon^2 8M]x^2.$$

Take $\alpha > \varepsilon_0^2 8M$, then $L(w) \geq 0$.

(2) For $|x| < 1$,

$$L(w) \geq e^{\alpha t/\varepsilon}\frac{m}{L^2}[(\alpha + q)(x^2 + pt) + \varepsilon p - \varepsilon^2 8M] \geq e^{\alpha t/\varepsilon}\frac{m}{L^2}[p - \varepsilon 8M]\varepsilon.$$

Take $p > \varepsilon_0 8M$, then $L(w) \geq 0$.

Besides,

$$w\big|_{t=0} = u\big|_{t=0} + \frac{m}{L^2}x^2 \geq 0,$$

$$w\big|_{\pm L} = u\big|_{\pm L} + e^{\alpha t/\varepsilon}\frac{m}{L^2}(L^2 + pt) \geq -m + m = 0.$$

Hence, by the maximum theorem in a bounded domain, we have $w \geq 0$ in D_L, i.e.

$$u + e^{\alpha t/\varepsilon}\frac{m}{L^2}(x^2 + pt) \geq 0.$$

Letting $L \to +\infty$, we get $u \geq 0$.

Stage 2. Consider the inhomogeneous Cauchy problem

$$\begin{cases} L(R_N) = H(x, t, \varepsilon), \\ R_N(x, 0, \varepsilon) = 0. \end{cases} \tag{22}$$

We introduce the function

$$w = \pm R_N + \frac{M_1 t}{\varepsilon} + m.$$

Then,

$$L(w) = \pm H + M_1 + q(t)\left(\frac{M_1 t}{\varepsilon} + m\right) \geq 0, \ \ w\big|_{t=0} = m \geq 0.$$

From the result of Stage 1, it follows that $w \geq 0$ in $D = (-\infty, +\infty) \times [0, T]$ $\forall \varepsilon \in (0, \varepsilon_0]$, i.e. $\pm R_N + \frac{M_1 t}{\varepsilon} + m \geq 0$. Consequently, $|R_N| \leq \frac{M_1 t}{\varepsilon} + m \leq \frac{M_3}{\varepsilon}$, $M_3 > 0$. We write the remainder

$$R_N = u_N + \varepsilon R_{N+1}.$$

Then, $|R_N| \leq |u_N| + \varepsilon\frac{M_4}{\varepsilon} \leq M_2$, $M_4 > 0$. ▷ □

4. Appendix

Lemma 1 (on the solvability of iterative problems). *Let the equation be given*

$$t^{m/n} Z(x, t) = F(x, t) \tag{23}$$

and conditions are met

$$F(x,t) \in C^{\infty}(R \times [0,T]).$$

Then, Equation (23) is solvable in the class of smooth functions if and only if

$$\frac{\partial F^k(x,0)}{\partial t} = 0, \ \ k = \overline{0, \left[\frac{m}{n}\right]}.$$

Proof of Lemma 1. We expand $F(x,t)$ by Maclaurin's formula in t:

$$F(x,t) = F(x,0) + \dot{F}(x,0)t + \ldots + F^{[m/n]}(x,0)t^{[m/n]} + t^{[m/n]+1}f(x,t).$$

Then, Equation (23) has the form

$$t^{m/n}Z(x,t) = F(x,0) + \dot{F}(x,0)t + \ldots + F^{[m/n]}(x,0)t^{[m/n]} + t^{[m/n]+1}f(x,t).$$

Necessity, Let Equation (23) have a solution. For $t = 0$, we have $0 = F(x,0)$:

$$t^{m/n}Z(x,t) = F(x,t) - F(x,0).$$

Dividing the equation by t if $\left[\frac{m}{n}\right] > 1$:

$$t^{\frac{m}{n}-1}Z(x,t) = \frac{F(x,t) - F(x,0)}{t}.$$

For $t = 0$, we get $0 = \dot{F}(x,0)$.

Continuing this process to step $\left[\frac{m}{n}\right]$, we get $F^k(x,0) = 0 \ \forall k = \overline{0, \left[\frac{m}{n}\right]}$.

Adequacy: Let be $F(x,0) = \dot{F}(x,0) = \ldots = F^{[m/n]}(x,0) = 0$. Then, Equation (23) has the form:

$$t^{m/n}Z(x,t) = t^{[m/n]+1}f(x,t),$$

hence the decision $Z(x,t) = t^{1-\{\frac{m}{n}\}}f(x,t)$. □

5. Conclusions

The novelty of the article lies in the construction by the regularization method of an asymptotic solution of a singularly perturbed Cauchy problem for a parabolic equation in the presence of a "simple" rational turning point of the limit operator. The nature of this "simple" turning point affects the structure of the functions describing the singular dependence of the solution on the parameter ε. The asymptotic expansion of the constructed solution is justified using the maximum principle. This approach can be applied both in the study of applied problems containing turning points and in the construction of numerical algorithms for solving problems with spectral features.

Funding: This research received no external funding.

Conflicts of Interest: The author declare no conflict of interest.

References

1. Lomov, S.A. *Introduction to the General Theory of Singular Perturbations*; Nauka: Moscow, Russia, 1981.
2. Butuzov, V.F.; Nefedov, N.N.; Schneider, K.R. Singularly Perturbed Problems in Case of Exchange of Stabilities. *J. Math. Sci.* **2004**, *121*, 1973–2079. [CrossRef]
3. Safonov, V.F.; Bobodzhanov, A.A. *Course of Higher Mathematics. Singularly Perturbed Equations and the Regularization Method*; Izdatelstvo MPEI: Moscow, Russia, 2012.
4. Butuzov, V.F.; Gromova, E.A. Singularly perturbed parabolic problem in the case of intersecting roots of the degenerate equation. In *Proceedings of the Steklov Institute of Mathematics (Supplementary Issues)*; Springer: Berlin/Heidelberg, Germany, 2003; Volume 1, pp. 37–44.

5. Bobochko, V.N. An Unstable Differential Turning Point in the Theory of Singular Perturbations. *Russ. Math.* **2005**, *49*, 6–14.
6. Tursunov, D.A.; Kozhbekov, K.G. Asymptotics of the Solution of Singularly Perturbed Differential Equations with a Fractional Turning Point. *Izvestiya Irkutskogo gos. universiteta* **2017**, *21*, 108–121.
7. Eliseev, A.G. Regularized solution of the singularly perturbed Cauchy problem in the presence of an irrational simple turning point. *Differ. Equ. Control. Process.* **2020**, *2*, 15–32.
8. Eliseev, A.G. On the Regularized Asymptotics of a Solution to the Cauchy Problem in the Presence of a Weak Turning Point of the Limit Operator. *Axioms* **2020**, *9*, 86. [CrossRef]
9. Eliseev, A.G.; Lomov, S.A. The theory of singular perturbations in the case of spectral singularities of the limit operator. *Sb. Math.* **1986**, *131*, 544–557. [CrossRef]
10. Eliseev, A.G.; Ratnikova, T.A. A singularly perturbed Cauchy problem in the presence of a rational "simple" turning point for the limit operator. *Differ. Equ. Control. Process.* **2019**, *3*, 63–71.
11. Eliseev, A.; Ratnikova, T. Regularized Solution of Singularly Perturbed Cauchy Problem in the Presence of Rational "Simple" Turning Point in Two-Dimensional Case. *Axioms* **2019**, *8*, 124. [CrossRef]
12. Ilyin, A.M.; Kalashnikov, A.S.; Oleinik, O.A. Linear equations of the second order of parabolic type. *UMN* **1962**, *17*, 3–116. [CrossRef]

axioms

MDPI

Article

Complete Controllability Conditions for Linear Singularly Perturbed Time-Invariant Systems with Multiple Delays via Chang-Type Transformation

Olga Tsekhan

Faculty of Economics and Management, Yanka Kupala State University of Grodno, 230023 Grodno, Belarus; tsekhan@grsu.by

Received: 19 February 2019; Accepted: 23 May 2019; Published: 3 June 2019

Abstract: The problem of complete controllability of a linear time-invariant singularly-perturbed system with multiple commensurate non-small delays in the slow state variables is considered. An approach to the time-scale separation of the original singularly-perturbed system by means of Chang-type non-degenerate transformation, generalized for the system with delay, is used. Sufficient conditions for complete controllability of the singularly-perturbed system with delay are obtained. The conditions do not depend on a singularity parameter and are valid for all its sufficiently small values. The conditions have a parametric rank form and are expressed in terms of the controllability conditions of two systems of a lower dimension than the original one: the degenerate system and the boundary layer system.

Keywords: time delay system; multiple commensurate delays; singular perturbation; decomposition; Chang transformation; complete controllability; robust sufficient condition

1. Introduction

We consider a singularly-perturbed linear time-invariant system with a small multiplier for the derivatives and with non-small commensurate delays in the slow state variables (SPLTISD).

Singularly-perturbed controlled systems (SPS) occur as models in automatic control theory, nonlinear oscillation theory, quantum mechanics, gas dynamics, biology, chemical kinetics, and others (see, e.g., the references in [1–3]). Time-delay systems arise from inherent time delays in the components of the systems or from the deliberate introduction of time delays into the systems for control purposes. Time delays occur often in systems in engineering, biology, chemistry, physics, ecology, economics, technology, the social sphere, etc. (see, e.g., the references in [4,5]).

Controllability is one of the basic properties of controlled systems. This property is well known from the mathematical theory of systems (Kalman) as the concept of the reachability of terminal states. This means that it is possible to control a dynamic system from an arbitrary initial state to an arbitrary final state using a set of admissible controls.

Time-delay systems can be represented by delay differential equations, which belong to the class of functional differential equations and are infinite-dimensional systems [6,7]. Due to this fact, the controllability concepts for systems with delay are more diverse, and their analysis is significantly more complicated than for systems without delay. Different types of controllability for infinite-dimensional systems were studied in the literature (see, e.g., [8–10] and the references therein). There are several concepts of time delay system controllability that are a direct generalization of the concept of controllability in the Kalman theory of systems: relative controllability (equivalently, R^n or Euclidean-space controllability) (rank criterion [11]), complete controllability (formulation of the problem: Krasovsky N.N., 1963; condition [12]), and pointwise controllability [13,14]).

Controllability in the full state space (approximate, functional controllability [15,16]) is quite a restrictive concept [16]. For instance, many delay systems do not satisfy one of the necessary conditions of approximate controllability, even though they often possess other good properties such as stabilizability and spectral controllability. This suggests that from the controllability point of view, the full state space is "too big", and in the above-mentioned cases of controllability, the state (unlike the classical works of Kalman) is not based on the concept of minimality. Therefore, one should search for a "smaller space" in which a controllability would be characterized by less restrictive conditions and would be related to stabilizability and spectral controllability. The concept of F-controllability, weaker than the approximate controllability in the state space, has been introduced (see [16] and the references therein). Works on the study of systems with delay on the basis of an approach of the space of minimal states also appeared (see [17] and the references therein).

The functional controllability of systems with delay according to the approach, based on the notion of time delay systems' state minimality [17], turns out to be equivalent to the complete controllability, which solves one of the most difficult problems of controllability for systems with delay: the problem of complete damping of such systems in a finite time (total quieting, controllability to null function, null controllability). The spectral criterion of complete controllability for a system with delay (without parameter) was proven in [12,16,18] (see also the references in [17]). In [12,15,19] and a number of other papers, these conditions were associated with parametric rank Kalman-type conditions.

Studying the controllability of SPS without delay is well known (see, e.g., the reviews of [1,20] and the references therein). The controllability of SPS with delay has been studied much less (see [21–27], reviews [1,20], and the references therein).

To check a proper type of controllability for a given SPS, corresponding controllability conditions can be directly applied for any specified value of a small parameter of singular perturbation. However, the stiffness, as well as a possible high dimension of the SPS, can considerably complicate this application. Therefore, for example, with the direct use of rank controllability criteria [12,16,18], the controllability matrix of such systems has a large dimension and is ill-conditioned. Correct checking of the rank of such matrices can be complicated. Moreover, such an application depends on the value of the parameter, i.e., it is not robust with respect to this parameter, while in most of real-life problems, this value is unknown [28].

One of the approaches independent of the singular perturbation parameter controllability analysis of SPS is based on the time scale separation concept (see, e.g., [29]). Using this concept, the complete controllability of SPS without delays was analyzed in the works [29–31]. Parameter-free conditions of complete Euclidean space controllability, robust with respect to the small parameter, were obtained for linear singularly-perturbed time-invariant system with a single pointwise non-small delay in the state variables in [21], for linear SPS with point-wise and distributed small delays (of the order of the small parameter) in the state variables in [25–27,32,33], and for a linear singularly-perturbed neutral-type system with a single non-small point-wise delay in [34].

The problem of functional controllability for SPS with delay has been investigated much less. In [28], a singularly-perturbed linear time-dependent controlled system with multiple point-wise and distributed state delays was considered (the delays were small in the fast state variable and non-small in the slow state variable). It has been established that the approximate state-space controllability of two parameter-free subsystems (the slow and fast ones), associated with the original system, yields the approximate state-space controllability of the original system robustly with respect to the parameter of singular perturbation for all its sufficiently small values. In [35], the conditions of controllability in the $L_2^2[t_1 - h, t_1] \times R^2$ space for linear stationary SPS with delay was obtained on the basis of the state-space method.

One of the realizations of the time scale separation concept is Chang's transformation with a nondegenerate change of variables (for linear singularly-perturbed continuous-time varying systems without delay, introduced in [36]). Generalizations of Chang-type transformation for linear SPS slowly varying in time were proposed in [37,38] and on systems with many time scales in [39]. For SPS with delay, Chang-type transformations were constructed in [21,23,40,41]. In [40], the existence of a continuous function on the small parameter linear transformation for partial decomposition of SPS with distributed and concentrated non-small delays in slow and fast variables was proven. As a result, in the transformed system, there is a connection between fast and slow variables only through variables with delay. Constructed in [21,23], Chang-type transformation for a linear stationary SPS with a constant (non-small) delay in the state was performed by a linear operator with a finite number of delay operators and resulted in the original SPS with one delay in the state and without delay in the control to split the subsystems: slow with many delays in the state and control and heterogeneity depending on the initial conditions and fast with one delay in the state variable and with heterogeneity. In [41], the change of variables for linear time-invariant SPS of functional-differential equations with small concentrated and distributed delay in fast variables was constructed. The transformation in [42] generalizes the transformation in [36] to linear stationary SPS with concentrated delay in the slow variable. Unlike [21,23], this transformation was constructed in the form of an asymptotic series, and the obtained slow and fast subsystems did not contain inhomogeneities with the exception of control components.

In [29,43] and other papers, Chang's transformation was applied to split the original system with fast and slow variables into two independent subsystems and to obtain controllability conditions for SPS without delay. The result of the non-degenerate transformation [21,23] was used to study the relative controllability of the original SPS with delay. In [44], the sufficient conditions for complete controllability based on Chang's transformation [42] of linear stationary SPS with the single delay were obtained (without detailed proof).

In this paper, the problem of complete controllability of a linear time-invariant singularly-perturbed system with multiple commensurate non-small delays in the slow state variables on the basis of the time scale separation concept is considered. The main differences of this work from [28] are in the property under investigation (complete controllability) and in the method used for the investigation (the method of non-degenerate variable transformation is evolved in this work). The non-degenerate change of variables was developed in [42], where decoupling transformation in the form of asymptotic series was constructed for a singular perturbed system with single non-small delay. The exact separation is performed by means of non-degenerate transformation of the original system. Two much simpler subsystems than the original SPS parameter-free ones are associated with the original system. They are $O(\mu)$, close to the decoupled subsystems, the slow and fast ones. It is established that the complete controllability of the slow and fast subsystems yields the complete controllability of the original system.

Parameter-free sufficient conditions of complete controllability of the singularly-perturbed system with non-small delay are obtained. The conditions are valid for all sufficiently small values of the parameter of singular perturbation, i.e., robustly with respect to this parameter, and have a rank form.

The paper is organized as follows. In the second section, the problem is formulated, the main definitions dependent on the parameter criterion of complete controllability of the considered system are presented. Criteria of complete controllability for the fast and slow subsystems, associated with the original one, are presented in Section 3. Section 4 is devoted to Chang-type decoupling transformation in the form of asymptotic series for singular perturbed systems with non-small delay. Section 5 is devoted to the main result. An illustrative example is presented in Section 6. The discussion and conclusions are placed in Sections 7 and 8, respectively.

The following main notations and notions are applied in the paper:

- $'$ means the transposition;
- I_n denotes the identity matrix of dimension n;
- C is the set of complex numbers;
- R is the set of real numbers;
- $p \triangleq \frac{d}{dt}$ is a differential operator;

- e^{-ph} is a delay operator: $e^{-ph}z(t) \triangleq z(t-h)$.

2. Singularly-Perturbed Linear Time-Invariant System with Delays and Its Complete Controllability: Definitions

2.1. Singularly-Perturbed Linear Time-Invariant System with Delays

Consider the singularly-perturbed linear time-invariant system with multiple commensurate delays in the slow state variables (SPLTISD):

$$\dot{x}(t) = \sum_{j=0}^{l} A_{1j}x(t-jh) + A_2y(t) + B_1u(t),\ x \in R^{n_1},\ y \in R^{n_2}, \tag{1}$$

$$\mu\dot{y}(t) = \sum_{j=0}^{l} A_{3j}x(t-jh) + A_4y(t) + B_2u(t),\ u \in R^{r},\ t \geq 0, \tag{2}$$

with the initial conditions:

$$x(0) = x_0,\ y(0) = y_0,\ x_0 \in R^{n_1},\ y_0 \in R^{n_2},\ x(\theta) = \varphi(\theta), \theta \in [-lh, 0). \tag{3}$$

Here, $0 < h$ is a given constant, x is a slow variable, y is a fast variable, u is a control, $u(t) \in U$, U is a set of piecewise continuous $t \geq 0$ vector functions, μ is a small parameter, $\mu \in (0, \mu^0], \mu^0 \ll 1$, $A_{ij}, i = 1,3, j = \overline{0,l}, A_k, k = 2,4, B_j, j = 1,2$, are constant matrices of appropriate dimensions, and $\phi(\theta)$, $\theta \in [-lh, 0)$, is a piecewise continuous n_1 vector function. Assume that det $A_4 \neq 0$.

For a given $\mu > 0$, using the notations $p \triangleq \frac{d}{dt}$, a differential operator, e^{-ph}, and a delay operator, $e^{-ph}z(t) \triangleq z(t-h)$, introduce the following matrix-valued operators that depend on the parameter:

$$\mathsf{A}\left(\mu, e^{-ph}\right) = \begin{pmatrix} \mathsf{A}_1\left(e^{-ph}\right) & A_2 \\ \frac{\mathsf{A}_3\left(e^{-ph}\right)}{\mu} & \frac{A_4}{\mu} \end{pmatrix},\quad B(\mu) = \begin{pmatrix} B_1 \\ \frac{B_2}{\mu} \end{pmatrix},\quad \mu > 0, \tag{4}$$

where:

$$\mathsf{A}_i\left(e^{-ph}\right) \triangleq \sum_{j=0}^{l} A_{ij}e^{-jph},\ i = 1,3. \tag{5}$$

Introduce also the vector $z = (x', y')'$. Using the above notations, we can rewrite SPLTISD (1)–(2) in the equivalent operator form:

$$pz(t) = \mathsf{A}\left(\mu, e^{-ph}\right) z(t) + B(\mu)\, u(t),\ u \in R^{r},\ t \geq 0. \tag{6}$$

From (6), we obtain the characteristic equation of the system (1)–(2):

$$w\left(\mu, \lambda, e^{-\lambda h}\right) \triangleq \det\left[\lambda I_{n_1+n_2} - \mathsf{A}\left(\mu, e^{-\lambda h}\right)\right] = 0. \tag{7}$$

For any fixed $\mu \in (0, \mu^0]$, by:

$$\sigma(\mu) = \left\{\lambda(\mu) \in C : w\left(\mu, \lambda, e^{-\lambda h}\right) = 0\right\} \tag{8}$$

denote the spectrum (set of the eigenvalues) of the SPLTISD (1).

From the known properties of the delay system spectrum [6,7] follows the characterization of the SPLTISD spectrum (8) for any given $\mu \in (0, \mu^0]$.

Characterization 1. *For a given $\mu \in (0, \mu^0]$, the following statements are true:*

(a) *the spectrum $\sigma(\mu)$ of the SPLTISD (1)–(2) consists of a finite or countable set of complex numbers;*
(b) *the real part of all SPLTISD (1)–(2) eigenvalues is bounded above by some real value γ;*
(c) *any vertical strip of the complex plane with $a \le Re\, z \le b$ contains a finite number of SPLTISD eigenvalues;*
(d) *any two subsets of the set $\sigma(\mu)$ are separated on the complex plane by a vertical strip of nonzero width.*

2.2. Definition and Dependent on the μ Criterion of Split Complete Controllability

Similar to [12], let us introduce the following definition.

Definition 1. *For a given $\mu \in (0, \mu^0]$, the SPLTISD (1)–(2) is said to be completely controllable if for any fixed initial conditions (3), there exist a time moment $t_1 < +\infty$ $(t_1 > (n_1 + n_2)h)$ and a piecewise continuous control $u(t)$, $t \in [0, t_1]$, such that for this control and the corresponding solution $(x(t,\mu), y(t,\mu))$, $t \ge 0$, of the system (1)–(2) with the initial conditions (3), the following identities are valid:*

$$(x(t,\mu), y(t,\mu)) \equiv 0,\; u(t) \equiv 0,\; t \ge t_1.$$

Definition 2. *If there exists a number $\mu^* > 0$ for which SPLTISD (1)–(2) is completely controllable for any $\mu \in (0, \mu^*]$, we say that complete controllability is robust with respect to $\mu \in (0, \mu^*]$.*

For $\mu > 0, \lambda \in C$, we introduce the following matrix-valued function:

$$\mathsf{N}\left(\mu, \lambda, e^{-\lambda h}\right) \triangleq \left[\lambda I_{n_1+n_2} - \mathsf{A}\left(\mu, e^{-\lambda h}\right), B(\mu)\right]. \tag{9}$$

The following criterion of the SPLTISD complete controllability for a fixed $\mu \in (0, \mu^0]$ follows from [12].

Theorem 1. *For a given $\mu \in (0, \mu^0]$, the SPLTISD (1)–(2) is completely controllable if and only if:*

$$rank\, \mathsf{N}\left(\mu, \lambda, e^{-\lambda h}\right) = n_1 + n_2, \quad \forall \lambda \in \sigma(\mu). \tag{10}$$

2.3. Objective of the Paper

Our objective in this paper is the following. On the basis of a non-degenerate change of variables in the original system, we prove the approximation of the original SPLTISD (1)–(2) by two independent small parameter subsystems of lower dimension and obtain parametric rank-type sufficient conditions for complete controllability of the original singularly-perturbed system in terms of the complete controllability of these subsystems. The conditions do not depend on the parameter and are robust with respect to μ for all its sufficiently small values.

3. Subsystems of SPLTISD

3.1. Slow and Fast Subsystems of SPLTISD

With the $n_1 + n_2$-dimensional system (1)–(3) is associated two independent μ subsystems: the slow and the fast ones. The slow subsystem, the degenerate system (DS), has the form:

$$\dot{x}_s(t) = \sum_{j=0}^{l} A_{sj} x_s(t - jh) + B_s u_s(t),\; x_s \in R^{n_1},\; t \ge 0, \tag{11}$$

$$x_s(0) = x_0, x_s(\theta) = \phi(\theta), \theta \in [-lh, 0), \tag{12}$$

where:

$$A_{sj} \triangleq A_{1j} - A_2 A_4^{-1} A_{3j},\; j = \overline{0,l},\; B_s \triangleq B_1 - A_2 A_4^{-1} B_2, \tag{13}$$

and $u_s(t) \in R^r$ (u_s is a control). Introducing the following matrix-valued operator:

$$\mathsf{A}_s\left(e^{-ph}\right) \triangleq \mathsf{A}_1\left(e^{-ph}\right) - A_2 A_4^{-1} \mathsf{A}_3\left(e^{-ph}\right), \tag{14}$$

we can rewrite the degenerate system (11) in the operator form:

$$px_s(t) = \mathsf{A}_s\left(e^{-ph}\right) x_s(t) + B_s u_s(t), \; x_s \in R^{n_1}, \; t \geq 0, \tag{15}$$

$$x_s(0) = x_0, x_s(\theta) = \phi(\theta), \theta \in [-lh, 0). \tag{16}$$

The degenerate system (11) is a linear stationary n_1-dimensional system with multiple commensurate delays. It is obtained from (1)–(2) by setting there formally $\mu = 0$, expressing $y_s(t) = A_4^{-1}\left[\mathsf{A}_3\left(e^{-ph}\right) x_s(t) + B_2 u_s(t)\right]$ from (2) and substituting it into (1).

The characteristic equation for the DS (11) is:

$$w_s\left(\lambda, e^{-\lambda h}\right) \triangleq \det\left[\lambda I_{n_1} - \mathsf{A}_s\left(e^{-\lambda h}\right)\right] = 0, \tag{17}$$

the spectrum of the DS (11):

$$\sigma_s = \left\{\lambda \in \mathsf{C} : w_s\left(\lambda, e^{-\lambda h}\right) = 0\right\} \tag{18}$$

is a finite or countable set of complex numbers.

Since the DS (11) is a system with delay, properties similar to the properties from the characterization 1 are valid for the spectrum (18).

The fast subsystem, the boundary layer system (BLS), has the form:

$$\frac{dy_f(\tau)}{d\tau} = A_4 y_f(\tau) + B_2 u_f(\tau), \, y_f \in R^{n_2}, \, \tau = \frac{t}{\mu} \geq 0, \tag{19}$$

$$y_f(0) = y_0 - A_4^{-1}\left[\mathsf{A}_3\left(e^{-\lambda h}\right)\phi(0) + B_2 u_s(0)\right]. \tag{20}$$

Here, $y_f(\tau) = y(\mu\tau) - y_s(\mu\tau), u_f(\tau) = u(\mu\tau) - u_s(\mu\tau)$.

The boundary layer system (19) is a linear stationary n_2-dimensional system without delay and is derived from Equation (2) for the fast state variable y in the following way: (i) the terms containing the slow state variable x are removed from Equation (2); (ii) the transformation of variables $t = \mu\tau, y(\mu\tau) = y_f(\tau), u(\mu\tau) = u_f(\tau)$ is done in the resulting equation, where τ, y_f and u_f are new independent variables (the stretched time), state and control, respectively.

The characteristic equation for the BLS (19) is:

$$w_f(\lambda) \triangleq \det\left[\lambda I_{n_2} - A_4\right] = 0. \tag{21}$$

The spectrum of the BLS (19) is the finite set of complex numbers:

$$\sigma_f = \left\{\lambda \in \mathsf{C} : w_f(\lambda) = 0\right\}. \tag{22}$$

Similar to [12,45], we introduce the following definitions.

Definition 3. *The DS (11) is said to be completely controllable if for any fixed initial conditions (16), there exists a time moment $t_1 < +\infty$ and a piecewise continuous control $u_s(t)$, $t \in [0, t_1]$ such that for this control and corresponding solution $x_s(t), t \geq 0$, of the system (11) with the initial conditions (16), the following identities are true:*

$$x_s(t) \equiv 0, \; u_s(t) \equiv 0, \; t \geq t_1.$$

Definition 4. *The BLS (19) is said to be completely controllable if for any fixed initial conditions (20), there exist a time moment $\tau_1 < +\infty$ and a piecewise continuous control $u_f(\tau)$, $\tau \in [0, \tau_1]$ such that for this control and corresponding solution $y_f(\tau), \tau \geq 0$, of the system (14) with the initial conditions (20), the following equality is true:*

$$y_f(\tau_1) = 0.$$

Note that the subsystems (11) and (19) have smaller dimensions than the original SPLTISD (1)–(2) and do not depend on a small parameter μ.

The main objective of the article is to obtain the conditions of complete controllability of the SPLTISD (1)–(2) (Definition 1) in terms of complete controllability of its subsystems (11) and (19) (Definitions 3 and 4), robust with respect to μ for all its sufficiently small values (Definition 2).

3.2. Controllability of Subsystems

Define the matrix-valued functions:

$$\mathsf{N}_s\left(\lambda, e^{-\lambda h}\right) \triangleq \left[\lambda I_{n_1} - \mathsf{A_s}\left(e^{-\lambda h}\right), B_s\right],\ \lambda \in C, \tag{23}$$

$$N_f(\lambda) \triangleq [\lambda I_{n_2} - A_4, B_2],\ \lambda \in C. \tag{24}$$

Applying the conditions of complete controllability from [12] to DS (11) and BLS (19), we obtain that the following theorems are valid.

Theorem 2. *The DS (11) is completely controllable if and only if the following condition is valid:*

$$rank\,\mathsf{N}_s\left(\lambda, e^{-\lambda h}\right) = n_1 \quad \forall\lambda \in \sigma_s. \tag{25}$$

Theorem 3. *The BLS (19) is completely controllable if and only if the following condition is valid:*

$$rank\,N_f(\lambda) = n_2 \quad \forall\lambda \in \sigma_f. \tag{26}$$

Along with Conditions (25) and (26), we formulate some more applicable conditions for the complete controllability of the subsystems, which simplify the procedure for checking this property. To do this, we define the matrix-valued function:

$$\mathsf{P_s}(z) \triangleq \left[B_s, \mathsf{A_s}(z)\,B_s, ..., \mathsf{A_s}^{n_1-1}(z)\,B_s\right],\ z \in C, \tag{27}$$

and the matrix:

$$\mathsf{P_f} \triangleq \left[B_2, \mathsf{A_4}B_2, ..., \mathsf{A_4}^{n_2-1}B_2\right]. \tag{28}$$

The following theorem follows from the application to the subsystems (11) and (19) of the results from [19] about the connection of the ranks of matrices (27) and (23).

Theorem 4. *Let for some $\lambda \in C$:*

$$rank\ \mathsf{P}_s\left(e^{-\lambda h}\right) = n_1. \tag{29}$$

Then, for this λ:

$$rank\,\mathsf{N}_s\left(\lambda, e^{-\lambda h}\right) = n_1. \tag{30}$$

If we apply the well-known criterion of the controllability of a linear stationary system [45] to the BLS (19), we obtain the following theorem:

Theorem 5. *The following equality:*

$$rank\, \mathsf{N}_f(\lambda) = n_2\ \forall\lambda \in C \tag{31}$$

holds if and only if:

$$rank\ \mathsf{P}_f = n_2. \tag{32}$$

In order to formulate the conditions of subsystems' controllability, that do not require the computation of all eigenvalues from (18) and (22), let us define the following set of complex numbers:

$$Z_s = \{z \in C : rank\ \mathsf{P_s}(z) < n_1\} \tag{33}$$

the set of numbers for which the matrix $\mathsf{P_s}(z)$ does not have full rank by rows.

Let $\pi_s(z)$ be the greatest common divisor of all minors of order n_1 of the function matrix $P_s(z)$ (27). Taking into account (5), (14), (13), and (27), we have that $\pi_s(z)$ is a polynomial of degree l_s no higher than n_1 and can be represented as:

$$\pi_s(z) = \sum_{i=0}^{l_s} k_i z,\ k_{l_s} \neq 0. \tag{34}$$

Since the set Z_s (33) coincides with the set of all roots of the polynomial $\pi_s(z)$, then:

$$Z_s = \{z \in C : \pi_s(z) = 0\}$$

and the set of Z_s contains the finite numbers of elements.

Along with (33), let us define the set of complex numbers:

$$\Lambda_s = \left\{\lambda \in C : e^{-\lambda h} = z,\ z \in Z_s\right\}, \tag{35}$$

associated with Z_s. The elements $z \in Z_s$, $\lambda_k \in \Lambda_s$, are connected by the relation:

$$h\,\lambda_k = \ln|z| + i\,(\arg z + 2\pi k)\,,\ z \in Z_s,\ i = \sqrt{-1},\ k = 0, \pm 1, \pm 2, \ldots. \tag{36}$$

Since the set Z_s consists of a finite number of elements, from the connection (36) between the sets Λ_s and Z_s, the validity follows:

Characterization 2. *There are real numbers $\alpha, \gamma,\ \alpha < \gamma$, such that $\alpha < Re\,\lambda \leq \gamma\ \forall\lambda \in \Lambda_s$.*

Define also the set:

$$\Omega_s \triangleq \sigma_s \cap \Lambda_s.$$

By virtue of the connection (36) between the elements of the sets (35) and (33) from Theorem 4, it follows that the conditions (25) of complete controllability of DS (11) are sufficient to check only for $\lambda \in \Omega_s$, and the following theorem holds [19].

Theorem 6. *Let:*

(1) *there exists $z \in C$ that:*

$$rank\ \mathsf{P}_s(z) = n_1; \tag{37}$$

(2)

$$rank\,\mathsf{N}_s(\lambda, z) = n_1\ \forall\lambda \in \Omega_s,\ z \in Z_s. \tag{38}$$

Then, the DS (11) is completely controllable.

From [45] follows:

Theorem 7. *The BLS (19) is completely controllable if and only if the following condition is satisfied:*

$$rank\, P_f = n_2. \tag{39}$$

4. Decoupling Transformation for the SPLTISD

Similar to [42] for asymptotic decomposition of the SPLTISD, we introduce the change of variables:

$$\begin{pmatrix} x(t) \\ y(t) \end{pmatrix} = \mathsf{G}\left(\mu, e^{-ph}\right) \begin{pmatrix} \xi(t) \\ \eta(t) \end{pmatrix}, \quad \xi(t) \in R^{n_1},\ \eta(t) \in R^{n_2},\ t \in T, \tag{40}$$

$$\mathsf{G}\left(\mu, e^{-ph}\right) = \begin{pmatrix} I_{n_1} & \mu \mathsf{H}\left(\mu, e^{-ph}\right) \\ -\mathsf{L}\left(\mu, e^{-ph}\right) & I_{n_2} - \mu \mathsf{L}\left(\mu, e^{-ph}\right) \mathsf{H}\left(\mu, e^{-ph}\right) \end{pmatrix}, \tag{41}$$

where $\mathsf{H}\left(\mu, e^{-ph}\right)$ and $\mathsf{L}\left(\mu, e^{-ph}\right)$ are the matrix-valued operators, depending on the parameter μ. They are the solutions of the following matrix-valued functional equations (in order to reduce the records, where this does not lead to ambiguous understanding, we omit the arguments $\left(\mu, e^{-\lambda h}\right)$ of the matrix-valued operators $\mathsf{H}\left(\mu, e^{-ph}\right)$, $\mathsf{L}\left(\mu, e^{-ph}\right)$):

$$\begin{aligned} &\mathsf{A}_3\left(e^{-ph}\right) - A_4\mathsf{L} + \mu\,\mathsf{L}\,\mathsf{A}_1\left(e^{-ph}\right) - \mu\,\mathsf{L}\,A_2\mathsf{L} = 0, \\ &\mu\left(\mathsf{A}_1\left(e^{-ph}\right) - A_2\mathsf{L}\right)\mathsf{H} - \mathsf{H}\left(A_4 + \mu\,\mathsf{L}\,A_2\right) + A_2 = 0. \end{aligned} \tag{42}$$

Notice that:

$$\det \mathsf{G}\left(\mu, e^{-ph}\right) \equiv 1, \quad \mathsf{G}^{-1}\left(\mu, e^{-ph}\right) = \begin{pmatrix} I_{n_1} - \mu\,\mathsf{HL} & -\mu\,\mathsf{H} \\ \mathsf{L} & I_{n_2} \end{pmatrix}. \tag{43}$$

By $O(\mu)$, we denote any vector function $f(t,\mu)$, $t \in [t_1, t_2]$, with the following property: there exist positive constants μ^* and c such that the Euclidean norm $|f(t,\mu)|$ satisfies the inequality $|f(t,\mu)| \le c\mu$ for all $\mu \in (0, \mu^*]$ and all $t \in [t_1, t_2]$.

Lemma 1. *Suppose that* $\det A_4 \neq 0$*. Then, there exists a* $\mu^* > 0$ *such that for all* $\mu \in [0, \mu^*]$*, there is a continuous function depending on the* μ *solution* $\mathsf{L}\left(\mu, e^{-ph}\right)$, $\mathsf{H}\left(\mu, e^{-ph}\right)$ *of Equation (42) that could be represented in asymptotic series form:*

$$\begin{aligned} \mathsf{L}\left(\mu, e^{-ph}\right) &= \sum_{i=0}^{k} \mu^i \mathsf{L}^i\left(e^{-ph}\right) + O\left(\mu^{k+1}\right), \\ \mathsf{H}\left(\mu, e^{-ph}\right) &= \sum_{i=0}^{k} \mu^i \mathsf{H}^i\left(e^{-ph}\right) + O\left(\mu^{k}\right), \end{aligned} \tag{44}$$

where:

$$\begin{aligned} &\mathsf{L}^0\left(e^{-ph}\right) = A_4^{-1}\mathsf{A}_3\left(e^{-ph}\right), \ \mathsf{L}^1\left(e^{-ph}\right) = A_4^{-2}\mathsf{A}_3\left(e^{-ph}\right)\mathsf{A}_0\left(e^{-ph}\right), \\ &\mathsf{A}_0\left(e^{-ph}\right) = \mathsf{A}_1\left(e^{-ph}\right) - A_2A_4^{-1}\mathsf{A}_3\left(e^{-ph}\right), \ H^0 = A_2A_4^{-1}. \end{aligned} \tag{45}$$

Proof. It is easy to prove the decomposition (44) according to the scheme of the proof of [29]. Continuity is proven as in [40]. For the SPLTISD with a simple delay, see [42]. □

The next corollary follows from Lemma 1 if we substitute (44) into (42) and equate the coefficients of equal powers of μ in the resulting equations.

Corollary 1. *Let* $\det A_4 \neq 0$. *A solution of matrix equations (42) can be found with any degree of accuracy in the form of (44), where terms of the asymptotic series (44) can be found according to the following iterative scheme (in order to reduce the records, we omit the arguments* $e^{-\lambda h}$ *of the matrix-valued operators* $\mathsf{L}^k\left(e^{-ph}\right)$, $\mathsf{H}^k\left(e^{-ph}\right)$):

$$\begin{aligned}
&\mathsf{L}^{k+1} = A_4^{-1}\left(\mathsf{L}^k \mathsf{A}_1\left(e^{-ph}\right) - \sum_{j=0}^{k} \mathsf{L}^{k-j} A_2 \mathsf{L}^j\right),\ \mathsf{L}^0\left(e^{-ph}\right) = A_4^{-1}\mathsf{A}_3\left(e^{-ph}\right),\\
&\mathsf{H}^{k+1} = A_4^{-1}\left(\mathsf{A}_1\left(e^{-ph}\right)\mathsf{H}^k - A_2\sum_{i=0}^{k} \mathsf{L}^i \mathsf{H}^{k-i} - \sum_{i=0}^{k} \mathsf{H}^i \mathsf{L}^{k-i} A_2\right),\ \mathsf{H}^0 = A_2 A_4^{-1}.
\end{aligned} \tag{46}$$

By using the SPLTISD (1)–(2) matrix parameters and the matrix-valued functions $\mathsf{L}\left(\mu, e^{-ph}\right)$, $\mathsf{H}\left(\mu, e^{-ph}\right)$, we introduce the matrix-valued functions:

$$\begin{aligned}
\mathsf{A}_\xi\left(\mu, e^{-ph}\right) &\triangleq \mathsf{A}_1\left(e^{-ph}\right) - A_2\mathsf{L}\left(\mu, e^{-ph}\right),\\
\mathsf{B}_\xi\left(\mu, e^{-ph}\right) &\triangleq B_1 - \mathsf{H}\left(\mu, e^{-ph}\right)B_2 - \mu\mathsf{H}\left(\mu, e^{-ph}\right)\mathsf{L}\left(\mu, e^{-ph}\right)B_1,\\
\mathsf{A}_\eta\left(\mu, e^{-ph}\right) &\triangleq A_4 + \mu\,\mathsf{L}\left(\mu, e^{-ph}\right)A_2,\\
\mathsf{B}_\eta\left(\mu, e^{-ph}\right) &\triangleq B_2 + \mu\mathsf{L}\left(\mu, e^{-ph}\right)B_1.
\end{aligned} \tag{47}$$

Note here that due to Lemma 1, similar to [40], it is easy to prove that matrices from (47) continuously depend on μ for $[0, \mu^*]$.

From (14), (13), (44), and (45), we have:

$$\begin{aligned}
\mathsf{A}_\xi\left(\mu, e^{-ph}\right) &= \mathsf{A}_s\left(e^{-\lambda h}\right) + O\left(\mu\right),\\
\mathsf{B}_\xi\left(\mu, e^{-ph}\right) &= B_s + O\left(\mu\right),\\
\mathsf{A}_\eta\left(\mu, e^{-ph}\right) &= A_4 + O\left(\mu\right),\\
\mathsf{B}_\eta\left(\mu, e^{-ph}\right) &= B_2 + O\left(\mu\right).
\end{aligned} \tag{48}$$

As a result of the application to the system (1)–(2) of the transformation (40), taking into account (43) and (47), the SPLTISD (1)–(2) goes into the equivalent system with separated motions:

$$\dot{\xi}\left(t\right) = \mathsf{A}_\xi\left(\mu, e^{-ph}\right)\xi\left(t\right) + \mathsf{B}_\xi\left(\mu, e^{-ph}\right)u\left(t\right), \xi\left(t\right) \in R^{n_1}, \tag{49}$$

$$\mu\dot{\eta}\left(t\right) = \mathsf{A}_\eta\left(\mu, e^{-ph}\right)\eta\left(t\right) + \mathsf{B}_\eta\left(\mu, e^{-ph}\right)u\left(t\right), \eta\left(t\right) \in R^{n_2},\ t > 0. \tag{50}$$

Due to (48), the decoupled system (49) and (50) is $O\left(\mu\right)$-close to the DS (11) and the BLS (19).

The decomposition (49) and (50) allows us to prove the separation (at sufficiently small μ) of the SPLTISD spectrum $\sigma\left(\mu\right)$ (8) into two disjoint parts with "slow" and "fast" eigenvalues, as well as the approximation of the SPLTISD spectrum $\sigma\left(\mu\right)$ (8) elements by the eigenvalues of σ_s (18) and σ_f (22).

Let us define:

$$w_\xi\left(\mu, \lambda, e^{-\lambda h}\right) \triangleq \left[\lambda I_{n_1} - \mathsf{A}_\xi\left(\mu, e^{-\lambda h}\right)\right]. \tag{51}$$

Due to (48) and (17), we have:

$$w_\xi\left(\mu, \lambda, e^{-\lambda h}\right) = w_s\left(\lambda, e^{-\lambda h}\right) + O\left(\mu\right). \tag{52}$$

Theorem 8. *For sufficiently small* $\mu \in \left(0, \mu^0\right]$, *the spectrum* $\sigma\left(\mu\right)$ *(8) of the SPLTISD (1)–(2) is separated into two disjoint parts:*

$$\sigma\left(\mu\right) = \sigma_x\left(\mu\right) \cup \sigma_y\left(\mu\right),\quad \sigma_x\left(\mu\right) \cap \sigma_y\left(\mu\right) = \varnothing. \tag{53}$$

The "slow" part:

$$\sigma_x(\mu) = \left\{\lambda \in C : \det\left[\lambda I_{n_1} - A_\xi\left(\mu, e^{-\lambda h}\right)\right] = 0\right\} \tag{54}$$

consists of elements that for sufficiently small μ are the functions $\lambda(\mu)$ that continuously depend on μ and tend to the elements of the DS (11) spectrum (18) as $\mu \to 0$:

$$\lim_{\mu \to 0} \lambda_i(\mu) = \lambda_{si} \in \sigma_s. \tag{55}$$

The fast part:

$$\sigma_y(\mu) = \left\{\lambda \in C : \det\left[\lambda I_{n_2} - A_\eta\left(\mu, e^{-\lambda h}\right)\right] = 0\right\} \tag{56}$$

consists of n_2 elements that tend to infinity, with the rate μ^{-1}, and are of the form $\frac{\lambda_i(\mu)}{\mu}$, where:

$$\lim_{\mu \to 0} \lambda_i(\mu) = \lambda_{fi} \in \sigma_f. \tag{57}$$

If in the spectrum σ_s (18) of the DS (11), there are not multiple values and in the spectrum σ_f (22) of BLS (19) there are not multiple values (it is allowed $\sigma_s \cap \sigma_f \neq \emptyset$), then the eigenvalues of the SPLTISD (1)–(2) are approximated as:

$$\lambda_i(\mu) = \lambda_{si} + O(\mu), \ \lambda_{si} \in \sigma_s, \ \forall \lambda_i(\mu) \in \sigma_x, \tag{58}$$

$$\lambda_i(\mu) = \frac{\lambda_{fi} + O(\mu)}{\mu}, \ \lambda_{fi} \in \sigma_f, \ \forall \lambda_i(\mu) \in \sigma_y. \tag{59}$$

Proof. The separation (53) and the representation (58) and (59) of the SPLTISD spectrum can be proven according to the scheme from [29]. The continuity of $\lambda(\mu) \in \sigma_x(\mu)$ follows from the continuous dependence of the roots of a quasi-polynomial with respect to its coefficients and Lemma 1. □

Note that from Theorem 8 follows the continuity and, therefore, boundedness on $\mu \in [0, \mu^0]$ the functions $\lambda(\mu) \in \sigma_x(\mu)$.

Let us define:

$$N_\xi\left(\mu, \lambda, e^{-\lambda h}\right) \triangleq \left[\ \lambda I_{n_1} - A_\xi\left(\mu, e^{-\lambda h}\right), \ \ B_\xi\left(\mu, e^{-\lambda h}\right)\ \right]. \tag{60}$$

Due to (48) and (23), the following equality is true:

$$N_\xi\left(\mu, \lambda, e^{-\lambda h}\right) = N_s\left(\lambda, e^{-\lambda h}\right) + O(\mu), \tag{61}$$

For $\mu \geq 0, z \in C$, we introduce the following matrix function:

$$P_\xi(\mu, z) \triangleq \left[B_\xi(\mu), A_\xi(\mu, z) B_\xi(\mu), ..., A_\xi^{n_1+n_2-1}(\mu, z) B_\xi(\mu)\right]. \tag{62}$$

For a given $\mu \geq 0$, define the following set of complex numbers:

$$Z_\xi(\mu) = \{z \in C : \ rank\ P_\xi(\mu, z) < n_1\}. \tag{63}$$

Let $\pi_\xi(\mu, z)$ be the greatest common divisor of all minors of the order n_1 of the function matrix $P_\xi(\mu, z)$. Taking into account (48), (27), (62), and (34), $\pi_\xi(\mu, z)$ can be represented as:

$$\pi_\xi(\mu, z) = \sum_{i=0}^{l_s} k_i(\mu) z, \ k_i(\mu) = k_i + O(\mu). \tag{64}$$

Since the set Z_ξ (63) coincides with the set of roots of the polynomial $\pi_\xi(\mu, z)$ (64), then:

$$Z_\xi = \{z \in C : \pi_\xi(\mu, z) = 0\}.$$

By virtue of the continuous dependence of the roots of a polynomial with respect to its coefficients for sufficiently small $\mu > 0$, the elements $z(\mu) \in Z_\xi(\mu)$ are continuous and, therefore, bounded functions of μ.

For a given $\mu \in [0, \mu^0]$, define the set of complex numbers:

$$\Lambda_\xi(\mu) = \left\{\lambda \in C : e^{-\lambda h} = z,\, z \in Z_\xi(\mu)\right\},$$

associated with $Z(\mu)$. The elements $z(\mu) \in Z_\xi(\mu)$, $\lambda_k \in \Lambda_\xi(\mu)$, are connected by:

$$h\,\lambda_k(\mu) = \ln|z(\mu)| + i\,(\arg z(\mu) + 2\pi k),\; z \in Z_\xi(\mu),\; i = \sqrt{-1},\; k = 0, \pm 1, \pm 2, \ldots. \tag{65}$$

By virtue of the connection (65) and the continuity of the functions $z(\mu) \in Z_\xi(\mu)$, the functions $Re\lambda_k(\mu) = \frac{1}{h}\ln|z(\mu)|$ are also continuous for sufficiently small $\mu > 0$.

5. Complete Controllability of a Singularly-Perturbed Linear Time-Invariant System with Delays

5.1. Auxiliary Results

Lemma 2. *Let δ and α be two real numbers, $\delta < \alpha$, and:*

$$w_s\left(\lambda, e^{-\lambda h}\right) \neq 0\; \forall \lambda \in C : \delta \leq Re\,\lambda \leq \alpha. \tag{66}$$

Then, there exists a positive number μ^*, such that for all $\mu \in [0, \mu^*]$, the following inequalities hold:

$$w_\xi\left(\mu, \lambda, e^{-\lambda h}\right) \neq 0\; \forall \lambda \in C : \delta \leq Re\,\lambda \leq \alpha. \tag{67}$$

Proof. (By contradiction) Let the statement of the lemma be wrong. Then, two sequences $\{\mu_i\}$ and $\{\lambda_i\}$, $i = 1, 2, \ldots$ exist such that:

(a) $\mu_i > 0$, $i = 1, 2, \ldots$;

(b) $\lim_{i\to+\infty} \mu_i = 0$;

(c) $\delta \leq Re\,\lambda_i \leq \alpha$, $i = 1, 2, \ldots$

(d)

$$w_\xi\left(\mu_i, \lambda_i, e^{-\lambda_i h}\right) = 0,\; i = 1, 2, \ldots \tag{68}$$

Two cases can be distinguished: (i) the sequence $\{\lambda_i\}$, $i = 1, 2, \ldots$ is bounded; (ii) the sequence $\{\lambda_i\}$, $i = 1, 2, \ldots$ is unbounded. Begin with Case (i). In this case, there exists a convergent subsequence of $\{\lambda_i\}$. For the sake of simplicity (but without loss of generality), we assume that this subsequence coincides with $\{\lambda_i\}$. Let $\bar{\lambda} \triangleq \lim_{i\to\pm\infty} \lambda_i$. Due to Assumption (c),

$$\delta \leq Re\,\bar{\lambda} \leq \alpha. \tag{69}$$

Calculating the limit of (68) for $i \to \infty$, we obtain that $w_s\left(\bar{\lambda}, e^{-\bar{\lambda} h}\right) = 0$. The latter along with (69) contradicts the assumption (66) of the lemma.

Proceed to Case (ii). In this case, there exists a subsequence of λ_i, which tends to infinity. For the sake of simplicity (but without loss of generality), we assume that this subsequence coincides with λ_i. Then, $\lim_{i\to+\infty} \lambda_i = \pm\infty$.

Using (51) and (48), Equation (68) can be rewritten in the form:

$$(-1)^{n_1}\lambda_i^{n_1} + \lambda_i^{n_1-1} f_1(\lambda_i, \mu_i) + \cdots + f_{n_1}(\lambda_i, \mu_i) = 0, \tag{70}$$

where $f_j(\lambda_i, \mu_i)$, $j = 1, \ldots, n_1$ are some functions of (λ_i, μ_i). The functions $f_j(\lambda_i, \mu_i)$ are bounded uniformly with respect to μ_i for all sufficiently large i.

Dividing Equation (70) by $\lambda_i^{n_1}$ and calculating the limit of the resulting one for $i \to +\infty$ yield the contradiction $(-1)^{n_1} = 0$. This contradiction and the contradiction obtained in Case (i) imply that the statement of the lemma holds. □

Lemma 3. *Let $\alpha \in R$ such that $\forall \lambda \in \Lambda_s : Re\,\lambda > \alpha$. Then, there exists a positive number $\bar{\mu}$ such that $Re\,\lambda\,(\mu) \geq \alpha$, $\forall \lambda\,(\mu) \in \Lambda_{\xi}\,(\mu)$, for all $\mu \in (0, \bar{\mu}]$.*

Proof. Let the statement of the lemma be wrong. Then, three sequences $\{\mu_i\}$, $\{\lambda_i\}$, and $\{z_i\}$, exist such that:

(a) $\mu_i > 0, i = 1, 2, \ldots;$

(b) $\lim_{i \to +\infty} \mu_i = 0;$

(c) $\lambda_i \in \Lambda_{\xi}\,(\mu_i)$,

(d) $Re\,\lambda_i < \alpha,\ i = 1, 2, \ldots,$

(e) $Re\,\lambda_i = ln|z_i|, i = 1, 2, \ldots$

(f)

$$\pi_{\xi}\,(\mu_i, z_i) = 0. \tag{71}$$

Since the sequence $\{z_i\}$ is bounded, there exists a convergent subsequence of $\{z_i\}$. For the sake of simplicity (but without loss of generality), we assume that this subsequence coincides with $\{z_i\}$. Let $\bar{z} \triangleq \lim_{i \to \pm\infty} z_i$, $Re\bar{\lambda} = ln|\bar{z}|$, $\bar{z} \in Z_s$. Due to Assumptions (c) and (d), in view of the continuous dependence of $Re\lambda(\mu) \in \Lambda_{\xi}$ on μ for $\mu \in [0, \mu^0]$, the following inequalities are satisfied:

$$Re\,\bar{\lambda} \leq \alpha,\ \bar{\lambda} \in \Lambda_s. \tag{72}$$

Calculating the limit of (71) for $i \to \infty$, we obtain that $\pi_s\,(\bar{z}) = 0$. The latter along with (72) contradicts the assumption $\forall \lambda \in \Lambda_s : Re\lambda > \alpha$ of the lemma. This contradiction implies that the statement of the lemma holds. □

Lemma 4. *Let $\alpha \in R$ such that $\forall \lambda \in \sigma_s : Re\lambda > \alpha$. Then, there exists a positive number $\hat{\mu}$ such that $Re\lambda\,(\mu) \geq \alpha$, $\forall \lambda\,(\mu) \in \sigma_x\,(\mu)$, for all $\mu \in (0, \hat{\mu}]$.*

Proof. Let the statement of the lemma be wrong. Then, two sequences $\{\mu_i\}$, $\{\lambda_i\}$ exist such that:

(a) $\mu_i > 0, i = 1, 2, \ldots;$

(b) $\lim_{i \to +\infty} \mu_i = 0;$

(c) $\lambda_i \in \sigma_x\,(\mu_i)$,

(d) $Re\,\lambda_i < \alpha,\ i = 1, 2, \ldots$

Let us note that the sequence $\{Re\lambda_i\}, i = 1, 2, \ldots$ is bounded. Therefore, there exists a convergent subsequence of $\{Re\lambda_i\}$. For the sake of simplicity (but without loss of generality), we assume that this corresponding subsequence coincides with $\{\lambda_i\}$. Let $Re\bar{\lambda} \triangleq \lim_{i \to \pm\infty} Re\lambda_i, \bar{\lambda} \in \sigma_s$. Due to Assumptions (c) and (d), in view of the continuous dependence of $\lambda(\mu) \in \sigma_x$ on μ for $\mu \in [0, \mu^0]$:

$$Re\,\bar{\lambda} \leq \alpha,\ \bar{\lambda} \in \Lambda_s. \tag{73}$$

Calculating the limit of (71) for $i \to \infty$, we obtain that $rankP_s\,(\bar{z}) < n_1$. The latter along with (69) contradicts the assumption (66) of the lemma. □

Lemma 5. *Let:*

$$rank\,\mathrm{N}_s\left(\lambda, e^{-\lambda h}\right) = n_1 \quad \forall \lambda \in \sigma_s, \tag{74}$$

Then, there exists a positive number μ^*, such that for all $\mu \in (0, \mu^*]$:

$$rank\, \mathsf{N}_\xi\left(\mu, \lambda, e^{-\lambda h}\right) = n_1 \quad \forall \lambda \in \sigma_x(\mu). \tag{75}$$

Proof. From the fact that σ_s is a countable set, the characterizations 2, it follows that there exist real numbers $\delta, \alpha, \gamma,\ \delta < \alpha < \gamma$, such that:

(a) $Re\,\lambda \leq \gamma\ \forall \lambda \in \sigma_s$;

(b) $\alpha < Re\,\lambda \leq \gamma\ \forall \lambda \in \Lambda_s$;

(c) $w_s\left(\lambda, e^{-\lambda h}\right) \neq 0\ \forall \lambda \in C : \delta \leq Re\,\lambda \leq \alpha$.

By $\sigma_s^{<\delta},\ \sigma_s^{>\alpha}$, denote the following subset of σ_s:

$$\sigma_s^{<\delta} = \{\lambda \in \sigma_s :\ Re\,\lambda < \delta\},\ \sigma_s^{>\alpha} = \{\lambda \in \sigma_s :\ \alpha < Re\,\lambda \leq \gamma\}.$$

Therefore,

$$\sigma_s = \sigma_s^{<\delta} \cup \sigma_s^{>\alpha}.$$

In view of Theorem 4, (33) and (35), the condition (74) can be violated only for $\lambda \in \Lambda_s$,. Similarly, we can prove that for a given $\mu > 0$, the condition (75) can be violated only for $\lambda \in \Lambda_\xi$. Due to Lemmas 4 and 3 for all sufficiently small $\mu > 0$, it is true that $Re\lambda(\mu) \geq \alpha,\ \forall \lambda(\mu) \in \sigma_x^{>\alpha}(\mu)$, and $\forall \lambda(\mu) \in \Lambda_\xi(\mu)$. Due to Lemma 2 for all sufficiently small $\mu > 0$, it is true that $Re\lambda(\mu) \leq \delta,\ \forall \lambda(\mu) \in \sigma_x^{<\delta}(\mu)$. Therefore, $\sigma_x^{<\delta}(\mu) \cup \Lambda_\xi(\mu) = \emptyset$, and the condition (75) is true $\forall \lambda \in \sigma_x^{<\delta}(\mu)$ for all sufficiently small $\mu > 0$.

Then, similar to [29] (p. 75), it is proven that if for some λ, the condition (74) is true, then for the same λ, the condition (75) is true for all sufficiently small $\mu > 0$.

Due to the above-mentioned property (a), the characterization 1 (c), and the continuity of the functions $\lambda(\mu) \in \sigma_\xi(\mu)$, it is possible to choose such μ^*, such that (74) follows (75). □

5.2. Split Controllability: Parameter-Free Sufficient Conditions

Theorem 9. *Let the DS (11) be completely controllable, i.e., the conditions (37) and (38) are fulfilled, and the BLS is completely controllable, i.e., the condition (39) is fulfilled. Then, there exists a $\mu^* \in (0, \mu^0]$ such that the SPLTISD (1)–(2) is completely controllable for all $\mu \in (0, \mu^*]$.*

Proof. For a given $\mu \in (0, \mu^0]$, the SPLTISD (1)–(2) is completely controllable if and only if:

$$rank\, \mathsf{N}\left(\mu, \lambda, e^{-\lambda h}\right) = n_1 + n_2, \quad \forall \lambda \in \sigma(\mu). \tag{76}$$

Consider the matrix-valued function:

$$\mathsf{N}_{\xi\eta}\left(\mu, \lambda, e^{-\lambda h}\right) \triangleq \mathsf{G}^{-1} N\left(\mu, \lambda, e^{-\lambda h}\right) diag\{\mathsf{G}, E_r\}, \tag{77}$$

that by virtue of continuity with μ of the matrices (48), (9) for sufficiently small $\mu > 0$ can be extended by continuity at $\mu = 0$.

By using (9), (43), and (47), it is easy to make sure that for $\mu \geq 0$:

$$\mathsf{N}_{\xi\eta}\left(\mu, \lambda, e^{-\lambda h}\right) = \begin{pmatrix} \lambda I_{n_1} - \mathsf{A}_\xi\left(\mu, e^{-\lambda h}\right) & 0_{n_1 \times n_2} & B_\xi\left(\mu, e^{-\lambda h}\right) \\ 0_{n_2 \times n_1} & \mu\lambda I_{n_2} - \mathsf{A}_\eta\left(\mu, e^{-\lambda h}\right) & B_\eta\left(\mu, e^{-\lambda h}\right) \end{pmatrix}. \tag{78}$$

Due to the invariance of the spectrum and preserving the matrix rank under nondegenerate transformations, it is determined from (76) that the SPLTISD (1)–(2) is completely controllable at a fixed $\mu > 0$ if and only if:

$$rank\, \mathsf{N}_{\xi\eta}\left(\mu, \lambda, e^{-\lambda h}\right) = n_1 + n_2, \quad \forall \lambda \in \sigma(\mu). \tag{79}$$

Due to (48) and (78), the condition (79) has a view: $\forall\lambda \in \sigma(\mu)$:

$$rank\begin{pmatrix} \lambda I_{n_1} - \mathsf{A}_s\left(e^{-\lambda h}\right) + O(\mu) & 0_{n_1\times n_2} & B_s + O(\mu) \\ 0_{n_2\times n_1} & \mu\lambda I_{n_2} - A_4 + O(\mu) & B_2 + O(\mu) \end{pmatrix} = n_1 + n_2. \quad (80)$$

Let us show that the condition (80) is fulfilled for all sufficiently small $\mu > 0$ if DS (11) and BLS (19) are completely controllable.

Let DS (11) be completely controllable, i.e., (25) is true. Then, by Lemma 5, the following condition:

$$rank\,\mathsf{N}_{\xi}\left(\mu, \lambda, e^{-\lambda h}\right) = n_1, \quad \forall\lambda \in \sigma_x(\mu) \quad (81)$$

is true for all sufficiently small $\mu > 0$.

Since for all sufficiently small $\mu > 0$, the sets $\sigma_x(\mu)$ (54) and $\sigma_y(\mu)$ (56) have no elements in common, then:

$$rank\left[\mu\lambda I_{n_2} - \mathsf{A}_{\eta}\left(\mu, e^{-\lambda h}\right)\right] = n_2, \forall\lambda = \lambda(\mu) \in \sigma_x(\mu) = \sigma(\mu) \setminus \sigma_y(\mu) \quad (82)$$

for all sufficiently small $\mu > 0$. From the conditions (81) and (82), it follows that (80) is true for all $\lambda \in \sigma_x(\mu)$ for all sufficiently small $\mu > 0$.

Let BLS (19) is completely controllable, i.e., (26) is true. Then, since for all $\lambda(\mu) = \frac{1}{\mu}\lambda_i(\mu) \in \sigma_y(\mu)$, it is true that $\lambda_i(\mu) \underset{\mu\to 0}{\to} \lambda_{fi} \in \sigma_f$ (see Theorem 8), and due to the finiteness of the set σ_f (22) in view of the preservation of the full rank of a matrix under small regular perturbation, we have that the condition:

$$rank\begin{pmatrix} \mu\lambda I_{n_2} - A_4 + O(\mu) & B_2 + O(\mu) \end{pmatrix} = n_2, \quad \forall\lambda(\mu) \in \sigma_y(\mu)$$

is true for all sufficiently small $\mu > 0$. Since for all sufficiently small $\mu > 0$, the sets $\sigma_x(\mu)$ and $\sigma_x(\mu)$ have no elements in common, then:

$$rank\left[\lambda I_{n_1} - \mathsf{A}_{\xi}\left(\mu, e^{-\lambda h}\right)\right] = n_2, \ \forall\lambda = \lambda(\mu) \in \sigma_y(\mu) = \sigma(\mu) \setminus \sigma_x(\mu)$$

for all sufficiently small $\mu > 0$. From the last conditions, it follows that the condition (80) is true for all sufficiently small $\mu > 0$ for all $\lambda(\mu) \in \sigma_y(\mu)$.

Combining the above results, we have that if DS (11) is completely controllable and BLS (19) is completely controllable, then the condition (80) is fulfilled for all sufficiently small $\mu > 0$ for all $\lambda(\mu) \in \sigma(\mu)$.

Applying Theorems 6 and 7, we are convinced of the validity of the statement of Theorem 9. □

According to the proven theorem, the complete controllability of the slow and fast subsystems yields the complete controllability of the original system for all sufficiently small values of the parameter of singular perturbation.

Note that the conditions (37)–(39) do not depend on the small parameter; they have a ranked form; they are expressed through the matrix parameters of the SPLTISD and guarantee the preservation of its complete controllability for all sufficiently small values of the parameter $\mu > 0$. A similar statement for SPS without delay was proven in [29] and with a single delay in [44].

It is not difficult to verify the condition (37). In addition, for a given $\mu \in (0, \mu^0]$, the condition (38) may be violated only for λ_k that also are the roots of the DS (11) characteristic Equation (18), so the verification of this condition (38) is necessary only for the λ_k view of (36) that comprise the roots of the polynomial $w_s(\lambda, z)$ for $z \in Z_s$.

6. Example

In this section, we consider an illustrative example. Consider the following system, a particular case of the SPLTISD (1)–(2),

$$\begin{aligned} \dot{x}_1(t) &= -x_1(t) + 2x_2(t) - x_2(t-1) - y(t), \\ \dot{x}_2(t) &= -x_1(t) + 2x_2(t) + x_1(t-1) - x_2(t-1), \\ \mu\dot{y}(t) &= -x_1(t-1) - y(t) + u(t), \end{aligned} \tag{83}$$

with the parameters $n_1 = 2,\ n_2 = r = 1,\ l = 1,\ h = 1$ and matrices:

$$\begin{aligned} &A_{10} = \begin{pmatrix} -1 & 2 \\ -1 & 2 \end{pmatrix},\ A_{11} = \begin{pmatrix} 0 & -1 \\ 1 & -1 \end{pmatrix},\ A_2 = \begin{pmatrix} -1 \\ 0 \end{pmatrix},\ B_1 = \begin{pmatrix} 0 \\ 0 \end{pmatrix}, \\ &A_{30} = \begin{pmatrix} 0 & 0 \end{pmatrix},\ A_{31} = \begin{pmatrix} -1 & 0 \end{pmatrix},\ A_4 = (-1),\ B_2 = (1). \end{aligned} \tag{84}$$

The characteristic Equation (7) for the system (83) is:

$$w\left(\mu, \lambda, e^{-\lambda}\right) = \frac{1}{\mu}\left(\lambda(\lambda-1)(1+\mu\lambda) - \mu\lambda \mathrm{e}^{-\lambda}\left(2 - \mathrm{e}^{-\lambda} - \lambda\right)\right) = 0$$

and for sufficient small $\mu > 0$ has the roots (8):

$$\sigma(\mu) = \left\{0, 1 + O(\mu), -\frac{1}{\mu} + O(\mu)\right\}.$$

The DS (11) for SPLTISD (83) has the form:

$$\begin{aligned} \dot{x}_{s1}(t) &= -x_{s1}(t) + x_{s1}(t-1) + 2x_{s2}(t) - x_{s2}(t-1) - u_s(t), \\ \dot{x}_{s2}(t) &= -x_{s1}(t) + x_{s1}(t-1) + 2x_{s2}(t) - x_{s2}(t-1), \end{aligned} \tag{85}$$

and the matrix parameters (13) for DS (85) have the form:

$$A_{s0} = \begin{pmatrix} -1 & 2 \\ -1 & 2 \end{pmatrix},\ A_{s1} = \begin{pmatrix} 1 & -1 \\ 1 & -1 \end{pmatrix},\ B_s = \begin{pmatrix} -1 \\ 0 \end{pmatrix}.$$

The BLS (19) for SPLTISD (83) has the form:

$$\frac{dy_f(\tau)}{d\tau} = -y_f(\tau) + u_f(\tau). \tag{86}$$

The characteristic Equation (17) for the DS (85) is:

$$w_s = \lambda^2 - \lambda + 2e^{-\lambda}(1 - e^{-\lambda}) = 0,$$

and the characteristic Equation (21) for the BLS (86) is:

$$w_f = \lambda + 1 = 0.$$

The spectra (18) of the DS (85) and the BLS (86) for (83): $\sigma_s = \{0, 1\}, \sigma_f = \{-1\}$.

Since the matrices (23) and (24) for (1)–(2) and (84):

$$N_s\left(\lambda, e^{-\lambda h}\right) = \begin{bmatrix} \lambda + 1 - e^{-\lambda} & -1 - e^{-\lambda} & -1 \\ 2 - e^{-\lambda} & \lambda - 2 + e^{-\lambda} & 0 \end{bmatrix},\quad N_f(\lambda) = [\lambda + 1, 1]$$

have a full rank for all $\lambda \in \sigma_s$, all $\lambda \in \sigma_f$, respectively, then according to Theorem 2, the DS for (83) is completely controllable and the BLS for (1)–(2) and (84) is completely controllable.

Then, with according to Theorem 9, there exists a $\mu^* > 0$ such that the SPLTISD (1)–(2) and (84) is completely controllable for all $\mu \in (0, \mu^*]$.

Let us show the validity of the conditions (37)–(39) for (1)–(2) and (84). We have:

$$P_s(z) = \begin{bmatrix} -1 & 1-z \\ 0 & 1-z \end{bmatrix}, \quad \pi_s(z) = -1 + z, \quad Z_s = \{1\}, \quad P_f = [1, \ -1].$$

It is obvious that (37) and (39) for SPLTISD (83) are valid. Since among the roots of the polynomial $w_s(\lambda, 1) = \lambda(\lambda - 1)$, there are no numbers of the form $\lambda_k = \ln|1| + i \cdot 2\pi k$, $i = \sqrt{-1}$, $k = 0, \pm 1, \pm 2, \ldots$, so $\Omega_s = \emptyset$, then we conclude that (38) is also fulfilled.

Thus, all the conditions of Theorem 9 are fulfilled for the system (83). This confirms the above conclusion about the complete robust controllability of the SPLTISD (83) with respect to $\mu > 0$ for all sufficiently small values of this parameter.

7. Discussion

The rank condition of complete controllability [12] is also known as a condition of spectral controllability and observability [46], and related to various structural properties of the system, for example realization, modal control, completeness, etc. Therefore, the results of this work can be used to obtain the conditions of similar properties for a singularly-perturbed system (1)–(2), robust with respect to a small parameter and expressed in the form of rank parametric conditions for systems of lower dimensions than the original system.

For complete controllable systems with delay, we can design static feedback controllers, providing an arbitrary finite spectrum of a closed system [47,48]. In particular, by choosing a spectrum, a closed system can be made asymptotically stable. Based on the decoupling transformation for the original singularly-perturbed system, it is possible to construct a stabilizing feedback in the form of a composite regulator that combines the stabilizing regulators of its slow and fast subsystems of lower dimensions (see, e.g., [49]).

8. Conclusions

In this paper, a singularly-perturbed linear time-invariant controlled system with multiple commensurate time delays in the slow state variables was considered. For this system, the complete controllability, robust with respect to a small parameter μ, was studied. This study is based on the Chang-type transformation of the original system, which decouples the original singularly-perturbed system into two $O(\mu)$-close to μ-free subsystems, slow and fast subsystems of smaller dimensions than the original.

Based on the above-mentioned Chang-type transformation of the original singularly-perturbed system, μ-free verifiable parametric rank-type sufficient conditions for the complete controllability of this system were established. These conditions, being μ-free, provide the complete controllability of the original singularly-perturbed system with delay for all sufficiently small values of $\mu > 0$, i.e., robustly with respect to this parameter of singular perturbation.

Funding: The work is partially supported by the State research program "Convergence-2020" of the Republic of Belarus: Task 1.3.02.

Conflicts of Interest: The author declares no conflict of interest.

References

1. Dmitriev, M.G.; Kurina, G.A. Singular perturbations in control problems. *Autom. Remote Control* **2006**, *67*, 1–43. [CrossRef]

2. Vasil'eva, A.B.; Dmitriev, M.G. Singular perturbations in optimal control problems. *J. Soviet Math.* **1986**, *34*, 1579–1629. (In Russian) [CrossRef]
3. Zhang, Y.; Naidu D.S.; Cai, C.; Zou, Y. Singular perturbations and time scales in control theories and applications: An overview 2002–2012. *Int. J. Inf. Syst. Sci.* **2014**, *9*, 1–36.
4. Michiels, W.; Niculescu, S.I. Stability and stabilization of time delay systems: An eigenvalue based approach. In *Advances in Design and Control*; SIAM: Philadelphia, PA, USA, 2007.
5. Niculescu, S.I. *Delay Effects on Stability: A Robust Control Approach*; Springer: New York, NY, USA, 2001.
6. Bellman, R.; Cooke, K. *Differential Difference Equations*; Academic Press: New York, NY, USA, 1963.
7. Hale, J. *Theory of Functional Differential Equations*; Springer: New York, NY, USA, 1977.
8. Klamka, J. Controllability of dynamical systems. In *Notes in Computer Science*; Springer: New York, NY, USA, 2008; pp. 156–168.
9. Klamka, J. Controllability of dynamical systems. A survey. *Bull. Pol. Acad. Sci. Tech.* **2013**, *61*, 335–342. [CrossRef]
10. Richard, P. Time-delay systems: An overview of some recent advacesand open problems. *Automatica* **2003**, *39*, 1667–1694. [CrossRef]
11. Kirillova, F.M.; Churakova S.V. On the problem of controllability of linear systems with delay. *Differ. Equ.* **1967**, *3*, 436–445. (In Russian)
12. Marchenko, V.M. To controllability of linear systems with aftereffect. *Dokl. AN SSSR* **1977**, *236*, 1083–1086. (In Russian)
13. Marchenko, V.M. On the theory of point-wise controllability of systems with delay. *Vestnik BSU Ser. I* **1977**, *3*, 27–30. (In Russian)
14. Marchenko, V.M. Multipoint controllability of systems with delay. *Differ. Equ.* **1978**, *14*, 1324–1327. (In Russian)
15. Mantius, A.; Triggiani, R. Function space controllability of linear retarded systems: A derivation from abstract operator conditions. *SIAM J. Control Optim.* **1978**, *16*, 599–645. [CrossRef]
16. Mantius, A. F-Controllability and observability of linear retarded systems *Appl. Math. Optim.* **1982**, *9*, 73–95. [CrossRef]
17. Marchenko, V.M. A brief rewiew of the development of qualitative control theory in Belarus. *Cybern. Syst. Anal.* **2002**, *38*, 597–607. [CrossRef]
18. Miniuk, S.A. On complete controllability of linear time delay systems. *Differ. Equ.* **1972**, *8*, 254–259. (In Russian)
19. Minyuk, S.A.; Lyakhovets, S.N. On a problem of controllability for systems with maltiple delays. *Bull. Byeloruss. SSR Ser. Fiz.-Math. Sci.* **1980**, *2*, 12–17. (In Russian)
20. Kurina, G.A. Singular perturbations of control problems with equations of state not solved for the derivative (a survey). *J. Comput. Syst. Sci. Int.* **1993**, *31*, 17–45. (In Russian)
21. Kopeikina, T.B. Controllability of singularly-perturbed linear systems with delay. *Differ. Equ.* **1989**, *25* 1055–1064.
22. Kopeikina, T.B.; Tsekhan, O.B. On the controllability of linear nonstationary singularly-perturbed systems in state space. *Tech. Cybern.* **1993**, *3*, 40–46.
23. Kopeikina, T.B. Some approaches to the investigation controllability of singularly-perturbed dynamic systems. *Syst. Sci.* **1995**, *21*, 17–36.
24. Kopeikina, T.B. Defining equations in the problem of controllability of stationary singularly-perturbed systems with delay. *Trudy BSTU. Ser. VI Phys.-Math. Inform. Sci.* **2009**, *XVII*, 11–13. (In Russian)
25. Glizer, V.Y. Euclidean space controllability of singularly-perturbed linear systems with state delay. *Syst. Control Lett.* **2001**, *43*, 181–191. [CrossRef]
26. Glizer, V.Y. Controllability of singularly-perturbed linear time-dependent systems with small state delay. *Dyn. Control* **2001**, *11*, 261–281. [CrossRef]
27. Glizer, V.Y. Controllability of nonstandard singularly-perturbed systems with small state delay. *IEEE Trans. Autom. Control* **2003**, *48*, 1280–1285. [CrossRef]
28. Glizer, V.Y. Approximate state-space controllability of linear singularly-perturbed systems with two scales of state delays. *Asymptot. Anal.* **2018** *107*, 73–114. [CrossRef]
29. Kokotovic, P.V.; Khalil, H.K.; O'Reilly, J. *Singular Perturbation Methods in Control: Analysis and Design*; Academic Press: London, UK, 1986.

30. Sannuti, P. On the controllability of singularly-perturbed systems. *IEEE Trans. Autom. Control* **1977**, *22*, 622–624. [CrossRef]
31. Sannuti, P. On the controllability of some singularly-perturbed nonlinear systems. *J. Math. Anal. Appl.* **1978**, *64*, 579–591. [CrossRef]
32. Glizer, V.Y. Novel controllability conditions for a class of singularly-perturbed systems with small state delays. *J. Optim. Theory Appl.* **2008**, *137*, 135–156. [CrossRef]
33. Glizer, V.Y. Controllability conditions of linear singularly-perturbed systems with small state and input delays. *Math. Control Signals Syst.* **2016**, *28*, 1–29. [CrossRef]
34. Kopeikina, T.B. Unified method of investigating controllability and observability problems of time-variable differential systems. *Funct. Differ. Equ.* **2006**, *13*, 463–481. (In Russian)
35. Tsekhan, O.B. About the conditions of controllability in the space $L_2^2[t_1 - h, t_1] \times R^2$ of linear stationary singularly-perturbed systems of second order with delay *Vestnik Grdu of Ya. Ser. 2* **2012**, *3*, 62–77. (In Russian)
36. Chang, K. Singular perturbations of a general boundary value problem. *SIAM J. Math. Anal.* **1972**, *3*, 520–526. [CrossRef]
37. Yang, X.; Zhu, J.J. A Generalization of Chang Transformation for Linear Time-Varying Systems. In Proceedings of the IEEE Conference on Decision and Control (CDC), Atlanta, GA, USA, 15–17 December 2010; pp. 6863–6869.
38. Yang, X.; Zhu, J.J. Chang transformation for decoupling of singularly-perturbed linear slowly time-varying systems. In Proceedings of the 51st IEEE Conference on Decision and Control, Maui, Hi, USA, 10–13 December 2012; pp. 5755–5760.
39. Kurina, G.A. Complete controllability of singularly-perturbed systems with slow and fast modes. *Math. Notes* **1992**, *52*, 1029–1033. (In Russian) [CrossRef]
40. Magalhaes, L.T. Exponential estimates for singularly-perturbed linear functional differential equations. *J. Math. Anal. Appl.* **1984**, *103*, 443–460. [CrossRef]
41. Fridman, E. Decoupling transformation of singularly-perturbed systems with small delays and its applications. *Z. Angew. Math. Mech. Berlin* **1996**, *76*, 201–204.
42. Tsekhan, O.B. Decoupling transformation for linear stationary singularly-perturbed system with delay and its applications to analysis and control of spectrum. *Vesnik GrDU Imja Janki Kupaly. Ser. 2 Matjematyka* **2017**, *7*, 50–61. (In Russian)
43. Kokotovic, P.V.; Haddad, A.H. Controllability and time-optimal control of systems with slow and fast modes. *IEEE Trans. Autom. Control* **1975**, *20*, 111–113. [CrossRef]
44. Tsekhan, O.B. Sufficient conditions for complete controllability based on the decomposition of linear stationary singularly-perturbed system with delay. *Vesnik GrDU Imja Janki Kupaly. Ser. 2 Matjematyka* **2017**, *7*, 51–65. (In Russian)
45. Gabasov, R.; Kirillova, F. *The Qualitative Theory of Optimal Processes*; M. Dekker: New York, NY, USA, 1976.
46. Bhat, K.; Koivo, H. Modal characterizations of controllability and observability for time delay systems. *IEEE Trans. Autom. Control* **1976**, *21*, 292–293. [CrossRef]
47. Manitius, A.Z.; Olbrot, A.W. Finite spectrum assignment problem for systems with delays. *IEEE Trans. Autom. Control* **1979**, *24*, 541–553. [CrossRef]
48. Metel'skii, A.V. Feedback control of the spectrum of differential-difference system. *Autom. Remote Control* **2015**, *76*, 560–572. [CrossRef]
49. Dmitriev, M.; Makarov, D. Composite regulator in a linear time variant control system. *J. Comput. Syst. Sci. Intern.* **2014**, *53*, 777–787. [CrossRef]

axioms

MDPI

Article

Cheap Control in a Non-Scalarizable Linear-Quadratic Pursuit-Evasion Game: Asymptotic Analysis

Vladimir Turetsky [1,*,†] and Valery Y. Glizer [2,†]

1 Department of Mathematics, Ort Braude College of Engineering, 51 Snunit Str., P.O. Box 78, Karmiel 2161002, Israel
2 The Galilee Research Center for Applied Mathematics, Ort Braude College of Engineering, 51 Snunit Str., P.O. Box 78, Karmiel 2161002, Israel; valgl120@gmail.com
* Correspondence: turetsky1@braude.ac.il
† These authors contributed equally to this work.

Abstract: In this work, a finite-horizon zero-sum linear-quadratic differential game, modeling a pursuit-evasion problem, was considered. In the game's cost function, the cost of the control of the minimizing player (the minimizer/the pursuer) was much smaller than the cost of the control of the maximizing player (the maximizer/the evader) and the cost of the state variable. This smallness was expressed by a positive small multiplier (a small parameter) of the square of the L_2-norm of the minimizer's control in the cost function. Parameter-free sufficient conditions for the existence of the game's solution (the players' optimal state-feedback controls and the game value), valid for all sufficiently small values of the parameter, were presented. The boundedness (with respect to the small parameter) of the time realizations of the optimal state-feedback controls along the corresponding game's trajectory was established. The best achievable game value from the minimizer's viewpoint was derived. A relation between solutions of the original cheap control game and the game that was obtained from the original one by replacing the small minimizer's control cost with zero, was established. An illustrative real-life example is presented.

Keywords: linear-quadratic differential game; cheap control; singular (degenerate) differential game; pursuit-evasion game

Citation: Turetsky, V.; Glizer, V.Y. Cheap Control in a Non-Scalarizable Linear-Quadratic Pursuit-Evasion Game: Asymptotic Analysis. *Axioms* **2022**, *11*, 214. https://doi.org/10.3390/axioms11050214

Academic Editor: Boualem Djehiche

Received: 27 March 2022
Accepted: 27 April 2022
Published: 5 May 2022

1. Introduction

A cheap control problem is an extremal control problem where a control cost of at least one of the decision makers is much smaller than a state cost in at least one cost function of the problem. Cheap control problems appear in many topics of optimal control and differential game theories. For example, such problems appear in the following topics: (1) regularization of singular optimal controls (see, e.g., [1–4]); (2) limitation analysis for optimal regulators and filters (see, e.g., [5–7]); (3) extremal control problems with high gain control in dynamics (see, e.g., [8,9]); (4) inverse optimal control problems (see, e.g., [10]); (5) robust optimal control of systems with uncertainties/disturbances (see, e.g., [11,12]); (6) guidance problems (see, e.g., [13,14]).

The Hamilton boundary-value problem and the Hamilton–Jacobi–Bellman–Isaacs equation, associated with the cheap control problem by solvability (control optimality) conditions, are singularly perturbed because of the smallness of the control cost.

In the present paper, we considered one class of cheap control pursuit-evasion differential games. Cheap control differential games have been studied in a number of works in the literature (see, e.g., [4,11,12,15,16] and references therein). In most of these studies, the case where a state cost appeared in the integral part of the cost function was treated. This feature allowed (subject to some additional condition on the state cost) the use of the boundary function method [17] for an asymptotic analysis of the corresponding singularly perturbed Hamilton–Jacobi–Bellman–Isaacs equation. Moreover, the time realization of the optimal state-feedback control with the small cost had an impulse-like behaviour, meaning it was unbounded as the control cost tended to zero. To the best of our knowledge, cheap control games, where the time realization of the state-feedback optimal control with the small cost remains bounded as this cost tends to zero, were considered only in a few works and only for specific problem settings. Thus in [13], a pursuit-evasion problem, modeled by a linear-quadratic zero-sum differential game with time-invariant four-dimensional dynamics and scalar controls of the players, was considered. In this game, the control cost of the pursuer was assumed to be small. Moreover, the integral part of the game's cost function did not contain the state cost. By a linear state transformation, this cheap control game was converted to a scalar linear-quadratic cheap control game. In this scalar game, the time realization of the optimal state-feedback pursuer's control against a bang–bang evader's control was analyzed. Sufficient conditions for the boundedness of this time realization for all sufficiently small values of the pursuer's control cost were derived. In [14], a similar problem was solved in the case where the control costs of both the pursuer and evader were small and had the same order of smallness. In [11], a more general pursuit-evasion problem was studied. This problem was modeled by a linear-quadratic zero-sum differential game with time-dependent six-dimensional dynamics. The controls of both the pursuer and evader were scalar. The costs of these controls were small and had the same order of smallness. The state cost was absent in the integral part of the game's cost function. This game also allowed a transformation to a scalar linear-quadratic cheap control game. In this scalar game, the time realization of the optimal state-feedback pursuer's control against an open-loop bounded evader's control was analyzed. Sufficient conditions, guaranteeing that the time realization satisfied given constraints for all sufficiently small values of the controls' costs, were obtained. In [12], a robust tracking problem, modeled by a linear-quadratic zero-sum differential game with time-dependent n-dimensional ($n \geq 1$) dynamics, was analyzed. The controls of both minimizing and maximizing players were vector-valued. The costs of these controls were small and had the same order of smallness. For this game, the limit behaviour of the state-dependent part of the cost function, generated by the optimal state-feedback control of the minimizing player (the minimizer) and any L_2-bounded open-loop control of the maximizing player (the maximizer), was studied. Sufficient conditions, providing the tendency to zero of this part of the cost function as the small controls' costs approached zero (the exact tracking), were derived. Subject to these conditions, necessary conditions for the boundedness of the time realization of the optimal state-feedback minimizer's control for all sufficiently small values of the controls' costs were obtained.

In the present work, we studied a much more general cheap control linear-quadratic zero-sum differential game than those in [11,13,14]. For this game, an asymptotic analysis of its solution was carried out in the case where the small control's cost of the minimizer tended to zero. In particular, the asymptotic behavior of the time realizations of both players' optimal state-feedback controls along the corresponding (optimal) trajectory of the game was analyzed. The boundedness of these time realizations was established for all sufficiently small values of the minimizer's control cost. Moreover, in contrast to the results of the work [12], the conditions for such boundedness were sufficient and they were not restricted by any other specific conditions, such as the exact tracking in [12].

Also in the present work, we considered one more linear-quadratic zero-sum differential game. This game was obtained from the original cheap control game by replacing the small control cost of the minimizer with zero. This new game was called a degenerate game and was similar to the continuous/discrete time system obtained from a singularly perturbed system by replacing a small parameter of singular perturbation with zero. The relation between the original cheap control game and the degenerate game was established.

This paper is organised as follows. In Section 2, the problems of the paper (the cheap control differential game and the degenerate differential game) are rigorously formulated, main definitions and some preliminary results are presented and the objectives of the paper are stated. In Section 3, the solution of the cheap control differential game is obtained and the asymptotic analysis of this solution is carried out. Section 4 is devoted to deriving the solution of the degenerate differential game. In addition, some relations between the solution of the cheap control differential game and the degenerate differential game are established in this section. In Section 5, based on the theoretical results of the paper, one interception problem in 3D space was studied. Conclusions of the paper are presented in Section 6.

2. Preliminaries and Problem Statement

Consider the controlled system

$$\dot{x} = A(t)x + B(t)u + C(t)v, \; x(t_0) = x_0, \; t \in [t_0, t_f], \tag{1}$$

where $x \in \mathbb{R}^n$, $u \in \mathbb{R}^r$ and $v \in \mathbb{R}^s$ are the state, the pursuer's control and the evader's control, respectively; t_0 is an initial time moment; t_f is a final time moment; the matrix-valued functions $A(t)$, $B(t)$ and $C(t)$ of appropriate dimensions are continuous for $t \in [t_0, t_f]$. The controls $u(t)$ and $v(t)$ are assumed to be measurable bounded functions for $t \in [t_0, t_f]$.

The target set is a linear manifold

$$T_x = \{x \in \mathbb{R}^n : \; Dx + d = 0\}, \tag{2}$$

where D is a prescribed $m \times n$-matrix ($m < n$) and $d \in \mathbb{R}^m$ is a prescribed vector. The objective of the pursuer is to steer the system onto a target set at $t = t_f$, whereas the evader desires to avoid hitting the target set by exploiting feedback strategies $u(t, x)$ and $v(t, x)$, respectively.

Let us consider the set $\mathcal{U}_x$ of all functions $u = u(t, x) : [0, t_f] \times \mathbb{R}^n \to \mathbb{R}^r$, which are measurable w.r.t. $t \in [0, t_f]$ for any fixed $x \in \mathbb{R}^n$ and satisfy the local Lipschitz condition w.r.t. $x \in \mathbb{R}^n$ uniformly in $t \in [0, t_f]$. Similarly, we consider the set $\mathcal{V}_x$ of all functions $v = v(t, x) : [0, t_f] \times \mathbb{R}^n \to \mathbb{R}^s$, which are measurable w.r.t. $t \in [0, t_f]$ for any fixed $x \in \mathbb{R}^n$ and satisfy the local Lipschitz condition w.r.t. $x \in \mathbb{R}^n$ uniformly in $t \in [0, t_f]$.

Definition 1. *Let us denote by U_x the set of all functions $u(t, x) \in \mathcal{U}_x$ satisfying the following conditions: (1_{ux}) the initial-value problem (1) for $u(t) = u(t, x)$ and any fixed $v(t) \in L_2([0, t_f], \mathbb{R}^s)$ has the unique absolutely continuous solution $x_u(t)$, $t \in [0, t_f]$; (2_{ux}) $u(t, x_u(t)) \in L_2([0, t_f], \mathbb{R}^r)$.*

Also, let us denote by V_x the set of all functions $v(t, x) \in \mathcal{V}_x$ satisfying the following conditions: (1_{vx}) the initial-value problem (1) for $v(t) = v(t, x)$ and any fixed $u(t) \in L_2([0, t_f], \mathbb{R}^r)$ has the unique absolutely continuous solution $x_v(t)$, $t \in [0, t_f]$; (2_{vx}) $v(t, x_v(t)) \in L_2([0, t_f], \mathbb{R}^s)$.

In what follows, the set U_x is called the set of all admissible state-feedback controls (strategies) of the pursuer, while the set V_x is called the set of all admissible state-feedback controls (strategies) of the evader.

Below, two differential games modeling this conflict situation are formulated.

2.1. Cheap Control Differential Game

The first is the Cheap Control Differential Game (CCDG) with the dynamics (1) and the cost function

$$\widetilde{J}_{\alpha\beta}(u, v) = |Dx(t_f) + d|^2 + \alpha \int_{t_0}^{t_f} |u(t)|^2 dt - \beta \int_{t_0}^{t_f} |v(t)|^2 dt, \tag{3}$$

where $|x|$ denotes the Euclidean norm of the vector x; $\alpha, \beta > 0$ are the penalty coefficients for the players' control expenditure, and α is assumed to be small. The objectives of the pursuer and the evader were to minimize and to maximize the cost function (3) by $u(\cdot) \in U_x$ and $v(\cdot) \in V_x$, respectively.

The CCDG (1), (3) is a zero-sum linear-quadratic differential game (see, e.g., [18–22]).

Definition 2. *Let $u(t,x)$, $(t,x) \in [t_0, t_f] \times \mathbb{R}^n$, be any given admissible pursuer strategy, i.e., $u(\cdot) \in U_x$. Then, the value*

$$\widetilde{J}^u_{\alpha\beta}(u(\cdot); t_0, x_0) = \sup_{v(t) \in L_2([t_0,t_f],\mathbb{R}^s)} \widetilde{J}_{\alpha\beta}(u(\cdot), v(t)), \tag{4}$$

calculated along the corresponding trajectories of the system (1), is called the guaranteed result of the strategy $u(\cdot)$ in the CCDG.

The value

$$\widetilde{J}^{u*}_{\alpha\beta}(t_0, x_0) = \inf_{u(\cdot) \in U_x} \widetilde{J}^u_{\alpha\beta}(u(\cdot); t_0, x_0) \tag{5}$$

is called the upper value of the CCDG.

If the infimum value (5) is attained for $\widetilde{u}^0_{\alpha\beta}(t,x) \in U_x$, i.e.,

$$\inf_{u(\cdot) \in U_x} \widetilde{J}^u_{\alpha\beta}(u(\cdot); t_0, x_0) = \min_{u(\cdot) \in U_x} \widetilde{J}^u_{\alpha\beta}(u(\cdot); t_0, x_0)$$

and

$$\widetilde{u}^0_{\alpha\beta}(t,x) = \arg\min_{u(\cdot) \in U_x} \widetilde{J}^u_{\alpha\beta}(u(\cdot); t_0, x_0), \tag{6}$$

the strategy $\widetilde{u}^0_{\alpha\beta}(t,x)$ is called the optimal strategy of the pursuer in the CCDG.

Definition 3. *Let $v(t,x)$, $(t,x) \in [t_0, t_f] \times \mathbb{R}^n$, be any given admissible evader strategy, i.e., $v(\cdot) \in V_x$. Then, the value*

$$\widetilde{J}^v_{\alpha\beta}(v(\cdot); t_0, x_0) = \inf_{u(t) \in L_2([t_0,t_f],\mathbb{R}^r)} \widetilde{J}_{\alpha\beta}(u(t), v(\cdot)), \tag{7}$$

calculated along the corresponding trajectories of the system (1), is called the guaranteed result of the strategy $v(\cdot)$ in the CCDG.

The value

$$\widetilde{J}^{v*}_{\alpha\beta}(t_0, x_0) = \sup_{v(\cdot) \in V_x} \widetilde{J}^v_{\alpha\beta}(v(\cdot); t_0, x_0) \tag{8}$$

is called the lower value of the CCDG.

If the supremum value (8) is attained for $\widetilde{v}^0_{\alpha\beta}(t,x) \in V_x$, i.e.,

$$\sup_{v(\cdot) \in V_x} \widetilde{J}^v_{\alpha\beta}(v(\cdot); t_0, x_0) = \max_{v(\cdot) \in V_x} \widetilde{J}^v_{\alpha\beta}(v(\cdot); t_0, x_0)$$

and

$$\widetilde{v}^0_{\alpha\beta}(t,x) = \arg\max_{v(\cdot) \in V_x} \widetilde{J}^v_{\alpha\beta}(v(\cdot); t_0, x_0), \tag{9}$$

the strategy $\widetilde{v}^0_{\alpha\beta}(t,x)$ is called the optimal strategy of the evader in the CCDG.

Definition 4. *If*

$$\widetilde{J}^{u*}_{\alpha\beta}(t_0, x_0) = \widetilde{J}^{v*}_{\alpha\beta}(t_0, x_0) \triangleq \widetilde{J}^0_{\alpha\beta}(t_0, x_0), \tag{10}$$

then it is said that the CCDG has the game value $\widetilde{J}^0_{\alpha\beta}$.

2.2. Singular (Degenerate) Differential Game

In this game the dynamics were the same as in the CCDG, i.e., (1), while the cost function of this game was obtained from (3) by replacing α with zero:

$$\widetilde{J}_{\beta}(u,v) = |Dx(t_f) + d|^2 - \beta \int_{0}^{t_f} |v(t)|^2 dt. \tag{11}$$

The differential game (1), (11) is called the Singular Differential Game (SDG).

Remark 1. *The sets of all admissible state-feedback controls (strategies) of the pursuer and the evader in the SDG are the same as in the CCDG, i.e., U_x and V_x, respectively. The guaranteed results $\widetilde{J}^u_{\beta}(u(\cdot);t_0,x_0)$ and $\widetilde{J}^v_{\beta}(v(\cdot);t_0,x_0)$ of any given strategies $u(\cdot) \in U_x$ and $v(\cdot) \in V_x$ in the SDG are defined similarly to (4) and (7), respectively. Namely,*

$$\widetilde{J}^u_{\beta}(u(\cdot);t_0,x_0) = \sup_{v(t)\in L_2([t_0,t_f],\mathbb{R}^s)} \widetilde{J}_{\beta}(u(\cdot),v(t)), \tag{12}$$

$$\widetilde{J}^v_{\beta}(v(\cdot);t_0,x_0) = \inf_{u(t)\in L_2([t_0,t_f],\mathbb{R}^r)} \widetilde{J}_{\beta}(u(t),v(\cdot)). \tag{13}$$

The upper $\widetilde{J}^{u*}_{\beta}(t_0,x_0)$ and lower $\widetilde{J}^{v*}_{\beta}(t_0,x_0)$ values of the SDG are defined similarly to (5) and (8), respectively. Namely,

$$\widetilde{J}^{u*}_{\beta}(t_0,x_0) = \inf_{u(\cdot)\in U_x} \widetilde{J}^u_{\beta}(u(\cdot);t_0,x_0), \tag{14}$$

$$\widetilde{J}^{v*}_{\beta}(t_0,x_0) = \sup_{v(\cdot)\in V_x} \widetilde{J}^v_{\beta}(v(\cdot);t_0,x_0). \tag{15}$$

If

$$\widetilde{J}^{u*}_{\beta}(t_0,x_0) = \widetilde{J}^{v*}_{\beta}(t_0,x_0) \triangleq \widetilde{J}^{*}_{\beta}(t_0,x_0), \tag{16}$$

then $\widetilde{J}^{*}_{\beta}(t_0,x_0)$ is called the value of the SDG.

Definition 5. *The sequence of state-feedback controls $\{\widetilde{u}_{\beta,k}(\cdot)\}$, $\widetilde{u}_{\beta,k}(\cdot) \in U_x$, $(k = 1,2,...)$, is called minimizing in the SDG if*

$$\lim_{k\to\infty} \widetilde{J}^u_{\beta}(\widetilde{u}_{\beta,k}(\cdot);t_0,x_0) = \widetilde{J}^{u*}_{\beta}(t_0,x_0). \tag{17}$$

If there exists $\widetilde{u}^*_{\beta}(t,x) \in U_x$, for which the upper value of the SDG is attained, this state-feedback control is called an optimal state-feedback control of the pursuer in the SDG:

$$\widetilde{u}^*_{\beta}(t,x) = \arg\min_{u(\cdot)\in U_x} \widetilde{J}^u_{\beta}(u(\cdot);t_0,x_0). \tag{18}$$

Definition 6. *The sequence of state-feedback controls $\{\widetilde{v}_{\beta,k}(\cdot)\}$, $\widetilde{v}_{\beta,k}(\cdot) \in V_x$, $(k = 1,2,...)$, is called maximizing in the SDG if*

$$\lim_{k\to\infty} \widetilde{J}^v_{\beta}(\widetilde{v}_{\beta,k}(\cdot);t_0,x_0) = \widetilde{J}^{v*}_{\beta}(t_0,x_0). \tag{19}$$

If there exists $\widetilde{v}^*_{\beta}(t,x) \in V_x$, for which the lower value of the SDG is attained, this state-feedback control is called an optimal state-feedback control of the evader in the SDG:

$$\widetilde{v}^*_{\beta}(t,x) = \arg\max_{v(\cdot)\in V_x} \widetilde{J}^v_{\beta}(v(\cdot);t_0,x_0). \tag{20}$$

Remark 2. *Since the cost function (11) of the SDG does not contain a quadratic control cost of u, its solution (if it exists) cannot be obtained either by the Isaacs's MinMax principle or by the Bellman–Isaacs equation method (see [23]). This justified calling this game singular. The CCDG could be considered as a singularly perturbed SDG, whereas the SDG was a degenerate CCDG.*

2.3. Reduction of the Games

Let $\Phi(t,\tau)$ be the transition matrix of the homogeneous system $\dot{x} = A(t)x$. By applying the state transformation

$$z = D\Phi(t_f, t)x + d, \tag{21}$$

the system (1) is reduced to

$$\dot{z} = H_1(t)u + H_2(t)v, \quad z(t_0) = z_0, \quad t \in [t_0, t_f], \tag{22}$$

where $m \times r$ and $m \times s$ matrices $H_1(t)$ and $H_2(t)$ are

$$H_1(t) = D\Phi(t_f, t)B(t), \quad H_2(t) = D\Phi(t_f, t)C(t), \tag{23}$$

$$z_0 = D\Phi(t_f, t_0)x_0 + d. \tag{24}$$

Due to (21), for the reduced system (22), the cost functions (3) and (11) of the CCDG and SDG become

$$J_{\alpha\beta} = |z(t_f)|^2 + \alpha \int_{t_0}^{t_f} |u(t)|^2 dt - \beta \int_{t_0}^{t_f} |v(t)|^2 dt, \tag{25}$$

and

$$J_{\beta} = |z(t_f)|^2 - \beta \int_{t_0}^{t_f} |v(t)|^2 dt, \tag{26}$$

respectively.

The games (22), (25) and (22), (26) are called the Reduced Cheap Control Differential Game (RCCDG) and the Reduced Singular Differential Game (RSDG), respectively.

Let us consider the set $\mathcal{U}_z$ of all functions $u = u(t,z) : [0, t_f] \times \mathbb{R}^m \to \mathbb{R}^r$, which are measurable w.r.t. $t \in [0, t_f]$ for any fixed $z \in \mathbb{R}^m$ and satisfy the local Lipschitz condition w.r.t. $z \in \mathbb{R}^m$ uniformly in $t \in [0, t_f]$. Similarly, we consider the set $\mathcal{V}_z$ of all functions $v = v(t,z) : [0, t_f] \times \mathbb{R}^m \to \mathbb{R}^s$, which are measurable w.r.t. $t \in [0, t_f]$ for any fixed $z \in \mathbb{R}^m$ and satisfy the local Lipschitz condition w.r.t. $z \in \mathbb{R}^m$ uniformly in $t \in [0, t_f]$.

Definition 7. *Let us denote by U_z the set of all functions $u(t,z) \in \mathcal{U}_z$ satisfying the following conditions: (1_{uz}) the initial-value problem (22) for $u(t) = u(t,z)$ and any fixed $v(t) \in L_2([0,t_f], \mathbb{R}^s)$ has the unique absolutely continuous solution $z_u(t)$, $t \in [0, t_f]$; (2_{uz}) $u(t, z_u(t)) \in L_2([0,t_f], \mathbb{R}^r)$.*

In addition, let us denote by V_z the set of all functions $v(t,z) \in \mathcal{V}_z$ satisfying the following conditions: (1_{vz}) the initial-value problem (22) for $v(t) = v(t,z)$ and any fixed $u(t) \in L_2([0,t_f], \mathbb{R}^r)$ has the unique absolutely continuous solution $z_v(t)$, $t \in [0, t_f]$; (2_{vx}) $v(t, z_v(t)) \in L_2([0,t_f], \mathbb{R}^s)$.

In what follows, the set U_z is called the set of all admissible state-feedback controls (strategies) of the pursuer in both games RCCDG and RSDG, while the set V_z is called the set of all admissible state-feedback controls (strategies) of the evader in both games RCCDG and RSDG.

Remark 3. *Based on Definition 7, the guaranteed results $J^u_{\alpha\beta}(u(\cdot); t_0, z_0)$ and $J^v_{\alpha\beta}(v(\cdot); t_0, z_0)$ of any given strategies $u(\cdot) \in U_z$ and $v(\cdot) \in V_z$ in the RCCDG are defined similarly to (4) and (7), respectively. The upper $J^{u*}_{\alpha\beta}(t_0, z_0)$ and lower $J^{v*}_{\alpha\beta}(t_0, z_0)$ values of the RCCDG are defined similarly to (5) and (8), respectively. The optimal state-feedback controls of the pursuer $u^0_{\alpha\beta}(t,z)$ and the* *evader $v^0_{\alpha\beta}(t,z)$, $(t,z) \in [0,t_f] \times \mathbb{R}^m$, are defined similarly to (6) and (9), respectively. The value of the RCCDG $J^0_{\alpha\beta}(t_0,z_0)$ is defined similarly to (10).*

Remark 4. *Based on Definition 7, the guaranteed results $J^u_\beta(u(\cdot);t_0,z_0)$ and $J^v_\beta(v(\cdot);t_0,z_0)$ of any given strategies $u(\cdot) \in U_z$ and $v(\cdot) \in V_z$ in the RSDG are defined similarly to (12) and (13), respectively. The upper $J^{u*}_\beta(t_0,z_0)$ and lower $J^{v*}_\beta(t_0,z_0)$ values of the RSDG are defined similarly to (14) and (15), respectively. The minimizing sequence $\{u_{\beta,k}(\cdot)\}$, $u_{\beta,k}(\cdot) \in U_z$, $(k=1,2,...)$, and the optimal state-feedback control $u^*_\beta(t,z)$ of the pursuer in the RSDG are defined similarly to (17) and (18), respectively. The maximizing sequence $\{v_{\beta,k}(\cdot)\}$, $v_{\beta,k}(\cdot) \in V_z$, $(k=1,2,...)$, and the optimal state-feedback control $v^*_\beta(t,z)$ of the evader in the RSDG are defined similarly to (19) and (20), respectively. The value of the RSDG $J^*_\beta(t_0,z_0)$ is defined similarly to (16).*

Remark 5. *If $u^0_{\alpha\beta}(t,z)$ and $v^0_{\alpha\beta}(t,z)$ are the optimal strategies of the pursuer and the evader in the RCCDG, then the strategies*

$$u^0_{\alpha\beta}\Big(t, D\Phi(t_f,t)x+d\Big) \quad \text{and} \quad v^0_{\alpha\beta}\Big(t, D\Phi(t_f,t)x+d\Big), \tag{27}$$

are optimal strategies of the pursuer and the evader in the CCDG.

If $\{u_{\beta,k}(t,z)\}_{k=1}^{+\infty}$ and $\{v_{\beta,k}(t,z)\}_{k=1}^{+\infty}$ are the minimizing sequence and the maximizing sequence in the RSDG, then the sequences

$$\Big\{u_{\beta,k}\Big(t, D\Phi(t_f,t)x+d\Big)\Big\}_{k=1}^{+\infty} \quad \text{and} \quad \Big\{v_{\beta,k}\Big(t, D\Phi(t_f,t)x+d\Big)\Big\}_{k=1}^{+\infty} \tag{28}$$

are minimizing and maximizing sequences in the SDG. Moreover, if $u^*_\beta(t,z)$ and $v^*_\beta(t,z)$ are the optimal strategies of the pursuer and the evader in the RSDG, then the strategies

$$u^*_\beta\Big(t, D\Phi(t_f,t)x+d\Big) \quad \text{and} \quad v^*_\beta\Big(t, D\Phi(t_f,t)x+d\Big), \tag{29}$$

are optimal strategies of the pursuer and the evader in the SDG.

2.4. Objectives of the Paper

In this paper, we investigated the asymptotic behaviour of the solution to the RCCDG and the relation between the RCCDG and the RSDG solutions. In particular, the objectives of the paper were:

(1) to establish the boundedness of the time realizations $u^0_{\alpha\beta}(t) = u^0_{\alpha\beta}\Big(t, z^0_{\alpha\beta}(t)\Big)$, $v^0_{\alpha\beta}(t) = v^0_{\alpha\beta}\Big(t, z^0_{\alpha\beta}(t)\Big)$ of the RCCDG optimal strategies along the corresponding trajectory $z^0_{\alpha\beta}(t)$ of (22) for $\alpha \to 0$;

(2) to establish the best achievable RCCDG value from the pursuer's point of view:

$$J^0_{\text{best}}(t_0,z_0) \triangleq \inf_{\alpha \in (0,\alpha^0]} J^0_{\alpha\beta}(t_0,z_0), \tag{30}$$

where $\alpha^0 > 0$ is some sufficiently small number;

(3) to obtain the RSDG value, and establish the limiting relation between the values of the RCCDG and the RSDG:

$$\lim_{\alpha \to 0} J^0_{\alpha\beta}(t_0,z_0) = J^*_\beta(t_0,z_0); \tag{31}$$

(4) to construct the RSDG pursuer's minimizing sequence $\Big\{u_{\beta,k}(\cdot)\Big\}_{k=1}^{+\infty}$ and the evader's optimal state-feedback control $v^*_\beta(\cdot)$ based on the RCCDG solution.

3. The RCCDG Solution and Its Asymptotic Properties

By virtue of [19–22], we obtained the RCCDG solution:

$$J^0_{\alpha\beta}(t_0, z_0) = z_0^T R_{\alpha\beta}(t_0) z_0, \tag{32}$$

$$u^0_{\alpha\beta}(t, z) = -\frac{1}{\alpha} H_1^T(t) R_{\alpha\beta}(t) z, \tag{33}$$

$$v^0_{\alpha\beta}(t, z) = \frac{1}{\beta} H_2^T(t) R_{\alpha\beta}(t) z, \tag{34}$$

where the matrix-valued function $R_{\alpha\beta}(t)$ is the solution of the Riccati matrix differential equation

$$\dot{R} = R Q_{\alpha\beta}(t) R, \quad R(t_f) = I_m, \quad t \in [t_0, t_f], \tag{35}$$

$$Q_{\alpha\beta}(t) = \frac{1}{\alpha} H_1(t) H_1^T(t) - \frac{1}{\beta} H_2(t) H_2^T(t), \tag{36}$$

H^T denotes a transposed matrix and I_m is the unit $m \times m$-matrix.

The solution of (35) is readily obtained:

$$R_{\alpha\beta}(t) = S^{-1}_{\alpha\beta}(t), \quad t \in [t_0, t_f], \tag{37}$$

if and only if the matrix

$$S_{\alpha\beta}(t) = I_m + \int_t^{t_f} Q_{\alpha\beta}(\tau) d\tau \tag{38}$$

is invertible for all $t \in [t_0, t_f]$.

Thus, the RCCDG is solvable if and only if

$$\det(S_{\alpha\beta}(t)) \neq 0, \quad t \in [t_0, t_f]. \tag{39}$$

Condition S. The system (22) is controllable with respect to $u(t)$ at any interval $[t, t_f]$, $t \in [t_0, t_f)$.

Remark 6. *By using the t-dependent controllability gramians*

$$G_1(t) = \int_t^{t_f} H_1(\tau) H_1^T(\tau) d\tau, \quad t \in [t_0, t_f), \tag{40}$$

Condition S can be rewritten [18] as

$$\det G_1(t) > 0, \quad t \in [t_0, t_f). \tag{41}$$

The following statement is a direct consequence of (Theorem 3.1 [24]).

Proposition 1. *Let Condition S hold. Then, for any $\beta > 0$ there exists $\tilde{\alpha} = \tilde{\alpha}(\beta)$ such that the condition (39) holds for all $\alpha > 0$ satisfying*

$$\alpha \leq \tilde{\alpha}. \tag{42}$$

Let $z^0_{\alpha\beta}(t)$ denote the optimal motion of (22) for $u = u^0_{\alpha\beta}(t, z)$, $v = v^0_{\alpha\beta}(t, z)$.

Proposition 2. *Let Condition S hold. Then, there exists the bounded limit function*

$$\tilde{z}(t) = \lim_{\alpha \to 0} z^0_{\alpha\beta}(t), \quad t \in [t_0, t_f], \tag{43}$$

which is independent of β. Moreover

$$\lim_{\alpha\to 0} z^0_{\alpha\beta}(t_f) = \tilde{z}(t_f) = 0. \tag{44}$$

Proof. Let $\alpha > 0$ satisfy (42). By substituting the optimal strategies (33) and (34) into the system (22), due to (36), (37) and (38), the dynamics become

$$\dot{z} = -Q_{\alpha\beta}(t)R_{\alpha\beta}(t)z. \tag{45}$$

Define

$$y \triangleq R_{\alpha\beta}(t)z = \left(I_m + \int_t^{t_f} Q_{\alpha\beta}(\tau)d\tau\right)^{-1} z. \tag{46}$$

Then,

$$\dot{y} = -\left(I_m + \int_t^{t_f} Q_{\alpha\beta}(\tau)d\tau\right)^{-1} (-Q_{\alpha\beta}(t)) \left(I_m + \int_t^{t_f} Q_{\alpha\beta}(\tau)d\tau\right)^{-1} z +$$
$$\left(I_m + \int_t^{t_f} Q_{\alpha\beta}(\tau)d\tau\right)^{-1} (-Q_{\alpha\beta}(t)) \left(I_m + \int_t^{t_f} Q_{\alpha\beta}(\tau)d\tau\right)^{-1} z = 0, \tag{47}$$

yielding

$$y(t) \equiv c = \text{const}, \quad t \in [t_0, t_f]. \tag{48}$$

For $t = t_0$,

$$y(t_0) = c = \left(I_m + \int_{t_0}^{t_f} Q_{\alpha\beta}(\tau)d\tau\right)^{-1} z_0. \tag{49}$$

Thus, due to (46) and (48), the solution $z^0_{\alpha\beta}(t)$ of (45) is

$$z^0_{\alpha\beta}(t) = \left(I_m + \int_t^{t_f} Q_{\alpha\beta}(\tau)d\tau\right)\left(I_m + \int_{t_0}^{t_f} Q_{\alpha\beta}(\tau)d\tau\right)^{-1} z_0. \tag{50}$$

Due to (36) and (40),

$$z^0_{\alpha\beta}(t) = \left(I_m + \frac{1}{\alpha}G_1(t) - \frac{1}{\beta}\int_t^{t_f} H_2(\tau)H_2^T(\tau)d\tau\right)\left(I_m + \frac{1}{\alpha}G_1(t_0) - \frac{1}{\beta}\int_{t_0}^{t_f} H_2(\tau)H_2^T(\tau)d\tau\right)^{-1} z_0. \tag{51}$$

By factoring $\dfrac{1}{\alpha}$ out of both matrices, (51) becomes

$$z^0_{\alpha\beta}(t) = \left(\alpha I_m - \frac{\alpha}{\beta}\int_t^{t_f} H_2(\tau)H_2^T(\tau)d\tau + G_1(t)\right)\left(\alpha I_m - \frac{\alpha}{\beta}\int_t^{t_f} H_2(\tau)H_2^T(\tau)d\tau + G_1(t_0)\right)^{-1} z_0. \tag{52}$$

Since the gramian $G_1(t_0)$ is non-singular, the limit (43) is readily calculated for $t \in [t_0, t_f]$:

$$\lim_{\alpha\to 0} z^0_{\alpha\beta}(t) = G_1(t)G_1^{-1}(t_0)z_0 \triangleq \tilde{z}(t). \tag{53}$$

For $t = t_f$, (51) is

$$z^0_{\alpha\beta}(t_f) = \alpha\left(\alpha I_m - \frac{\alpha}{\beta}\int\limits_{t_0}^{t_f} H_2(\tau)H_2^T(\tau)d\tau + G_1(t_0)\right)^{-1} z_0, \tag{54}$$

and

$$\lim_{\alpha\to 0} z^0_{\alpha\beta}(t_f) = 0. \tag{55}$$

Since $G_1(t_f) = 0$, (53) yields

$$\tilde{z}(t_f) = 0. \tag{56}$$

Equations (55) and (56) prove (44). This completes the proof of the proposition. □

Proposition 3. *Let Condition S hold. Then the time realizations $u^0_{\alpha\beta}(t) = u^0_{\alpha\beta}(t, z^0_{\alpha\beta}(t))$, $v^0_{\alpha\beta}(t) = v^0_{\alpha\beta}(t, z^0_{\alpha\beta}(t))$ of the optimal strategies (33)–(34) are bounded for $\alpha \to 0$.*

Proof. By substituting (50) into (33), by using (36) and (40), and by factoring $\frac{1}{\alpha}$ out of the matrix, the time realization of the RCCDG optimal minimizer's strategy is

$$u^0_{\alpha\beta}(t) = -H_1^T(t)\left(\alpha I_m + G_1(t_0) - \frac{\alpha}{\beta}\int\limits_{t_0}^{t_f} H_2(\tau)H_2^T(\tau)(\tau)d\tau\right)^{-1} z_0. \tag{57}$$

Thus, for any $\beta > 0$, there exists the bounded limit function

$$\lim_{\alpha\to 0} u^0_{\alpha\beta}(t) = -H_1(t)G_1^{-1}(t_0)z_0 \triangleq \tilde{u}(t), \;\; t \in [t_0, t_f]. \tag{58}$$

Similarly, the time realization the RCCDG optimal maximizer's strategy is

$$v^0_{\alpha\beta}(t) = \frac{\alpha}{\beta}H_2^T(t)\left(\alpha I_m + G_1(t_0) - \frac{\alpha}{\beta}\int\limits_{t_0}^{t_f} H_2(\tau)H_2^T(\tau)(\tau)d\tau\right)^{-1} z_0. \tag{59}$$

yielding

$$\lim_{\alpha\to 0} v^0_{\alpha\beta}(t) = 0 \triangleq \tilde{v}(t), \;\; t \in [t_0, t_f]. \tag{60}$$

□

Proposition 4. *Let Condition S hold. Then the feedback strategies (33) and (34) are well defined for $\alpha = 0$ for all $(t, z) \in [t_0, t_f) \times \mathbb{R}^m$.*

Proof. Similarly to (57), by factoring $\frac{1}{\alpha}$ from the gain of the strategy (33),

$$u^0_{\alpha\beta}(t, z) = -H_1^T(t)\left(\alpha I_m + G_1(t) - \frac{\alpha}{\beta}\int\limits_{t}^{t_f} H_2(\tau)H_2^T(\tau)(\tau)d\tau\right)^{-1} z, \tag{61}$$

which is well defined for $\alpha = 0$, $(t, z) \in [t_0, t_f) \times \mathbb{R}^m$:

$$\lim_{\alpha\to 0} u^0_{\alpha\beta}(t, z) = \tilde{K}(t)z \triangleq \tilde{u}(t, z), \tag{62}$$

where

$$\tilde{K}(t) = -H_1^T(t)G_1^{-1}(t). \tag{63}$$

Similarly to (59),

$$v^0_{\alpha\beta}(t,z) = \frac{\alpha}{\beta}H_2^T(t)\left(\alpha I_m + G_1(t) - \frac{\alpha}{\beta}\int_t^{t_f} H_2(\tau)H_2^T(\tau)(\tau)d\tau\right)^{-1} z, \tag{64}$$

yielding

$$\lim_{\alpha\to 0} v^0_{\alpha\beta}(t,z) = 0 \triangleq \tilde{v}(t,z), \tag{65}$$

for all $(t,z) \in [t_0,t_f) \times \mathbb{R}^m$. □

Remark 7. *Due to (40), the gain (63) of the limit feedback $\tilde{u}(t,z)$ is infinite for $t \to t_f$:*

$$\lim_{t\to t_f} ||\tilde{K}(t)|| = \infty, \tag{66}$$

where $||\cdot||$ is the Euclidean norm of a matrix.

Remark 8. *The limit motion $\tilde{z}(t)$ given in (53) is generated by the limit feedback strategies $\tilde{u}(t,z)$ and $\tilde{v}(t,z))$ given in (62) and (65), respectively. Moreover, their time realizations along $\tilde{z}(t)$ are equal to $\tilde{u}(t)$ and $\tilde{v}(t)$ given in (58) and (60), respectively:*

$$\tilde{u}(t,\tilde{z}(t)) = \tilde{u}(t), \quad \tilde{v}(t,\tilde{z}(t)) = \tilde{v}(t). \tag{67}$$

Proposition 5. *Let Condition S hold. Then for any $\beta > 0$, the RCCDG game value satisfies*

$$\lim_{\alpha\to 0} J^0_{\alpha\beta}(t_0,z_0) = 0. \tag{68}$$

Moreover, all the terms of the optimal cost function (25) tend to zero for $\alpha \to 0$:

$$\lim_{\alpha\to 0} |z^0_{\alpha\beta}(t_f)|^2 = 0, \tag{69}$$

$$\lim_{\alpha\to 0}\left(\alpha\int_{t_0}^{t_f} |u^0_{\alpha\beta}(t)|^2 dt\right) = 0, \tag{70}$$

$$\lim_{\alpha\to 0}\left(\beta\int_{t_0}^{t_f} |v^0_{\alpha\beta}(t)|^2 dt\right) = 0. \tag{71}$$

Proof. By factoring $\frac{1}{\alpha}$ from the matrix $R_{\alpha\beta}(t)$,

$$J^0_{\alpha\beta}(t_0,z_0) = \alpha z_0^T\left(\alpha I_m + G_1(t_0) - \frac{\alpha}{\beta}\int_{t_0}^{t_f} H_2(\tau)H_2^T(\tau)(\tau)d\tau\right)^{-1} z_0. \tag{72}$$

Since the matrix $G_1(t_0)$ is non-singular, (72) directly leads to (68).

The limiting Equation (69) is the consequence of (55); (70) holds, because, due to Proposition 3, the limit time realization of the minimizer's optimal strategy is bounded; (71) follows from (60). □

Corollary 1. *Let Condition S hold. Then,*

$$J^0_{\text{best}}(t_0, z_0) = 0. \tag{73}$$

Proof. First of all, let us note that, due to Remark 6, the matrix $G_1(t_0)$ is positive definite. Therefore, using (72), we can conclude the following. There exists a positive number $\alpha^0 \leq \tilde{\alpha}$ such that, for all $\alpha \in (0, \alpha^0]$,

$$J^0_{\alpha\beta}(t_0, z_0) \geq 0. \tag{74}$$

This inequality, along with the equality (68), directly yields the statement of the corollary. □

4. RSDG Solution

Lemma 1. *Let Condition S hold. Then, there exists a positive number $\alpha_0 < \tilde{\alpha}$, such that for all $\alpha \in (0, \alpha_0]$ the guaranteed result $J^u_\beta(u^0_{\alpha\beta}(\cdot); t_0, z_0)$ of the pursuer's state-feedback control $u^0_{\alpha\beta}(t, z)$ in the RSDG satisfies the inequality*

$$0 \leq J^u_\beta(u^0_{\alpha\beta}(\cdot); t_0, z_0) \leq a\alpha, \tag{75}$$

where $a > 0$ is some value independent of α.

Proof. First of all, let us remember that $u^0_{\alpha\beta}(t, z)$ is the optimal pursuer's control in the RCCDG, and this control is given by Equation (33). Taking into account Remark 4 and Equation (26), the guaranteed result of this control in the RSDG is calculated as follows:

$$\begin{aligned} J^u_\beta(u^0_{\alpha\beta}(\cdot); t_0, z_0) &= \sup_{v(t)\in L_2\left([t_0,t_f],\mathbb{R}^s\right)} J_\beta(u^0_{\alpha\beta}(\cdot), v(\cdot)) \\ &= \sup_{v(t)\in L_2\left([t_0,t_f],\mathbb{R}^s\right)} \left(|z(t_f|^2 - \beta \int_{t_0}^{t_f} |v(t)|^2 dt \right) \end{aligned} \tag{76}$$

along trajectories of the system

$$\dot{z} = H_1(t) u^0_{\alpha\beta}(t, z) + H_2(t) v(t), \quad t \in [t_0, t_f], \quad z(0) = z_0. \tag{77}$$

For any $v(t) \in L_2([t_0, t_f], \mathbb{R}^s)$, we have the inequality

$$|z(t_f)|^2 - \beta \int_{t_0}^{t_f} |v(t)|^2 dt \leq |z(t_f|^2 + \alpha \int_{t_0}^{t_f} \left|u^0_{\alpha\beta}(t, z)\right|^2 dt - \beta \int_{t_0}^{t_f} |v(t)|^2 dt \tag{78}$$

along trajectories of the system (77). Therefore,

$$\begin{aligned} 0 \leq \sup_{v(t)\in L_2\left([t_0,t_f],\mathbb{R}^s\right)} \left(|z(t_f|^2 - \beta \int_{t_0}^{t_f} |v(t)|^2 dt \right) \leq \\ \sup_{v(t)\in L_2\left([t_0,t_f],\mathbb{R}^s\right)} \left(|z(t_f|^2 + \alpha \int_{t_0}^{t_f} \left|u^0_{\alpha\beta}(t, z)\right|^2 dt - \beta \int_{t_0}^{t_f} |v(t)|^2 dt \right). \end{aligned} \tag{79}$$

Since $u^0_{\alpha\beta}(t, z)$ is the optimal state-feedback control in the RCCDG, then using the form of the cost function in this game (see Equation (25)) and the definition of the value in this game (see Remark 3), we directly have

$$\sup_{v(t)\in L_2\left([t_0,t_f],\mathbb{R}^s\right)} \left(|z(t_f|^2 + \alpha \int_{t_0}^{t_f} \left|u^0_{\alpha\beta}(t, z)\right|^2 dt - \beta \int_{t_0}^{t_f} |v(t)|^2 dt \right) = J^0_{\alpha\beta}(t_0, z_0). \tag{80}$$

Remember that $J^0_{\alpha\beta}(t_0, z_0)$ is the RCCDG value given by Equation (32).

Further, using Equations (76), (80) and the inequality (79), we obtain immediately

$$0 \leq J^u_\beta(u^0_{\alpha\beta}(\cdot); t_0, z_0) \leq J^0_{\alpha\beta}(t_0, z_0). \tag{81}$$

Now, the statement of the lemma directly follows from Equation (72) and the inequality (81). □

Consider the following admissible state-feedback control of the maximizing player (the evader) in the RSDG:

$$\bar{v}^0(t, z) \equiv 0, \quad (t, z) \in [t_0, t_f] \times \mathbb{R}^m. \tag{82}$$

Lemma 2. *Let Condition S hold. Then, the guaranteed result $J^v_\beta(\bar{v}^0(\cdot); t_0, z_0)$ of $\bar{v}^0(t, z)$ in the RSDG is*

$$J^v(\bar{v}^0(\cdot); t_0, z_0) = 0. \tag{83}$$

Proof. Substituting $v(t) = \bar{v}^0(t, z)$ into the system (22) and the cost function (26) yields the following system and cost function:

$$\dot{z} = H_1(t)u, \ \ z(t_0) = z_0, \ \ t \in [t_0, t_f], \tag{84}$$

$$\bar{J}(u(\cdot)) = J_\beta(u(\cdot), \bar{v}^0(\cdot)) = |z(t_f)|^2. \tag{85}$$

Therefore, $J^v(\bar{v}^0(\cdot); t_0, z_0)$ is the infimum value with respect to $u(t) \in L_2\Big([t_0, t_f], \mathbb{R}^r\Big)$ of the cost function (85) along trajectories of the system (84), i.e.,

$$J^v(\bar{v}^0(\cdot); t_0, z_0) = \inf_{u(\cdot) \in L_2\left([t_0, t_f], \mathbb{R}^r\right)} \bar{J}(u(\cdot)). \tag{86}$$

The optimal control problem (84) and (85) is singular (see, e.g., [3]), and the value (86) can be derived similarly to this work. To do this, first, we replaced approximately the singular problem (84) and (85) with the regular optimal control problem consisting of the system (84) and the new cost function

$$\bar{J}_\alpha(u(\cdot)) \stackrel{\triangle}{=} |z(t_f)|^2 + \alpha \int_{t_0}^{t_f} |u(t)|^2 dt \tag{87}$$

to be minimized by $u(\cdot) \in L_2\Big([t_0, t_f], \mathbb{R}^r\Big)$ along trajectories of the system (84). In (87), $\alpha > 0$ is a small parameter of the regularization.

For any given $\alpha > 0$, the problem in (84), (87) is a linear-quadratic optimal control problem. By virtue of the results of [25], we directly have that the solution (the optimal control) of this problem is $\bar{u}^0_\alpha(t) = -(1/\alpha)H_1^T(t)\bar{R}_\alpha(t)\bar{z}_\alpha(t)$, and the optimal value of its function has the form

$$\bar{J}^0_\alpha = \bar{J}_\alpha(\bar{u}^0_\alpha(\cdot)) = z_0^T \bar{R}_\alpha(t_0) z_0, \tag{88}$$

where the $m \times m$-matrix-valued function $\bar{R}_\alpha(t)$ is the solution of the terminal-value problem

$$\dot{\bar{R}}_\alpha = \frac{1}{\alpha}\bar{R}_\alpha H_1(t)H_1^T(t)\bar{R}_\alpha, \quad t \in [t_0, t_f], \quad \bar{R}_\alpha(t_f) = I_m, \tag{89}$$

the vector-valued function $\bar{z}_\alpha(t)$ is the solution of the initial-value problem

$$\dot{\bar{z}}_\alpha = -\frac{1}{\alpha}H_1(t)H_1^T(t)\bar{R}_\alpha(t)\bar{z}_\alpha, \quad t \in [t_0, t_f], \quad z(t_0) = z_0. \tag{90}$$

Using Remark 6, we obtain the unique solution of the problem (89) as follows:

$$\bar{R}_\alpha(t) = \left(I_m + \frac{1}{\alpha}G_1(t)\right)^{-1}, \quad t \in [t_0, t_f], \tag{91}$$

where the $m \times m$-matrix-valued function $G_1(t)$ is given in Remark 6 (see (40) for $t \in [t_0, t_f]$).

Substituting (91) into (88), we obtain after some rearrangement

$$\bar{J}_\alpha^0 = \alpha z_0^T \left(\alpha I_m + G_1(t_0)\right)^{-1} z_0, \tag{92}$$

yielding the following inequality for all sufficiently small $\alpha > 0$:

$$0 \le \bar{J}_\alpha^0 \le c\alpha, \tag{93}$$

where $c > 0$ is some value independent of α.

Using Equation (88) and inequality (93), we obtain for all sufficiently small $\alpha > 0$:

$$0 \le \inf_{u(\cdot)\in L_2([t_0,t_f],\mathbb{R}^r)} \bar{J}(u(\cdot)) \le \bar{J}(\bar{u}_\alpha^0(\cdot)) \le \bar{J}_\alpha(\bar{u}_\alpha^0(\cdot)) = \bar{J}_\alpha^0 \le c\alpha,$$

yielding

$$0 \le \inf_{u(\cdot)\in L_2([t_0,t_f],\mathbb{R}^r)} \bar{J}(u(\cdot)) \le c\alpha.$$

The latter implies immediately

$$\inf_{u(\cdot)\in L_2([t_0,t_f],\mathbb{R}^r)} \bar{J}(u(\cdot)) = 0$$

which, along with Equation (86), proves the statement of the lemma. □

Theorem 1. *Let Condition S hold. Then, the RSDG value $J_\beta^*(t_0, z_0)$ exists and*

$$J_\beta^*(t_0, z_0) = 0. \tag{94}$$

Proof. Let $J_\beta^{u*}(t_0, z_0)$ and $J_\beta^{v*}(t_0, z_0)$ be the upper and lower values of the RSDG, respectively. Then, due to the definitions of these values (see Remark 4), we have

$$J_\beta^{u*}(t_0, z_0) \le J_\beta^u(u_{\alpha\beta}^0(\cdot); t_0, z_0), \quad \alpha \in (0, \alpha_0], \tag{95}$$

$$J_\beta^v(\bar{v}^0(\cdot); t_0, z_0) \le J_\beta^{v*}(t_0, z_0), \tag{96}$$

$$J_\beta^{v*}(t_0, z_0) \le J_\beta^{u*}(t_0, z_0). \tag{97}$$

Now, using the equality (83) and the inequalities (75), (95)–(97) yield

$$\begin{aligned} 0 = J_\beta^v(\bar{v}^0(\cdot); t_0, z_0) &\le J_\beta^{v*}(t_0, z_0) \le J_\beta^{u*}(t_0, z_0) \\ &\le J_\beta^u(u_{\alpha\beta}^0(\cdot); t_0, z_0) \le a\alpha, \quad \alpha \in (0, \alpha_0]. \end{aligned} \tag{98}$$

The latter implies

$$0 \le J_\beta^{v*}(t_0, z_0) \le J_\beta^{u*}(t_0, z_0) \le a\alpha, \quad \alpha \in (0, \alpha_0]. \tag{99}$$

From (99), for $\alpha \to 0$, we directly have $J_\beta^{v*}(t_0, z_0) = J_\beta^{u*}(t_0, z_0) = 0$, which proves the theorem. □

Corollary 2. *Let Condition S hold. Then,*

$$J^0_{\text{best}}(t_0, z_0) = J^*_\beta(t_0, z_0). \tag{100}$$

Proof. The statement of the corollary directly follows from Theorem 1 and Equation (73). □

Corollary 3. *Let Condition S hold. Then, the limit Equality (31) is valid.*

Proof. The statement of the corollary is a direct consequence of Equations (68) and (94). □

By $\{\alpha_k\}_{k=1}^{+\infty}$, we denote a sequence of numbers, satisfying the following conditions: (I) $\alpha_k \in (0, \alpha_0]$, $(k = 1, 2, ...)$; (II) $\lim_{k \to +\infty} \alpha_k = 0$.

Theorem 2. *Let Condition S hold. Then, the sequence of the pursuer's state-feedback controls $\{u^0_{\alpha_k\beta}(t, z)\}_{k=1}^{+\infty}$ is the minimizing sequence in the RSDG. The state-feedback control $\bar{v}^0(t, z)$, given by (82), is the optimal evader's strategy in the RSDG.*

Proof. From the chain of the equality and the inequalities (98) we obtain

$$\lim_{k \to +\infty} J^u_\beta(u^0_{\alpha_k\beta}(\cdot); t_0, z_0) = J^{u*}_\beta(t_0, z_0), \tag{101}$$

meaning the validity of the first statement of the theorem.

Similarly, we have

$$J^v_\beta(\bar{v}^0(\cdot); t_0, z_0) = J^{v*}_\beta(t_0, z_0), \tag{102}$$

which implies the validity of the second statement of the theorem. □

Remark 9. *It should be noted that the optimal evader's strategy $\bar{v}^0(t, z)$ in the RSDG coincides with the limit (as $\alpha \to 0$) of the optimal evader's strategy in the RCCDG for all $(t, z) \in [t_0, t_f) \times \mathbb{R}^m$ (see Proposition 4 and Equation (65)). Also, it should be noted that the limit (as $k \to +\infty$) of the minimizing sequence $\{u^0_{\alpha_k\beta}(t, z)\}_{k=1}^{+\infty}$ in the RSDG is $\bar{u}(t, z)$ for all $(t, z) \in [t_0, t_f) \times \mathbb{R}^m$ (see Proposition 4 and Equations (62) and (63)). However, the function $\bar{u}(t, z)$ does not belong to the set $\mathcal{U}_z$. Therefore, this function does not belong to the set U_z, i.e., it is not an admissible pursuer's state-feedback control in the RSDG.*

5. Example: Interception Problem in Three-Dimensional Space

5.1. Engagement Model and Its Reduction

Consider the engagement in 3D space of two flying vehicles (the interceptor or the pursuer and the target or the evader), which has similar geometry to that considered in [26,27]. In contrast to [26,27], we assumed that both the pursuer and the evader have first-order dynamics controllers. Two mutually perpendicular control channels could have different time constants: τ_{p_1}, τ_{p_2} for the pursuer's controller and τ_{e_1}, τ_{e_2} for the evader's one.

The equations of motion were written down in the line-of-sight coordinate system where the axis X was the initial line-of-sight, the plane XY was the collision plane determined by the initial line-of-sight and the target's velocity vectors and the plane XZ was normal to XY.

Let (X_p, Y_p, Z_p) and (X_e, Y_e, Z_e) be the coordinates of the interceptor (the pursuer) and the target (the evader), respectively. The relative separations in the Y and Z-directions were $Y = Y_p - Y_e$ and $Z = Z_p - Z_e$. By linearization along the initial line-of-sight, the equations of motion were written down in the form (1) where the state vector was

$$x = (Y, \dot{Y}, \ddot{Y}_p, \ddot{Y}_e, Z, \dot{Z}, \ddot{Z}_p, \ddot{Z}_e)^T, \tag{103}$$

the players' control vectors (lateral acceleration commands) were $u = (u_1, u_2)^T$ (for the pursuer) and $v = (v_1, v_2)^T$ (for the evader); the final time t_f was the time of achieving the zero distance between the players along the axis X. The matrices in (1) were

$$A(t) \equiv \begin{bmatrix} 0 & 1 & 0 & 0 & 0 & 0 & 0 & 0 \\ 0 & 0 & 1 & -1 & 0 & 0 & 0 & 0 \\ 0 & 0 & -1/\tau_{p_1} & 0 & 0 & 0 & 0 & 0 \\ 0 & 0 & 0 & -1/\tau_{e_1} & 0 & 0 & 0 & 0 \\ 0 & 0 & 0 & 0 & 0 & 1 & 0 & 0 \\ 0 & 0 & 0 & 0 & 0 & 0 & 1 & -1 \\ 0 & 0 & 0 & 0 & 0 & 0 & -1/\tau_{p_2} & 0 \\ 0 & 0 & 0 & 0 & 0 & 0 & 0 & -1/\tau_{e_2} \end{bmatrix}, \tag{104}$$

$$B(t) \equiv \begin{bmatrix} 0 & 0 \\ 0 & 0 \\ 1/\tau_{p_1} & 0 \\ 0 & 0 \\ 0 & 0 \\ 0 & 0 \\ 0 & 1/\tau_{p_2} \\ 0 & 0 \end{bmatrix}, \quad C(t) \equiv \begin{bmatrix} 0 & 0 \\ 0 & 0 \\ 0 & 0 \\ 1/\tau_{e_1} & 0 \\ 0 & 0 \\ 0 & 0 \\ 0 & 0 \\ 0 & 1/\tau_{e_2} \end{bmatrix}. \tag{105}$$

In the pursuit problem, the target set was $x_1 = Y = 0, x_5 = Z = 0$, meaning that in (2),

$$D = \begin{bmatrix} 1 & 0 & 0 & 0 & 0 & 0 & 0 & 0 \\ 0 & 0 & 0 & 0 & 1 & 0 & 0 & 0 \end{bmatrix}, \quad d = \begin{bmatrix} 0 \\ 0 \end{bmatrix}. \tag{106}$$

Thus, in this example, $n = 8, r = s = m = 2$.

The transition matrix of the homogeneous system was readily obtained as

$$\Phi(t_f, t) = \begin{bmatrix} \Phi_1(t_f, t, \tau_{p_1}, \tau_{e_1}) & O_4 \\ O_4 & \Phi_1(t_f, t, \tau_{p_2}, \tau_{e_2}) \end{bmatrix}, \tag{107}$$

where O_4 is the zero 4×4 matrix,

$$\Phi_1(t_f, t, \tau_p, \tau_e) = \begin{bmatrix} 1 & t_f - t & -h(t, \tau_p) & h(t, \tau_e) \\ 0 & 1 & -\tau_p\left(1 - e^{-\vartheta(t,\tau_p)}\right) & \tau_e\left(1 - e^{-\vartheta(t,\tau_e)}\right) \\ 0 & 0 & e^{-\vartheta(t,\tau_p)} & 0 \\ 0 & 0 & 0 & e^{-\vartheta(t,\tau_e)} \end{bmatrix}, \tag{108}$$

$$\vartheta(t, \tau) \triangleq \frac{t_f - t}{\tau}, \tag{109}$$

$$h(t, \tau) \triangleq \tau^2\left(e^{-\vartheta(t,\tau)} + \vartheta(t, \tau) - 1\right). \tag{110}$$

Then, by applying the transformation (21) with D and d as in (106), the original system was reduced to the two-dimensional system of the form (22), where

$$H_1(t) = \begin{bmatrix} -h(t, \tau_{p_1}) & 0 \\ 0 & -h(t, \tau_{p_2}) \end{bmatrix}, \quad H_2(t) = \begin{bmatrix} h(t, \tau_{e_1}) & 0 \\ 0 & h(t, \tau_{e_2}) \end{bmatrix}. \tag{111}$$

Explicitly, the system (22) became

$$\begin{aligned} \dot{z}_1 &= -h(t, \tau_{p_1})u_1 + h(t, \tau_{e_1})v_1, \quad z_1(t_0) = z_{0_1}, \quad t \in [t_0, t_f], \\ \dot{z}_2 &= -h(t, \tau_{p_2})u_2 + h(t, \tau_{e_2})v_2 \quad z_2(t_0) = z_{0_2}, \quad t \in [t_0, t_f]. \end{aligned} \tag{112}$$

5.2. Reduced Cheap Control Game

In this example, the RCCDG cost function (25) is

$$J_{\alpha\beta} = z_1^2(t_f) + z_2^2(t_f) + \alpha \int_{t_0}^{t_f} \left[u_1^2(t) + u_2^2(t)\right] dt - \beta \int_{t_0}^{t_f} \left[v_1^2(t) + v_2^2(t)\right] dt. \tag{113}$$

Due to (111), the gramian (40) is calculated as

$$G_1(t) = \begin{bmatrix} \int_t^{t_f} h^2(\eta, \tau_{p_1}) d\eta & 0 \\ 0 & \int_t^{t_f} h^2(\eta, \tau_{p_2}) d\eta \end{bmatrix}, \tag{114}$$

and

$$\det G_1(t) = \left(\int_t^{t_f} h^2(\eta, \tau_{p_1}) d\eta \right) \left(\int_t^{t_f} h^2(\eta, \tau_{p_2}) d\eta \right). \tag{115}$$

For all $\tau > 0$, we have that $h(t, \tau) > 0$, $t \in [t_0, t_f)$, and $h(t_f, \tau) = 0$. Therefore, the condition (41), and, consequently, Condition S hold.

Due to the symmetry of the matrices (111), the matrix (37) is also symmetric:

$$R_{\alpha\beta}(t) = \begin{bmatrix} r_{\alpha\beta_1}(t) & 0 \\ 0 & r_{\alpha\beta_2}(t) \end{bmatrix}, \tag{116}$$

where

$$r_{\alpha\beta_i}(t) = \frac{1}{1 + \frac{1}{\alpha} \int_t^{t_f} h^2(\eta, \tau_{p_i}) d\eta - \frac{1}{\beta} \int_t^{t_f} h^2(\eta, \tau_{e_i}) d\eta}, \quad i = 1, 2. \tag{117}$$

Thus, the RCCDG is solvable if

$$1 + \frac{1}{\alpha} \int_t^{t_f} h^2(\eta, \tau_{p_i}) d\eta - \frac{1}{\beta} \int_t^{t_f} h^2(\eta, \tau_{e_i}) d\eta > 0, \quad t \in [t_0, t_f], \ i = 1, 2. \tag{118}$$

Similarly to [24], it is proved that the solvability condition (118) yields the value $\tilde{\alpha}$ in (42) as

$$\tilde{\alpha} = \min\{\tilde{\alpha}_1, \tilde{\alpha}_2\}, \tag{119}$$

where

$$\tilde{\alpha}_i = \tilde{\alpha}_i(\beta) = \begin{cases} \mu_i(\beta)\beta, & \beta < \int_{t_0}^{t_f} h^2(\eta, \tau_{e_i}) d\eta, \\ +\infty, & \beta \geq \int_{t_0}^{t_f} h^2(\eta, \tau_{e_i}) d\eta, \end{cases} \quad i = 1, 2, \tag{120}$$

$$\mu_i(\beta) = \frac{1}{\max_{t \in [t_0, \bar{t}_i]} F_i(t, \beta)}, \quad i = 1, 2, \tag{121}$$

$$F_i(t,\beta) = \frac{\int_t^{\bar{t}_i(\beta)} h^2(\eta, \tau_{e_i})d\eta}{\int_t^{t_f} h^2(\eta, \tau_{p_i})d\eta}, \quad i = 1,2, \tag{122}$$

the moments $\bar{t}_i(\beta) \in (t_0, t_f)$, $i = 1,2$, satisfy

$$\int_{\bar{t}_i(\beta)}^{t_f} h^2(\eta, \tau_{e_i})d\eta = \beta, \quad i = 1,2. \tag{123}$$

By using (32)–(34) and (116), the solution of the game (112) and (113) is

$$J^0_{\alpha\beta}(t_0, z_0) = r_1(t_0)z^2_{0_1} + r_2(t_0)z^2_{0_2}, \tag{124}$$

$$u^0_{\alpha\beta}(t,z) = -\frac{1}{\alpha}\Big(h(t,\tau_{p_1})r_1(t)z_1, h(t,\tau_{p_2})r_2(t)z_2\Big)^T, \tag{125}$$

$$v^0_{\alpha\beta}(t,z) = \frac{1}{\beta}\Big(h(t,\tau_{e_1})r_1(t)z_1, h(t,\tau_{e_2})r_2(t)z_2\Big)^T. \tag{126}$$

Let us consider the numerical example for $t_0 = 0$ s, $t_f = 3$ s, $\beta = 0.1$, $\tau_{p_1} = \tau_{p_2} = 0.1$ s, $\tau_{e_1} = 0.15$ s, $\tau_{e_2} = 0.2$ s. For these parameters,

$$\beta = 0.1 < \int_{t_0}^{t_f} h^2(\eta, \tau_{e_1})d\eta = \int_0^3 h^2(\eta, 0.15)d\eta = 0.1737, \tag{127}$$

$$\beta = 0.1 < \int_{t_0}^{t_f} h^2(\eta, \tau_{e_2})d\eta = \int_0^3 h^2(\eta, 0.2)d\eta = 0.293. \tag{128}$$

In this example, the moments, defined by (123), are $\bar{t}_1 = 0.4792$ s, $\bar{t}_2 = 0.8443$ s (see Figure 1).

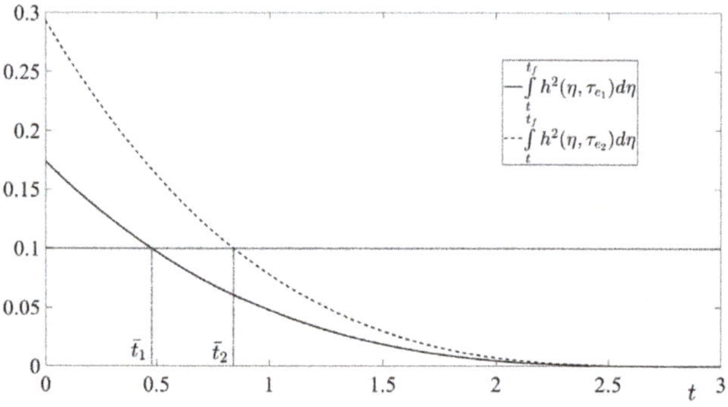

Figure 1. Moments $\bar{t}_i(\beta)$.

In Figure 2, the functions $F_i(t,\beta)$, given by (122), are shown for $t \in [t_0, \bar{t}_i]$, $i = 1,2$. It is seen that these functions were decreasing. Therefore,

$$\mu_i = \frac{1}{F_1(0,\beta)} = 1.1035, \quad \mu_2 = \frac{1}{F_2(0,\beta)} = 0.4214. \tag{129}$$

Due to (119) and (120), $\tilde{\alpha} = \beta \min\{\mu_1, \mu_2\} = 0.04214$.

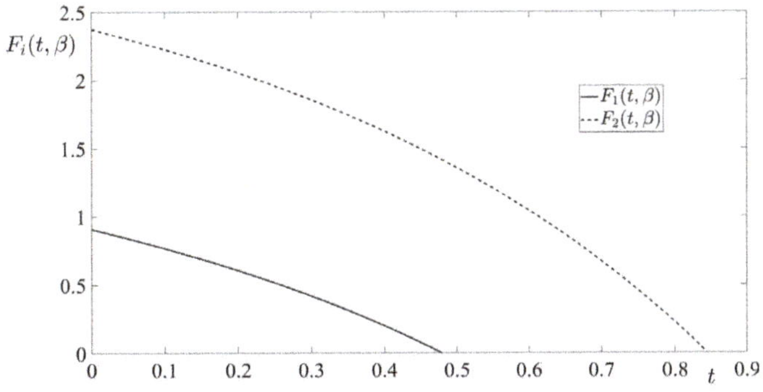

Figure 2. Functions $F_i(t, \beta)$.

In Figures 3 and 4, the components of the optimal trajectories $z^0_{\alpha\beta}(t)$ are shown for decreasing values of $\alpha < \tilde{\alpha}$, along with the components of the corresponding limiting function $\tilde{z}(t)$. It is clearly seen that the optimal trajectories tended to $\tilde{z}(t)$ for $\alpha \to 0$, and $z^0_{\alpha\beta}(t_f)$ tended to zero.

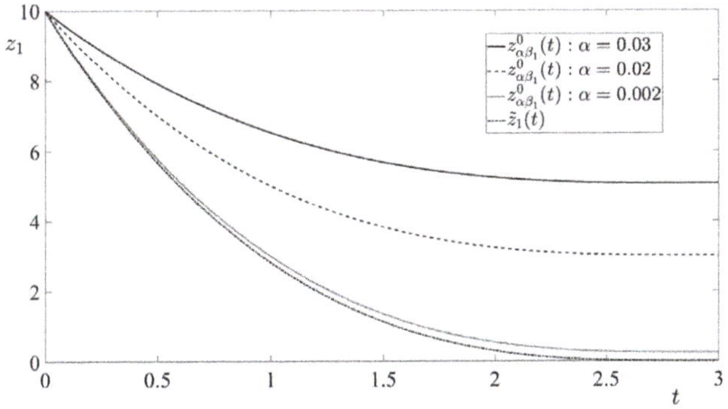

Figure 3. Trajectories $z^0_{\alpha\beta_1}(t)$ and limiting function $\tilde{z}_1(t)$.

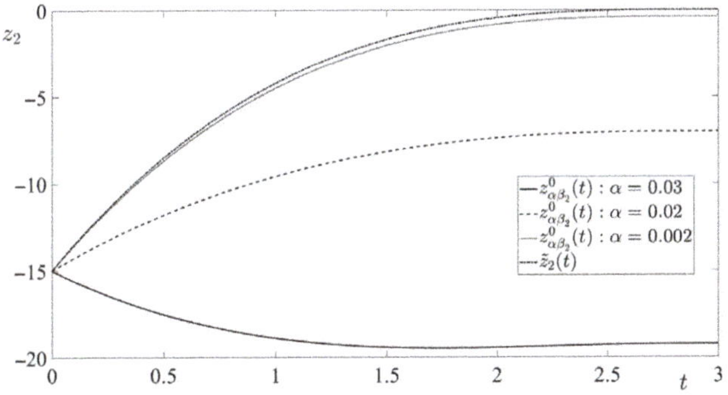

Figure 4. Trajectories $z^0_{\alpha\beta_2}(t)$ and limiting function $\tilde{z}_2(t)$.

The respective components of time realizations of the optimal strategies $u^0_{\alpha\beta}(\cdot)$ and $v^0_{\alpha\beta}(\cdot)$, along with the components of the corresponding limiting functions $\tilde{u}(t)$ and $\tilde{v}(t)$, are depicted in Figures 5–8, respectively. It is seen that the time realizations of the optimal strategies tended to the corresponding limiting functions for $\alpha \to 0$, remaining bounded.

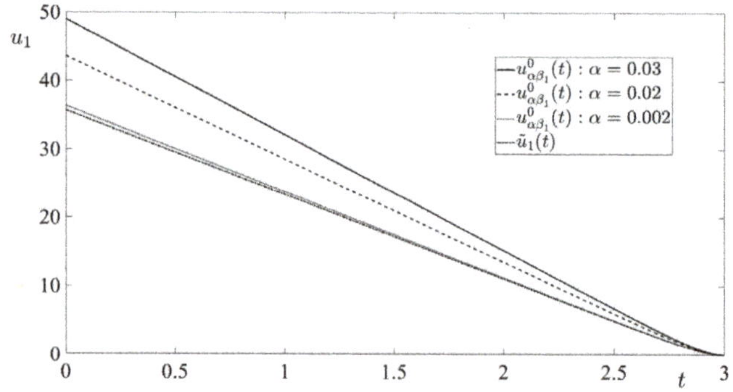

Figure 5. Time realizations $u^0_{\alpha\beta_1}(t)$ and limiting function $\tilde{u}_1(t)$.

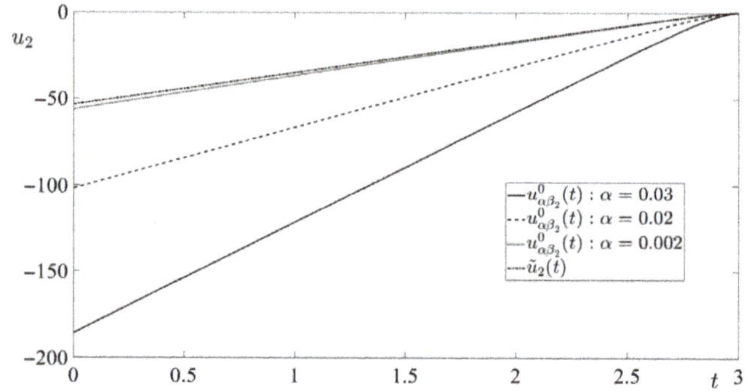

Figure 6. Time realizations $u^0_{\alpha\beta_2}(t)$ and limiting function $\tilde{u}_2(t)$.

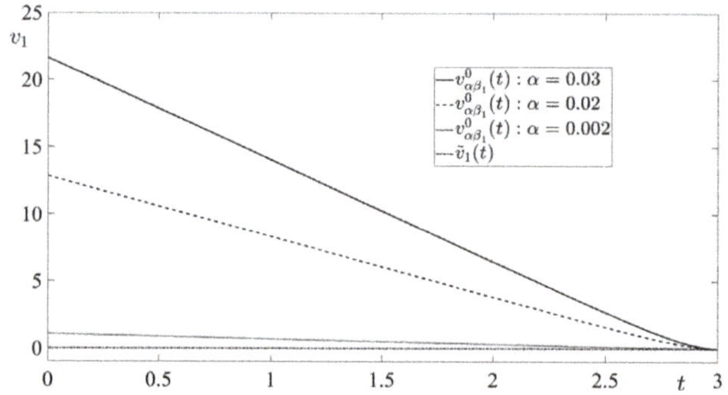

Figure 7. Time realizations $v^0_{\alpha\beta_1}(t)$ and limiting function $\tilde{v}_1(t)$.

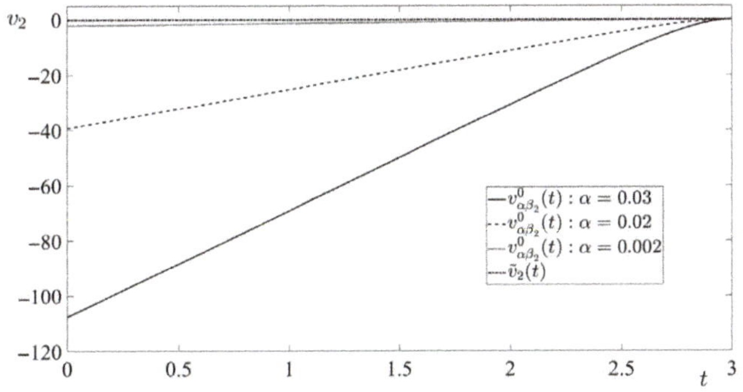

Figure 8. Time realizations $v^0_{\alpha\beta_2}(t)$ and limiting function $\tilde{v}_2(t)$.

The game value $J^0_{\alpha\beta}(t_0, z_0)$ is depicted in Figure 9 as a function of α. It is seen that it tended to zero for $\alpha \to 0$.

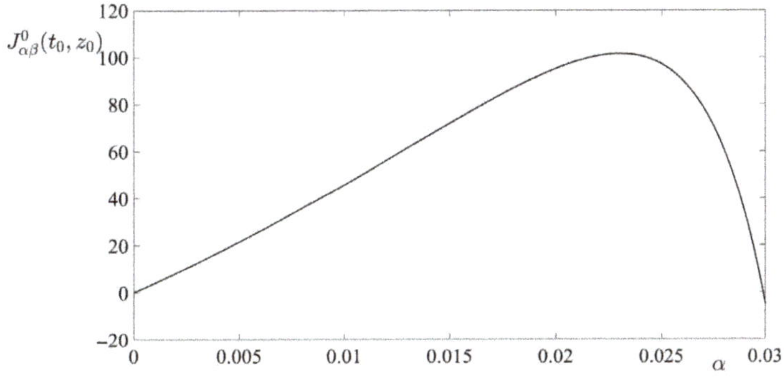

Figure 9. The game value.

The respective terminal and integral terms of the cost function are shown in Figures 10 and 11, respectively. It is seen that all components of the optimal cost tended to zero for $\alpha \to 0$.

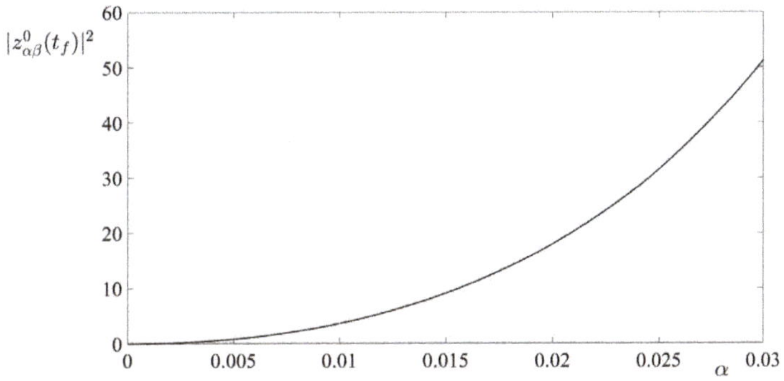

Figure 10. The terminal term of the cost function.

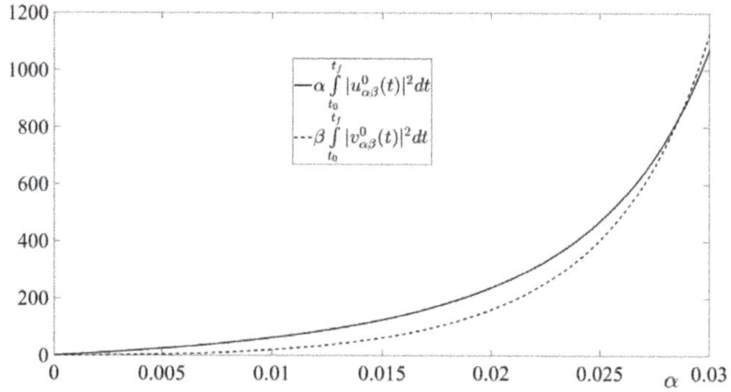

Figure 11. Integral terms of the cost function.

Remark 10. *From Equation (125), it was seen that the small control cost of the interceptor yielded the high gain in its optimal state-feedback control. This important feature of the interceptor's optimal state-feedback control increased considerably the ability of the interceptor to capture the target. One more important feature of the interceptor's optimal state-feedback control was that the time realization of this control along the optimal interception's trajectory and, especially, the trajectory itself, were bounded while the small parameter α tended to zero. Both aforementioned features of the interceptor's state-feedback control, obtained by solution of the cheap control game, were extremely important in various real-life situations of a capture of a maneuverable flying target by a maneuverable flying interceptor. It should be noted that if the small control cost of the interceptor tended to zero, the ability of the interceptor to capture the target increased tending to the best achievable result, which was the zero-miss distance at the end of the interception.*

6. Conclusions

In this paper, a pursuit-evasion problem, modeled by a finite-horizon linear-quadratic zero-sum differential game, was considered. In the game's cost function, the penalty coefficient for the minimizing player's control expenditure was a small value $\alpha > 0$. Thus, the considered game was a zero-sum differential game with a cheap control of the minimizing player. By the proper state transformation, the initially formulated game was converted to a smaller Euclidean dimension differential game, called the reduced game. This game, also was a cheap control game and it was treated in the sequel of the paper. Due to the game's solvability conditions, the solution of the reduced cheap control game was converted to the solution of the terminal-value problem for the matrix Riccati differential equation. Sufficient condition for the existence of the solution to this terminal-value problem in the entire interval of the game's duration was presented, and the solution of this terminal-value problem was obtained. Using this solution, the value of the reduced cheap control game, as well as the optimal state-feedback controls of the minimizing player (the pursuer) and the maximizing player (the evader), were derived. The trajectory of the game, generated by the optimal players' state-feedback controls, (the optimal trajectory), was obtained. The limits of the optimal trajectory, as well as of the time realizations of the players' optimal state-feedback controls along the optimal trajectory, for $\alpha \to 0$ were calculated. By this calculation, the boundedness of the optimal trajectory and the corresponding time realizations of the players' optimal state-feedback controls for $\alpha \to 0$ were shown. The limit of the game value for $\alpha \to 0$ also was calculated, yielding the best achievable game value from the pursuer's viewpoint. Along with the cheap control game, its degenerate version was considered. This version was obtained from the cheap control game by setting there formally $\alpha = 0$, yielding the new zero-sum linear-quadratic pursuit-evasion game. This new game was singular, because it could not be solved either by the Isaacs's MinMax principle or by the Bellman–Isaacs equation method. For this singular game, the notion of the pursuer's minimizing sequence of state-feedback controls (instead of the pursuer's optimal state-feedback control) was proposed. It was established that the α-dependent pursuer's optimal state-feedback control in the cheap control game constituted the pursuer's minimizing sequence of state-feedback controls (as $\alpha \to 0$) in the singular game. It was shown that the limit of this minimizing sequence was not an admissible pursuer's state-feedback control in the singular game. However, the evader's optimal state-feedback control and the value of the singular game coincided with the limits (for $\alpha \to 0$) of the evader's optimal state-feedback control and the value, respectively, of the cheap control game. Based on the theoretical results of the paper, the interception problem in 3D space, modeled by a zero-sum linear-quadratic game with the eight-dimensional dynamics, was studied. Similarly to the theoretical part of the paper, the case of the small penalty coefficient $\alpha > 0$ for the pursuer's (interceptor's) control expenditure in the cost function was considered. By proper linear state transformation, the original cheap control game was reduced to the new cheap control game with the two-dimensional dynamics. The asymptotic behaviour of the solution to this new game for $\alpha \to 0$ was analyzed.

Author Contributions: The authors contributed equally to this article. All authors have read and agreed to the published version of the manuscript.

Funding: This research received no external funding.

Data Availability Statement: Not applicable.

Conflicts of Interest: The authors declare no conflict of interest.

References

1. Bell, D.J.; Jacobson, D.H. *Singular Optimal Control Problems;* Academic Press: Cambridge, MA, USA, 1975.
2. Kurina, G.A. A degenerate optimal control problem and singular perturbations. *Soviet Math. Dokl.* **1977**, *18*, 1452–1456.
3. Glizer, V.Y. Stochastic singular optimal control problem with state delays: Regularization, singular perturbation, and minimizing sequence. *SIAM J. Control Optim.* **2012**, *50*, 2862–2888. [CrossRef]
4. Shinar, J.; Glizer, V.Y.; Turetsky, V. Solution of a singular zero-sum linear-quadratic differential game by regularization. *Int. Game Theory Rev.* **2014**, *16*, 1–32. [CrossRef]
5. Kwakernaak, H.; Sivan, R. The maximally achievable accuracy of linear optimal regulators and linear optimal filters. *IEEE Trans. Autom. Control* **1972**, *17*, 79–86. [CrossRef]
6. Braslavsky, J.H.; Seron, M.M.; Mayne, D.Q.; Kokotović, P.V. Limiting performance of optimal linear filters. *Automatica* **1999**, *35*, 189–199. [CrossRef]
7. Seron, M.M.; Braslavsky, J.H.; Kokotović, P.V.; Mayne, D.Q. Feedback limitations in nonlinear systems: From Bode integrals to cheap control. *IEEE Trans. Autom. Control* **1999**, *44*, 829–833. [CrossRef]
8. Kokotović, P.V.; Khalil, H.K.; O'Reilly, J. *Singular Perturbation Methods in Control: Analysis and Design;* Academic Press: London, UK, 1986.
9. Young, K.D.; Kokotović, P.V.; Utkin, V.I. A singular perturbation analysis of high-gain feedback systems. *IEEE Trans. Autom. Control* **1977**, *22*, 931–938. [CrossRef]
10. Moylan, P.J.; Anderson, B.D.O. Nonlinear regulator theory and an inverse optimal control problem. *IEEE Trans. Autom. Control* **1973**, *18*, 460–465. [CrossRef]
11. Turetsky, V.; Glizer, V.Y. Robust solution of a time-variable interception problem: A cheap control approach. *Int. Game Theory Rev.* **2007**, *9*, 637–655. [CrossRef]
12. Turetsky, V.; Glizer, V.Y.; Shinar, J. Robust trajectory tracking: Differential game/cheap control approach. *Int. J. Systems Sci.* **2014**, *45*, 2260–2274. [CrossRef]
13. Turetsky, V.; Shinar, J. Missile guidance laws based on pursuit—Evasion game formulations. *Automatica* **2003**, *39*, 607–618. [CrossRef]
14. Turetsky, V. Upper bounds of the pursuer control based on a linear-quadratic differential game. *J. Optim. Theory Appl.* **2004**, *121*, 163–191. [CrossRef]
15. Petersen, I.R. Linear-quadratic differential games with cheap control. *Syst. Control Lett.* **1986**, *8*, 181–188. [CrossRef]
16. Glizer, V.Y. Asymptotic solution of zero-sum linear-quadratic differential game with cheap control for the minimizer. *NoDEA Nonlinear Diff. Equ. Appl.* **2000**, *7*, 231–258. [CrossRef]
17. Vasil'eva, A.B.; Butuzov, V.F.; Kalachev, L.V. *The Boundary Function Method for Singular Perturbation Problems;* SIAM Books: Philadelphia, PA, USA, 1995.
18. Bryson, A.; Ho, Y. *Applied Optimal Control;* Hemisphere: New York, NY, USA, 1975.
19. Zhukovskii, V.I. Analytic design of optimum strategies in certain differential games. I. *Autom. Remote Control* **1970**, *4*, 533–536.
20. Krasovskii, N.N.; Subbotin, A.I. *Game-Theoretical Control Problems;* Springer: New York, NY, USA, 1988.
21. Basar, T.; Olsder, G.J. *Dynamic Noncooperative Game Theory;* Academic Press: London, UK, 1992.

22. Petrosyan, L.A.; Zenkevich, N.A. *Game Theory*; World Scientific Publishing Company: Singapore, 2016.
23. Isaacs, R. *Differential Games*; John Wiley: New York, NY, USA, 1965.
24. Turetsky, V. Robust route realization by linear-quadratic tracking. *J. Optim. Theory Appl.* **2016**, *170*, 977–992. [CrossRef]
25. Kalman, R.E. Contributions to the Theory of Optimal Control. *Bol. Soc. Mat. Mex.* **1960**, *5*, 102–119.
26. Shinar, J.; Gutman, S. Three-Dimensional Optimal Pursuit and Evasion with Bounded Controls. *IEEE Trans. Autom. Control* **1980**, *25*, 492–496. [CrossRef]
27. Shinar, J.; Medinah, M.; Biton, M. Singular surfaces in a linear pursuit-evasion game with elliptical vectograms. *J. Optim. Theory Appl.* **1984**, *43*, 431–458. [CrossRef]

axioms

MDPI

Article

Non-Resonant Non-Hyperbolic Singularly Perturbed Neumann Problem

Robert Vrabel

Institute of Applied Informatics, Automation and Mechatronics, Slovak University of Technology in Bratislava, Bottova 25, 917 01 Trnava, Slovakia; robert.vrabel@stuba.sk

Abstract: In this brief note, we study the problem of asymptotic behavior of the solutions for non-resonant, singularly perturbed linear Neumann boundary value problems $\varepsilon y'' + ky = f(t)$, $y'(a) = 0$, $y'(b) = 0$, $k > 0$, with an indication of possible extension to more complex cases. Our approach is based on the analysis of an integral equation associated with this problem.

Keywords: singular perturbation; linear ordinary differential equation; Neumann boundary value problem

MSC: 34E15; 34B05

Citation: Vrabel, R. Non-Resonant Non-Hyperbolic Singularly Perturbed Neumann Problem. *Axioms* **2022**, *11*, 394. https://doi.org/10.3390/axioms11080394

Academic Editor: Valery Y. Glizer

Received: 3 July 2022
Accepted: 10 August 2022
Published: 11 August 2022

1. Introduction

In this paper, we are dealing with the singularly perturbed linear problem

$$\varepsilon y'' + ky = f(t), \quad k > 0, \quad 0 < \varepsilon \ll 1, \quad f \in C^3([a,b]), \tag{1}$$

with the Neumann boundary condition

$$y'(a) = 0, \quad y'(b) = 0. \tag{2}$$

The analysis of the differential equations under consideration is complicated by the fact that all roots of characteristic equations of this differential equation are located on the imaginary axis; that is, the differential equation is not hyperbolic. For the singularly perturbed dynamical systems, the dynamics near a normally hyperbolic critical manifold are well-known; see [1–5] for a geometric approach to the singular perturbation theory, Refs. [6–9] for the lower and upper solution method and [10] for applications in control theory. However, if the condition of normal hyperbolicity of a critical manifold is not fulfilled, then the problem of existence and asymptotic behavior (as $\varepsilon \to 0^+$) of solutions is hard to solve in general, and leads to the principal technical difficulties in nonlinear cases; see, for example [11]. Thus, the considerations below may be instructive and helpful for the analyses of this class of problems. The calculations that will follow (and thus, the main result formulated in Theorem 1 below) can also be applied to nonlinear differential equations, where the right-hand side of (1), (2) will have the function $f(t,y)$ instead of $f(t)$, but in this case it will be necessary to guarantee that the set of solutions $y_\varepsilon(t)$, $\varepsilon \to 0^+$, of such problems also belong to the space $C^3([a,b])$, and are uniformly bounded together with their second and third derivatives on the interval $[a,b]$ (Remark 2). The uniform boundedness of the first derivatives follows from the boundary conditions imposed on the solutions (2), and uniform boundedness of the second derivatives.

Despite these difficulties, we will prove that there are an infinite number of sequences $\{\varepsilon_n\}_{n=0}^{\infty}$, $\varepsilon_n \to 0^+$, such that $y_{\varepsilon_n}(t)$ converge uniformly to $u(t)$ on $[a,b]$ for $\varepsilon_n \to 0^+$, where y_{ε_n} is a solution of the Problem (1), (2) with $\varepsilon = \varepsilon_n$ and u represents the critical manifold for our system, that is, a solution of the reduced problem $ky = f(t)$ obtained from Equation (1) for $\varepsilon = 0$.

Henceforth, in this paper, for the values of parameter ε, we consider the closed intervals J_n only, defined as

$$J_n \triangleq \left[k\left(\frac{b-a}{(n+1)\pi-\lambda}\right)^2, k\left(\frac{b-a}{n\pi+\lambda}\right)^2\right], \quad n = 0, 1, 2, \dots,$$

where $\lambda > 0$ is an arbitrarily small but fixed constant ($\lambda \ll \pi/2$), which guarantees the existence and uniqueness to the solutions of (1), (2); that is, a non-resonant case.

Example 1. *As an academic example, let us consider the linear problem*

$$\varepsilon y'' + ky = e^t, \quad t \in [a,b], \quad k > 0, \quad 0 < \varepsilon \ll 1,$$
$$y'(a) = 0, \quad y'(b) = 0,$$

and its solution

$$y_\varepsilon(t) = \frac{-e^a \cos\left[\sqrt{\frac{k}{\varepsilon}}(b-t)\right] + e^b \cos\left[\sqrt{\frac{k}{\varepsilon}}(t-a)\right]}{\sqrt{\frac{k}{\varepsilon}}(k+\varepsilon)\sin\left[\sqrt{\frac{k}{\varepsilon}}(b-a)\right]} + \frac{e^t}{k+\varepsilon}.$$

Hence, for every sequence $\{\varepsilon_n\}_{n=0}^{\infty}$, $\varepsilon_n \in J_n$, the solution of the problem under consideration satisfies

$$y_{\varepsilon_n}(t) = \frac{e^t}{k+\varepsilon_n} + O(\sqrt{\varepsilon_n})$$

and thus, the solutions converge uniformly on the interval $[a,b]$ to the solution $u(t) = e^t/k$ of the reduced problem for $n \to \infty$. The second term on the right-hand side denotes the convenient Big–O notation. For better illustration, Figure 1 graphically shows the solutions for different values of the parameter ε. The MATLAB code for Figure 1 is below, in Listing 1.

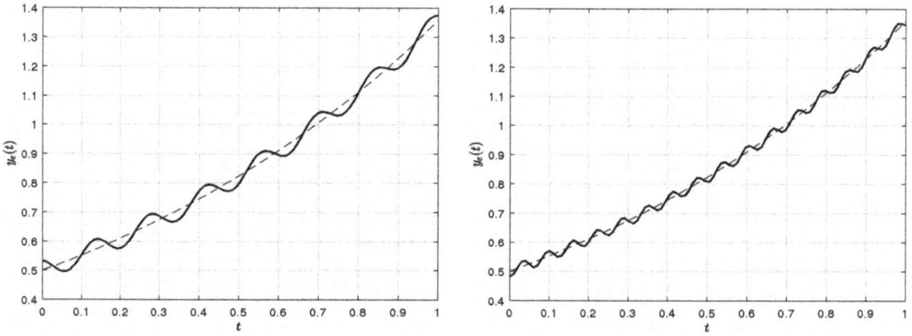

Figure 1. Solutions of the Neumann boundary value problem from Example 1 on the interval $[0,1]$ for $k = 2$ and $\varepsilon = 0.001$ (**left**) and $\varepsilon = 0.0002$ (**right**). A dashed line is used to draw the function $u(t) = e^t/k$, the solution of the reduced problem.

Listing 1. MATLAB code for Figure 1.

```
%bvp5cNeumann.m
format long;
a = 0;
b = 1;
k = 2;
eps = 0.0002;
ode = @(x,y) [y(2) ; (-k*y(1) + exp(x))/eps];
bc = @(ya,yb)[ya(2); yb(2)]; %Neumann BC
solinit = bvpinit(linspace(a,b,50),[1 0]);
sol = bvp5c(ode,bc,solinit);
x = linspace(a,b);
y = deval(sol,x);
X=x'; Y=y(1,:)';
%[X Y]
plot(x,Y,'linewidth',1.5);
hold on
plot(x,exp(x)/k, '--');
hold on
grid on
xlabel('$t$','interpreter','latex');
ylabel('$y_{\varepsilon}(t)$','interpreter','latex');
%print('figure1','-deps')
```

The main result of this note is the following theorem generalizing the Example 1 to all right-hand sides $f(t)$.

2. Main Result

Theorem 1. *For all $f \in C^3([a,b])$ and for every sequence $\{\varepsilon_n\}_{n=0}^{\infty}$, $\varepsilon_n \in J_n$ there exists a unique sequence of the solutions $\{y_{\varepsilon_n}\}_{n=0}^{\infty}$ of the Problem (1), (2) satisfying*

$$y_{\varepsilon_n} \to u \text{ uniformly on } [a,b] \text{ for } n \to \infty.$$

More precisely,

$$y_{\varepsilon_n}(t) = \frac{f(t)}{k} + O(\sqrt{\varepsilon_n}) \text{ on } [a,b]$$

for $n \to \infty$ ($\Rightarrow \varepsilon_n \to 0^+$) and, if $f'(a) = f'(b) = 0$, then on $[a,b]$, the following asymptotics for $n \to \infty$ hold:

$$y_{\varepsilon_n}(t) = \frac{f(t)}{k} + O(\varepsilon_n) \text{ and } y'_{\varepsilon_n}(t) = \frac{f'(t)}{k} + O(\sqrt{\varepsilon_n}).$$

Proof. First, we show that the function

$$y_{\varepsilon}(t) = \frac{\cos\left[\sqrt{\frac{k}{\varepsilon}}(t-a)\right] \int\limits_a^b \cos\left[\sqrt{\frac{k}{\varepsilon}}(b-s)\right] \frac{f(s)}{\varepsilon} ds}{\sqrt{\frac{k}{\varepsilon}} \sin\left[\sqrt{\frac{k}{\varepsilon}}(b-a)\right]} + \int\limits_a^t \frac{\sin\left[\sqrt{\frac{k}{\varepsilon}}(t-s)\right] \frac{f(s)}{\varepsilon}}{\sqrt{\frac{k}{\varepsilon}}} ds \tag{3}$$

is a solution of (1), (2). Differentiating (3) twice, taking into consideration the relation

$$\frac{\mathrm{d}}{\mathrm{d}t}\int_a^t H(t,s)f(s)\mathrm{d}s = \int_a^t \frac{\partial H(t,s)}{\partial t}f(s)\mathrm{d}s + H(t,t)f(t),$$

we obtain that

$$y'_\varepsilon(t) = -\frac{\sqrt{\frac{k}{\varepsilon}}\sin\left[\sqrt{\frac{k}{\varepsilon}}(t-a)\right]\int_a^b \cos\left[\sqrt{\frac{k}{\varepsilon}}(b-s)\right]\frac{f(s)}{\varepsilon}\mathrm{d}s}{\sqrt{\frac{k}{\varepsilon}}\sin\left[\sqrt{\frac{k}{\varepsilon}}(b-a)\right]} + \int_a^t \frac{\sqrt{\frac{k}{\varepsilon}}\cos\left[\sqrt{\frac{k}{\varepsilon}}(t-s)\right]\frac{f(s)}{\varepsilon}}{\sqrt{\frac{k}{\varepsilon}}}\mathrm{d}s, \tag{4}$$

$$y''_\varepsilon(t) = -\frac{\left(\sqrt{\frac{k}{\varepsilon}}\right)^2\cos\left[\sqrt{\frac{k}{\varepsilon}}(t-a)\right]\int_a^b \cos\left[\sqrt{\frac{k}{\varepsilon}}(b-s)\right]\frac{f(s)}{\varepsilon}\mathrm{d}s}{\sqrt{\frac{k}{\varepsilon}}\sin\left[\sqrt{\frac{k}{\varepsilon}}(b-a)\right]} - \int_a^t \frac{\left(\sqrt{\frac{k}{\varepsilon}}\right)^2\sin\left[\sqrt{\frac{k}{\varepsilon}}(t-s)\right]\frac{f(s)}{\varepsilon}}{\sqrt{\frac{k}{\varepsilon}}}\mathrm{d}s + \frac{f(t)}{\varepsilon}. \tag{5}$$

From (5) and (3), after a little algebraic rearrangement, we get

$$y''_\varepsilon = \frac{k}{\varepsilon}(-y_\varepsilon) + \frac{f(t)}{\varepsilon},$$

that is, y_ε is a solution of differential Equation (1), and from (4), it is easy to verify that this solution of (1) satisfies the boundary condition (2).

Let $t_0 \in [a,b]$ be arbitrary, but fixed. Let us denote by I_1 and I_2 the integrals

$$I_1 \triangleq \int_a^b \cos\left[\sqrt{\frac{k}{\varepsilon}}(b-s)\right]\frac{f(s)}{\varepsilon}\mathrm{d}s$$

and

$$I_2 \triangleq \int_a^{t_0} \sin\left[\sqrt{\frac{k}{\varepsilon}}(t_0-s)\right]\frac{f(s)}{\varepsilon}\mathrm{d}s.$$

Then

$$y_\varepsilon(t_0) = \frac{\cos\left[\sqrt{\frac{k}{\varepsilon}}(t_0-a)\right]I_1}{\sqrt{\frac{k}{\varepsilon}}\sin\left[\sqrt{\frac{k}{\varepsilon}}(b-a)\right]} + \frac{I_2}{\sqrt{\frac{k}{\varepsilon}}}.$$

Integrating I_1 and I_2 by parts we obtain that

$$I_1 = \sqrt{\frac{\varepsilon}{k}}\sin\left[\sqrt{\frac{k}{\varepsilon}}(b-a)\right]\frac{f(a)}{\varepsilon} + \int_a^b \sqrt{\frac{\varepsilon}{k}}\sin\left[\sqrt{\frac{k}{\varepsilon}}(b-s)\right]\frac{f'(s)}{\varepsilon}\mathrm{d}s,$$

$$I_2 = \frac{\sqrt{\frac{\varepsilon}{k}}f(t_0)}{\varepsilon} - \sqrt{\frac{\varepsilon}{k}}\cos\left[\sqrt{\frac{k}{\varepsilon}}(t_0-a)\right]\frac{f(a)}{\varepsilon} - \int_a^{t_0} \sqrt{\frac{\varepsilon}{k}}\cos\left[\sqrt{\frac{k}{\varepsilon}}(t_0-s)\right]\frac{f'(s)}{\varepsilon}\mathrm{d}s.$$

Thus,

$$y_\varepsilon(t_0) = \frac{f(t_0)}{k} + \frac{\cos\left[\sqrt{\frac{k}{\varepsilon}}(t_0 - a)\right]}{\sin\left[\sqrt{\frac{k}{\varepsilon}}(b - a)\right]} \int_a^b \sin\left[\sqrt{\frac{k}{\varepsilon}}(b - s)\right] \frac{f'(s)}{k} \mathrm{d}s$$
$$- \int_a^{t_0} \cos\left[\sqrt{\frac{k}{\varepsilon}}(t_0 - s)\right] \frac{f'(s)}{k} \mathrm{d}s.$$

Now, we estimate the difference $y_\varepsilon(t_0) - \frac{f(t_0)}{k}$. We have

$$\left| y_\varepsilon(t_0) - \frac{f(t_0)}{k} \right| \le \frac{1}{k \sin \lambda} \left| \int_a^b \sin\left[\sqrt{\frac{k}{\varepsilon}}(b - s)\right] f'(s) \mathrm{d}s \right|$$
$$+ \frac{1}{k} \left| \int_a^{t_0} \cos\left[\sqrt{\frac{k}{\varepsilon}}(t_0 - s)\right] f'(s) \mathrm{d}s \right|. \tag{6}$$

The integrals in (6) converge to zero for $\varepsilon = \varepsilon_n \in J_n$ as $n \to \infty$. Indeed, with respect to the assumption imposed on f we may integrate by parts in (6). Thus,

$$\int_a^b \sin\left[\sqrt{\frac{k}{\varepsilon}}(b - s)\right] f'(s) \mathrm{d}s = \left[\sqrt{\frac{\varepsilon}{k}} \cos\left[\sqrt{\frac{k}{\varepsilon}}(b - s)\right] f'(s) \right]_a^b$$
$$- \int_a^b \sqrt{\frac{\varepsilon}{k}} \cos\left[\sqrt{\frac{k}{\varepsilon}}(b - s)\right] f''(s) \mathrm{d}s$$
$$\le \sqrt{\frac{\varepsilon}{k}} \left(|f'(a)| + |f'(b)| + \left| \int_a^b \cos\left[\sqrt{\frac{k}{\varepsilon}}(b - s)\right] f''(s) \mathrm{d}s \right| \right)$$
$$\le \sqrt{\frac{\varepsilon}{k}} \left\{ |f'(a)| + |f'(b)| + \sqrt{\frac{\varepsilon}{k}} (|f''(a)| + (b - a)\mu_2) \right\} \tag{7}$$

and

$$\int_a^{t_0} \cos\left[\sqrt{\frac{k}{\varepsilon}}(t_0 - s)\right] f'(s) \mathrm{d}s = \left[-\sqrt{\frac{\varepsilon}{k}} \sin\left[\sqrt{\frac{k}{\varepsilon}}(t_0 - s)\right] f'(s) \right]_a^{t_0}$$
$$+ \int_a^{t_0} \sqrt{\frac{\varepsilon}{k}} \sin\left[\sqrt{\frac{k}{\varepsilon}}(t_0 - s)\right] f''(s) \mathrm{d}s$$
$$\le \sqrt{\frac{\varepsilon}{k}} \left(|f'(a)| + \left| \int_a^{t_0} \sin\left[\sqrt{\frac{k}{\varepsilon}}(t_0 - s)\right] f''(s) \mathrm{d}s \right| \right)$$
$$\le \sqrt{\frac{\varepsilon}{k}} \left\{ |f'(a)| + \sqrt{\frac{\varepsilon}{k}} (\mu_1 + |f''(a)| + (b - a)\mu_2) \right\}, \tag{8}$$

where $\mu_1 = \max\limits_{t \in [a,b]} |f''(t)|$ and $\mu_2 = \max\limits_{t \in [a,b]} |f'''(t)|$.

Substituting (7) and (8) into (6), we obtain the a priori estimate of solutions of the problem (1), (2) for all $t_0 \in [a,b]$ in the form

$$\begin{aligned}
&\left|y_\varepsilon(t_0) - \frac{f(t_0)}{k}\right| \\
&\leq \frac{1}{k\sin\lambda}\sqrt{\frac{\varepsilon}{k}}\left\{|f'(a)| + |f'(b)| + \sqrt{\frac{\varepsilon}{k}}(|f''(a)| + (b-a)\mu_2)\right\} \\
&+ \frac{1}{k}\sqrt{\frac{\varepsilon}{k}}\left\{|f'(a)| + \sqrt{\frac{\varepsilon}{k}}(\mu_1 + |f''(a)| + (b-a)\mu_2)\right\}.
\end{aligned} \tag{9}$$

Because the right-hand side of the inequality (9) is independent of t_0, the convergence is uniform on $[a,b]$.

Analogously, using (4), for $y'_\varepsilon(t_0)$, we obtain for all $t_0 \in [a,b]$ the estimate

$$\begin{aligned}
&\left|y'_\varepsilon(t_0) - \frac{f'(t_0)}{k}\right| \\
&\leq \frac{1}{k\sin\lambda}\left\{|f'(a)| + |f'(b)| + \sqrt{\frac{\varepsilon}{k}}(|f''(a)| + (b-a)\mu_2)\right\} \\
&+ \frac{1}{k}\left\{|f'(a)| + \sqrt{\frac{\varepsilon}{k}}(|f''(a)| + (b-a)\mu_2)\right\},
\end{aligned} \tag{10}$$

where the constant on the right-hand side does not depend on $t_0 \in [a,b]$. Theorem 1 is proved. □

Remark 1. *We conclude that in the case when $f'(a) = f'(b) = 0$,—that is, the solution $u = f(t)/k$ of a reduced problem satisfies the prescribed boundary conditions (2)—the convergence rate of the solutions of (1), (2) to the function u on the interval $[a,b]$ is even faster; namely, $O(\varepsilon_n)$ for $\varepsilon_n \in J_n$, as follows from (9).*

For example, the Neumann boundary value problem $\varepsilon y'' + ky = \cos t$, $t \in [0,\pi]$, (2) $k > 0$, $\varepsilon = \varepsilon_n \in J_n$, $n = 0,1,2,\ldots$, has solution $y_\varepsilon(t) = \cos t/(k-\varepsilon)$ satisfying

$$\left|y_\varepsilon(t_0) - \frac{\cos(t_0)}{k}\right| = \frac{\varepsilon|\cos(t_0)|}{k|k-\varepsilon|} = O(\varepsilon)$$

for all $t_0 \in [0,\pi]$ as $\varepsilon \to 0^+$. Note here that $\varepsilon \in J_n \Rightarrow k/\varepsilon \neq 1$.

Remark 2. *As follows from the proof of Theorem 1, the boundedness of the set*

$$\{|y_{\varepsilon_n}(t)| + |y'_{\varepsilon_n}(t)| + |y''_{\varepsilon_n}(t)| + |y'''_{\varepsilon_n}(t)|,\ t \in [a,b],\ \varepsilon_n \in J_n,\ n = 0,1,2\ldots\}$$

implies $|y_{\varepsilon_n}(t) - u(t)| = O(\sqrt{\varepsilon_n})$ for $n \to \infty$ uniformly on $[a,b]$ for the solutions y_{ε_n} of the nonlinear Neumann problem

$$\varepsilon_n y'' + ky = f(t,y), \quad k > 0, \quad f \in C^3([a,b]\times\mathbb{R}), \quad \varepsilon_n \in J_n,$$

where u is a solution of the reduced problem $ky = f(t,y)$ defined on $[a,b]$. In the proof we just replace $f'(s)$ with $\frac{\partial f}{\partial s} + \frac{\partial f}{\partial y}y'_\varepsilon(s)$, and so on.

3. Conclusions

In this paper, we dealt with a standard problem in the field of singular perturbations, namely the asymptotic behavior of the solutions when the parameter ε reaches zero, and the relation of this limit to the solution of the reduced problem ($\varepsilon = 0$).

The problem, namely (1), (2) which we analyze in the paper looks seemingly simple, but our approach represents a possible way of analyzing singularly perturbed problems when the critical manifold (solution of the reduced problem) is not normally hyperbolic (the roots of the characteristic equation are located on the imaginary axis). The investigation of this type of problem is still far from complete, and this article represents a small contribution (perhaps rather an attempt) towards grasping it.

Funding: This publication has been published with the support of the Operational Program Integrated Infrastructure within project "Výskum v sieti SANET a možnosti jej d'alšieho využitia a rozvoja", code ITMS 313011W988, co-financed by the European Regional Development Fund (ERDF).

Institutional Review Board Statement: Not applicable.

Informed Consent Statement: Not applicable.

Data Availability Statement: Not applicable.

Acknowledgments: The author thanks the editors and the anonymous reviewers for their insightful comments, which improved the quality of the paper.

Conflicts of Interest: The author declares no conflict of interest.

References

1. Jones, C.K.R.T. Geometric Singular Perturbation Theory. In *Dynamical Systems, Part of the Lecture Notes in Mathematics*; Springer: Heidelberg, Germany, 2006; Volume 1609, 44–118.
2. Fenichel, N. Geometric singular perturbation theory for ordinary differential equations. *J. Differ. Equ.* **1979**, *31*, 53–98. [CrossRef]
3. Wiggins, S. *Normally Hyperbolic Invariant Manifolds in Dynamical Systems*; Springer Science+Business Media: New York, NY, USA, 1994.
4. Kuehn, C. *Multiple Time Scale Dynamics*; Springer: Cham, Switzerland; Heidelberg, Germany; New York, NY, USA; Dordrecht, The Netherlands; London, UK, 2015.
5. Riley, J.W. *Fenichel's Theorems with Applications in Dynamical Systems*; University of Louisville: Louisville, KY, USA, 2012.
6. De Coster, C.; Habets, P. *Two-Point Boundary Value Problems: Lower and Upper Solutions*; Elsevier Science: Amsterdam, The Netherlands, 2006.
7. Chang K.W.; Howes, F.A. *Nonlinear Singular Perturbation Phenomena: Theory and Applications*; Springer: New York, NY, USA, 1984.
8. Vrabel, R. Upper and lower solutions for singularly perturbed semilinear Neumann's problem. *Math. Bohem.* **1997**, *122*, 175–180. [CrossRef]
9. Cabada, A.; Lopez-Somoza, L. Lower and Upper Solutions for Even Order Boundary Value Problems. *Mathematics* **2019**, *7*, 878. [CrossRef]
10. Kokotovic, P.; Khalil, H.K.; O'Reilly, J. *Singular Perturbation Methods in Control, Analysis and Design*; Academic Press: London, UK, 1986.
11. Vrabel, R. Singularly perturbed semilinear Neumann problem with non-normally hyperbolic critical manifold. *Electron. J. Qual. Theory Differ. Equ.* **2010**, 9, 1–11. [CrossRef]

MDPI
St. Alban-Anlage 66
4052 Basel
Switzerland
Tel. +41 61 683 77 34
Fax +41 61 302 89 18
www.mdpi.com

Axioms Editorial Office
E-mail: axioms@mdpi.com
www.mdpi.com/journal/axioms

www.ingramcontent.com/pod-product-compliance
Lightning Source LLC
LaVergne TN
LVHW070451100526
838202LV00014B/1702
9783036560298